Units

Symbol	Name	Measure of	Conversion Factors
A	ampere	Electrical Current	1 C/sec
Å	Ångström	Length	10^{-10} m, 0.1 nm
Bq	becquerel	Radioactivity	1 disintegration/sec, 60 dpm
C	coulomb	Electrical Charge	1 A sec
°C	centigrade degree	Temperature	°K − 273
Ci	curie	Radioactivity	3.7×10^{10} Bq
cm	centimeter	Length	10^{-2} m, 10^7 nm
cpm	counts/min	Radioactivity	dpm × counting efficiency[a]
d	dalton	Molecular Mass	1.66×10^{-24} g ($^1/_{12}$ mass of carbon atom)
dpm	disintegrations/min	Radioactivity	0.016 Bq, cpm/counting efficiency[a]
g	gram	Mass	6.02×10^{23} daltons
°K	Kelvin	Temperature	°C + 273
kb	kilobase	Nucleotides	1000 bases or base pairs
kcal	kilocalorie	Energy	4.18 kilojoules
kd	kilodalton	Molecular Mass	1000 d
l	liter	Volume	1000 ml
m	meter	Length	100 cm, 10^9 nm
M	molar	Concentration	moles solute per liter of solution
μg	microgram	Mass	10^{-6} g
min	minute	Time	60 sec
ml	milliliter	Volume	1 cm³
mole	mole	Number	6.02×10^{23} molecules
mV	millivolt	Electrical Potential	10^{-3} volts
nm	nanometer	Length	10^{-9} m, 10 Å
sec	second	Time	3,600 sec/hour; 86,400 sec/day
V	volt	Electrical Potential	1000 mV

[a] See table of radioactive isotopes (inside back cover) for efficiency of counting of specific isotopes.

Cover Problem (see Problem 13–25, page 152)

MOLECULAR BIOLOGY OF
THE CELL

THE PROBLEMS BOOK
Revised Edition

MOLECULAR BIOLOGY OF
THE CELL

THE PROBLEMS BOOK
Revised Edition

John Wilson • Tim Hunt

Garland Publishing, Inc.
New York & London

John Wilson received his Ph.D. from the California Institute of Technology (in 1971) and is currently a Professor of Biochemistry and Molecular Genetics at Baylor College of Medicine in Houston, Texas. He is coauthor of *Biochemistry: A Problems Approach* and of *Immunology*. His laboratory studies genetic recombination and genome rearrangements in mammalian cells.

Tim Hunt received his Ph.D. from the University of Cambridge (in 1968), where he taught biochemistry and cell biology from 1965 until 1990. He is now at the Imperial Cancer Research Fund Laboratories in England, where his laboratory studies the control of the cell cycle. He is coauthor (with Andrew Murray) of *The Cell Cycle: An Introduction.*

© 1989, 1994 by John Wilson and Tim Hunt

ISBN 0-8153-1621-6

Published by Garland Publishing, Inc.
717 Fifth Avenue, New York, NY 10022

Printed in the United States of America

15 14 13 12 11 10 9 8 7 6 5 4 3

for

Gavin Borden
1939–1991

Preface to the Revised Edition

In writing the first edition of *The Problems Book*, we had three aims. First, we wanted to help students recall the specialized language used by biologists by providing sets of simple fill-in-the-blank questions. Second, we tried to test the student's understanding of that language by providing a set of statements that might be either true or false. Finally, we provided a more challenging set of questions that were based on real experiments, which served, we hoped, to illustrate the actuality of thinking and research in cell and molecular biology. Many of these questions were based on classic experiments that provided new insights into biological processes, firm footholds in new landscapes of discovery. We have preserved these aims in the new edition, which has, however, been completely reorganized to correspond to the structure of the third edition of *Molecular Biology of the Cell* by Bruce Alberts, Dennis Bray, Julian Lewis, Martin Raff, Keith Roberts, and James D. Watson.

In revising our book and bringing it up to date, we were pleasantly surprised to find that for the most part, our choice of illustrations from the primary literature had stood the test of time. Classic experiments do stand the test of time, and the modern scientist must understand the past if he or she is to make present progress. In a few cases, however, subsequent research proved our chosen examples wrong or the conclusions drawn from the experiments misleading. These problems are omitted from the present book. A new feature starts on page 453, where we have gathered the literature references in the text into a list keyed to their associated problems. A similarly cross-referenced author list can be found on page 461. This should allow teachers to locate particular problems more readily, as well as making it easy for our colleagues to see if their work features in the book.

The last four years have seen many technical advances, particularly in genetic engineering, that are explored in Chapter 7. In addition, great progress has been made in certain fields, of which signal transduction and the cell cycle are two notable examples. To accomodate this progress, we have added nearly 60 new problems to bring the book up to date. The majority of these refer to experiments done in the interval between the two editions, but not all. In one case, recent research traced back to brilliant experiments done more than fifty years ago by the late Barbara McClintock. These experiments on plant chromosomes shed unanticipated light on the origins of chromosomal aberrations that cause cancer; surprisingly, they are not described in general genetics texts. In another, we lead the reader through the logic of Theodor Boveri's pioneering studies on chromosomes, which were performed at the turn of the century.

Our secret aim has been to try to answer the questions that should repeatedly surface during the reading of a textbook: How do they know that? And how did they find it out? Such things are rarely obvious, especially to people who have never worked in a real research laboratory, but it is in fact the whole point, for Science is a way of knowing. Our problems provide examples of practically all the ways there are of finding things out in biology, or so we believe. We have thus tried to provide a kind of running commentary on the third edition of *Molecular Biology of the Cell*.

In reacquainting ourselves with our old problems, we recognized for ourselves what our friends had repeatedly told us, and often urged us to amend. Many of these problems are hard. Yet we have let them stand, partly because properly understanding life is hard, and partly because, as we renewed our acquaintance with our old problems, we remembered what they stood for, and again could find no way to make them easier. Correctly analyzed and properly understood, we think that, in fact, only a very few of these problems present real difficulty. But they will all benefit, we believe, from discussion, whether between fellow students or in a class led by a teacher.

Once again, we have opted to include a mixture of answered and unanswered problems. The answered problems will allow readers to check their reasoning quickly and to verify their mastery of basic concepts. The unanswered questions allow readers to test their prowess on their own—in the absence of easily accessed answers. Unanswered problems are marked with an asterisk (*); they are answered in the *Instructor's Answer Manual* (available to instructors on request from Garland Publishing). We selected with care the answers to omit, but the choices were not easy. We left in the answers to many of the more difficult problems and to problems that made interesting points that might be overlooked. We imagine that the reader who makes a serious attempt to solve a problem before consulting the answer will gain most. We hope that both the answered and unanswered problems will also prove useful to instructors, as a source of homework assignments, as a framework for class discussions, and perhaps as the basis for exam questions.

We would love to hear from our readers, whether students or teachers, especially in cases where we have gotten things wrong or have stated them badly. You can FAX John Wilson at (713) 795-5487 or Tim Hunt at [44] 71 269 3804, or you can write to us c/o Ruth Adams, Garland Publishing, 717 Fifth Avenue, New York, NY 10022.

Preface to the First Edition

The Problems Book is designed as a companion volume to the second edition of *Molecular Biology of the Cell* by Bruce Alberts, Dennis Bray, Julian Lewis, Martin Raff, Keith Roberts, and James Watson. The intent of the book is to engage the reader actively in exploring the principles of cell biology. We firmly believe that principles must be examined, questioned, and played with in order to be fully understood. We hope to complement the elegant presentation in MBOC by encouraging a more questioning attitude toward cell biology. Textbooks make things as plain as possible and do not normally interrogate the reader. Yet, if the reader is not challenged, some of the richness and depth is lost. Accordingly, we present three different kinds of questions—simple review questions, more challenging thought questions, and problems based on experiments. These are integrated with the main text to provide a kind of running commentary that should help the reader learn and, above all, understand.

Our most important aim is to introduce readers to the experimental foundation of cell and molecular biology in order to increase their appreciation of the link between the behavior of molecules and the biology of cells. Experiments are the only way we have to find out how nature works, and our knowledge is only as sophisticated as our experiments permit. In the real world of research biology, half the battle is knowing what question to ask; the other half is finding a way to answer it. In biology, new knowledge rarely comes from simply sitting back and thinking. Unfortunately, no book can replace the actual doing of experiments, but we have tried to capture some of the flavor of the process, especially the thinking that leads from observation to interpretation. Most of our problems are based on real experiments. Some are from classic papers in cell biology, some are from studies that provide the crucial experimental support for our current concepts, and some are taken from papers at the frontiers—ones that were hot off the press when the problems were devised. All are meant to involve the reader actively in the thinking behind the experiments that form the basis of our knowledge of cell biology.

Our greatest difficulty was finding papers that could be turned into suitable problems. The authors of MBOC were always helpful in pointing out important papers, as were our friends and colleagues. But papers that make really good problems are rare. The majority—even many excellent ones that measurably advance our understanding—are difficult to cast in problem form. In general, the papers we selected were simplified considerably in abstracting data for the problems. We limited each problem to a few essential points, and we provided just the information necessary to reason through to the answer. Real life in the lab is not like this, of course: it rarely happens that one can look at a single Petri dish and know that the genetic code uses nonoverlapping triplets. In our efforts to simplify, we also handled data in a way that may occasionally require a word of apology to the authors. We tried carefully not to damage the sense of logic of the work, however, and hope that we will stimulate a closer reading of the many excellent papers we used.

Composing these problems took us a very long time and an enormous amount of work, but it was almost always stimulating and enjoyable. We very much hope that some of the fun we had is transmitted to our readers. Our object is to explore and explain, not to baffle and bamboozle.

Acknowledgments

This edition of *The Problems Book* builds on the foundation of its predecessor, and we wish to thank once again all those whose help we acknowledged in the first edition. In addition, we thank Angus MacNicol for the photograph in Figure 7–25, Dr. Nasser Hajibagheri and the staff of the ICRF Electron Microscopy Unit for specially providing the picture of *S. pombe* on page 230, and Keith Roberts for help with the chapter openers. We are greatly indebted to Alastair Ewing for checking the problems this second time around. We also thank Emily Preece for looking after us royally in London, as well as Miranda Robertson, who, with Ruth Adams, provided support well above and beyond what is normally expected of publishing executives. Libby Borden generously shared her country house and garden with us during the course of revision, and we are extremely grateful to her for this and many other kindnesses. Sadly, Gavin Borden died suddenly of cancer before we even embarked on this new edition, but the memory of that extraordinary man continually refreshed and spurred our efforts, as the Dedication of this volume attests.

Contents

Contents

Problems

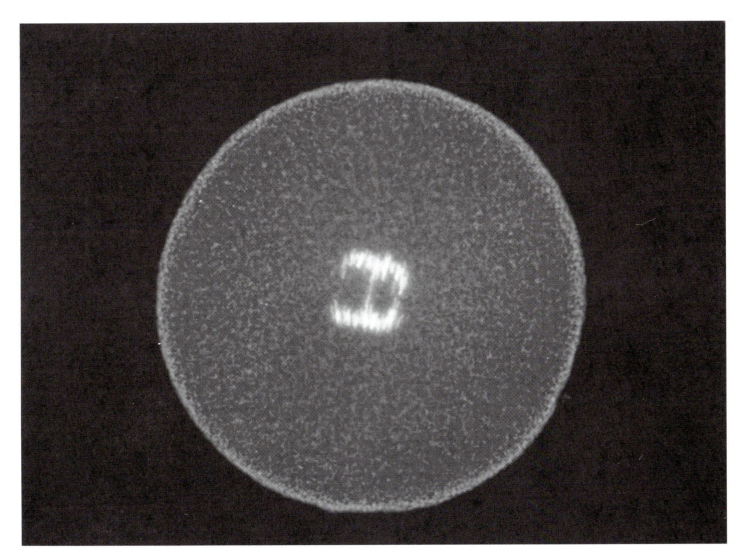

Anaphase at first cleavage of a sea urchin egg. The chromosomes are stained with a flourescent dye that binds DNA.

Unsolved problem: What molecule(s) hold the chromosomes together in pairs before the metaphase-anaphase transition? What is the signal, and what is the process by which the attachment is severed? How is it that the individual chromatids part from one another and move towards each spindle pole in such perfect synchrony?

Basic Genetic Mechanisms

6

RNA and Protein Synthesis
(MBOC 223–242)

6–1   Fill in the blanks in the following statements.

A. _____ copies a stretch of DNA into RNA in a process known as _____.

B. RNA synthesis begins at a _____ in the DNA and ends at a _____.

C. Conserved nucleotide sequences that are found in all examples of a particular type of regulatory region in DNA are called _____.

D. The _____ in a tRNA molecule is designed to base-pair with a complementary sequence of three nucleotides, the _____, in an mRNA molecule.

E. Enzymes called _____ couple each amino acid to its appropriate tRNA molecule to create an _____ molecule.

F. A _____ contains two binding sites for tRNA molecules: the _____, or P-site, holds the tRNA molecule that is linked to the growing end of the polypeptide chain, and the _____, or A-site, holds the incoming tRNA molecule that is charged with an amino acid.

G. Peptide bond formation is catalyzed by _____, a catalytic activity that is thought to be mediated by the major _____ molecule in the large ribosomal subunit.

H. Proteins called release factors bind to _____ codons in the A-site on the ribosome, causing peptidyl transferase to hydrolyze the bond that links the nascent polypeptide to the tRNA molecule.

I. An RNA sequence can be translated in any of three _____, each of which will specify a completely different polypeptide chain.

J. The initiation process of protein synthesis is complicated, involving a number of steps catalyzed by proteins called _____.

K. In all cells, a special _____ molecule, recognizing the _____ codon AUG and carrying the amino acid _____, provides the amino acid that begins a protein chain.

6–2   Indicate whether the following statements are true or false. If a statement is false, explain why.

___ A. Binding to the promoter orients RNA polymerase so that it transcribes the adjacent gene; however, the choice of template strand is dictated by additional protein factors.

1

___ B. In any one region of the DNA double helix, only one DNA strand is usually used as a template.

___ C. Bacterial cells use one type of RNA polymerase to transcribe all classes of RNA, whereas eucaryotic cells use three different types of RNA polymerase.

___ D. Modified nucleotides, which are especially common in tRNA molecules, are produced by covalent modification of the standard nucleotides before incorporation into RNA transcripts.

___ E. If the anticodon of a tRNA^{Tyr} were modified by a single-base change to recognize a Ser codon and then added to a cell-free system, the resulting protein would have tyrosine at all positions normally occupied by serine.

___ F. Each aminoacyl-tRNA linkage is activated for addition of the next amino acid to the growing polypeptide chain rather than for its own addition.

___ G. Wobble base-pairing occurs between the first position in the codon and the third position in the anticodon.

___ H. The primary function of the small ribosomal subunit is to bind mRNA and tRNAs, whereas the large ribosomal subunit catalyzes peptide bond formation.

___ I. Overall, the synthesis of proteins, which uses four high-energy phosphate bonds per added amino acid (4/codon), consumes less total energy than the transcription of DNA into RNA, which uses two high-energy phosphate bonds per added nucleotide (6/codon).

___ J. Because AUG serves as the start codon for protein synthesis, methionine is found only at the N-terminus of proteins.

___ K. Inserting a delay between the binding of a charged tRNA to the ribosome and its utilization in synthesis increases fidelity by giving improperly base-paired tRNAs an opportunity to diffuse off the ribosome.

___ L. Many antibiotics used in modern medicine selectively inhibit bacterial protein synthesis by exploiting the structural and functional differences between procaryotic and eucaryotic ribosomes.

6–3 One strand of a section of DNA isolated from *E. coli* reads

5′-GTAGCCTACCCATAGG-3′

A. Suppose mRNA is transcribed from this DNA using the complementary strand as a template. What will be the sequence of the mRNA?

B. What peptide would be made if translation started exactly at the 5′ end of this mRNA? (Assume no start codon is required, as is true under certain test tube conditions.) When tRNA^{Ala} leaves the ribosome, what tRNA will be bound next? When the amino group of alanine forms a peptide bond, what bonds, if any, are broken, and what happens to tRNA^{Ala}?

C. How many different peptides are encoded in this mRNA? Would the same peptides be made if the other strand of the DNA served as the template for transcription?

D. Suppose this stretch of DNA is transcribed as indicated in part A, but you do not know which reading frame is used. Could this DNA be from the beginning of a gene? The middle? The end?

*6–4 A few amino acids are removed from the C-terminal end of the beta-lactamase enzyme from *B. lichenformis* after it is synthesized. The sequence of the C-terminus of the enzyme can be deduced by comparing it to a mutant in which the reading frame is shifted by insertion or by deletion of a nucleotide. The amino acid sequence of the purified wild-type enzyme and that from the frameshift mutant are given below from amino acid residue 263 to the C-terminal end.

wild-type: N M N G K
mutant: N M I W Q I C V M K D

A. What was the mutational event that gave rise to the frameshift mutant?

---

*Problems with an asterisk (*) are answered in the Instructor's Manual.

B. Deduce the number of amino acids in the synthesized form of the wild-type enzyme and, as far as possible, the actual sequence of the wild-type enzyme.

6–5 You are studying protein synthesis in *Tetrahymena*, which is a unicellular ciliate. You have good news and bad news. The good news is that you have the first bit of protein and nucleic acid sequence data for the C-terminus of a *Tetrahymena* protein, which is shown as follows:

I   M   Y   K   Q   V   A   Q   T   Q   L   *

AUU AUG UAU AAG UAG GUC GCA UAA ACA CAA UUA UGA GAC UUA

The bad news is that you have been unable to translate purified *Tetrahymena* mRNA in a reticulocyte lysate, which is a standard system for analyzing protein synthesis *in vitro*. The mRNA looks good by all criteria, but the translation products are mostly small polypeptides (Figure 6–1, lane 1).

To figure out what is wrong, you do a number of control experiments using a pure mRNA from tobacco mosaic virus (TMV) that encodes a 116-kd protein. TMV mRNA alone is translated just fine in the *in vitro* system, giving a major band at 116 kd—the expected product—and a very minor band about 50 kd larger (Figure 6–1, lane 2). When *Tetrahymena* RNA is added, there is a significant increase in a higher molecular weight product (Figure 6–1, lane 3). When some *Tetrahymena* cytoplasm (minus the ribosomes) is added, the TMV mRNA now gives almost exclusively the higher molecular weight product (Figure 6–1, lane 4); furthermore, much to your delight, the previously inactive *Tetrahymena* mRNA now appears to be translated (Figure 6–1, lane 4). You confirm this by leaving out the TMV mRNA (Figure 6–1, lane 5).

A. What is unusual about the sequence data for the *Tetrahymena* protein?
B. How do you think the minor higher molecular weight band is produced from pure TMV mRNA in the reticulocyte lysate?
C. Explain the basis for the shift in proportions of the major and minor TMV proteins upon addition of *Tetrahymena* RNA alone and in combination with *Tetrahymena* cytoplasm. What *Tetrahymena* components are likely to be required for efficient translation of *Tetrahymena* mRNA?
D. Comment on the evolutionary implications of your results.

6–6 Consider the following experiment on the coordinated synthesis of the α and β chains of hemoglobin. Rabbit reticulocytes were labeled with ³H-lysine for 10 minutes, which is very long relative to the time required for synthesis of a single globin chain. The ribosomes, with attached nascent chains, were then isolated by centrifugation to give a preparation free of soluble (finished) globin chains. The nascent globin chains were digested with trypsin, which gives peptides ending in C-terminal lysine or arginine. The peptides were then separated by HPLC, and their radioactivity was measured. A plot of the radioactivity in each peptide versus the position of the lysine residues in the chains is shown in Figure 6–2.

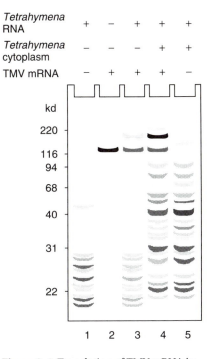

Tetrahymena RNA	+	−	+	+	+
Tetrahymena cytoplasm	−	−	−	+	+
TMV mRNA	−	+	+	+	−

Figure 6–1 Translation of TMV mRNA in the presence and absence of various components from *Tetrahymena* (Problem 6–5).

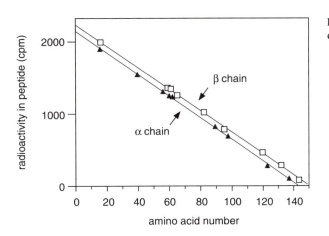

Figure 6–2 Synthesis of α- and β-globin chains (Problem 6–6).

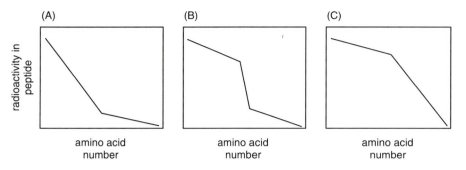

Figure 6–3 Hypothetical curves for globin synthesis with a roadblock to ribosome movement at the midpoint of the mRNA (Problem 6–6). These schematic diagrams are analogous to the graph in Figure 6–2.

A. Do these data allow you to decide which end of the globin chains (N- or C-terminus) is synthesized first? How?

B. In what ratio are the two globin chains produced? Can you estimate the relative numbers of α- and β-globin mRNA molecules from these data?

C. How long does a protein chain stay attached to the ribosome once the termination codon has been reached?

D. It was once suggested that heme is added to nascent globin chains during their synthesis and, furthermore, that ribosomes must wait for insertion of heme before they can proceed. The straight lines in Figure 6–2 indicate that ribosomes do not pause significantly, and heme is now thought to be added after synthesis. From among the graphs shown in Figure 6–3, choose the one that would result if there were a significant roadblock to ribosome movement halfway down the mRNA.

*6–7 Rates of peptide chain growth can be estimated from data such as those shown in Figure 6–4. In this experiment, a single TMV mRNA species encoding a 116,000 dalton protein was translated in a rabbit-reticulocyte lysate in the presence of ^{35}S-methionine. Samples were removed at one-minute intervals and subjected to electrophoresis on polyacrylamide gels, and the translation products were visualized by autoradiography. As is apparent in the figure, the largest detectable proteins get larger with time.

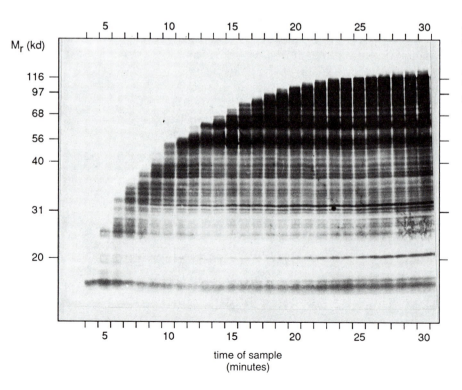

Figure 6–4 Time course of synthesis of a TMV protein in a rabbit-reticulocyte lysate (Problem 6–7). No radioactivity was detected during the first 3 minutes because the short chains ran off the bottom of the gel.

A. Is the rate of synthesis linear with time? One simple way to evaluate linearity of synthesis in this experiment is to plot the molecular mass (M_r) of the standards, which are shown on the left in Figure 6–4, against the time at which the largest peptide in a sample is equal to that molecular mass.

B. What is the rate of protein synthesis (in amino acids/minute) in this experiment? Assume the average molecular mass of an amino acid is 110 daltons.

C. Why does the autoradiograph have so many bands in it rather than just a few bands that get larger as time passes; that is, why does the experiment produce the "actual" result rather than the "theoretical" result shown in Figure 6–5? Can you think of a way to manipulate the experimental conditions to produce the theoretical result?

6–8 Termination codons in bacteria are decoded by one of two proteins. Release factor 1 (RF1) recognizes UAG and UAA, whereas RF2 recognizes UGA and UAA. The molecular details of how these proteins recognize stop codons and catalyze chain termination are unknown. However, the genes for RF1 and RF2 recently have been cloned and sequenced. For RF2, a comparison of the nucleotide sequence of the gene with the amino acid sequence of the protein revealed a startling surprise, which is contained within the sequences shown before the map of the gene in Figure 6–6. Sequences of the gene and protein were checked carefully to rule out any artifacts.

A. What is the surprise?

B. What hypothesis concerning the regulation of expression of the RF2 gene is suggested by it?

6–9 The overall accuracy of protein synthesis is very difficult to measure, partly because it is extremely accurate and mistakes, therefore, are very rare and partly because cells tend to destroy their mistakes very quickly. Technically there are problems, too. For example, how exactly do you identify a protein that, by definition, is not like the one you know how to purify and identify?

One ingenious approach to this problem used flagellin (molecular weight 40,000), which is the sole protein in bacterial flagella. Flagellin offers two advantages. First, flagella (hence flagellin) can be sheared off bacteria and purified by differential centrifugation. Second, flagellin contains no cysteine, thereby allowing a sensitive measure of the misincorporation of cysteine into the protein.

To radioactively label cysteine, bacteria were grown in the presence of $^{35}SO_4^{2-}$ (specific activity 5.0×10^3 cpm/pmol) for exactly one generation with excess unlabeled methionine in the growth medium. (No detectable radioactivity enters methionine under these conditions.) The resulting flagellin was purified and run on an SDS polyacrylamide gel. The 8 µg of flagellin recovered from the gel were found to contain 300 cpm of ^{35}S radioactivity.

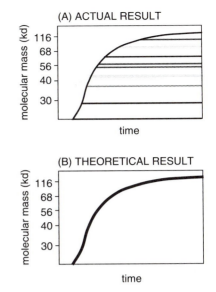

Figure 6–5 Actual (A) and theoretical (B) outcomes of protein synthesis experiment (Problem 6–7).

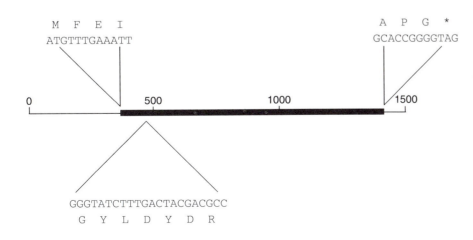

Figure 6–6 Schematic representation of the gene for RF2 (Problem 6–8). The coding sequence is shown as a thick line, with sequences at the start and finish shown for reference.

A. Of the flagellin molecules that were synthesized during the labeling period, what fraction contain cysteine? Assume that the mass of flagellin doubles during the labeling period and that the specific activity of cysteine in flagellin is equal to the specific activity of the $^{35}SO_4^{2-}$ used to label the cells.

B. In flagellin, cysteine is misincorporated at the arginine codons CGU and CGC. In terms of anticodon-codon interaction, what mistake is made during misincorporation of cysteine for arginine?

C. Given that there are 18 arginines in flagellin and that all arginine codons are equally represented, what is the frequency of misreading of each sensitive (CGU and CGC) arginine codon?

D. Assuming that the error frequency per codon calculated above applies to all amino acid codons equally, estimate the percentage of molecules that is correctly synthesized for proteins 100, 1000, and 10,000 amino acids in length. (The probability of not making a mistake is one minus the probability of making a mistake.)

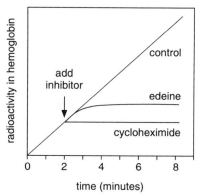

Figure 6–7 Effects of the inhibitors edeine and cycloheximide on protein synthesis in reticulocyte lysates (Problem 6–10).

*6–10 You have isolated an antibiotic, named edeine, from a bacterial culture. Edeine inhibits protein synthesis but has no effect on either DNA synthesis or RNA synthesis. When added to a reticulocyte lysate, edeine stops protein synthesis after a short lag, as shown in Figure 6–7, where it is contrasted with cycloheximide, which stops protein synthesis immediately. Analysis of the edeine-inhibited lysate by centrifugation in sucrose density gradients showed that no polyribosomes remained by the time protein synthesis had stopped. Instead, all the globin mRNA accumulated in an abnormal 40S peak, which contained equimolar amounts of the small ribosomal subunit and initiator tRNA.

A. What step in protein synthesis does edeine inhibit?

B. Why is there a lag between addition of edeine and cessation of protein synthesis? What determines the length of the lag?

C. Would you expect the polyribosomes to disappear if you added cycloheximide at the same time as edeine?

DNA Repair
(MBOC 242–251)

6–11 Fill in the blanks in the following statements.

A. Most spontaneous changes in DNA are quickly erased by a correction process called _____; only rarely do DNA maintenance procedures fail and allow a permanent sequence change, which is called a _____.

B. The genes encoding the _____, which are discarded during blood clotting, accumulate mutations without being selected against.

C. Two common spontaneous changes in DNA are _____, which results from disruption on the N-glycosyl linkages of adenine and guanine to deoxyribose, and _____, which converts cytosine to uracil.

D. DNA repair involves three steps: recognition and removal of the altered portion of the DNA strand by enzymes called _____, resynthesis of the excised region by _____, and sealing the remaining nick by the enzyme _____.

E. One major repair pathway, called _____, involves a battery of enzymes called _____, each of which recognizes an altered base in DNA and catalyzes its removal.

F. The _____ pathway is capable of removing almost any type of DNA damage that creates a large change in the DNA double helix.

G. In *E. coli* any block to DNA replication caused by DNA damage produces a signal that induces the _____, which allows replication through the block, hence giving the cell a chance for survival.

6–12   Indicate whether the following statements are true or false. If a statement is false, explain why.

___ A.   The fibrinopeptides, which are discarded from fibrinogen when it is activated to form fibrin during blood clotting, are especially useful for estimating mutation rates because they apparently have no direct function.

___ B.   Estimates of mutation rate based on amino acid differences in the same protein in different species always will be underestimates of the actual mutation rate because some mutations compromise the function of the protein and vanish from the population under selective pressure.

___ C.   Since histone H4 proteins are virtually identical in all species, one would expect histone H4 genes in different species to be virtually identical as well.

___ D.   Observed mutation rates, although extremely low, nevertheless limit the number of essential genes in any organism to about 60,000.

___ E.   There are a variety of DNA repair mechanisms, but all of them depend on the existence of two copies of the genetic information, one in each chromosome of a diploid organism.

___ F.   Spontaneous depurination and the removal of a deaminated C by uracil DNA glycosylase both leave an identical intermediate, which is the substrate recognized by AP endonuclease.

___ G.   Only the first step in DNA repair is catalyzed by enzymes that are unique to the repair process; the later steps are catalyzed by enzymes that play more general roles in DNA metabolism.

___ H.   The principal function of the SOS response in *E. coli* is to increase cell survival by introducing compensating mutations near the site of the original DNA damage.

___ I.   All the spontaneous deamination products of the usual four DNA bases are recognizable as unnatural when they occur in DNA.

6–13   Several genes in *E. coli*, including *uvrA*, *uvrB*, *uvrC*, and *recA*, are involved in repair of UV damage. Strains of *E. coli* that are defective in any one of these genes are much more sensitive to killing by UV light than are nonmutant (wild-type) cells, as shown for *uvrA* and *recA* strains in Figure 6–8A. Individual mutations in different genes can be combined in pairs to make all the possible

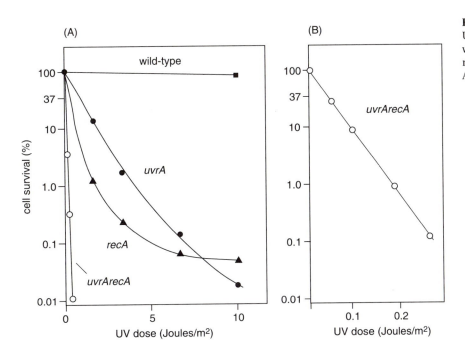

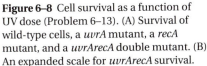

Figure 6–8 Cell survival as a function of UV dose (Problem 6–13). (A) Survival of wild-type cells, a *uvrA* mutant, a *recA* mutant, and a *uvrArecA* double mutant. (B) An expanded scale for *uvrArecA* survival.

double mutants. The sensitivity of the double mutants varies much more than that of the single mutants. The combinations of *uvr* mutants with one another show little increase in sensitivity relative to the *uvr* single mutants. The combination of *recA* with any of the *uvr* mutations, however, gives a strain that is exquisitely sensitive to UV light, as shown for *uvrArecA* on an expanded scale in Figure 6–8B.

A. Why do combinations of a *recA* mutation with a *uvr* mutation give an extremely UV-sensitive strain of bacteria, whereas combinations of mutations in different *uvr* genes are no more UV sensitive than the individual mutations?

B. According to the Poisson distribution, when a population of bacteria receives an overall average of one lethal "hit," 37% (e^{-1}) will survive because they actually receive no hits. For the double mutant *uvrArecA*, a dose of 0.04 J/m² gives 37% survival (Figure 6–8B). Calculate how many pryimidine dimers constitute a lethal hit for the *uvrArecA* strain, given that *E. coli* has 4×10^6 base pairs in its genome, which is 50% GC, and that exposure of DNA to UV light at 400 J/m² converts 1% of the total pyrimidine pairs (TT, TC, CT, plus CC) to pyrimidine dimers.

*6–14 In addition to killing bacteria, UV light causes mutations. You have measured the UV-induced mutation frequency in wild-type *E. coli* and in strains defective in either the *uvrA* gene or the *recA* gene. The results are shown in Table 6–1. Surprisingly, these strains differ markedly in their mutability by UV.

A. Assuming that the *recA* and *uvrA* gene products participate in different pathways for repair of UV damage, decide which pathway is more error prone. Which pathway predominates in wild-type cells?

B. The error-prone pathway is thought to result from misincorporation of nucleotides opposite a site of unrepaired damage. When forced to, DNA polymerases tend to incorporate an adenine nucleotide opposite a site with ambiguous coding properties, such as pyrimidine dimer. Is this so-called "A rule" a good strategy for dealing with UV damage? Calculate the frequency of base changes (mutations) using the A rule versus random incorporation (each nucleotide with equal probability) for *E. coli* where pyrimidine dimers are approximately 60% TT, 30% TC and CT, and 10% CC.

*6–15 The SOS pathway in *E. coli* represents an emergency response to DNA damage. As illustrated in Figure 6–9, under normal conditions the SOS set of damage-inducible genes is turned off by the LexA repressor, which also partially represses its own synthesis and that of RecA. In response to DNA damage, a signal (thought to be single-stranded DNA) activates RecA, which then mediates the cleavage of LexA. In the absence of LexA, all genes are maximally expressed. The SOS response increases cell survival in the face of DNA damage and transiently increases the mutation rate, hence variability, in the bacterial population. Although indispensable during an emergency, the constant expression of the SOS genes would be very deleterious.

One aspect of the regulation of the SOS response appears paradoxical: the expression of LexA (the repressor of the SOS response) is substantially increased during an SOS response. If the object of the response is maximal

Table 6–1 Frequency of UV-induced Mutations in Various Strains of *E. coli* (Problem 6–14)

Strain	Survival (%)	Mutations/10¹⁰ Survivors
Wild-type	100	400
recA	10	1
uvrA	10	40,000

UNINDUCED

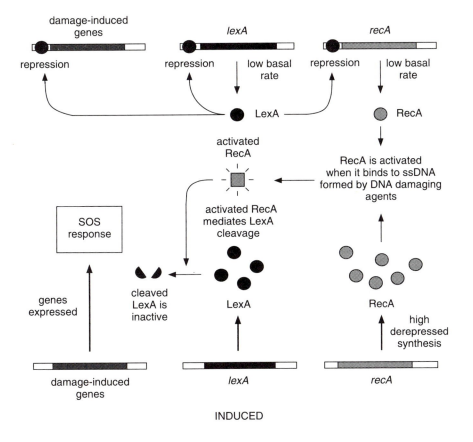

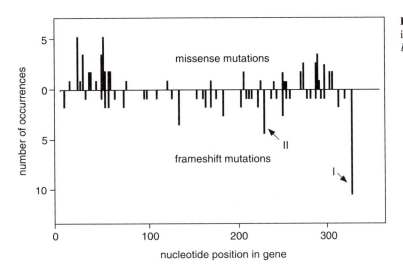

expression of the damage-induced genes, why should the repressor be expressed at high levels? Put another way, why is LexA not expressed at a low rate all the time? Do you see any advantage that the induced expression of LexA offers for the regulation of the SOS response?

6–16   The patterns of UV-induced mutations occurring in the *E. coli lacI* gene have been extensively analyzed. Figure 6–10 shows the total number of independently isolated missense (amino acid substitution) mutations (above the line) and frameshift (change in the reading frame) mutations (below the line). There

Figure 6–9 The SOS response in *E. coli* (Problem 6–15).

Figure 6–10 Pattern of UV-induced mutations in the *E. coli lacI* gene (Problem 6–16).

are almost equal numbers of mutations in each category. Missense mutations were identified by scoring for loss of function of the protein specified by the *lacI* gene (the lac repressor); the frameshift mutations were scored by a gene-fusion assay, which is independent of the function of the lac repressor.

A. Why do you think that there are so many more missense mutations at the ends of the gene than in the middle? Why do you think that the frameshift mutations are more or less evenly distributed across the gene (except for one or two "hot spots")?

B. DNA sequence analysis of the hot spot (labeled I in Figure 6–10) revealed that its wild-type sequence was TTTTTC and that the mutated sequence was TTTTC. The next most common mutation (labeled II in Figure 6–10) was the change from GTTTTC to GTTTC. Analysis of other frameshifts indicated that they resulted most commonly from the loss of one base; no insertions were found. Based on what you know of the nature of UV damage, can you suggest a molecular mechanism for the loss of a single base pair?

6–17 In addition to UvrABC endonuclease repair, recombinational repair, and SOS repair, bacteria have an even more potent repair system for dealing with pyrimidine dimers (one that is not discussed in the text). This phenomenon was discovered by careful observation to define an uncontrolled variable in an investigation of the effect of UV light on bacteria—not unlike the scenario described below.

You and your advisor are trying to isolate mutants in *E. coli* using UV light as the mutagenic agent. To get plenty of mutants, you find it necessary to use a dose of irradiation that kills 99.99% of the bacteria. You have been getting much more consistent results than your advisor, who also requires tenfold to a hundredfold higher doses of irradiation to achieve the same degree of killing. He wonders about the validity of your results since you always do your experiments at night after he has left. When he insists that you come in the morning to do the experiments in parallel, both of you are surprised when you get exactly the same results. You are a bit chagrined because the results are more like your advisor's, and you had confidently assumed you were better at doing lab work than your advisor, who is rarely seen in a lab coat these days. When you repeat the experiments in parallel at night, however, your advisor is surprised at the results, which are exactly as you had described them.

Now that you believe one another's observations, you make rapid progress. You find that you need a higher dose of UV light in the afternoon than in the morning to get the same degree of killing. Even higher doses are required on sunny days than on overcast days. Your laboratory faces west. What is the variable that has been plaguing your experiments?

*6–18 Mutagens such as N-methyl-N'-nitro-N-nitrosoguanidine (MNNG) and methyl nitrosourea (MNU) are potent DNA methylating agents and are extremely toxic to cells. Nitrosoguanidines are used in research as mutagens and clinically as drugs in cancer chemotherapy because they preferentially kill cells in the act of replication.

The original experiment that led to the discovery of the alkylation repair system in bacteria was designed to assess the long-term effects of exposure to low doses of MNNG (as in chemotherapy), in contrast to brief exposure to large doses (as in mutagenesis). Bacteria were placed first in a low concentration of MNNG (1 µg/ml) for 1.5 hours and then in fresh medium lacking MNNG. At various times during and after exposure to the low dose of MNNG, samples of the culture were treated with a high concentration (100 µg/ml) of MNNG for 5 minutes and then tested for viability and the frequency of mutants. As shown in Figure 6–11, exposure to low doses of MNNG temporarily increased the number of survivors and decreased the frequency of mutants among the survivors. As shown in Figure 6–12, this adaptive response to low doses of

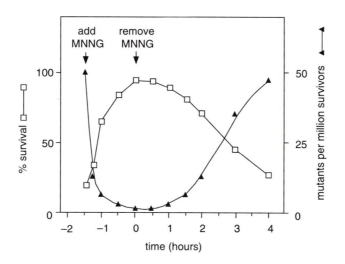

Figure 6–11 Adaptive response of *E. coli* to low doses of MNNG (Problem 6–18). MNNG at 1 μg/ml was present from −1.5 to 0 hours. Samples were removed at various times and treated briefly with 100 μg/ml MNNG to assess the number of survivors and the frequency of mutants.

MNNG was prevented if chloramphenicol (an inhibitor of protein synthesis) was included in the incubation.

A. Does the adaptive response of *E. coli* to low levels of MNNG require activation of preexisting protein or synthesis of new protein?

B. Why do you think the adaptive response of *E. coli* to a low dose of MNNG is so short-lived?

6–19 The nature of the mutagenic lesion introduced by MNNG and the mechanism of its removal from DNA were identified in the following experiments. To determine the nature of the mutagenic lesion, untreated bacteria and bacteria that had been exposed to low doses of MNNG were incubated with 50 μg/ml ^{3}H-MNNG for 10 minutes. Their DNA was isolated and then hydrolyzed to nucleotides, and the radioactive purines were then analyzed by paper chromatography as shown in Figure 6–13.

To examine the mechanism of removal of the mutagenic lesion, the enzyme responsible for removal was first purified. The kinetics of removal were studied by incubating different amounts of the enzyme (molecular weight, 19,000) with DNA containing 0.26 pmol of the mutagenic base, which was radioactively labeled with ^{3}H. At various times samples were taken, and the DNA was analyzed to determine how much of the mutagenic base remained (Figure 6–14). When the experiment was repeated at 5°C instead of 37°C, the initial rates of removal were slower, but exactly the same end points were achieved.

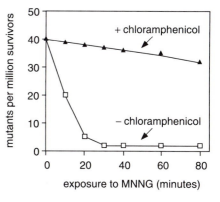

Figure 6–12 Effects of chloramphenicol on the adaptive response to low doses of MNNG (Problem 6–18). After different times of exposure of bacteria to 1 μg/ml MNNG, samples were removed and treated with 100 μg/ml MNNG to measure susceptibility to mutagenesis.

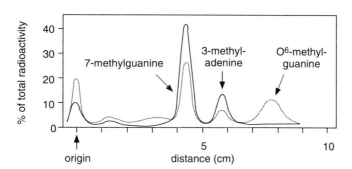

Figure 6–13 Chromatographic separation of labeled methylated purines in the DNA of untreated bacteria and bacteria treated with low doses of MNNG (Problem 6–19). The solid line indicates methylated purines from the DNA of untreated bacteria; the dashed line shows the results from MNNG-treated bacteria.

A. Which methylated purine is responsible for the mutagenic action of MNNG?
B. What is peculiar about the kinetics of removal of the methyl group from the mutagenic base? Is this peculiarity due to an unstable enzyme?
C. Calculate the number of methyl groups that are removed by each enzyme molecule. Does this calculation help to explain the peculiar kinetics?

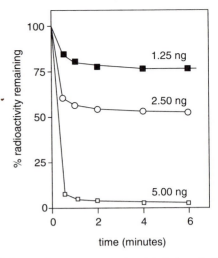

DNA Replication
(MBOC 251–262)

6–20 Fill in the blanks in the following statements.

A. The enzyme responsible for DNA synthesis in both replication and repair is called _____.
B. The active region of a chromosome involved in replication is a Y-shaped structure called a _____.
C. The enzyme that seals nicks in the DNA helix during DNA synthesis and repair is called _____.
D. During DNA replication, the daughter strand that is synthesized continuously is called the _____, and the strand that is synthesized discontinuously is known as the _____.
E. If DNA polymerase adds an incorrect nucleotide to the 3′ terminus, a separate catalytic domain containing a 3′-to-5′ _____ activity removes the mismatched base.
F. Initiation of DNA synthesis on the lagging strand requires short _____ made by an enzyme called _____, which uses ribonucleotide triphosphates as substrates.
G. The unwinding of the DNA helix at the replication fork is catalyzed by a _____, which uses the energy from ATP hydrolysis to move unidirectionally along DNA.
H. _____, which aid DNA unwinding, bind to single-stranded DNA in such a way that the bases are still available for templating reactions.
I. The DNA primase molecule is linked directly to the DNA helicase to form a unit called a _____, which moves down the lagging strand at the fork, synthesizing RNA primers as it goes.
J. If DNA polymerase makes a mistake, thus creating a pair of incorrectly hydrogen-bonded bases, the mistake is corrected by a special _____ system that uses methylation to distinguish new strands from old.
K. For bacteria and yeasts, as well as for several viruses that grow in eucaryotic cells, replication bubbles have been shown to form at special DNA sequences called _____.
L. _____ can be viewed as reversible nucleases that create either a transient single-strand break (type I) or a transient double-strand break (type II).

6–21 Indicate whether the following statements are true or false. If a statement is false, explain why.

__ A. In *E. coli*, where the replication fork moves forward at 500 nucleotide pairs per second, the DNA ahead of the fork rotates at nearly 3000 revolutions per minute.
__ B. Semiconservative replication means that the parental DNA strands serve as templates for the synthesis of the new progeny DNA strands, so the new double-stranded DNA molecules are composed of one old and one new strand.
__ C. When read in the same direction (5′ to 3′), the sequence of nucleotides in the newly synthesized DNA strand is the same as in the parental template strand.

Figure 6–14 Removal of ^{3}H-labeled methyl groups from DNA by purified methyltransferase enzyme (Problem 6–19). The quantities of purified enzyme are indicated.

___ D. A 5′-to-3′ synthesis of DNA means that growth occurs by addition of dNTPs to the exposed 3′-OH group, with expulsion of inorganic pyrophosphate.

___ E. DNA synthesis occurs in the 5′-to-3′ direction on the leading strand and in the 3′-to-5′ direction on the lagging strand.

___ F. If DNA polymerization occurred in the 3′-to-5′ direction, it would require that the growing end of the chain terminate in a 5′-triphosphate or that 3′-deoxynucleoside triphosphates act as precursors.

___ G. Loss of the 3′-to-5′ proofreading exonuclease activity of DNA polymerase in *E. coli* should slow the rate of DNA synthesis but not affect its fidelity.

___ H. Single-strand binding proteins at the replication fork hold the two strands of DNA apart by covering the bases and thus preventing base-pairing.

___ I. The mismatch proofreading system in *E. coli* can distinguish the parental strand from the progeny strand as long as one or both are methylated, but not if both strands are unmethylated.

___ J. In *E. coli*, yeasts, and eucaryotic viruses, new rounds of DNA replication are initiated at a specific site, which often contains several copies of a short sequence that binds a complex of initiator proteins.

___ K. Topoisomerase I does not require ATP to break and to rejoin DNA strands because the energy of the phosphodiester bond is stored transiently in a phosphotyrosine linkage in the enzyme's active site.

___ L. Topoisomerase II mutants in yeast can replicate their DNA but cannot separate their chromosomes at mitosis.

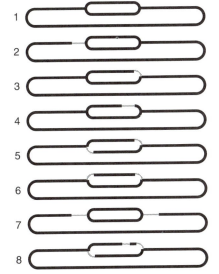

6–22 The DNA fragment in Figure 6–15 is double stranded at each end but single stranded in the middle. The polarity of the top strand is indicated.

A. Is the indicated phosphate (P) on the bottom strand at the 5′ end or the 3′ end of the fragment to which it is attached?

B. How would you expect the gap to be filled in by DNA repair processes inside a cell?

C. How many pieces will the bottom strand contain if the gap is filled in a test-tube reaction containing only deoxyribonucleoside triphosphates and DNA polymerase?

*6–23** The electron microscope is an important tool for studying DNA replication because it is possible to observe the replication fork directly and, for small DNA molecules, to observe the entire replicating structure. In addition, by using appropriate techniques of sample preparation, one can distinguish double-stranded DNA from single-stranded DNA.

A series of hypothetical replicating molecules are illustrated schematically in Figure 6–16, with regions of single-stranded DNA shown as thin lines. In an important early electron microscope study of bacteriophage lambda replication, some of these structures were observed commonly and others were never observed.

A. Draw a diagram of a replication structure with two forks moving in opposite directions. Label the ends of all strands (5′ or 3′), and indicate the leading and lagging strands at each replication fork.

B. Based on your knowledge of DNA replication, pick out the four structures in Figure 6–16 that were most commonly observed.

6–24 DNA polymerase I possesses a 3′-to-5′ proofreading exonuclease activity in addition to its polymerizing activity. This activity functions during proofreading to remove mismatched bases from the terminus of the newly polymerized DNA strand. To examine this activity, you prepare an artificial substrate with one poly(dA) strand and one poly(dT) strand that contains a few ^{32}P-labeled dT residues followed by a few ^{3}H-labeled dC residues at its 3′ end as shown in

Figure 6–16 Hypothetical structures of replicating DNA molecules (Problems 6–23).

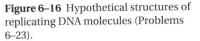

Figure 6–15 A DNA fragment with a single-stranded gap on the bottom strand (Problem 6–22).

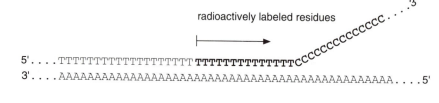

radioactively labeled residues

Figure 6–17 Artificial substrate for studying proofreading by DNA polymerase I of *E. coli* (Problem 6–24). Bold letters indicate nucleotides that are radioactively labeled.

5'....TTTTTTTTTTTTTTTTTTTTTT **TTTTTTTTTTTTTTC**CCCCCCCCCCCCC....3'
3'....AAAAAAAAAAAAAAAAAAAAAAAAAAAAAAAAAAAAAAAAAAAAAAAAAAA....5'

Figure 6–17. You measure the loss of the labeled dT and dC residues either without any dTTP present, so that no DNA synthesis is possible, or with dTTP present, so that DNA synthesis can occur. The results are shown in Figure 6–18.

A. Why were the T and C residues labeled with different isotopes?
B. Why did it take longer for the T residues to be removed in the absence of dTTP than the C residues?
C. Why were none of the T residues removed in the presence of dTTP, whereas the C residues were lost regardless of whether or not dTTP was present?
D. Would you expect different results in Figure 6–18B if you had included dCTP and dTTP?

*6–25 Does RNA priming occur at specific sites or at random sites on the template? The M13 viral DNA is an ideal template for studying this question because it has no 3'-OH group to confuse the issue. To answer this question, the M13 circle was completely copied in the presence of the bacteriophage T4 DNA polymerase, the T4 primosome (a complex of a helicase and an RNA primase), rNTPs, and dNTPs. The double-stranded circular products were then digested with a restriction enzyme that makes a double-strand cut at a unique site. The digestion products were denatured and analyzed on a high-resolution DNA sequencing gel. Many discrete bands were observed. If the digestion products were treated with RNase before electrophoresis, they all became five nucleotides shorter as judged by their faster migration on the sequencing gels.

Knowing that each product ended at the unique restriction site, it was possible to deduce the template sequence near their 5' ends from their length. (The complete sequence of M13 is known.) Some of these template sequences are shown on the left side of Figure 6–19. The DNA sequence at each of the corresponding 5' ends of the product strands was determined after removal of the priming ribonucleotides. These sequences are shown on the right side of Figure 6–19, on the same line as the region of complementary sequence in the DNA template.

From these data, deduce the start site for the RNA primer on each template sequence. What is the likely signal for starting the RNA primase reaction?

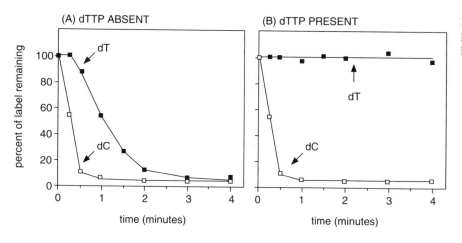

Figure 6–18 Proofreading by DNA polymerase I in the absence (A) and presence (B) of dTTP (Problem 6–24).

site	M13 template sequences	DNA sequences linked to RNA primer
	5' 3'	5' 3'
1	A T C C T T G C G T T G A A A T	A G G A T
2	T C T T G T T T G C T C C A G A	C A A G A
3	A T T C T C T T G T T T G C T C	A G A A T
4	A C A T G C T A G T T T T A C G	C A T G T
5	A T T G A C A T G C T A G T T T	T C A A T
6	A T C T T C C T G T T T T T T G G	A A G A T
7	A A A T A T T T G C T T A T A C	T A T T T
8	C T A G A A C G G T T A C C C T	T C T A G

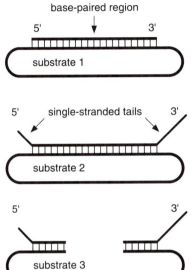

Figure 6–19 M13 template sequences (*left*) and the corresponding DNA sequences produced at each site (*right*) during the priming reaction (Problem 6–25). The RNA primers were removed from these DNA chains prior to sequencing.

6–26 The *dnaB* gene of *E. coli* encodes a helicase (DnaB) that unwinds DNA at the replication fork. Its properties have been studied using artificial substrates like those shown in Figure 6–20. The experimental approach is to incubate the substrates under a variety of conditions and then subject a sample to electrophoresis on agarose gels. The short single strand will move slowly if it is still annealed to the longer DNA strand, but it will move much faster if it has been unwound and detached. The migration of the short single strand can be followed selectively by making it radioactive and examining its position in the gel by autoradiography.

The results of several experiments are shown in Figure 6–21. Substrate 1, the hybrid without tails, was not unwound by DnaB (Figure 6–21, lanes 1 and 2). However, when either substrate with tails was incubated at 37°C with DnaB and ATP, a significant amount of small fragment was released by unwinding (lanes 6 and 10). For substrate 3 only the 3′ half-fragment was unwound (lane 10). All unwinding was absolutely dependent on ATP hydrolysis.

Unwinding was considerably enhanced by adding single-stranded DNA-binding protein (SSB) (compare lanes 5 and 6 and lanes 9 and 10). Interestingly, SSB had to be added about 3 minutes after DnaB; otherwise it inhibited unwinding.

Figure 6–20 Substrates used to test the properties of DnaB (Problem 6–26).

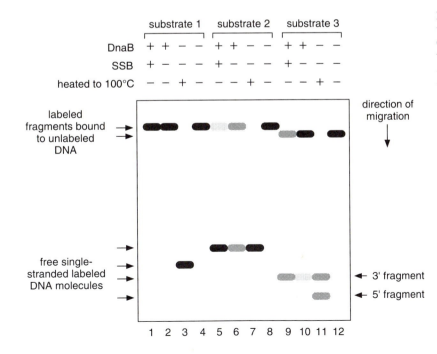

Figure 6–21 Results of several experiments to measure unwinding by DnaB (Problem 6–26). Only the single-stranded fragments were radioactively labeled. Their positions are shown by the bands in this schematic diagram.

A. Why is ATP hydrolysis required for unwinding?
B. In what direction does DnaB move along the long single-stranded DNA? Is this direction more consistent with its movement on the leading strand or on the lagging strand at the replication fork?
C. Why might SSB inhibit unwinding if it is added before DnaB but stimulate unwinding if added after DnaB?

6–27 The laboratory you joined is studying the life cycle of an animal virus that uses a circular, double-stranded DNA as its genome. Your project is to define the location of the origin(s) of replication and to determine whether replication proceeds in one or both directions away from an origin (unidirectional or bidirectional). To accomplish your goal, you isolated replicating molecules, cleaved them with a restriction enzyme that cuts the viral genome at one site to produce a linear molecule from the circle, and examined the resulting molecules in the electron microscope. Some of the molecules that you have observed are illustrated schematically in Figure 6–22. (Note that it is impossible to tell one end of a DNA molecule from the other in the electron microscope.)

You must present your conclusions to the rest of the lab tomorrow. How will you answer the questions that your advisor has posed for you? (1) Is there a single, unique origin of replication or several origins? (2) Is replication unidirectional or bidirectional?

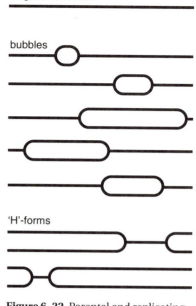

Figure 6–22 Parental and replicating forms of an animal virus (Problem 6–27).

*6–28 Conditionally lethal mutations are very useful in genetic and biochemical analyses of complex processes such as DNA replication. Temperature-sensitive (ts) mutations, which are one form of conditional lethal mutation, allow growth at one temperature (for example, 30°C) but not at a higher temperature (for example, 42°C).

A large number of temperature-sensitive replication mutants have been isolated in *E. coli*. These mutant bacteria are defective in DNA replication at 42°C but not at 30°C. If the temperature of the medium is raised from 30°C to 42°C, these mutants stop making DNA in one of two characteristic ways. The "quick-stop" mutants halt DNA synthesis immediately, whereas the "slow-stop" mutants stop DNA synthesis only after many minutes.

A. Predict which of the following proteins, if temperature sensitive, would display a quick-stop phenotype and which would display a slow-stop phenotype. In each case explain your prediction.

  1. DNA topoisomerase I
  2. A replication initiator protein
  3. Single-strand binding protein
  4. DNA helicase
  5. DNA primase
  6. DNA ligase

B. Cell-free extracts of the mutants show essentially the same patterns of replication as the intact cells. Extracts from quick-stop mutants halt DNA synthesis immediately at 42°C, whereas extracts from slow-stop mutants do not stop DNA synthesis for several minutes after a shift to 42°C. Suppose extracts from a temperature-sensitive DNA helicase mutant and a temperature-sensitive DNA ligase mutant were mixed together at 42°C. Would you expect the mixture to exhibit a quick-stop phenotype, a slow-stop phenotype, or a nonmutant phenotype?

Genetic Recombination
(MBOC 263–273)

6–29 Fill in the blanks in the following statements.

A. In _____, genetic exchange occurs between homologous DNA sequences, most commonly between two copies of the same chromosome.

B. At the site of exchange, a strand of one DNA molecule has become base-paired to a strand of the second DNA molecule to create a _____ between the two double helices.

C. Two single-stranded, complementary DNA molecules come together to form a fully double-stranded helix by _____, which is thought to start with a slow _____ step.

D. The _____ is required for chromosome pairing in *E. coli*; it binds to single-stranded DNA and promotes its pairing with homologous, double-stranded DNA.

E. Once synapsis has occurred, the short heteroduplex region where the strands from two different DNA molecules have begun to pair is enlarged through a reaction called _____.

F. A central intermediate in general recombination is the _____, which is also called a _____, after its discoverer.

G. Occasionally during recombination between two slightly different copies of the same gene (alleles), one allele is replaced by the other in a process known as _____.

H. Mobile DNA sequences and some viruses enter and leave a target chromosome by _____ genetic recombination.

6–30 Indicate whether the following statements are true or false. If a statement is false, explain why.

__ A. General recombination requires long regions of homologous DNA on both partners in the exchange, whereas site-specific recombination requires only short, specific nucleotide sequences, which in some cases need be present on only one of the exchanging partners.

__ B. General recombination involves the physical exchange of DNA segments, which entails the breaking and rejoining of phosphodiester bonds in the DNA backbone.

__ C. The RecA protein combines a site-specific, single-strand nicking activity with an ATP-dependent DNA helicase function that can unravel single-stranded "whiskers" off a duplex DNA molecule.

__ D. The SSB protein of *E. coli* melts out short hairpin helices in single strands by binding to the sugar-phosphate backbone and holding the bases in an exposed position.

__ E. The RecA protein binds both single-stranded and double-stranded DNA so that it can catalyze synapsis between them.

__ F. The cross-strand exchange contains two distinct pairs of strands (crossing strands and noncrossing strands), which cannot be interconverted without breaking the phosphodiester backbone of at least one strand.

__ G. Gene conversion is the process whereby fungi occasionally change sex; normally, equal numbers of male and female spores are produced by a mating event, but occasionally the ratio is 1:3 or 3:1.

__ H. All known mechanisms of gene conversion require a limited amount of DNA synthesis.

__ I. The integration of the lambda genome into the *E. coli* genome is catalyzed by a site-specific topoisomerase (called lambda integrase), which recognizes short, specific DNA sequences on both chromosomes.

6–31 Using Figure 6–23A as a guide, draw the products of a crossover recombination event between the homologous regions of the molecules represented in Figure 6–23B. In the figure the DNA duplex is represented schematically by a single line and the targets for homologous recombination are represented by arrows.

*6–32 Specific DNA sequences known as Chi sites locally stimulate RecBCD-mediated homologous recombination in *E. coli*. Presumably, interaction between

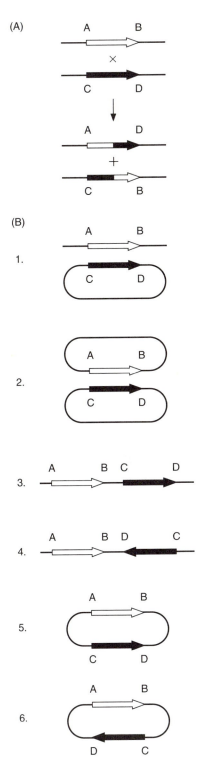

Figure 6–23 A variety of recombination substrates (Problem 6–31).

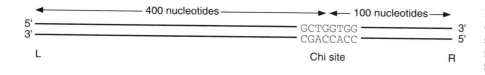

Figure 6–24 A linear DNA fragment containing a Chi site (Problem 6–32). The sequence of the Chi site is shown. L and R indicate the left and right ends of the fragment.

the RecBCD protein and the Chi site stimulates a rate-limiting step in the recombination pathway. To study this interaction in detail, RecBCD was purified and incubated with a linear, double-stranded DNA fragment containing a Chi site (Figure 6–24).

Different samples of the linear DNA were first labeled specifically at the 5′ end on the left (5′L), at the 5′ end on the right (5′R), at the 3′ end on the left (3′L), and the 3′ end on the right (3′R). Each sample was then incubated in a reaction buffer containing RecBCD. As a control, a separate aliquot of labeled DNA was incubated in the reaction buffer without RecBCD. After 1 hour of reaction the DNA was denatured by boiling, and the resulting single strands were separated by electrophoresis through a polyacrylamide gel. The pattern of radioactively labeled DNA fragments is shown in Figure 6–25. As a further control, a sample of 3′R that had been incubated with RecBCD was run on the gel without first denaturing it (Figure 6–25).

A. What is the evidence that RecBCD cuts the DNA at the Chi site? Does it cut one or both strands? If you decide it cuts only one strand, specify the strand and indicate your reasoning.

B. What is the evidence that RecBCD can act as a helicase—that is, what is the evidence that it can separate the strands of duplex DNA?

C. How might the action of RecBCD stimulate homologous recombination in the neighborhood of a Chi site?

6–33   Bacteriophage T4 encodes a single-stranded DNA-binding protein (SSB protein) that is important for recombination and DNA replication. T4 mutants with a temperature-sensitive mutation in the gene that encodes the SSB protein rapidly cease recombination and DNA replication when the temperature is raised.

The T4 SSB protein is an elongated monomeric protein with a molecular weight of 35,000. It binds tightly to single-stranded, but not double-stranded, DNA. Binding saturates at a 1:12 weight ratio of DNA to protein. However, the binding of SSB protein to DNA shows a peculiar property that is illustrated in Figure 6–26. In the presence of excess single-stranded DNA (10 μg), virtually no binding is detectable at 0.5 μg SSB protein (Figure 6–26A), whereas almost quantitative binding is seen at 7.0 μg SSB protein (Figure 6–26B).

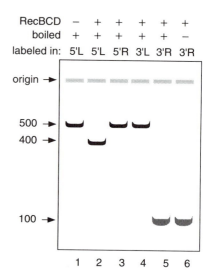

Figure 6–25 Results of incubating the RecBCD protein with Chi-containing DNA fragments, which were labeled at the ends of defined strands (Problem 6–32). Numbers adjacent to bands indicate the length of the labeled fragment in nucleotides. All samples were denatured before electrophoresis, except the unboiled sample in lane 6.

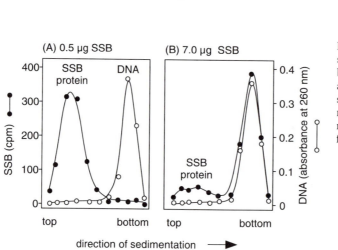

Figure 6–26 Binding of T4 SSB protein to single-stranded DNA (Problem 6–33). The binding of SSB protein to DNA was analyzed by centrifugation through sucrose gradients, on which the much more massive DNA sediments more rapidly than protein and is consequently found closer to the bottom of the gradient.

A. At saturation, what is the ratio of nucleotides of single-stranded DNA to molecules of SSB protein? (The average mass of a single nucleotide is 330 d.)

B. At the point at which binding of SSB protein to DNA reaches saturation, are adjacent monomers of SSB protein likely to be in contract? Assume that a monomer of SSB protein extends for 12 nm along the DNA upon binding and that the spacing of bases in single-stranded DNA after binding SSB protein is the same as in double-stranded DNA (that is, 10 nucleotides per 3.4 nm).

C. Why do you think that the binding of SSB protein to single-stranded DNA depends so strongly on the amount of SSB protein, as shown in Figure 6–26?

*6–34 RecA protein catalyzes both the initial pairing step of recombination and subsequent branch migration in *E. coli*. It promotes recombination by binding to single-stranded DNA and catalyzing the pairing of such coated single strands to homologous double-stranded DNA. One assay for the action of RecA is the formation of double-stranded DNA circles from a mixture of double-stranded linears and homologous single-stranded circles, as illustrated in Figure 6–27. This reaction proceeds in two steps: circles pair with linears at an end and then branch migrate until a single-stranded linear DNA is displaced.

One important question about the RecA reaction is whether branch migration is directional. This question has been studied in the following way. Single-stranded circles, which where uniformly labeled with ^{32}P, were mixed with unlabeled double-stranded linears in the presence of RecA. As the single-stranded DNA pairs with the linear DNA, it becomes sensitive to cutting by restriction enzymes, which do not cut single-stranded DNA. By sampling the reaction at various times, digesting the DNA with a restriction enzyme, and separating the labeled fragments by electrophoresis, you obtain the pattern shown in Figure 6–28.

A. By comparing the time of appearance of labeled fragments with the restriction map of the circular DNA in Figure 6–28, deduce which end (5′ or 3′) of the minus strand of the linear DNA the circular plus strand invades. Also deduce the direction of branch migration along the minus strand. (The linear double-stranded DNA was cut at the boundary between fragments a and c on the restriction map.)

B. Calculate the rate of branch migration, given that the length of this DNA is 7 kb.

C. What would you expect to happen if the linear, double-stranded DNA carried an insertion of 500 nonhomologous nucleotides between restriction fragments e and a?

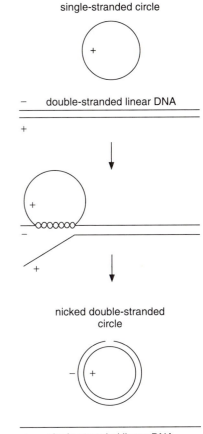

Figure 6–27 Strand assimilation assay for the RecA protein (Problem 6–34). The single-stranded (+) circle is complementary to the minus strand of the linear duplex and identical to the plus strand of the duplex.

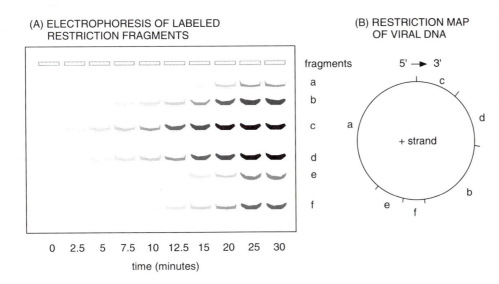

Figure 6–28 Electrophoretic separation (A) of labeled restriction fragments as a function of time of incubation with RecA (Problem 6–34). The restriction map (B) is shown for reference on the circle; clockwise around the circle is the 5′-to-3′ direction.

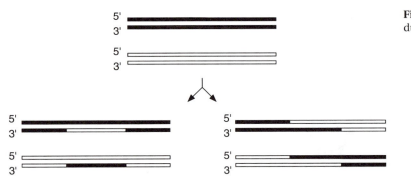

Figure 6–29 Parental and recombinant duplexes (Problem 6–35).

6–35 Two homologous parental duplexes and two sets of potential recombination products are illustrated in Figure 6–29. Diagram a Holliday junction between the parental duplexes that could generate the indicated recombination products. Label the left end of each strand in the Holliday junction 5′ or 3′ so that the relationship to the parental and recombinant duplexes is clear. Indicate which strands need to be cut to generate each set of recombination products. Finally, draw the recombinants as they would look after one round of replication.

*6–36 When plasmid DNA is extracted from *E. coli* and examined under the electron microscope, the majority of the plasmids are monomeric circles, but there are a variety of other forms, including dimeric and trimeric circles. In addition, about 1% of the molecules appear as figure-8 forms, in which the two loops are equal (Figure 6–30A).

You suspect that the figure 8s are recombination intermediates in the formation of a dimer from two monomers (or two monomers from a dimer). However, to rule out the possibility that they represent twisted dimers or touching monomers, you cut the DNA sample with a restriction enzyme, which cuts at a single site in the monomer, and then examine the molecules. After cutting, only two forms are seen: 99% of the DNA molecules are linear

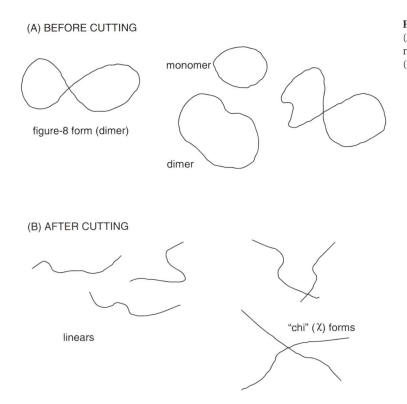

(A) BEFORE CUTTING

figure-8 form (dimer)

monomer

dimer

(B) AFTER CUTTING

linears

"chi" (χ) forms

Figure 6–30 Plasmid molecules observed (A) before digestion with a single-cut restriction enzyme and (B) after digestion (Problem 6–36).

monomers, and 1% are χ forms (Figure 6–30B). You note that the χ forms have an interesting property: the two longer arms are the same length, as are the two shorter arms. In addition, the sum of the lengths of a long arm and a short arm is equal to the length of the monomer plasmid. However, the position of the crossover point is completely random.

Unsure of yourself and feeling you are probably missing some hidden artifact, you show your pictures to a friend. She points out that your observations prove you are looking at recombination intermediates, which arose by random pairing at homologous sites.

A. Is your friend correct? What is her reasoning?
B. How would your observations differ if you repeated your experiments in a strain of *E. coli* that carried a nonfunctional *recA* gene?
C. What would the χ forms have looked like if the figure 8s were intermediates in a site-specific recombination between the monomers?
D. What would the χ forms have looked like if the figure 8s were intermediates in a totally random, nonhomologous recombination between the monomers?

Viruses, Plasmids, and Transposable Genetic Elements
(MBOC 273–287)

6–37 Fill in the blanks in the following statements.

A. The viruses that infect bacteria are called _____.
B. Virus multiplication is lethal, if the host cell must break open (_____) to allow the progeny virus to escape.
C. The protein shell that surrounds a viral genome is called the _____.
D. The outermost shell of _____ viruses is a typical membrane that is usually acquired in the process of budding from the plasma membrane.
E. Since normal cells do not have enzymes that copy RNA into RNA, viruses with RNA genomes must encode an _____, or a _____, to replicate.
F. The nucleocapsid of Semliki forest virus is surrounded by a closely apposed lipid bilayer that contains only three types of polypeptide chains, or _____, each of which is encoded by the viral RNA.
G. _____ bacteria carry a dormant but potentially active viral genome in their chromosomes: The integrated viral genome is known as the _____, a term that also is used to describe analogous viral genomes present in mammalian cells.
H. Animal cells that have been converted from their normal state to a cancerous one by a viral infection are said to have undergone a virus-mediated _____.
I. The enzyme _____, which transcribes RNA chains into complementary DNA molecules, accounts for the permanent genetic changes caused by _____.
J. _____ move from place to place in the host genome using their own site-specific recombination enzymes, known as _____.
K. The yeast Ty1 element is an example of a _____, whose transposition requires the synthesis of a complete RNA transcript, which is then copied into a DNA double helix and subsequently is integrated into a new chromosomal location.
L. Independently replicating elements, called _____, can replicate indefinitely outside the host chromosome.
M. _____, which are disease agents in plants, are small, single-stranded, circular RNA molecules that do not code for any protein.

6–38 Indicate whether the following statements are true or false. If a statement is false, explain why.

___ A. Since bacteriophage T4 encodes at least 30 different enzymes involved in genome replication and transcription, it is completely independent of the host DNA and RNA polymerases, though it must, of course, rely on the host protein synthesis machinery.

___ B. Small DNA viruses such as SV40 and φX174 rely entirely on the host replicative machinery to replicate their DNA.

___ C. Negative-strand viruses do not contain genes that code for proteins.

___ D. To follow the life cycle of an enveloped virus is to take a tour through the cell.

___ E. When bacteriophage lambda infects a suitable *E. coli* host cell, it usually carries out a lytic infection, releasing several hundred progeny viruses; more rarely, it integrates into the host chromosome, producing a lysogenic bacterium carrying the proviral lambda chromosome.

___ F. Infection by retroviruses often leads simultaneously to the nonlethal release of progeny virus and a permanent genetic change in the infected cell that makes it cancerous.

___ G. Viruses that reproduce by budding through their host cell membranes cause cancer by the changes that budding makes to the cell surface.

___ H. One simple way to classify a new virus as an RNA or a DNA virus is to see whether its growth is inhibited by actinomycin D, which blocks DNA-dependent RNA synthesis but not RNA-dependent replicases: if viral growth is inhibited by actinomycin D, it must be a DNA virus.

___ I. Transposases recognize sufficiently extensive sequences surrounding the integration sites so that the transposon does not become integrated into the middle of a gene, for gene disruption could be lethal to the cell.

___ J. The capacity of plasmids to replicate indefinitely without being part of a host chromosome distinguishes them from transposable elements.

___ K. Large viruses are more likely than small viruses to have overlapping genes because they have so many more genes.

___ L. Viroids are unusual because they encode no proteins and yet are able to replicate and to cause severe diseases in plants.

6–39 The one-step growth curve, which measures the increase in virus number during a single cycle of infection, was extremely important for defining the basic parameters of the interactions of phages with bacteria. Consider the following one-step growth curve with bacteriophage T4.

A small sample (0.1 ml) of a T4 phage suspension (10^{10} phage/ml) was mixed with a 100-ml culture of *E. coli* (10^7 bacteria/ml) and incubated at 37°C. At 5-minute intervals after mixing, duplicate samples of the infected bacteria were shaken with chloroform to kill the bacteria. One sample was treated with the enzyme lysozyme to break open the bacteria; the other was not treated. Neither chloroform nor lysozyme kills T4. The concentration (titer) of phages in all samples was measured, with the results shown in Figure 6–31.

A. What was the initial titer of T4 immediately after it was mixed with the bacterial culture?

B. Why is the phage titer at 5 minutes less than it was initially?

C. Explain why the phage titer rises more slowly in the samples that were not treated with lysozyme but ultimately reaches the same level as the lysozyme-treated samples.

D. Calculate how many phages were produced on average per infected bacterium.

*6–40 Bacterial and viral infections are fundamentally different and demand different strategies for treatment. Antibacterial drugs kill bacterial cells selectively without harming the host. However, a virus subverts the host cell's metabolic machinery to its own needs. As a consequence, agents directed at cellular components would harm infected and uninfected cells alike. Successful treatment of viral infections requires drugs that selectively block virus-specific processes but do not damage normal host cells.

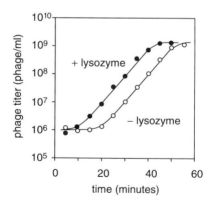

Figure 6–31 One-step growth curve of bacteriophage T4 (Problem 6–39).

Acyclovir (acycloguanosine) is one example of a clinically useful antiviral drug; it is a potent inhibitor of the replication of herpes virus. Acyclovir is phosphorylated to acyclo-GTP by the thymidine kinase encoded in the herpes virus genome but not by the normal cellular thymidine kinase. Acyclo-GTP is then incorporated into the replicating viral DNA by the DNA polymerase encoded in the herpes virus genome. Incorporation of acyclo-GTP blocks replication because it lacks the requisite 3′-OH needed for addition of the next nucleotide to the growing chain.

Acyclovir-resistant mutants of herpes virus have been isolated from patients treated with acyclovir. Replication of these herpes virus mutants is not affected by acyclovir. Propose two different explanations for how herpes virus might mutate so as to become resistant to acyclovir.

6–41 When a small number of T4 phages is mixed with a large excess of E. coli strain B and spread in a thin layer of soft agar over the surface of a petri dish containing a deep layer of nutrient agar, the bacteria grow up to form a continuous "lawn." However, in the places where a phage-infected bacterium happens to land, the phages multiply and kill the surrounding bacteria, causing the formation of a clear, round "plaque" in the cloudy lawn. One readily observable kind of virus mutation alters the plaque morphology. The "r" mutants of bacteriophage T4 were first noticed because they formed much larger plaques on lawns of E. coli B (r stands for rapid lysis). One class of r mutants, the r_{II} mutants, do not form plaques on E. coli strain K. Although they initiate an infection in E. coli K, the infection is abortive and progeny viruses do not form. Their distinctive plaque morphology on E. coli B and their inability to grow on E. coli K were elegantly exploited early on to define the general nature of genetic mutation and to elucidate the triplet structure of the genetic code.

As a practical introduction to these classical studies, you have been given 8 r_{II} mutants to characterize. First, you perform a set of spot tests. You infect one plate of E. coli K with a high concentration of mutant 1 and a second plate with a high concentration of mutant 2, so that many of the bacteria on each plate are infected. You then put a drop of each mutant and wild-type T4 in a ring around the plate. As a control, you spot the mutants and wild-type T4 on uninfected E. coli K. After overnight incubation you observe the results shown in Figure 6–32.

These results certainly appear meaningful, but you are still puzzled. To determine what kind of phage is present in the clear spots, you collect some of the phages from the wild-type spot and the clear spot formed by mutant 5 and test their growth. The phages from the wild-type spot form normal plaques on both E. coli B and E. coli K, as you expected. The majority of phages from the mutant-5 spot still behave as mutants: they form r plaques on E. coli B but do not grow on E. coli K. However, some of the phages from the mutant-5 spot appear to be wild-type: they form normal plaques on both strains. The frequency of wild-type phages is too high to arise from back mutation (reversion), which does occur, but only at a frequency of 10^{-5} to 10^{-6}.

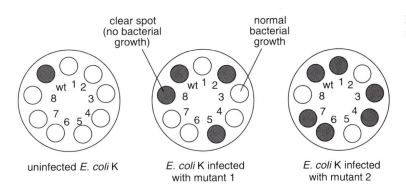

clear spot (no bacterial growth)

normal bacterial growth

uninfected E. coli K

E. coli K infected with mutant 1

E. coli K infected with mutant 2

Figure 6–32 Spot tests with various r_{II} mutants (Problem 6–41).

Table 6–2 Characteristics of Three RNA Viruses (Problem 6–42)

| | Protein Synthesis | Endogenous Polymerase | |
		RNA	DNA
Virus 1	−	+	−
Virus 2	+	−	−
Virus 3	+	−	+

A. Why do some mixtures of mutant phages grow in the spot test and others do not?

B. What pattern of growth would you expect if you repeated the spot test using *E. coli* K infected with mutant 3?

C. How did the small fraction of wild-type T4 arise in the clear spot formed by mutant 5?

*6–42 You have isolated three different viruses from animal cells. Each one contains a single-stranded RNA as its genome. To characterize them, you test them in two ways. First, you purify their RNA genomes and measure their abilities to serve as mRNA for cell-free protein synthesis. Second, you measure their endogenous polymerase activities by gently disrupting the intact viruses and then incubating them either with radioactive ribonucleoside triphosphates (NTPs) to check for RNA synthesis or radioactive deoxyribonucleoside triphosphates (dNTPs) to check for DNA synthesis. The results are shown in Table 6–2. From these results, briefly outline a life cycle for each virus, and give an example of a known virus with the same life cycle.

6–43 You are studying the procaryotic transposon Tn10 and have just figured out an elegant way to determine whether Tn10 replicates during transposition or moves directly without intervening DNA replication. Your idea is based on the key difference between these two mechanisms: both parental strands of the Tn10 will move if transposition is nonreplicative, whereas only one parental strand will move if transposition is replicative (Figure 6–33). You plan to mark the individual strands by annealing strands from two different Tn10s. To deliver these heteroduplex Tn10s efficiently into bacteria, you use Tn10-containing bacteriophage lambda DNA, which can be manipulated *in vitro* and then packaged into capsids to make infectious phage. You make sure to use a phage genome that is inactive due to other mutations, so that it cannot replicate or integrate and is ultimately destroyed; thus, the contribution of the phage genome can be ignored.

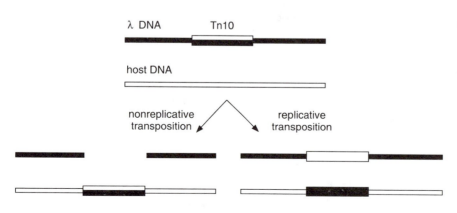

Figure 6–33 Replicative and nonreplicative transposition of a transposable element (Problem 6–43). The transposable element is shown as a heteroduplex, which is composed of two genetically different strands—one black and one white. During replicative transposition, one strand stays with the donor DNA and one strand is transferred to the recipient DNA. In nonreplicative transposition, the transposable element is cut out of the donor DNA and transferred entirely to the recipient DNA.

Your phage DNA carries slightly different versions of Tn10 inserted at the same site. Both Tn10s contain a gene for tetracycline resistance and a gene for lactose metabolism (*lacZ*), but in one, the *lacZ* gene is inactivated by a mutation. This difference provides a convenient way to follow the two Tn10s since *lacZ⁺* bacterial colonies (when incubated with an appropriate substrate) turn blue, but *lacZ⁻* colonies remain white. You denature and reanneal a mixture of the two bacteriophage DNAs, which produces an equal mixture of heteroduplexes and homoduplexes (Figure 6–34). You then package the mixture into phage capsids, infect *lacZ⁻* bacteria, and spread the infected bacteria onto petri dishes that contain tetracycline and the color-generating substrate.

Once the phage DNA is inside a bacterium, the transposon will move (at very low frequency) into the bacterial genome, where it confers tetracycline resistance on the bacterium. The rare bacterium that gains a Tn10 survives the selective conditions and forms a colony. When you score a large number of such colonies, you find that roughly 25% are white, 25% are blue, and 50% are mixed with one blue sector and one white sector.

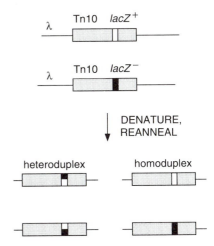

Figure 6–34 Formation of a mixture of heteroduplexes and homoduplexes by denaturing and reannealing two different Tn10-containing bacteriophage λ genomes (Problem 6–43). The phage DNA is shown as a thin line, and TN10 is shown as a box. The mutational difference between the two TN10s is indicated by the black and white segments.

A. Explain the source of each kind of bacterial colony and decide whether the results support a replicative or a nonreplicative mechanism of Tn10 transposition.

B. You performed these experiments using a recipient strain of bacteria that was incapable of repairing mismatches in DNA. How would the results differ if you used a bacterial strain that could repair mismatches?

C. Integration of bacteriophage lambda DNA into a bacterial genome by site-specific recombination occurs much more frequently than does transposition. How would the results differ if you used a mutant of bacteriophage lambda that could not replicate but could integrate?

*6–44 Bacteriophage φX174 shows an astonishing economy in the use of its limited coding capacity. It makes only a single mRNA. Yet one particular stretch of DNA encodes four completely different proteins that overlap as illustrated in Figure 6–35.

A. Which, if any, of the overlapping genes are translated left to right in Figure 6–35?

B. Which, if any, of the genes are translated in the same reading frame?

C. A mutation at a particular tyrosine codon (TAC) in the gene for protein K gives a rise to a stop codon (TAG). This mutation is accompanied by a glutamine (CAA) to glutamic acid (GAA) change in protein B. What is the mutational change in protein A? In protein C?

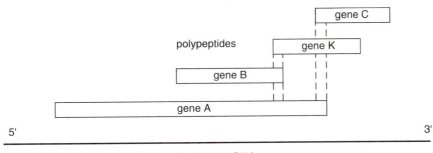

Figure 6–35 Four overlapping genes in φX174 along with the associated mRNA (Problem 6–44). Dashed lines indicate the two regions where three genes overlap.

So much has been written about genetic engineering and changing organisms in a directed way that many people may think that there is now nothing further to do. On the contrary, the subject is still in its infancy and, in fact, I claim that in one particular sense it does not yet exist. The essence of engineering is *design* and it is the implementation of a *plan* that distinguishes the engineer from the tinkerer and separates intentional human activity from the chance events of nature. Today, no-one can order a new organism like a unicorn, or a dragon, or a centaur, because nobody knows how to design one, let alone build it. Consider the centaur. It is a very old mythical beast and the man who first drew one simply cut off the torso of a man and stuck it onto the bottom of a decapitated horse. This is not genetic engineering but hopeful transplantation surgery, and you could not make real centaurs that way. (Caption from Sydney Brenner, Genes and Development. In Cellular Controls in Differentiation, ed. by Clive W. Lloyd and David A. Rees. New York: Academic Press, 1981.)

Recombinant DNA Technology

- The Fragmentation, Separation, and Sequencing of DNA Molecules
- Nucleic Acid Hybridization
- DNA Cloning
- DNA Engineering

The Fragmentation, Separation, and Sequencing of DNA Molecules

(MBOC 292–300)

7–1 Fill in the blanks in the following statements.

A. Many bacteria make _____, each of which recognizes a specific sequence of four to eight nucleotides in DNA.

B. By comparing the sizes of the DNA fragments produced from a particular genetic region after treatment with a combination of different restriction nucleases, a _____ of that region can be constructed showing the location of each restriction site in relation to neighboring restriction sites.

C. The technique of _____ allows one to identify specific sequences in DNA to which DNA-binding proteins attach.

7–2 Indicate whether the following statements are true or false. If a statement is false, explain why.

___ A. Bacteria that make a specific restriction nuclease for defense against viruses have evolved in such a way that their own genome does not contain the recognition sequence for the nuclease.

___ B. Comparison of restriction maps allows one to compare the same region of DNA in different individuals without having to determine the nucleotide sequences in detail.

___ C. Pulsed-field gel electrophoresis uses a strong electric field to separate very long DNA molecules, stretching them out so that they travel end-first through the gel at a rate that depends on their length.

___ D. Polynucleotide kinase can be used to label a DNA fragment by transferring a single ^{32}P-labeled nucleotide to the 5′ end of each DNA chain.

___ E. Although in principle there are six different reading frames in which a DNA sequence can be translated into protein, the correct one is generally recognizable as the only one lacking frequent stop codons.

___ F. A DNA footprint results when a DNA-binding protein covers the nucleotides at its binding site and protects their phosphodiester bonds from cleavage, leaving a prominent gap in the gel pattern.

7–3 The restriction enzymes BamHI and PstI cut their recognition sequences as shown in Figure 7–1 (next page).

A. Indicate the 5′ and 3′ ends of the cut DNA molecules.

B. How would the ends be modified if you incubated the cut molecules with DNA polymerase in the presence of all four dNTPs?

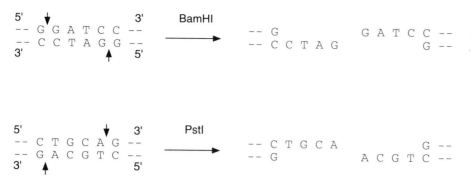

Figure 7–1 Cleavage at BamHI and PstI recognition sequences (Problem 7–3). Only the nucleotides that form the recognition sites are shown.

C. After the reaction in part B, could you still join the BamHI ends together by incubation with T4 DNA ligase? Could you still join the PstI ends together? (T4 DNA ligase will join blunt ends together as well as cohesive ends.)

D. Will joining of the ends in part C regenerate the BamHI site? Will it regenerate the PstI site?

*7–4   The restriction enzyme Sau3A recognizes the sequence 5′-GATC-3′ and cleaves on the 5′ side (to the left) of the G. (Since the top and bottom strands of most restriction sites read the same in the 5′-to-3′ direction, only one strand of the site needs to be shown.) The single-stranded ends produced by Sau3A cleavage are identical to those produced by BamHI cleavage (see Figure 7–1), allowing the two types of ends to be joined together by incubation with DNA ligase. (You may find it helpful to draw out the product of this ligation to convince yourself that it is true.)

A. What fraction of BamHI sites can be cut with Sau3A? What fraction of Sau3A sites can be cut with BamHI?

B. If two BamHI ends are ligated together, the resulting site can be cleaved again by BamHI. The same is true for two Sau3A ends. However, suppose you ligate a Sau3A end to a BamHI end. Can the hybrid site be cut with Sau3A? With BamHI?

C. What will be the average size of DNA fragments produced by digestion of chromosomal DNA with Sau3A? With BamHI?

7–5   You have purified two DNA fragments, which were generated by BamHI digestion of recombinant DNA plasmids. One fragment is 400 nucleotide pairs, and the other is 900 nucleotide pairs. You want to join them together as shown in Figure 7–2 to create a hybrid gene, which, if your speculations are right, will have amazing new properties.

You mix the two fragments together in the presence of DNA ligase and incubate them. After 30 minutes and again after 8 hours, you remove samples and analyze them by gel electrophoresis. You are surprised to find a complex pattern of fragments instead of the 1.3-kb recombinant molecule of interest (Figure 7–3A). You notice that with longer incubation the smaller fragments

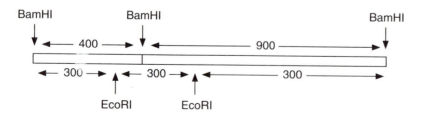

Figure 7–2 Final structure of hybrid gene (Problem 7–5).

*Problems with an asterisk (*) are answered in the Instructor's Manual.

diminish in intensity and the larger ones increase in intensity. If you cut the ligated mixture with BamHI, you regenerate the starting fragments (Figure 7–3A).

Puzzled, but undaunted, you purify the 1.3-kb fragment from the gel and check its structure by digesting a sample of it with BamHI. As expected, the original two bands are regenerated (Figure 7–3B). Just to be sure it is the structure you want, however, you digest another sample with EcoRI. You expected this digestion to generate two fragments 300 nucleotides in length and one fragment 700 nucleotides in length. Once again you are surprised by the complexity of the gel pattern (Figure 7–3B).

A. Why are there so many bands in the original ligation mixture?
B. Why are so many fragments produced by EcoRI digestion of the pure 1.3-kb fragment?

*7–6   The restriction enzyme EcoRI recognizes the sequence 5′-GAATTC-3′ and cleaves between the G and A to leave 5′ protruding single strands (like BamHI in Figure 7–1). PstI, on the other hand, recognizes the sequence 5′-CTGCAG-3′ and cleaves between the A and G to leave 3′ protruding single strands (see Figure 7–1). These two recognition sites are displayed on the helical representations of DNA in Figure 7–4.

A. For each restriction site indicate the position of cleavage on each strand of the DNA.
B. From the positions of the cleavage sites decide for each restriction enzyme whether you expect it to approach the recognition site from the major-groove side or from the minor-groove side.

7–7   Which, if any, of the restriction enzymes listed below will *definitely* cleave a segment of cDNA that encodes the peptide KIGPACF? (N can be any nucleotide.)

Restriction Enzyme	Recognition Sequence
AluI	AGCT
Sau96I	GGNCC
HindIII	AAGCTT

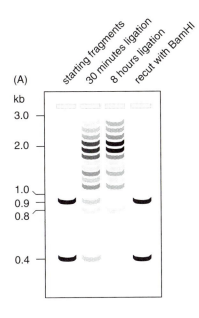

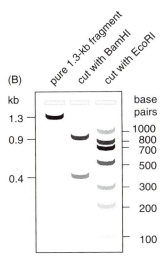

Figure 7–3 Ligation of pure DNA fragments (A) and diagnostic digestion of the purified 1.3-kb fragment (B) (Problem 7–5).

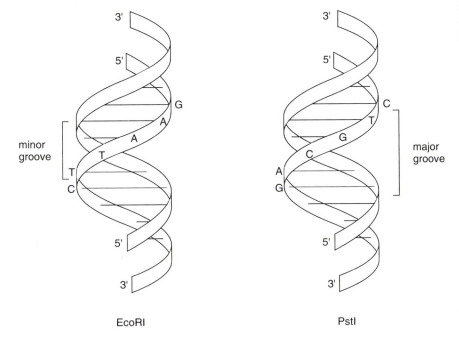

EcoRI   PstI

Figure 7–4 Restriction sites on helical DNA (Problem 7–6).

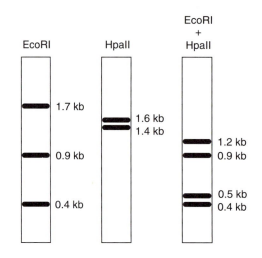

Figure 7–5 Sizes of DNA bands produced by digestion of a 3.0-kb BamHI fragment by EcoRI, HpaII, and a mixture of the two (Problem 7–8).

EcoRI
- 1.7 kb
- 0.9 kb
- 0.4 kb

HpaII
- 1.6 kb
- 1.4 kb

EcoRI + HpaII
- 1.2 kb
- 0.9 kb
- 0.5 kb
- 0.4 kb

7–8 You wish to make a restriction map of a 3.0-kb BamHI restriction fragment. You digest three samples of the fragment with EcoRI, HpaII, and a mixture of EcoRI and HpaII. You then separate the fragments by gel electrophoresis and visualize the DNA bands by staining with ethidium bromide (Figure 7–5). From these results, prepare a restriction map that shows the relative positions of the EcoRI and HpaII recognition sites and the distances in kilobases (kb) between them.

7–9 *Tetrahymena* is a ciliated protozoan with two nuclei. The smaller nucleus (the micronucleus) maintains a master copy of the cell's chromosomes; it participates in sexual conjugation, but not in day-to-day gene expression. The larger nucleus (the macronucleus) maintains a "working" copy of the cell's genome in the form of a large number of gene-sized double-stranded DNA fragments (minichromosomes), which are actively transcribed. The minichromosome that contains the ribosomal RNA genes is present in many copies; it can be separated from the other minichromosomes by gradient centrifugation and studied in detail.

When examined by electron microscopy, each ribosomal minichromosome is a linear structure 21 kb in length. Ribosomal minichromosomes also migrate at 21 kb when subjected to gel electrophoresis (Figure 7–6, lane 1). However, if the minichromosome is cut with the restriction enzyme BglII, the two fragments that are generated (13.4 kb and 3.8 kb) do not sum to 21 kb (Figure 7–6, lane 2). When the DNA is cut with other restriction enzymes, the sizes of the fragments always sum to less than 21 kb; moreover, the fragments in each digest add up to different overall lengths.

If the uncut minichromosome is first denatured and reannealed before it is run on a gel, the 21-kb fragment is replaced by a double-stranded fragment exactly half its length, 10.5 kb (Figure 7–6, lane 3). Similarly, if the BglII-cut minichromosome is denatured and reannealed, the 13.4-kb fragment is replaced by a double-stranded fragment half its length, 6.7 kb (Figure 7–6, lane 4).

Explain why the restriction fragments do not appear to add up to 21 kb. Why does the electrophoretic pattern change when the DNA is denatured and reannealed. What do you think might be the overall organization of sequences in the ribosomal minichromosome?

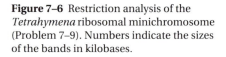

BglII digestion	−	+	−	+
denature and reanneal	−	−	+	+

kb
- 21.0
- 13.4
- 10.5
- 6.7
- 3.8

1 2 3 4

Figure 7–6 Restriction analysis of the *Tetrahymena* ribosomal minichromosome (Problem 7–9). Numbers indicate the sizes of the bands in kilobases.

*7–10 You have cloned a 4-kb segment of an important gene into a plasmid vector (Figure 7–7) and now wish to prepare a restriction map of the gene in preparation for other DNA manipulations. Your advisor left instructions on how to do it, but she is now on vacation, so you are on your own. You follow her instructions as outlined below.

1. Cut the plasmid with EcoRI.
2. Add a radioactive label to the EcoRI ends.
3. Cut the labeled DNA with BamHI.

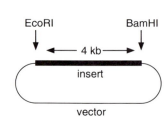

EcoRI   BamHI
←— 4 kb —→
insert

vector

Figure 7–7 Recombinant plasmid containing a cloned DNA segment (Problem 7–10).

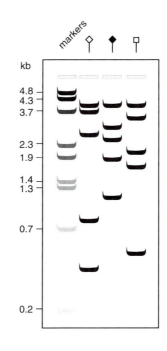

Figure 7–8 Autoradiogram showing the electrophoretic separation of the labeled fragments after partial digestion with the three restriction enzymes represented by the symbols (Problem 7–10). Numbers at the left indicate the sizes of a set of marker fragments (in kb).

4. Purify the insert away from the plasmid.
5. Digest the labeled insert briefly with a restriction enzyme so that on average each labeled molecule is cut about one time.
6. Repeat step 5 for several different restriction enzymes.
7. Run the partially digested samples side by side on an agarose gel.
8. Place the gel against x-ray film so that fragments with a radioactive end can expose the film to produce an autoradiogram.
9. Draw the restriction map.

Your biggest problem thus far has been step 5; however, by decreasing the amounts of enzyme and lowering the temperature, you were able to find conditions for partial digestion. You have now completed step 8, and your autoradiogram is shown in Figure 7–8.

Unfortunately, your advisor was not explicit about how to construct a map from the data in the autoradiogram. She is due back tomorrow. Will you figure it out in time?

7–11 A particularly clear example of a dideoxy sequencing gel is shown in Figure 7–9. Try reading it. As read from the bottom of the gel to the top, the sequence corresponds to the mRNA for a protein. Can you find the open reading frame in this sequence?

Nucleic Acid Hybridization
(MBOC 300–308)

7–12 Fill in the blanks in the following statements.

A. Complementary single strands of DNA will readily re-form double helices in a process called _____.
B. In the technique known as _____, intact RNA molecules are size fractionated by gel electrophoresis, transferred to a sheet of nitrocellulose (or nylon) paper, and hybridized to a radioactive DNA probe.
C. In the technique known as _____, DNA restriction fragments are separated by gel electrophoresis, transferred to nitrocellulose (or nylon) paper, and hybridized to a radioactive DNA probe.

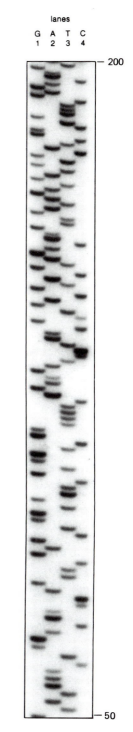

Figure 7–9 A dideoxy sequencing gel of a cloned DNA (Problem 7–11).

D. Differences in the size of DNA restriction fragments between individuals are known as _____.

E. _____ uses nucleic acid probes to determine the locations of specific nucleic acid sequences in cells and tissues.

7–13 Indicate whether the following statements are true or false. If a statement is false, explain why.

___ A. Hybridization reactions using DNA probes are so sensitive and selective that it is possible to determine how many copies of a particular DNA sequence are present in a cell's genome.

___ B. Hybridization of DNA probes to RNA molecules is useful for determining whether a cell is expressing a given gene, but it is not useful for determining transcription start and stop sites or the positions of introns.

___ C. If RNA or DNA molecules in a crude mixture are separated by electrophoresis before hybridization and if molecules of only one or a few sizes become labeled with the probe, one can be certain that the hybridization was specific.

___ D. If a variation in the DNA sequence is a rare one in the population, it is called a mutation; if it is a common one, it is called a polymorphism.

___ E. Genetic markers on different chromosomes are always genetically unlinked, while markers on the same chromosome are always genetically linked.

___ F. High-stringency hybridization using DNA oligonucleotides can be used in the prenatal diagnosis of genetic disease to detect a mutant gene that differs from the normal gene by a single nucleotide.

___ G. Because the members of a gene family are close relatives, they can usually be detected by high-stringency hybridization using one member as a probe.

___ H. The spatial resolution of *in situ* hybridization can be greatly improved by labeling the DNA probes chemically instead of radioactively and then using an antibody that specifically recognizes the chemical modification.

7–14 X-chromosome inactivation is a unique developmental regulatory mechanism that turns off the expression of genes all over one entire chromosome. In the case of the human X chromosome, inactivation involves over 150 million base pairs of DNA and several thousand genes. By randomly inactivating one of the two X chromosomes in normal females, X inactivation results in the expression of X-linked genes in such females being the same as that of the genes on the sole X chromosome in normal males. Although the mechanism of inactivation remains unknown, it is thought to be initiated at a specific region of the chromosome—the X-inactivation center—and then to spread to the rest of the chromosome.

Although most genes on the inactivated X chromosome are turned off, a few remain active. You think there may be a clue to the mechanism of X inactivation in the pattern of expression of X-linked genes. To test this idea, you isolate a large number of cDNAs for genes on the human X chromosome and examine their expression by Northern blotting. You compare expression in cells from normal males and females, in cells from individuals with abnormal numbers of X chromosomes, and in rodent:human hybrid cell lines that have retained either one inactive human X chromosome (X_i) or one active human X chromosome (X_a).

Among all your cDNAs you find three patterns of expression, as illustrated in Figure 7–10 for cDNAs A, B, and C.

A. For each pattern of expression decide whether the gene is expressed from the active X, the inactive X, or both. Which pattern do you expect to be the most common? Which pattern is the most surprising?

B. From the results with cells from abnormal individuals, formulate a rule as to how many chromosomes are inactivated and how many remain active during X inactivation.

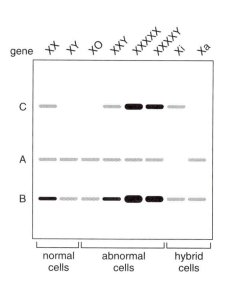

Figure 7–10 A Northern analysis of gene expression from cells with different numbers and types of X chromosomes (Problem 7–14). RNA from cells was run out on gels, blotted onto nitrocellulose, probed with a mixture of radioactive probes from cDNAs A, B, and C, and visualized by autoradiography. The positions of the RNA bands that correspond to genes A, B, and C were determined in a separate experiment.

*7–15 The DNA of certain animal viruses can be integrated into a cell's DNA as shown schematically in Figure 7–11. You want to know the structure of the viral genome as it exists in the integrated state. You digest samples of viral DNA and DNA from cells that contain the integrated virus with restriction enzymes that cut the viral DNA at known sites. Subsequently you separate the fragments by electrophoresis on agarose gels and visualize the bands that contain viral DNA by Southern blotting using radioactive viral DNA as a hybridization probe. You obtain the patterns shown in Figure 7–12.

From this information, in which of the five segments of the viral genome a–e did the integration event occur?

7–16 Many mutations that cause human genetic disease involve the substitution of one nucleotide for another. Oligonucleotide ligation provides a rapid way to detect such specific single-nucleotide differences. This assay uses pairs of oligonucleotides: one labeled with biotin and the other with a radioactive (or fluorescent) tag, as shown in Figure 7–13B for the detection of the mutation responsible for sickle-cell anemia (Figure 7–13A).

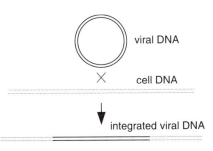

Figure 7–11 Integration of viral DNA into cell DNA (Problem 7–15).

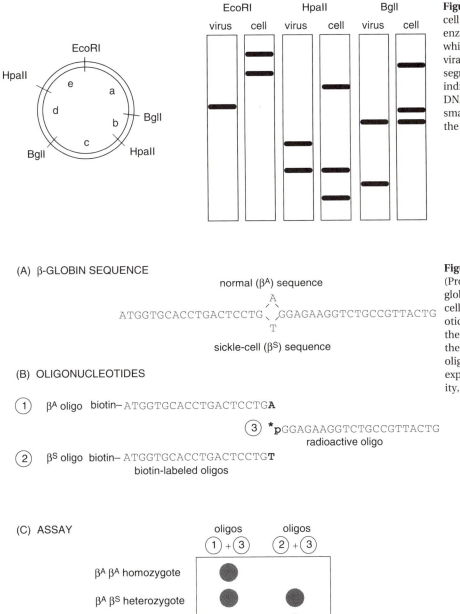

Figure 7–12 Southern blots of viral and cell DNA digested with various restriction enzymes (Problem 7–15). The sites at which the restriction enzymes cleave the viral DNA are shown at the left, and the segments defined by these sites are indicated by letters. Agarose gels separate DNA fragments on the basis of size—the smaller the fragment, the farther toward the bottom of the gel it moves.

Figure 7–13 Oligonucleotide ligation assay (Problem 7–16). (A) Sequence of the β-globin gene around the site of the sickle-cell (β^S) mutation. (B) Specific oligonucleotides for ligation assay. (C) Assay to detect the single-nucleotide difference between the β^A and β^S sequences. Biotinylated oligos are collected in a spot and then exposed to x-ray film to detect radioactivity, which turns the film black.

In the assay, pairs of oligonucleotides are hybridized to DNAs from individuals and incubated in the presence of DNA ligase. Biotinylated oligonucleotides are then bound to streptavidin on a solid support and any associated radioactivity is visualized by autoradiography, as shown in Figure 7–13C.

A. Do you expect the β^A and β^S oligonucleotides to hybridize to both β^A and β^S DNA?

B. How does this assay distinguish between β^A and β^S DNA?

*7–17 There has been a colossal snafu in the maternity ward at your local hospital. Four sets of male twins, born within an hour of each other, were inadvertently shuffled in the excitement occasioned by that unlikely event. You have been called in to set things right. As a first step you want to get the twins matched up. To that end you analyze a small blood sample from each infant using a hybridization probe that detects variable number tandem repeat (VNTR) polymorphisms located in widely scattered regions of the genome. The results are shown in Figure 7–14.

A. Which infants are brothers?

B. How will you match brothers to the correct parents?

7–18 RFLPs can be used to follow the inheritance of specific chromosomal regions. Under certain circumstances, RFLPs that are near a defective gene can be used to determine whether the deleterious gene is present in an unborn child.

A concerned couple has come to you for genetic counseling. Their second child died shortly after birth from an inherited genetic disease, and the mother is pregnant again. The deceased child was the second individual in the family lineage to be affected: a brother of the paternal grandmother of the child was the other. The couple wishes to know whether their unborn child has this genetic disease. You have studied the genetics of this disease and know that it is caused by an autosomal recessive gene. The chromosomal region, in which the disease gene is located, has several restriction-site polymorphisms within the population as a whole, as indicated in Figure 7–15A. The restriction-site polymorphisms in the population are indicated with a +/− designation.

Figure 7–14 DNA fingerprint analysis of shuffled twins (Problem 7–17).

Figure 7–15 Restriction map (A) of the chromosomal region carrying the disease gene and restriction patterns (B) from various family members (Problem 7–18). The polymorphic sites are labeled A, B, and C. In the geneology (B) circles represent females, squares represent males, and the triangle represents the unborn child.

(A) RESTRICTION MAP

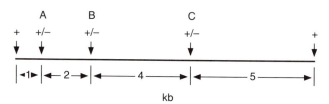

(B) GENEALOGY

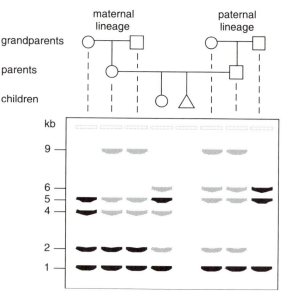

You isolate DNA samples from both parents, both sets of grandparents, and the unaffected child and characterize them by restriction mapping, as shown in Figure 7–15B. You are now ready to test the fetus. What restriction pattern will indicate to you that the fetus is likely to have inherited the disease?

*7–19   You wish to know whether the cDNA you have isolated and sequenced is the product of a unique gene or is made by a gene that is a member of a family of related genes. To address this question, you digest cell DNA with a restriction enzyme that cleaves the cell DNA but not the cDNA, separate the fragments by gel electrophoresis, and visualize bands using radioactive cDNA as a probe. The Southern blot shows two bands, one of which hybridizes more strongly to the probe than the other.

You interpret the stronger hybridizing band as the gene that encodes your cDNA and the weaker hybridizing band as a related gene. When you explain your result to your advisor, however, she cautions that you have not proven that there are two genes. She suggests that you repeat the Southern blot in duplicate, probing one with a radioactive segment from the 5′ end of the cDNA and the other with a radioactive segment from the 3′ end of the cDNA.

A. How might you get two hybridizing bands if the cDNA was the product of a unique gene?
B. What results would you expect from the experiment your advisor proposed if there is a single unique gene? If there are two genes?

DNA Cloning
(MBOC 308–318)

7–20   Fill in the blanks in the following statements.

A. In _____, a DNA fragment that contains a gene of interest is inserted into the purified DNA genome of a self-replicating genetic element—generally a virus or a plasmid.
B. The _____ used for gene cloning are small circular molecules of double-stranded DNA derived from larger plasmids that occur naturally in bacterial cells.
C. A plasmid that contains one fragment from cleaved genomic DNA is said to contain a _____; an extensive collection of such plasmids is said to constitute a _____.
D. A clone that contains a DNA copy of an mRNA is called a _____, and the entire collection of clones derived from one mRNA preparation constitutes a _____.
E. _____ can be used to enrich for a particular nucleotide sequence prior to cDNA cloning by hybridizing cDNA molecules from a cell type that makes the protein of interest with a large excess of mRNA molecules from a closely related cell type that does not make it.
F. Once one genetically mapped gene has been cloned, the clones in a genomic DNA library that correspond to neighboring genes can be identified using a technique called _____.
G. The technique called _____ allows the DNA from a selected region of a genome to be amplified a billionfold, provided that at least part of its nucleotide sequence is already known.

7–21   Indicate whether the following statements are true or false. If a statement is false, explain why.

___ A. If a number of bacterial cells have been successfully transfected, each will generally contain a different foreign DNA insert, which will be inherited by all of the progeny cells of that cell and will together form a small colony in a culture dish.

_ B. Genomic DNA libraries contain a random sample of all the DNA sequences in an organism, whereas a cDNA library contains a random sample of all the expressed genes in an organism.

_ C. By far the most important advantage of cDNA clones over genomic clones is that they contain the complete sequence of a gene.

_ D. cDNA libraries prepared after subtractive hybridization provide a powerful way to clone genes whose products are known to be restricted to a specific differentiated cell type.

_ E. The sequence, antigenicity, or ligand-binding properties of a protein of interest can be used to make a specific probe to find the corresponding gene in a DNA library.

_ F. It is relatively easy to differentiate between false-positive clones and authentic clones when the desired clone encodes a protein that has already been characterized by other means.

_ G. In chromosome walking one knows by DNA sequencing when the gene of interest has been reached.

_ H. Once an ordered set of genomic clones is available for an organism, one will then begin a "chromosome walk" by obtaining from the library all the clones covering the region of the genome that contains the mutant gene of interest.

_ I. Positional cloning, which starts with genetic linkage mapping to locate the gene in the genome, provides a straightforward and rapid method for isolating specific human genes.

_ J. If each cycle of PCR doubles the amount of DNA synthesized in the previous cycle, 10 cycles give a thousandfold amplification, 20 cycles give a millionfold amplification, and 30 cycles give a billionfold amplification.

7–22   You want to clone a DNA fragment that has KpnI ends into a vector that has BamHI ends. The problem is that BamHI and KpnI ends are not compatible: BamHI leaves a 5′ overhang and KpnI leaves a 3′ overhang (Figure 7–16). A friend suggests that you try to link them with an oligonucleotide "splint" as shown in Figure 7–16. It is not immediately clear to you that such a scheme will work because ligation requires an adjacent 5′ phosphate and 3′ hydroxyl. Although molecules that are cleaved with restriction enzymes have appropriate ends, oligonucleotides are synthesized with hydroxyl groups at both ends. Also, although the junction in Figure 7–16 is BamHI-KpnI, the other junction is KpnI-BamHI, and you are skeptical that the same oligonucleotide could splint both junctions.

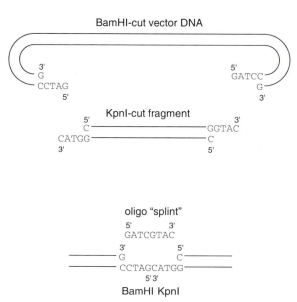

Figure 7–16 Scheme to use an oligo "splint" to link incompatible restriction ends (Problem 7–22).

A. Draw a picture of the KpnI-BamHI junction and the oligo splint that would be needed. Is this oligonucleotide the same or different from the one shown in Figure 7–16?

B. Draw a picture of the molecule after treatment with DNA ligase. Indicate which if any of the nicks will be ligated.

C. Will your friend's scheme work?

7–23 You are constructing a cDNA library in a high-efficiency cloning vector called λYES (Yeast–*E. coli* Shuttle). This vector is a clever combination of a bacteriophage lambda genome with a plasmid that can replicate in *E. coli* and yeast. It combines the advantages of viral and plasmid cloning vectors. cDNAs can be inserted into the plasmid portion of the vector, which can then be packaged into a virus coat *in vitro*. The packaged vector infects *E. coli* much more efficiently than plasmid DNA on its own. Once inside *E. coli* the plasmid sequence in λYES can be induced to recombine out of the lambda genome and replicate on its own. This allows cDNAs to be isolated on plasmids, which are ideal for subsequent manipulations.

To maximize the efficiency of cloning, both the vector and the cDNAs are prepared in special ways. The vector DNA is cut at its unique XhoI site (5'-C*TCGAG) and then incubated with DNA polymerase in the presence of dTTP. A preparation of double-stranded cDNAs with blunt ends is ligated to a double-stranded oligonucleotide adaptor composed of two oligonucleotides, 5'-CGAGATTTACC and 5'-GGTAAATC, each of which carries a phosphate at its 5' end. The vector DNA and cDNAs are then mixed and ligated together. This procedure turns out to be very efficient. Starting with 2 µg of vector and 0.1 µg of cDNA, you make a library consisting of 4×10^7 recombinant molecules.

A. Given that the vector is 43 kb long and the average size of the cDNAs is about 1 kb, estimate the ratio of vector molecules to cDNA molecules in the ligation mixture.

B. What is the efficiency with which vector molecules are converted to recombinant molecules in this procedure? (The average mass of a nucleotide pair is 660 d.)

C. Explain how the treatments of the vector molecules and the cDNAs allow them to be ligated together. Can the treated vector molecules be ligated together? Can the cDNAs be ligated together?

D. How does the treatment of the vector and cDNAs improve the efficiency of generating recombinant DNA molecules?

E. Can a cDNA be cut out of a recombinant plasmid with XhoI?

*7–24 It's midnight. Your friend has awakened you with yet another grandiose scheme. He has spent the last two years purifying a potent modulator of the immune response. Tonight he got the first 30 amino acids from a protein-sequenator run (Figure 7–17). He wants your help in cloning the gene so it can be expressed at high levels in bacteria. He argues that this protein, by stimulating the immune system, could be the ultimate cure for the common cold. He's already picked out a trade name—Immustim.

Even though he gets carried away at times, he is your friend, and you are intrigued by this idea. You promise to call him back in 15 minutes as soon as you have checked out the protein sequence. What two sets of 20-nucleotide-long oligonucleotides will you recommend to your friend as the best hybridization probes for screening a genomic DNA library?

```
          10        20        30
MFYWMIGRST EDWMPLYMKD FWAKHSLICE
```

Figure 7–17 The first 30 amino acids in your friend's protein-sequenator run (Problem 7–24).

```
5' -  GACCTGTGGAAGC  - - - - -  CATACGGGATTG  - 3'
3' -  CTGGACACCTTCG  - - - - -  GTATGCCCTAAC  - 5'
```

Figure 7–18 DNA to be amplified and potential PCR primers (Problem 7–25).

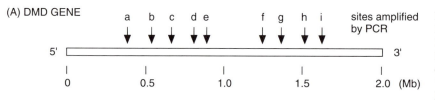

```
            primers                          primers

5' -  GACCTGTGGAAGC              5' -  CATACGGGATTG
5' -  CTGGACACCTTCG              5' -  GTATGCCCTAAC
5' -  CGAAGGTGTCCAG              5' -  GTTAGGGCATAC
5' -  GCTTCCACAGGTC              5' -  CAATCCCGTATG
```

7–25   You want to amplify the DNA between the two stretches of sequence shown in Figure 7–18. Of the listed primers choose the appropriate pair that will allow you to amplify the DNA by PCR.

*7–26   Duchenne's muscular dystrophy (DMD) is among the most common human genetic diseases, affecting approximately 1 in 3500 male births, and one-third of all new cases arise via new mutations. The DMD gene, which is located on the X chromosome, is greater than 2 million base pairs in length and contains at least 70 exons. Large deletions account for about 60% of all cases of the disease, and they tend to be concentrated around two regions of the gene.

   The very large size of the DMD gene complicates the analysis of mutations. One rapid approach, which can detect about 80% of all deletions, is termed multiplex PCR. It uses multiple pairs of PCR primers to amplify nine different segments of the gene in the two most common regions for deletions (Figure 7–19A). By arranging the PCR primers so that each pair gives a different size product, it is possible to amplify and analyze all nine segments in one PCR reaction. An example of multiplex PCR analysis of six unrelated DMD males is shown in Figure 7–19B.

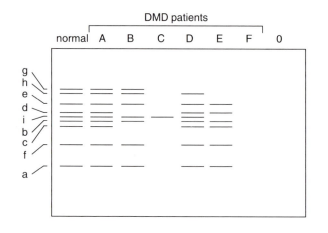

Figure 7–19 Multiplex PCR analysis of six DMD patients (Problem 7–26). (A) The DMD gene with the nine sites amplified by PCR indicated by arrows. (B) Agarose gel display of amplified PCR products. "Normal" indicates a normal male. The lane marked "0" shows a negative control with no added DNA.

A. Describe the deletions, if any, in each of the six DMD patients.

B. What additional control might you suggest to confirm your analysis of patient F?

7–27   You are working on an oncogene that appears from its sequence to encode a transcription factor. When expressed in bacteria, the protein produced by the gene is completely insoluble, which prohibits direct tests of its function, although it does allow you to raise good antibodies against it. Immunological staining shows that the protein is located in the nucleus, as expected for a transcription factor. If you could determine what DNA sequence it binds, you might be able to find genes that it regulates by searching existing sequence databases for natural sites in the promoters of known genes.

Your advisor, who is known for her clever imagination and green thumb when it comes to molecular biology, suggests that you might be able to use PCR to amplify rare DNA molecules that are bound by your protein. Her idea, as outlined in Figure 7–20, is (1) to synthesize a population of random-sequence oligonucleotides 26 nucleotides long, flanked by defined 25-nucleotide sequences that can serve as primer sites for PCR amplification (Figure 7–20A); (2) to add these oligos to a crude cell extract that contains the transcription factor, which can bind to oligos that contain the binding sites; (3) to isolate the oligonucleotides that are bound to the transcription factor using antibodies against the factor; and (4) to PCR amplify the selected oligonucleotides for sequence analysis.

You begin this procedure with 0.2 ng of single-stranded random-sequence oligo, which you convert to double-stranded DNA using one of the PCR primers. After four rounds of selection and amplification as outlined in Figure

(A) ORIGINAL RANDOM-SEQUENCE OLIGONUCLEOTIDE

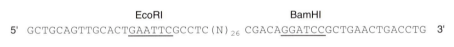

                EcoRI   BamHI

5' GCTGCAGTTGCACT$\underline{GAATTC}$GCCTC(N)$_{26}$ CGACA$\underline{GGATCC}$GCTGAACTGACCTG 3'

(B) SELECTION AND AMPLIFICATION PROTOCOL

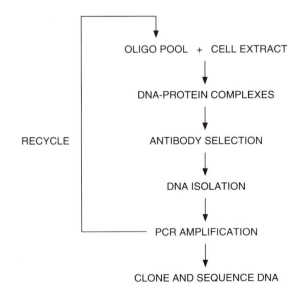

Figure 7–20 Selection and amplification of specific DNA sequences from a random pool (Problem 7–27). (A) Original random-sequence oligonucleotide used for selection. N stands for any nucleotide. The EcoRI and BamHI sites facilitate cloning of the selected DNA for sequence analysis. (B) Scheme for selecting and amplifying oligonucleotides that bind to a sequence-specific DNA-binding protein.

GAATTCGCCTCGAGCACATCATTGCCCATATATGGCACGACAGGATCC
GAATTCGCCTCTTCTAATGCCCATATATGGACTTGCTCGACAGGATCC
GGATCCTGTCGGTCCTTTATGCCCATATATGGTCATTGAGGCGAATTC

GAATTCGCCTCATGCCCATATATGGCAATAGGTGTTTCGACAGGATCC
GAATTCGCCTCTATGCCCATATAAGGCGCCACTACCCCGACAGGATCC

GAATTCGCCTCGTTCCCAGTATGCCCATATATGGACACGACAGGATCC
GGATCCTGTCGACACCATGCCCATATTTGGTATGCTCGAGGCGAATTC
GAATTCGCCTCATTTATGAACATGCCCTTATAAGGACCGACAGGATCC

GAATTCGCCTCTAATACTGCAATGCCCAAATAAGGAGCGACAGGATCC
GAATTCGCCTCATGCCCAAATATGGTCATCACCTACACGACAGGATCC

Underlined sequences correspond to PCR primer sites. Two sequences are shown starting at the BamHI end so that the binding site is oriented the same way in all sequences.

7–20B, you digest the isolated DNA with BamHI and EcoRI, clone the fragments into a plasmid, and sequence 10 individual clones (Table 7–1).

A. What is the consensus sequence to which your transcription factor binds?
B. Does the consensus binding sequence show any signs of symmetry; that is, is any portion of it palindromic? Does this help you decide whether the transcription factor binds to DNA as a monomer or as a dimer?
C. How many single-stranded oligonucleotides were present in the 0.2-ng sample with which you started the experiment? (The average mass of a nucleotide is 330 d.)
D. Assuming that the starting population of oligonucleotides was truly random in the 26 central nucleotides, is it likely that all possible 14-base-pair-long sequences were represented in the starting sample?

*7–28 A critical step in the technique of chromosome walking is to generate a probe from one end of a cloned genomic segment. This end-specific probe can then be used to search for additional clones containing genomic segments that overlap the original one. The difficulty is that the sequence of the inserted DNA is unknown, which precludes making primers for standard PCR. It is, of course, possible to subclone pieces from the end, but this procedure is tedious and time-consuming. One way around this problem is a clever adaptation of PCR, called inverse PCR because it uses primers that are oriented in the opposite direction from the usual arrangement.

The orientation of primers for inverse PCR is shown in Figure 7–21 for the generation of an end-specific probe from a large genomic insert in a YAC (yeast artificial chromosome). The two PCR primers, which point away from one

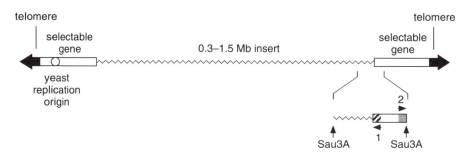

Figure 7–21 Arrangement of primers for generating an end-specific fragment by inverse PCR (Problem 7–28). The YAC cloning vector (boxes) flank the insert (jagged line). The nucleotide sequence of the vector is known.

another, are arranged so that one is adjacent to the inserted DNA and the other is adjacent to a known site for a frequently cutting restriction enzyme such as Sau3A (5'-GATC). The DNA is digested with Sau3A and then diluted sufficiently in the presence of DNA ligase so that fragments tend to circularize rather than ligate to other fragments. The DNA is then mixed with the primers and subjected to PCR.

A. Draw the structure of the PCR amplification product and show the positions and orientations of the PCR primers.
B. Will Sau3A cut the PCR product? Where?
C. Would the technique work if there were an additional Sau3A site located between the two primers on the YAC?
D. Why is circularization a critical step for inverse PCR?
E. Why is it necessary to use a restriction enzyme that cuts DNA frequently?

DNA Engineering
(MBOC 319–331)

7–29 Fill in the blanks in the following statements.

A. _____ can provide a powerful solution to the problem of defining the function of an unknown gene through the specific abnormalities in a mutant organism in which the gene is defective.
B. Recombinant DNA technology has made it possible to start with a particular gene, make mutations in it, and then create mutant cells or organisms, a process commonly referred to as _____.
C. Regions of a protein with special functions can be identified by making _____ with an easily detected reporter protein and then following the behavior of the reporter protein in a cell.
D. Fusing a protein to a short peptide of 8 to 12 amino acids that can be recognized by an antibody is called _____.
E. A _____ mutation is one in which a mutant gene eliminates the activity of its normal counterparts in the cell.
F. If a cloned gene is engineered so that the complementary DNA strand is transcribed instead of the usual strand, it will produce _____ molecules that have a sequence complementary to the normal RNA transcripts.
G. Animals that have been permanently altered so they pass foreign genes on to their progeny are called _____ organisms, and the foreign genes are called _____.

7–30 Indicate whether the following statements are true or false. If a statement is false, explain why.

__ A. PCR can be used both to amplify any natural nucleotide sequence and to redesign its two ends, so that any two naturally occurring DNA sequences can be amplified and spliced together very rapidly and efficiently.
__ B. The most efficient way to generate a pure RNA species is to overexpress it in cells and purify it.
__ C. The usual method for making large amounts of a protein whose gene is cloned is to transcribe and translate it *in vitro*.
__ D. The regulatory sequences for a gene of interest can be identified by attaching a reporter sequence to various fragments of DNA taken from upstream or downstream of the gene.
__ E. Mutants in which the abnormal protein is temperature sensitive are very useful, since in these mutants the abnormality can be switched on and off by changing the temperature.
__ F. Specific mutations can be introduced into cloned genes using synthetic DNA oligonucleotides containing the desired sequence change.

___ G. All of the important signals carried by proteins can be studied by fusing short amino acid sequences to a reporter protein.

___ H. Gene replacements occur by homologous recombination between the artificially introduced gene and the normal chromosomal gene.

___ I. Mutant proteins that display dominant-negative effects in cells generally function as monomeric species with abnormal activities that are deleterious to the cell.

___ J. Linear DNA fragments introduced into mammalian cells are rapidly ligated end to end by intracellular enzymes to form repeated arrays, which usually become integrated into a chromosome at apparently random sites.

___ K. The underlying principle of double drug selection to enrich for rare targeted recombinants in mammalian cells is to use one drug to select for cells that have integrated foreign DNA by any means and a second drug to kill cells that have integrated the DNA randomly.

___ L. Just as mutant mice can be derived by genetic manipulation of embryonic stem cells in culture, so transgenic plants can be created from single plant cells transfected with DNA in culture.

7–31   You want to clone a cDNA into an expression vector so you can make large amounts of the protein in *E. coli*. The cDNA is flanked by BamHI sites and you plan to insert it at the BamHI site in the vector. This is your first experience with cloning, so you carefully follow the procedures in the cloning manual.

The manual recommends that you cleave the vector DNA and then treat it with alkaline phosphatase to remove the 5′ phosphates. The next step is to mix the treated vector with the BamHI-cut cDNA fragment and incubate with DNA ligase. After ligation the DNA is mixed with bacterial cells that have been treated to make them competent to take up DNA. Finally, the mixture is spread onto culture dishes filled with a solid growth medium that contains an antibiotic, which kills all cells that have not taken up the vector. The vector allows cells to survive because it carries a gene for resistance to the antibiotic.

The cloning manual also suggests four controls.

Control 1.   Plate bacterial cells that have not been exposed to any vector onto the culture dishes.

Control 2.   Plate cells that have been transfected with vector that has not been cut.

Control 3.   Plate cells that have been transfected with vector that has been cut (but not treated with alkaline phosphatase) and then incubated with DNA ligase (in the absence of the cDNA fragment).

Control 4.   Plate cells that have been transfected with vector that has been cut and treated with alkaline phosphatase and then incubated with DNA ligase (in the absence of the cDNA fragment).

For your first attempt at the experiment and all its controls, you borrow a fellow student's competent cells, but all the plates have too many colonies to count (Table 7–2). For your second attempt, you use cells you have prepared yourself, but this time you get no colonies on any plate (Table 7–2). Ever the optimist, you try again, and this time you are rewarded with more encouraging results (Table 7–2). You pick 12 colonies from the experimental sample, prepare plasmid DNA from them, and digest the DNA with BamHI. Nine colonies yield a single band the same size as the linearized vector, but three colonies have, in addition, a fragment the size of the cDNA you wanted to clone. Success is sweet!

A. What do you think happened in the first experiment? What is the point of Control 1?

B. What do you think happened in the second experiment? What is the point of Control 2?

C. What is the point of controls 3 and 4?

Table 7–2 Results of Your cDNA Cloning Endeavor (Problem 7–31)

		Results of Experiments		
Preparation of Sample		1	2	3
Control 1	Cells alone	TMTC	0	0
Control 2	Uncut vector	TMTC	0	> 1000
Control 3	Omit phosphatase, omit cDNA	TMTC	0	435
Control 4	Omit cDNA	TMTC	0	25
Experimental sample		TMTC	0	34

TMTC = too many to count

D. Why does the cloning manual recommend treating the vector with alkaline phosphatase?

*7–32 From previous work, you suspect that the glutamine (Q) in the protein segment in Figure 7–22 plays an important role at the active site. Your advisor wants you to alter the protein in three ways: change the glutamine to lysine (K), change the glutamine to glycine (G), and delete the glutamine from the protein. You plan to accomplish these mutational alterations by hybridizing an appropriate oligonucleotide to the M13 viral DNA, such that when the oligonucleotide is extended around the single-stranded M13 circle by DNA polymerase, it will complete a strand that encodes the complement of the desired mutant protein. Design three 20-nucleotide-long oligonucleotides that could be hybridized to the cloned gene on single-stranded M13 viral DNA as the first step in effecting the mutational changes.

7–33 You want to express a rare human protein in bacteria so that you can make large quantities of it. To aid in its purification you decide to add a stretch of 6 histidines to either the N-terminus or the C-terminus of the protein. Such histidine-tagged proteins bind tightly to Ni^{2+} columns, but can be readily eluted with a solution of EDTA or imidazole. This procedure allows an enormous purification in one step.

The nucleotide sequence that encodes your protein is shown in Figure 7–23. Design a pair of PCR primers, each with 18 nucleotides of homology to the gene, that will amplify the coding sequence of the gene and add an initiation codon followed by six codons for histidine to the N-terminus. Design a pair of primers that will add six histidine codons followed by a stop codon to the C-terminus.

```
              L   R   D   P   Q   G   G   V   I
5' - CTTAGAGACCCGCAGGGCGGCGTCATC - 3'
```

Figure 7–22 Sequence of DNA and encoded protein (Problem 7–32).

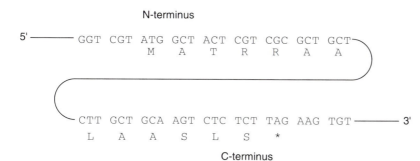

Figure 7–23 Nucleotide sequence around the N- and C-termini of the protein you want to modify (Problem 7–33). The encoded amino acid sequence is indicated below each codon using the one-letter code. The asterisk (*) indicates the stop codon. Only the top strand of the double-stranded DNA is shown.

*7–34   You have now cloned in an expression vector both versions of the histidine-tagged protein you created in the previous problem. Neither construct expresses particularly strongly in bacteria, but the product is soluble. You pass the crude extract over a Ni^{2+} affinity column, which binds histidine-tagged proteins specifically. After washing the column extensively, you elute your protein from the column using a solution containing EDTA, which chelates Ni^{2+} and strips the metal off the column, releasing your protein.

When you subject the eluted protein to electrophoresis and stain the gel for protein, you are pleased to find bands in the eluate that are not present when control bacteria are treated similarly. But you are puzzled to see that the construct tagged at the N-terminus gives a ladder of shorter proteins below the full-length protein, whereas the C-terminally tagged construct yields exclusively the full-length protein. The amount of full-length protein is about the same for each construct.

Offer an explanation for this difference between the two constructs.

*7–35   An adaptation of standard PCR, called recombinant PCR, allows virtually any two nucleotide sequences to be joined any way you want. Imagine, for example, that you want to combine the DNA-binding domain of protein A to the regulatory domain of protein B in order to test your conjectures about how these proteins work. The target domains in the cDNAs for the genes and the arrangement of PCR primers needed to join the domains are shown in Figure 7–24.

Recombinant PCR is usually carried out in two steps. In the first step PCR primers 1 and 2 are used to amplify the target segment of gene A, and, in a separate reaction, primers 3 and 4 are used to amplify the target sequence in gene B. In the second step the individually amplified products, separated from

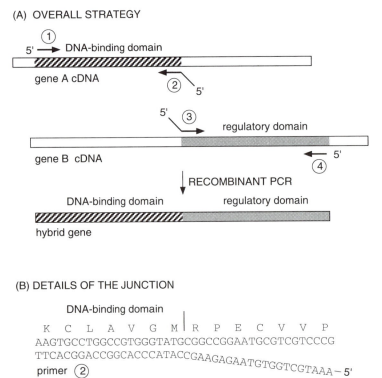

Figure 7–24 Recombinant PCR (Problem 7–35). (A) Overall strategy. (B) Details of the junction. Note that primers 2 and 3 are complementary to one another over their entire length.

Table 7–3 Injection of *raf-1* and NAF RNA into 2-Cell Embryos (Problem 7–36)

Injection	Total Survivors	Phenotype of Tadpole		
		Normal	**Truncated Tail**	**Other Abnormalities**
raf-1 RNA	94	75%	0%	25%
NAF RNA	80	40%	36%	24%
raf-1 and NAF RNA	93	73%	5%	22%
water	101	92%	0%	8%
uninjected	80	99%	0%	1%

their primers, are mixed together and amplified using primers 1 and 4. *Voila!* The desired hybrid gene is the major product.

A. Explain how recombinant PCR manages to link the two gene segments together.

B. Illustrate schematically the structure and arrangement of primers you would use to put the regulatory domain of gene B at the N-terminus of the hybrid protein.

7–36 The *raf-1* gene encodes a serine/threonine protein kinase that may be important in vertebrate development. In *Drosophila,* when the homologue of this gene is defective, embryos develop through the blastula stage normally, but from then on development is abnormal, leading to a truncated embryo that is lacking the posterior four segments. In vertebrates, fibroblast growth factor (FGF) stimulates mesoderm induction and posterior development, and it is thought that Raf-1 protein participates in the signaling pathway triggered by FGF. The Raf-1 protein consists of a serine/threonine kinase domain that is the carboxy-terminal half of the molecule and an amino-terminal regulatory domain that inhibits the kinase activity until a proper signal is received.

To test the role of Raf-1, you construct what you hope will be a dominant-negative *raf-1* cDNA by changing a key residue in the ATP-binding domain. This mutated cDNA encodes a kinase-defective protein, which you term NAF (not a functional Raf). You prepare RNA from the *raf-1* cDNA and from the NAF cDNA and inject them individually or as a mixture into both cells of the 2-cell frog embryo to determine their effects on frog development. These experiments are summarized in Table 7–3. The truncated-tail phenotype produced in some injections is very striking (Figure 7–25).

Figure 7–25 Normal and truncated-tail phenotype (Problem 7–36).

A. Do your experiments provide evidence in support of the hypothesis that Raf-1 participates in posterior development in vertebrates? If so, how do they do this?

B. Given the description of the Raf-1 protein, suggest a way that NAF might act as a dominant-negative protein.

C. Why does the coinjection of *raf-1* RNA with NAF RNA result in a low frequency of the truncated-tail phenotype?

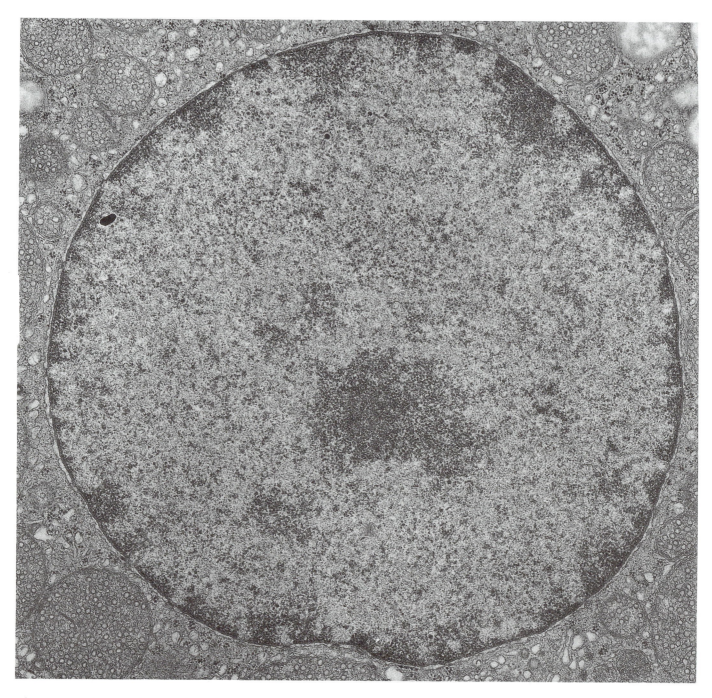

Thin-section electron micrograph of an adrenal cortex cell nucleus.
(Courtesy of Daniel S. Friend.)

The Cell Nucleus

- Chromosomal DNA and Its Packaging
- The Global Structure of Chromosomes
- Chromosome Replication
- RNA Synthesis and RNA Processing
- The Organization and Evolution of the Nuclear Genome

Chromosomal DNA and Its Packaging

(MBOC 336–346)

8–1 Fill in the blanks in the following statements.

A. Each DNA molecule is packaged in a separate _____, and the total genetic information stored in the chromosomes of an organism is said to constitute its _____.

B. A functional chromosome requires three DNA sequence elements: at least one _____ to permit the chromosome to be copied, one _____ to facilitate proper segregation of its two copies at mitosis, and two _____ to allow the chromosome to be maintained between cell generations.

C. Recombinant DNA methods allow yeast sequence elements to be added to human DNA molecules, which can then replicate in yeast cells as _____.

D. Each region of the DNA helix that produces a functional RNA molecule constitutes a _____.

E. In the genes of higher eucaryotes, short segments of coding DNA, called _____, are usually separated by long stretches of noncoding DNA, called _____.

F. The complex of the abundant structural proteins, the _____, and the _____ proteins with the nuclear DNA of eucaryotic cells is known as _____.

G. The five types of histones fall into two main groups: the _____ histones and the _____ histones.

H. The structure of eucaryotic chromosomes is dominated by a nucleoprotein particle, the _____, which plays a major role in packing and organizing all of the DNA in the cell nucleus.

I. Two copies each of H2A, H2B, H3, and H4 form a _____ around which the double-stranded DNA helix is wound twice.

J. Regions of DNA that lack nucleosomes are readily digested by trace amounts of deoxyribonuclease; these regions are known as _____.

8–2 Indicate whether the following statements are true or false. If a statement is false, explain why.

__ A. Each chromosome contains a single long DNA molecule.

__ B. A telomere allows a chromosome to be replicated precisely so that no nucleotides are lost from the end of the chromosome, thereby solving the end-replication problem.

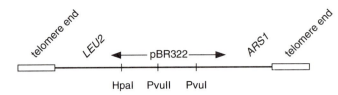

Figure 8–1 Structure of the intended linear chromosome with *Tetrahymena* telomeres flanking yeast and bacterial DNA (Problem 8–3). The sites of unique cutting by three restriction enzymes are indicated.

___ C. Population geneticists estimate from the observed mutation rate that only a small percentage of the mammalian genome can be involved in regulating or encoding essential proteins.

___ D. In genes from higher eucaryotes, introns are usually larger and more numerous than exons.

___ E. In a comparison between the DNAs of related organisms, such as humans and mice, conserved sequences represent functionally important exons and regulatory regions, and nonconserved sequences generally represent noncoding DNA.

___ F. Histones are relatively small proteins with a very high proportion of positively charged amino acids; the positive charge helps the histones bind tightly to DNA, regardless of its nucleotide sequence.

___ G. A nucleosome consists of about 146 nucleotide pairs wrapped in two turns around a histone octamer, which is a complex of eight nucleosomal histones.

___ H. While the majority of nucleosomes do not seem to be positioned precisely on DNA, striking examples of precise nucleosome positioning are known.

___ I. Nuclease-hypersensitive sites in chromatin are located in the linker DNA between nucleosomes.

___ J. In the living cell chromatin usually adopts the extended "beads-on-a-string" form.

*8–3 You think you may have devised a clever strategy for generating artificial chromosomes in yeast. You know that the ribosomal genes of *Tetrahymena*, when cut with the restriction enzyme BamHI, yield a 1.5-kb fragment that contains the telomere. You plan to attach these fragments to each end of a linear form of a yeast plasmid; you hope that the plasmid will then persist as a linear molecule—that is, an artificial chromosome.

As a source of plasmid DNA you use a circular yeast plasmid that contains a yeast origin of replication (ARS1), a selectable marker gene for growth in yeast (*LEU2*), and bacterial plasmid (pBR322) sequences (Figure 8–1). You linearize the 9-kb plasmid with BglII, which cuts the plasmid once. You then incubate the linear plasmid with the 1.5-kb fragments carrying the *Tetrahymena* telomere in the presence of DNA ligase and the two restriction enzymes, BglII and BamHI. When you analyze the ligation products, you find molecules of 10.5 kb and 12 kb in addition to the original components. You purify the 12-kb band, transform it into yeast, and select for yeasts that express the marker gene on the plasmid.

To test whether the plasmid is linear or circular, you prepare total DNA from one transformant; digest samples of it with the restriction enzymes HpaI, PvuII, and PvuI; separate the fragments by gel electrophoresis; and blot hybridize them to radioactively labeled pBR322 DNA. A diagram of the autoradiograph is shown in Figure 8–2.

A. How do the results of the analysis in Figure 8–2 distinguish between a linear and a circular form of the plasmid in the transformed yeast?

B. Explain how ligation of the DNA fragments in the presence of the restriction enzymes BamHI and BglII ensures that you get predominantly the construct you want. The recognition site for BglII is 5'-A'GATCT-3', where the asterisk (*) is the site of cutting, and the recognition site for BamHI is 5'-G'GATCC-3'.

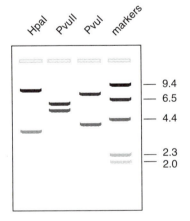

Figure 8–2 Autoradiograph of restriction analysis of plasmid structure (Problem 8–3). Marker DNAs of known fragment sizes (in kb) are shown at the right.

---

*Problems with an asterisk (*) are answered in the Instructor's Manual.

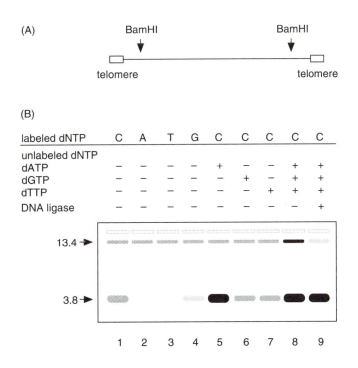

Figure 8–3 Structure of ribosomal minichromosome (A) and incorporation of radioactive nucleotides by DNA polymerase (B) (Problem 8–4). Ribosomal minichromosomes were incubated in the presence of various combinations of nucleotides, as indicated, and then cleaved with BamHI. The resulting fragments were separated by electrophoresis and visualized by autoradiography.

8–4   The telomeres at the ends of the *Tetrahymena* ribosomal minichromosomes have been studied extensively. The following observations give several important clues to their structure.

1. When ribosomal minichromosomes are incubated with DNA polymerase in the presence of ^{32}P-dCTP and the three other unlabeled dNTPs, the terminal 3.8-kb BamHI restriction fragments are much more extensively labeled than the central 13.4-kb fragment (Figure 8–3, lane 8). Incubations with single dNTPs give significant incorporation only with ^{32}P-dCTP (lane 1 versus lanes 2 to 4). Incorporation of dCTP in the presence of dATP is substantially greater than incorporation of dCTP alone (lane 5 versus lane 1). Incorporation of dCTP in the presence of dGTP or dTTP is no greater than dCTP alone (lanes 6 and 7).

2. Preincubation of the minichromosome with DNA ligase has little effect on incorporation into terminal fragments, but reduces incorporation into the central fragment by tenfold (Figure 8–3, lane 9).

3. In the presence of the other three unlabeled dNTPs, ^{32}P-dCTP is incorporated into the tandemly repeated sequence 5′-CCCCAA-3′.

4. Minichromosomes labeled with ^{32}P-dCTP alone, when denatured, give rise to single-stranded fragments that are composed of 2, 3, or 4 CCCCAA tandem repeats.

5. If the free 5′ phosphates (free means not in a phosphodiester bond) in minichromosomes are replaced with labeled phosphates and then treated so that all bonds to purine nucleotides are broken, the predominant labeled fragment is CCC.

6. If the minichromosome is cleaved with the restriction enzyme AluI, which cuts very near the ends of the minichromosome, a very broad band containing the terminal fragment is generated. The leading edge of the band corresponds to about 360 nucleotide pairs and the trailing edge corresponds to about 520 nucleotide pairs.

7. DNA synthesis that begins within the tandem CCCCAA repeats moves progressively toward the center of the minichromosome.

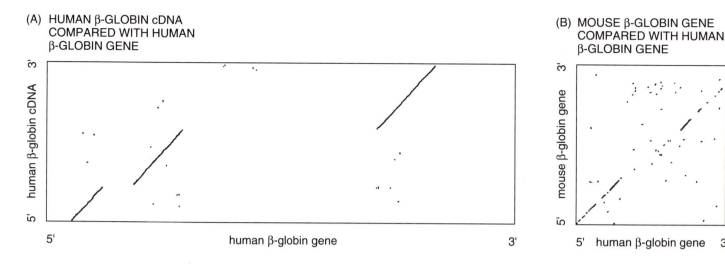

(A) HUMAN β-GLOBIN cDNA COMPARED WITH HUMAN β-GLOBIN GENE

human β-globin cDNA (3' to 5' on vertical axis)

human β-globin gene (5' to 3' on horizontal axis)

(B) MOUSE β-GLOBIN GENE COMPARED WITH HUMAN β-GLOBIN GENE

mouse β-globin gene (3' to 5' on vertical axis)

human β-globin gene (5' to 3' on horizontal axis)

Figure 8–4 Diagon plots comparing the human β-globin gene (A) with the human β-globin cDNA and (B) with the mouse β-globin gene (Problem 8–5). The 5' and 3' ends of the sequences are indicated. The human gene sequence is identical in the two plots. In order to accommodate the short β-globin cDNA sequence (549 nucleotides) and the sequence of the gene that encodes it (2052 nucleotides) in similar spaces, while maintaining proportional scales within each plot, the scale of (A) is about three times larger than that of (B).

A. Are the numbers of CCCCAA repeats in the telomere defined or variable? Give a minimum and maximum estimate of the number of repeats and explain your reasoning.

B. Are the CCCCAA repeats at the 5' ends of the individual strands of the minichromosome, or are they at the 3' ends? Explain your reasoning.

C. Incorporation of label by DNA synthesis or by replacement of 5' phosphates indicates that the nucleotides in the CCCCAA repeats are not all linked together by phosphodiester bonds. What do the various observations tell you about the nature of the single-strand interruptions in the CCCCAA repeats?

D. What is the spacing between the single-strand interruptions in the CCCCAA repeats?

E. Diagram as best you can from these data the structure of the telomeres on the *Tetrahymena* ribosomal minichromosome. (Note that these experiments do not define the structure of the very end of the chromosome, but only the structure of the CCCCAA repeats that make up the telomere.)

*8–5 A very useful graphic method for comparing nucleotide sequences is the so-called diagon plot. An example of this method is illustrated in Figure 8–4, where the human β-globin gene is compared to the human cDNA for β-globin (Figure 8–4A) and to the mouse β-globin gene (Figure 8–4B). These plots are generated by comparing blocks of sequence, in this case blocks of 11 nucleotides at a time. If 9 or more of the nucleotides match, a dot is placed on the diagram at the coordinates corresponding to the blocks being compared. A comparison of all possible blocks generates diagrams, such as the ones shown in Figure 8–4, in which sequence homologies show up as diagonal lines.

A. From the comparison of the human β-globin gene with the human β-globin cDNA (Figure 8–4A) deduce the positions of exons and introns in the β-globin gene.

B. Are the entire exons of the human β-globin gene homologous to the mouse β-globin gene (Figure 8–4B)? Identify and explain any discrepancies.

C. Is there any homology between the human and mouse β-globin genes that is outside the exons? If so, identify its location and offer an explanation for its preservation during evolution.

D. Have either of the genes undergone a change of intron length during their evolutionary divergence? How can you tell?

*8–6 You are studying chromatin structure in rat liver DNA. When you digest rat liver nuclei briefly with micrococcal nuclease, extract the DNA, and run it on an agarose gel, it forms a ladder of broad bands spaced at about 200-nucleotide intervals. If you use the enzyme DNase I instead, there is a much more continuous smear of DNA on the gels with only the haziest suggestion of a 200-nucleotide repeat. If you denature the DNase-I-treated DNA before fractionating it by gel electrophoresis, however, you find a new ladder of bands with a regular spacing of about 10 nucleotides.

You are puzzled by the different results with these two enzymes. When you describe the experiments to the rest of your research group, one colleague suggests that the difference derives from the steric properties of the DNA-binding sites on the two enzymes: micrococcal nuclease can only bind and cleave DNA that is free; DNase I can bind and cut free DNA and DNA that is bound to the surface of a nucleosome. Your colleague predicts that if DNA is bound to any surface and digested with DNase I, it will generate a 10-nucleotide ladder. You test this prediction by binding DNA to polylysine-coated plastic dishes and digesting with the two enzymes: micrococcal nuclease causes minimal digestion, but DNase I generates a 10-nucleotide ladder, verifying your friend's prediction.

A. Why does brief digestion of nuclei with micrococcal nuclease yield a ladder of bands spaced at intervals of about 200 nucleotides?

B. If you digested nuclei extensively with micrococcal nuclease, what pattern would you expect to see after fractionation of the DNA by gel electrophoresis?

C. Explain how your colleague's suggestion accounts for the generation of a 10-nucleotide ladder when nuclei are digested with DNase I.

*8–7   You have been sent the first samples of a newly discovered martian microorganism for analysis of its chromatin. The cells resemble earthly eucaryotes and are composed of similar molecules, including DNA, which is located within a nucleuslike structure in the cell. One member of your team has identified two basic histonelike proteins associated with the DNA in roughly an equal mass ratio with the DNA. You isolate nuclei from the cells and treat them with micrococcal nuclease for various times. You then extract the DNA and run it on an agarose gel alongside a similar digest of rat liver nuclei. As shown in Figure 8–5, the digest of rat-liver nuclei gives a standard ladder of nucleosomes, but the martian organism gives a smear of digestion products with a nuclease-resistant limit of about 300 nucleotides. As a control, you isolate the martian DNA free of all protein and digest it with micrococcal nuclease: it is completely susceptible, giving predominantly mono- and dinucleotides as the limit product.

Do these results suggest that the martian organism has nucleosomelike structures in its chromatin? If so, how are they spaced along the DNA?

8–8   The arrangement of nucleosomes around the centromere is of special interest because the centromere is the chromosome attachment site for microtubules. Thus, the usual arrangement of nucleosomes in chromatin might have to be altered to accommodate microtubule attachment. You are set up to address

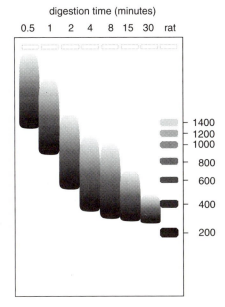

Figure 8–5 Micrococcal digest of chromatin from a martian organism (Problem 8–7). The results of digestion of rat-liver chromatin are shown on the right.

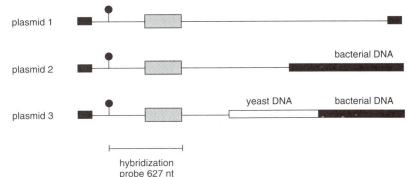

Figure 8–6 Diagram of the native chromosome and three plasmids around CEN3 (Problem 8–8). The native chromosome is linear; its true ends extend well beyond the position marked by the diagonal lines. The plasmids are shown here as linears for ease of comparison. The native yeast sequences around the centromere are shown as thin lines. Bacterial DNA sequences in the plasmids are shown as black rectangles. The yeast DNA in plasmid 3 that is shown as a white rectangle is a segment of yeast chromosomal DNA far removed from the centromere. The BamHI-cleavage site is shown as a closed circle to the left of the centromere. The location of the radioactive probe used in the hybridization is indicated at the bottom.

this question because you have cloned and sequenced about two kilobases of DNA surrounding the centromere (CEN3) of yeast chromosome III. As a result, you can test the arrangement of nucleosomes not only on the native chromosome but also on the plasmids into which you cloned various lengths of the native chromosomal DNA around the centromere (Figure 8–6).

You prepare chromatin from native yeasts and from yeasts that carry individual plasmids. You treat these chromatin samples briefly with micrococcal nuclease and then deproteinize the DNA and digest it to completion with the restriction enzyme BamHI, which cuts the DNA only once in the region of the centromere (Figure 8–6). The digested DNA is fractionated by gel electrophoresis and then analyzed by Southern blotting using a segment of radiolabeled DNA from the centromere as a hybridization probe (Figure 8–6). This procedure (called indirect end labeling) allows you to visualize all DNA fragments that include the DNA immediately to the right of the BamHI-cleavage site. As a control, you deproteinize a sample of chromatin to produce naked DNA, treat it with micrococcal nuclease, and then subject it to the same analysis. An autoradiogram of your results is shown in Figure 8–7.

A. If the digestion with BamHI is omitted, regular though less distinct sets of dark bands are apparent. Why does digestion with BamHI make the pattern so much clearer and easier to interpret?

B. Draw a diagram showing the micrococcal-nuclease-sensitive sites on the chromosomal DNA and the arrangement of nucleosomes along the chromosome. What is special about the centromeric region?

C. What is the purpose of including a naked DNA control in the experiment?

D. The autoradiogram in Figure 8–7 shows that the native chromosomal DNA yields a regularly spaced pattern of bands beyond the centromere; that is, the bands at 600 nucleotides and above are spaced at 160-nucleotide intervals. Does this regularity result from the lining up of nucleosomes at the centromere, like cars at a stop light? Or, is the regularity an intrinsic property of the DNA sequence itself? Explain how your results with plasmids 1, 2, and 3 decide the issue.

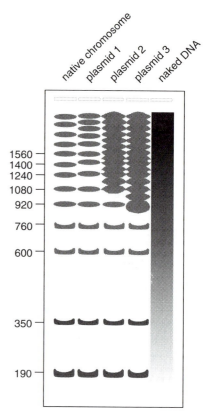

Figure 8–7 Results of micrococcal-nuclease digestion of DNA around CEN3 (Problem 8–8). Approximate lengths of DNA fragments in nucleotide pairs are indicated on the left of the autoradiogram.

The Global Structure of Chromosomes
(MBOC 346–356)

8–9   Fill in the blanks in the following statements.

A. The meiotically paired chromosomes in growing oocytes are known as _____ because they form unusually stiff and extended chromatin loops.

B. The precise side-to-side adherence of individual chromatin strands in _____ greatly elongates the chromosome axis and prevents tangling.

C. The regions on a polytene chromosome that are being actively transcribed are decondensed, forming distinctive _____.

D. _____ chromatin is unusually sensitive to digestion with nucleases, and its nucleosomes are thought to be altered in a way that makes their packing less condensed.

E. By light microscopy there are two types of chromatin in interphase nuclei of higher eucaryotic cells: a highly condensed form called _____ and all the rest, which is less condensed, called _____.

F. Chromosomes from nearly all cells are visible during mitosis, where they coil up to form much more condensed structures, called _____.

G. The display of the 46 human chromosomes at mitosis is called the human _____.

H. Certain fluorescent dyes appear to distinguish DNA that is rich in A-T nucleotide pairs (_____) from DNA that is rich in G-C nucleotide pairs

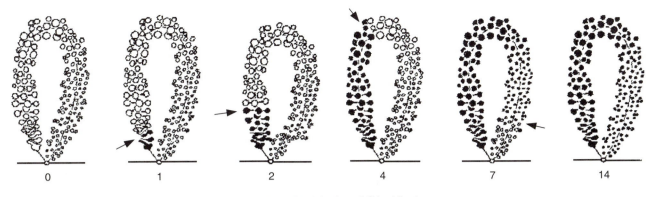

0	1	2	4	7	14
unlabeled		time (days after injection of ³H-uridine)			fully labeled

(_____), producing a striking and reproducible banding pattern along each mitotic chromosome.

Figure 8–8 Autoradiographs of a giant chromatin loop from a lampbrush chromosome of the newt (Problem 8–11). Arrows show forward progress of labeled regions around the loop at various times after injection of labeled uridine into the oocytes.

8–10 Indicate whether the following statements are true or false. If a statement is false, explain why.

__ A. In lampbrush chromosomes most of the chromatin is in the loops, which are actively transcribed, but some of the chromatin remains highly condensed in the chromomeres, which are transcriptionally inactive.

__ B. In several types of secretory cells of fly larvae, all the homologous chromosome copies remain side by side through several rounds of replication, thereby generating a single, giant polyploid chromosome.

__ C. Studies of chromosome puffs suggest that individual chromosome bands can decondense as a unit during transcription.

__ D. Classical genetic studies coupled with more recent molecular studies indicate that each band in a polytene chromosome probably corresponds to a single gene.

__ E. Treatment of the nuclei from different cells with an appropriate concentration of DNase I preferentially degrades DNA sequences that are actively transcribed in the particular cell type tested.

__ F. Nucleosomes in active chromatin selectively bind two closely related small chromosomal proteins, HMG 14 and HMG 17.

__ G. Transcriptionally inactive regions of chromosomes are condensed into a relatively nuclease-resistant form known as heterochromatin.

__ H. The coiling of chromosomes at mitosis, which reduces the length of DNA about 10,000-fold, is accompanied by the extensive phosphorylation of histone H1.

__ I. The bands observed after staining human mitotic chromosomes correspond to the bands seen in insect polytene chromosomes.

8–11 You are studying transcription in the lampbrush chromosomes of the newt *Triturus* by injecting ³H-uridine into the oocytes, waiting for various times, and detecting the radioactive RNA by autoradiography. Initially, you focus on the largest loops; they incorporate label progressively around the loop as shown in Figure 8–8. It takes about 14 days before the entire loop becomes labeled. When you compare this labeling pattern with that of the smaller loops, which are much more common in lampbrush chromosomes, you are surprised: the smaller loops show uniform labeling even at the shortest times of sampling. With increasing time after injection of radioactive uridine, this uniform labeling intensifies all around each loop.

Many loops in lampbrush chromosomes represent single transcription units, in which RNA polymerase initiates and terminates synthesis at the base of the loop as shown in Figure 8–9. Assuming that the loops you have examined are single transcription units, which pattern of loop labeling would you expect:

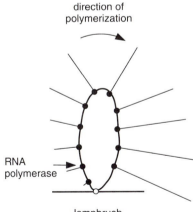

direction of
polymerization

RNA
polymerase

lampbrush
chromosome axis

Figure 8–9 Diagrammatic representation of a chromatin loop that is a single transcription unit (Problem 8–11). The progress of RNA polymerase molecules around the loop is illustrated along with the attendant growth of the nascent RNA chain associated with each polymerase.

The Global Structure of Chromosomes

progressive, as in the large loops (Figure 8–8), or uniform, as in the smaller loops?

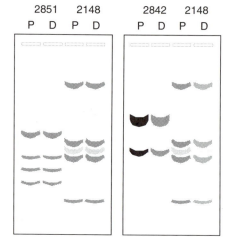

*8–12   The giant polytene chromosomes of *Drosophila melanogaster* have long been of interest to geneticists because their characteristic banding patterns provide a visible map of the genome. Bands apparently contain more DNA than interbands. Does this difference arise because DNA in bands is replicated more extensively than DNA in interbands? Or is all the DNA replicated to the same extent but folded in such a way that bands contain more DNA than interbands?

You are in a position to resolve this controversy because you have isolated a contiguous set of clones that span 315 kb of *Drosophila* DNA, including about 12 bands and interbands. You can use radiolabeled segments of these clones as hybridization probes to estimate the amount of corresponding DNA present in diploid tissues and polytene chromosomes. You isolate DNA from diploid tissues and polytene chromosomes, digest equal amounts of the DNA with combinations of restriction enzymes, separate the fragments by gel electrophoresis, and transfer them to nitrocellulose filters for hybridization analysis. In every case the restriction pattern is the same for the DNA from diploid tissues and polytene chromosomes, as illustrated for two examples in Figure 8–10. You measure the intensities of many specific restriction fragments and express the results as the ratio of the intensity of the fragment from polytene chromosomes to the intensity of the corresponding fragment from diploid tissues (Figure 8–11).

Do your results support differential replication or differential chromosome folding as the basis for the difference between bands and interbands?

Figure 8–10 Autoradiographs of blot-hybridization analysis of polytene and diploid DNA (Problem 8–12). P and D refer to polytene and diploid, respectively. Numbers at the top refer to cloned DNA segments used as probes: 2851 and 2842 are from the 315-kb region under analysis (see Figure 8–11); 2148 is from elsewhere in the genome and was used in all hybridizations to calibrate the amount of DNA added to the gels.

8–13   To understand the relationship between chromatin structure and gene expression, you are studying two types of chicken cells. Chicken red blood cells express large amounts of globin mRNAs, but chicken fibroblasts do not express globin at all. You want to know whether the globin DNA sequences exist in the same or a different chromatin structure in these two cell types. As probes of chromatin structure you use two nucleases: micrococcal nuclease and pancreatic deoxyribonuclease I (DNase I).

You prepare a ^{3}H-thymidine-labeled complementary DNA (cDNA) copy of the globin mRNAs from red blood cells. This globin cDNA is completely digested by a single-strand specific nuclease (S1 nuclease). However, if the cDNA is first annealed with a vast excess of chicken DNA, more than 90% of it is protected from subsequent digestion with S1 nuclease.

You assay the effects of micrococcal nuclease and DNase I on the capacity of DNA from red cells and fibroblasts to protect the cDNA. To preserve the natural organization of the chromatin, you treat isolated nuclei with the nucleases before extracting the DNA and assaying it. Digestion of red cell nuclei or fibroblast nuclei with micrococcal nuclease (so that about 50% of the DNA is degraded) yields DNA samples that still protect greater than 90% of the cDNA

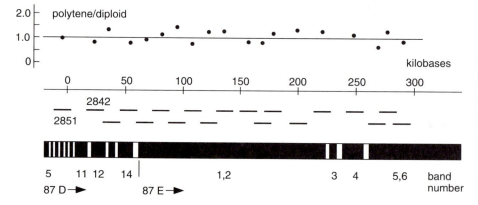

Figure 8–11 Relative amounts of DNA in diploid tissues and polytene chromosomes at different positions along the chromosome (Problem 8–12). The chromosomal segment covered by the cloned restriction fragments is shown at the bottom along with the cytological designations for the chromosome regions and bands. The cloned fragments are shown above the chromosomes, and the positions of 2851 and 2842 are indicated. The ratio of hybridization of each restriction fragment to DNA from polytene chromosomes and diploid tissues is given above each fragment.

Table 8–1 Protection of Globin cDNA and Total Red Cell DNA by Untreated and Nuclease-treated Chromatin Samples (Problem 8–13)

Excess DNA	Nuclease Treatment	Protected Globin cDNA	Protected Red Cell DNA
Total DNA	none	93%	95%
Red cell DNA	micrococcal nuclease	92%	94%
Red cell DNA	DNase I	25%	78%
Fibroblast DNA	micrococcal nuclease	91%	
Fibroblast DNA	DNase I	91%	
Nucleosome monomers	none	91%	94%
Nucleosome monomers	DNase I	25%	80%
Trypsin-treated nucleosome monomers	micrococcal nuclease	25%	83%

from subsequent digestion with S1 nuclease. Similarly, digestion of fibroblast nuclei with DNase I (so that less than 20% is degraded) yields DNA that protects greater than 90% of the cDNA. An identical digestion of red cell nuclei with DNase I, however, yields DNA that protects only about 25% of the cDNA. These results are summarized in Table 8–1.

You repeat some of these measurements using ^{3}H-thymidine-labeled total DNA from red blood cells in place of the globin cDNA. Annealing with total DNA or micrococcal-nuclease-digested DNA protects greater than 90% of the ^{3}H-labeled DNA, but annealing with DNase-I-digested DNA protects only 78% of the ^{3}H-labeled DNA (Table 8–1).

In a final set of experiments you first generate nucleosome monomers by digestion with micrococcal nuclease. DNA from the monomers protects more than 90% of globin cDNA. Treatment of the monomers with DNase I yields DNA that protects only 25% of globin cDNA. You then treat the monomers briefly with trypsin to remove 20 to 30 amino acids from the N-terminus of each histone molecule, redigest the modified nucleosomes with *micrococcal nuclease,* and isolate the DNA. This DNA protects 83% of total red cell DNA but only 25% of globin cDNA (Table 8–1).

A. Which nuclease—micrococcal nuclease or DNase I—digests chromatin that is being expressed (active chromatin)? How can you tell?

B. What fraction of red cell DNA is in active chromatin?

C. Does trypsin treatment of nucleosome monomers render a random population or a specific population of nucleosomes sensitive to micrococcal nuclease? How can you tell?

D. Is the alteration that distinguishes active chromatin from bulk chromatin a property of individual nucleosomes, or is it related to the way nucleosome monomers are packaged into higher-order structures within the cell nucleus?

Chromosome Replication
(MBOC 356–365)

8–14 Fill in the blanks in the following statements.

A. The DNA synthesis phase of the cell-division cycle is called the _____.

B. Multiple copies of SV40 _____ bind specifically to the SV40 origin of replication and act both as an initiator protein and a DNA helicase to open the DNA helix at that site, forming a _____.

C. Replication origins tend to be activated in clusters of perhaps 20 to 80 origins, which are called _____.

D. The problem of replicating the ends of chromosomes is solved using special DNA sequences, called _____, at the ends of chromosomes and an enzyme called _____ to extend them.

8–15 Indicate whether the following statements are true or false. If a statement is false, explain why.

__ A. ARS elements in yeast appear to contain multiple copies of an 11-nucleotide core consensus sequence clustered within a region of about 100 nucleotides.

__ B. Two distinct types of DNA polymerase are needed in eucaryotes: DNA polymerase alpha on the lagging strand and DNA polymerase delta on the leading strand.

__ C. Replication forks in bacteria and eucaryotic cells travel at the same rate, indicating that the packaging of DNA into chromatin does not hinder the replication process.

__ D. Chromosomal regions are replicated in large units, and different regions of each chromosome are replicated in a reproducible order.

__ E. The two X chromosomes in a female mammalian cell, only one of which is active, are replicated at the same time during the S phase.

__ F. Genes that are active in only a few cell types generally replicate early in the cells in which they are active and later in other types of cells.

__ G. Most G-C-rich bands (R bands) replicate during the first half of S phase, while most A-T-rich bands (G bands) replicate during the second half of S phase.

__ H. When an S-phase cell is fused with a G_2-phase cell, DNA synthesis is induced in the G_2-phase nucleus; when an S-phase cell is fused with a G_1-phase cell, however, the G_1 nucleus is not stimulated to synthesize DNA.

__ I. Like most proteins, histones are synthesized continuously throughout interphase, but they are deposited on DNA to make new chromatin only during S phase.

__ J. Telomerase synthesizes a new copy of the telomere repeat using an RNA template that is a component of the enzyme itself.

8–16 Autonomous replication sequences (ARS), which confer stability on plasmids in yeast, are thought to be origins of replication. Proving that an ARS is an origin of replication, however, is difficult, mainly because it is very hard to obtain enough well-defined replicating DNA molecules to analyze. This problem can be addressed using a two-dimensional gel-electrophoretic analysis that separates DNA molecules by mass in the first dimension and by shape in the second dimension. Because they have branches, replicating molecules migrate more

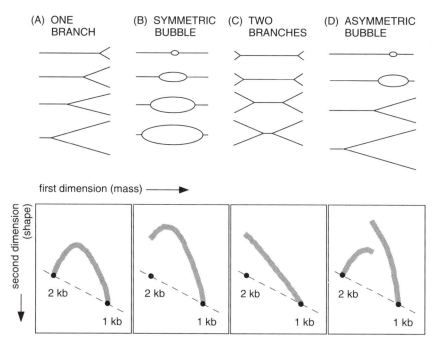

Figure 8–12 Two-dimensional gel patterns for molecules with (A) a single branch, (B) a symmetrically located replication bubble, (C) two branches, and (D) an asymmetrically located replication bubble (Problem 8–16). Intermediates in the replication of a hypothetical 1-kb fragment are shown at progressive states of replication at the top of the figure. The gel patterns that result from the continuum of such replication intermediates are shown below.

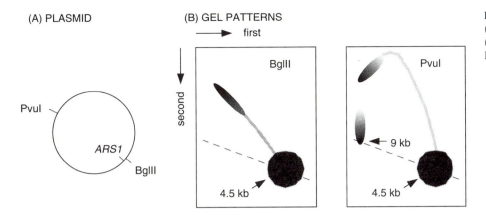

(A) PLASMID

(B) GEL PATTERNS

first

second

BglII

PvuI

9 kb

4.5 kb

4.5 kb

PvuI

ARS1

BglII

Figure 8–13 Structure of the ARS1 plasmid (A) and the two-dimensional gel patterns (B) resulting from cleavage with BglII and PvuI (Problem 8–16).

slowly in the second dimension than linear molecules of equal mass. By cutting replicating molecules with restriction enzymes, it is possible to generate a continuum of different branched forms that together give characteristic patterns on two-dimensional gels (Figure 8–12).

You apply this technique to the replication of a plasmid that contains ARS1. To maximize the fraction of plasmid molecules that are replicating, you synchronize a yeast culture and isolate DNA from cells in S phase. You then digest the DNA with BglII or PvuI, which cut the plasmid as indicated in Figure 8–13A. You separate the DNA fragments by two-dimensional electrophoresis and visualize the plasmid sequences by autoradiography after blot hydridization to radioactive plasmid DNA (Figure 8–13B).

A. What is the source of the intense spot of hybridization at the 4.5-kb position in both gels in Figure 8–13B?

B. Do the results of this experiment indicate that ARS1 is an origin of replication? Explain your answer.

C. There is a gap in the arc of hybridization in the PvuI gel pattern in Figure 8–13B. What is the basis for this discontinuity?

8–17 One important rule for eucaryotic DNA replication is that no chromosome or part of a chromosome should be replicated more than once per cell cycle. Eucaryotic viruses must evade or break this rule if they are to produce multiple copies of themselves during a single cell cycle. The animal virus SV40, for example, generates 100,000 copies of its genome during a single cycle of infection. In order to accomplish this feat, it synthesizes a special protein, termed T-antigen (because it was first detected immunologically). T-antigen binds to the SV40 origin of replication and in some way triggers initiation of DNA replication.

The mechanism by which T-antigen initiates replication has been investigated *in vitro*. When purified T-antigen, ATP, and a single-strand DNA-binding (SSB) protein were incubated with a circular plasmid DNA carrying the SV40 origin of replication, partially unwound structures, such as the one shown in Figure 8–14, were observed in the electron microscope. In the absence of any one of these components, no unwound structures were seen. Furthermore, no unwinding occurred in an otherwise identical plasmid that carried a 6-nucle-

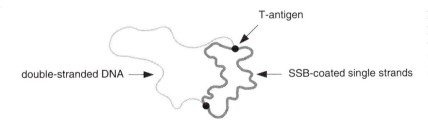

T-antigen

double-stranded DNA →

← SSB-coated single strands

Figure 8–14 A typical example of a plasmid molecule carrying an SV40 origin of replication after incubation with T-antigen, single-strand DNA-binding (SSB) protein, and ATP (Problem 8–17).

otide deletion at the origin of replication. If care was taken in the isolation of the plasmid so that it contained no nicks (that is, a covalently closed circular DNA), no unwound structures were observed unless topoisomerase I was also present in the mixture.

A. What activity in addition to site-specific DNA binding must T-antigen possess? How might this activity lead to initiation of DNA synthesis?

B. These experiments suggest that the structures observed in the electron microscope are unwound at the SV40 origin of replication. The location of the origin on the SV40 genome is precisely known. How might you use restriction enzymes to prove this point and to determine whether unwinding occurs in one or both directions away from the origin?

C. Why is there a requirement for topoisomerase I when the plasmid DNA is a covalently closed circle? (Topoisomerase I introduces single-strand breaks into duplex DNA and then rapidly recloses them, so that the breaks have only a transient existence.)

D. Draw an example of the kind of structure that would result if T-antigen repeatedly initiated replication at an SV40 origin that was integrated into a chromosome.

*8–18 You are investigating DNA synthesis in a line of tissue culture cells using a classic protocol. In this procedure ³H-thymidine is added to the cells, which incorporate it at replication forks. Then the cells are gently lysed in a dialysis bag to release the DNA. When the bag is punctured and the solution slowly drained, some of the DNA strands adhere to the walls and are stretched in the general direction of drainage. This method allows very long DNA strands to be isolated intact and examined; however, the stretching collapses replication bubbles so that daughter duplexes lie side by side. The support with its adhered DNA is fixed to a glass slide, overlaid with a photographic emulsion, and exposed for 3 to 6 months. The labeled DNA shows up as tracks of silver grains.

   You pretreat the cells to synchronize them at the beginning of S phase. In one experiment you release the synchronizing block and add ³H-thymidine immediately. After 30 minutes you wash the cells and change the medium so that the label is present at a third of its initial concentration. After an additional 15 minutes you prepare DNA for autoradiography. The results of this experiment are shown in Figure 8–15A. In the second experiment you release the synchronizing block and then wait 30 minutes before adding ³H-thymidine. After 30 minutes in the presence of ³H-thymidine, you once again change the medium to reduce the concentration of labeled thymidine and incubate the cells for an additional 15 minutes. The results of the second experiment are shown in Figure 8–15B.

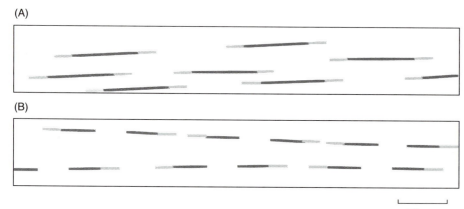

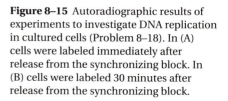

Figure 8–15 Autoradiographic results of experiments to investigate DNA replication in cultured cells (Problem 8–18). In (A) cells were labeled immediately after release from the synchronizing block. In (B) cells were labeled 30 minutes after release from the synchronizing block.

50 μm

A. Explain why in both experiments some regions of the tracks are dense with silver grains (dark), whereas others are less dense (light).

B. In the first experiment each track has a central dark section with light sections at each end. In the second experiment the dark section of each track has a light section at only one end. Explain the reason for the difference between the results in the two experiments.

C. Estimate the rate of fork movement (μm/min) in these experiments. Do the estimates from the two experiments agree? Can you use this information to estimate how long it would take to replicate the entire genome?

*8–19 The shell of a *Drosophila's* egg is made from more than 15 different chorion proteins, which are synthesized at a late stage in egg development by follicle cells surrounding the egg. The chorion genes are grouped in two clusters, one on chromosome 3 and the other on the X chromosome. In each cluster the genes are closely spaced with only a few hundred nucleotides separating adjacent genes. During egg development the number of copies of the chorion genes increases by overreplication of a segment of the surrounding chromosome. This amplification can be detected by preparing DNA from eggs at different stages of development, digesting the DNA samples with a restriction enzyme, and analyzing them by Southern blotting using a chorion cDNA as a probe. As shown in Figure 8–16, the number of chorion genes increases substantially between stages 8 and 12, whereas the copy number of a control gene that is far removed from the chorion gene clusters stays constant. The level of amplification around a chorion gene cluster can be determined using cloned probes covering the entire region. Measurements of the relative intensities of bands on autoradiographs such as the one in Figure 8–16 show that amplification of the chorion cluster on chromosome 3 is maximal in the region of the chorion genes but extends for nearly 50 kb on either side (Figure 8–17).

The DNA sequence responsible for amplification of the chorion cluster on chromosome 3 has been narrowed to a 510-nucleotide segment immediately upstream of one of the chorion genes. When this segment is moved to different places in the genome (using transposons to carry the DNA segment), those new sites are also amplified in follicle cells. No RNA or protein product seems to be synthesized from this amplification-control element.

A. Sketch what you think the DNA from an amplified cluster would look like under the electron microscope.

B. How many rounds of replication would be required to achieve a 60-fold amplification?

C. How do you think the 510-nucleotide amplification-control element promotes the overreplication of a chorion gene cluster?

8–20 Fertilized frog eggs are very useful for studying the cell-cycle regulation of DNA synthesis. Foreign DNA can be injected into the eggs and followed independently of chromosomal DNA replication. For example, in one study ^{3}H-labeled viral DNA was injected. The eggs were then incubated in a medium supplemented with ^{32}P-dCTP and nonradioactive bromodeoxyuridine triphosphate

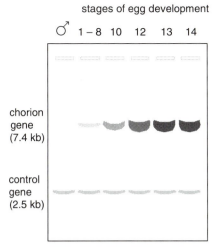

stages of egg development

♂ 1–8 10 12 13 14

chorion gene (7.4 kb)

control gene (2.5 kb)

Figure 8–16 Blot hybridization of a chorion gene and a control gene at various stages of egg development (Problem 8–19).

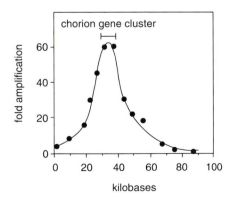

chorion gene cluster

fold amplification

kilobases

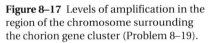

Figure 8–17 Levels of amplification in the region of the chromosome surrounding the chorion gene cluster (Problem 8–19).

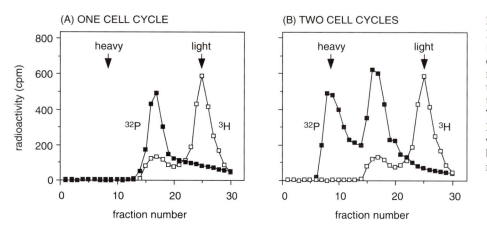

Figure 8–18 Density distribution of viral DNA after injection in fertilized frog eggs (Problem 8–20). (A) Results after one cell cycle. (B) Results after two cell cycles. The more dense end of the gradient is shown to the left ("heavy"), and the less dense end of the gradient is shown to the right ("light"). The ability to look specifically at the viral DNA depends on a technical trick: the eggs were heavily irradiated with UV light before the injection to block chromosomal replication.

(BrdUTP), which is a thymidine analogue that increases the buoyant density of DNA into which it is incorporated. Incubation was continued for long enough to allow one or two cell cycles to occur; then the viral DNA was extracted from the eggs and analyzed on CsCl density gradients, which separate DNA with 0, 1, or 2 BrdU-containing strands. Figure 8–18A and B show the density distribution of viral DNA after incubation for one and two cell cycles, respectively. If the eggs are bathed in cycloheximide (an inhibitor of protein synthesis) during the incubation, the results after incubation for one cycle or incubation for two cycles are both like those in Figure 8–18A.

A. Explain how the three density peaks in Figure 8–18 are related to replication of the injected DNA. Why is no ^{32}P radioactivity associated with the light peak, and why is no ^{3}H radioactivity associated with the heavy peak?

B. Does the injected DNA mimic the behavior that you would expect for the chromosomal DNA?

C. Why do you think that cycloheximide prevents the appearance of the most dense peak of DNA?

RNA Synthesis and RNA Processing

(MBOC 365–385)

8–21 Fill in the blanks in the following statements.

A. RNA synthesis, which is also called _____, is a highly selective process.

B. Transcription begins when an _____ molecule binds to a _____ sequence on the DNA double helix.

C. _____ transcribes the genes whose RNAs will be translated into proteins, _____ makes the large ribosomal RNAs, and _____ makes a variety of very small, stable RNAs.

D. RNA polymerase II transcripts in the nucleus are known as _____ molecules because one of the first characteristics used to distinguish them from other RNAs in the nucleus was the heterogeneity of their sizes.

E. The addition of a methylated G nucleotide to the 5′ end of the initial transcript forms the _____, which seems to protect the growing RNA from degradation and plays an important part in the initiation of protein synthesis.

F. The 3′ end of most polymerase II transcripts is defined by a modification, in which the growing transcript is cleaved at a specific site and a _____ is added by a separate polymerase to the cut 3′ end.

G. Modifications at the 5′ and 3′ ends of an RNA chain complete the formation of the _____.

H. Because the coding RNA sequences on either side of the intron are joined to each other after the intron sequence has been cut out, this reaction is known as _____.

I. Newly made RNA in eucaryotes appears to become immediately condensed into a series of closely spaced protein-containing particles, called _____ particles.

J. The small U RNAs in the cell nucleus are complexed with proteins to form _____.

K. The conserved sequences at the boundaries of an intron are called the _____ (donor site) and the _____ (acceptor site).

L. The large multicomponent ribonucleoprotein complex that carries out the splicing of the primary transcript is known as the _____.

M. Patients with _____ have an abnormally low level of hemoglobin—the oxygen-carrying protein in red blood cells.

N. The packaging of rRNAs with ribosomal proteins takes place in the nucleus in a large, distinct structure called the _____.

O. Each cluster of rRNA genes is known as a _____ region.

8–22   Indicate whether the following statements are true or false. If a statement is false, explain why.

__ A. In the RNA polymerase from *E. coli*, the initiation and elongation factors are permanent subunits of the enzyme, which allow it to recognize promoter sequences and extend the RNA chain.

__ B. RNA polymerases I, II, and III are each composed of multiple subunits, but none of the subunits are shared by all three polymerases.

__ C. Different RNA polymerase II start sites function with very different efficiencies, so that some genes are transcribed at much higher rates than others.

__ D. The 3′ end of most RNA polymerase II transcripts is defined by the termination of transcription, which releases a free 3′ end to which a poly-A tail is quickly added.

__ E. Only about 5% of the RNA synthesized by RNA polymerase II ever reaches the cytoplasm: the rest is degraded in the nucleus.

__ F. RNA splicing occurs in the cell nucleus, out of reach of the ribosomes, and RNA is exported to the cytoplasm only when processing is complete.

__ G. HnRNP particles and snRNPs resemble ribosomes in that each contains multiple polypeptide chains complexed to a stable RNA molecule.

__ H. Since introns are largely genetic "junk," they do not have to be removed precisely from the primary transcript during RNA splicing.

__ I. The major difference between group I and group II self-splicing introns is that the attacking nucleotide is free in group I introns but a part of the intron sequence in group II introns.

__ J. In most vertebrate cells, the clusters of genes encoding 28S rRNA are transcribed independently of the clusters of genes that encode 18S rRNA and 5.8S rRNA.

__ K. Ribosomal RNAs are produced in the nucleolus, a specialized region of the nucleus, and are then transported into the cytoplasm, where they are packaged with ribosomal proteins to form ribosomes.

__ L. Unlike cytoplasmic organelles, the nucleolus is not bounded by a membrane.

__ M. There is no nucleolus in a metaphase cell.

__ N. The extended chromosomes in interphase cells are thought to be extensively intertwined.

__ O. The proteins that constitute the nuclear matrix, or scaffold, can be shown to bind specific DNA sequences.

8–23   The trypanosome, which is the microorganism that causes sleeping sickness, can vary its surface glycoprotein coat and thus evade the immune defenses of its host. You are studying the synthesis of the variable surface glycoprotein (VSG) and have mapped the gene encoding this protein near the telomere of one of the chromosomes. However, you have been unable to locate the promoter, and your experiments suggest that it may be many thousands of nucleotides away from the VSG gene.

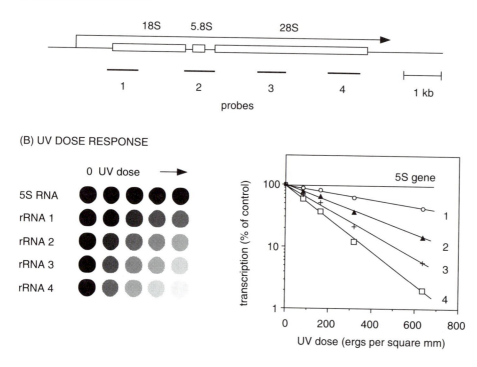

(A) TRANSCRIPTION MAP

(B) UV DOSE RESPONSE

Figure 8–19 Structure of the ribosomal RNA transcription unit (A) and UV sensitivity (B) of 5S RNA and ribosomal RNA transcription units (Problem 8–23). The positions of the hybridization probes along with a scale marker are indicated in (A) relative to the promoter (*left end of arrow*) for the transcription unit. Transcription as a function of UV dose is indicated in (B) in the form of a "dot blot" (*left*) and graph (*right*).

An old friend has repeatedly suggested that you use UV irradiation to map the promoter—a technique he has used successfully to map adenovirus transcription units. Since RNA polymerases cannot transcribe through pyrimidine dimers (the damage produced by UV irradiation), the sensitivity of transcription to UV irradiation is a measure of the distance between the start of transcription and the point where transcription is assayed. Since you have had no luck with your other approaches, you decide to try his suggestion.

To calibrate your system, you test transcription through the ribosomal RNA genes. The 5S RNA transcription unit is just over 100 nucleotides long, whereas the 18S, 5.8S, and 28S RNAs are part of a single transcription unit that is about 8 kb long (Figure 8–19A). You expose trypanosomes to increasing doses of UV irradiation, isolate their nuclei, and incubate them with ^{32}P-dNTPs. You then isolate RNA from the nuclei and hybridize it to cloned DNA corresponding to the 5S RNA gene and various parts of the ribosomal RNA transcription unit (Figure 8–19B). When the logarithm of the counts in each spot is plotted against the UV dose, the data give straight lines (Figure 8–19B). The slopes of these lines are proportional to the distance from the hybridization probe to the promoter.

When you repeat the experiment with a probe from the beginning of the VSG gene, you find that transcription is inactivated about seven times faster for the VSG gene than for probe 4 from the ribosomal transcription unit.

A. Why does RNA transcription increase in sensitivity to UV irradiation with increasing distance from the promoter?

B. Roughly how far is the VSG gene from its promoter? What assumption do you have to make in order to estimate this distance?

C. You have found another gene about 10 kb upstream of the VSG gene. Transcription through this gene is 20% less sensitive to UV irradiation than is transcription through the VSG gene. Could these two genes be transcribed from the same promoter?

*8–24   You are studying a DNA virus that makes a set of abundant proteins late in its infectious cycle. An mRNA for one of these proteins maps to a restriction fragment from the middle of the linear genome. To determine the precise

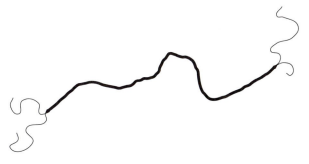

Figure 8–20 DNA-RNA hybrid between an mRNA and a restriction fragment from adenovirus (Problem 8–24).

location of this mRNA, you anneal it with the purified restriction fragment under conditions where only DNA-RNA hybrid duplexes are stable and DNA strands do not reanneal. When you examine the reannealed duplexes by electron microscopy, you see structures such as that in Figure 8–20. Why are there single-stranded tails at the ends of the DNA-RNA duplex region?

8–25 The 3′ ends of most eucaryotic mRNAs are established by cleavage of a precursor RNA followed by addition of a poly-A tail 200 to 300 nucleotides long. The sequence AAUAAA just 5′ of the polyadenylation site is the dominant signal for polyadenylation. The importance of this signal has been verified in many ways. One elegant approach makes use of chemical modification to interfere with specific protein interactions. In this case an RNA containing the signal sequences is treated with diethylpyrocarbonate, which reacts with A and G to give carboxyethylated derivatives. This treatment renders the modified sites highly sensitive to breakage by aniline under appropriate conditions. If the starting RNA molecules are labeled at one end and treated to contain roughly one modification per molecule, then cleavage with aniline will yield a series of fragments whose lengths correspond to the positions of A's and G's in the RNA (Figure 8–21, lane 1). (This method is exactly analogous to the chemical sequencing of DNA.)

To define critical A and G residues, the modified (but still intact) RNA molecules are incubated with an extract capable of cleavage and poly-adenylation. The RNA molecules are then separated into poly-A⁺ RNA and poly-A⁻ RNA, treated with aniline, and the fragments are analyzed by gel electrophoresis (lanes 2 and 3). In a second reaction EDTA is added to the extract to prevent addition of the poly-A tail (cleavage is unaffected), and the cleaved molecules are isolated, treated with aniline, and examined by electro-phoresis (lane 4).

A. At which end were the starting RNA molecules labeled?
B. Explain why the bands corresponding to the AAUAAA signal (bracket in Figure 8–21) are missing from the poly-A⁺ RNA and the cleaved RNA.
C. Explain why the band at the arrow (the normal nucleotide to which poly A is added) is missing in the poly-A⁺ RNA but is present in the cleaved RNA.
D. Which A and G nucleotides are important for cleavage, and which A and G nucleotides are important for addition of the poly-A tail?
E. What information might be obtained by labeling the RNA molecules at the other end?

8–26 Unlike most mRNAs, histone messages do not have poly-A tails at the 3′ end. Instead, they are cleaved from a longer precursor a few nucleotides to the 3′ side

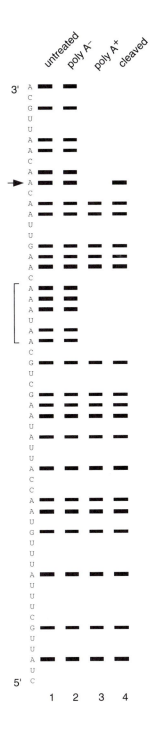

Figure 8–21 Autoradiographic analysis of experiments to define the purines important in cleavage and polyadenylation (Problem 8–25). The sequence of the precursor RNA is shown at the left with the 5′ end at the bottom and the 3′ end at the top. All RNAs were modified by reaction with diethylpyrocarbonate. RNA that was not treated with extract (untreated) is shown in lane 1. RNA that was treated with extract but not polyadenylated (poly A⁻) is shown in lane 2. RNA that was polyadenylated (poly A⁺) is shown in lane 3. RNA that was cleaved but not polyadenylated (cleaved) is shown in lane 4.

(A) HISTONE PRECURSOR RNA

human

5' ACCCAAAGGCUCUUUUCAGAGCCACCCAC▼UUAUUCCAACGAAAGUAGCUGUGAUAAUU 3'

sea urchin

5' AAACGGCUCUUUUCAGAGCCACC▼ACACCCCCAAGAAAGAUUCUCGUUAAA 3'

(B) DNA OLIGONUCLEOTIDES

human   5' ACGAAAGTAGCTGTG 3'

mouse   5' CGGAAAGAGCTGTT 3'

consensus   5' AAAGAAAGAGCTGGT 3'

(C) HUMAN U7 snRNA

5' m$_3$G-NNGUGUUACAGCUCUUUUAGAAUUUGUCUAGU   3'

Figure 8–22 Nucleotide sequences of histone precursor RNAs (A), DNA oligonucleotides (B), and snRNA (C) (Problem 8–26). In (A) the inverted repeat sequences capable of forming a stem-loop structure in the precursors are underlined with arrows. The site of cleavage is indicated by a vertical arrow. The conserved region is underlined in (A) and (B). In (C) m$_3$G is a trimethylated cap, which is characteristic of "U" RNAs. N refers to nucleotides whose identity is unknown.

of a stem-loop structure. Correct processing of the histone precursor depends, in addition, on a conserved sequence just beyond the cleavage site. From experiments with sea urchins it seems that this conserved sequence interacts with the RNA component of the U7 snRNP.

You are studying histone mRNA processing in mammalian cells and wonder if the same interactions define its 3' end. As shown in Figure 8–22A, there is a striking similarity between the 3' ends of histone mRNAs from sea urchin and human. You have shown that nuclear extracts from human cells correctly cleave a labeled synthetic histone mRNA precursor. Furthermore, if you pretreat extracts with a nuclease to digest RNA, the extract is no longer able to cleave histone precursor that is added subsequently. Activity can be restored to the treated extract by adding back a partially purified fraction containing snRNPs.

To define the processing reaction more thoroughly, you synthesize three DNA oligonucleotides corresponding to the region around the suspected site of snRNP interaction with histone precursors. One oligonucleotide matches the conserved sequence in human, one matches that in mouse, and one matches the consensus sequence derived from a comparison of all known conserved sequences in mammals (Figure 8–22B). When you preincubate these oligonucleotides with the extract in the presence of added RNase H (which cleaves RNA in a DNA-RNA hybrid), processing of the precursor was completely blocked by the mouse and consensus oligonucleotides, but the human oligonucleotide had no effect. The two inhibitory oligonucleotides also caused the disappearance of a 63-nucleotide snRNA from the extract. You manage to purify this snRNA and determine the sequence at its 5' end (Figure 8–22C).

A. Explain the design of the oligonucleotide experiment. What were you trying to accomplish by incubating the extract with a DNA oligonucleotide in the presence of RNase H?

B. Since you were using a human extract, you were surprised that the human oligonucleotide did not inhibit processing, whereas the mouse and consensus oligonucleotides did. Can you offer an explanation for this result?

*8–27 You have just gotten the computer to print out a whole set of DNA sequences for the β-globin gene family, and you take the thick file to the country to study for the weekend. When you look at the printout, you discover to your annoyance that you forgot to indicate where in the gene you are. You know that the sequences in Figure 8–23 come from one of the exon/intron or intron/exon boundaries and that the boundary lies down the dotted line. You know that introns always begin with the dinucleotide sequence GT and end with AG, but you realize that these particular sequences would fit *either* as the start *or* the finish of an intron (Figure 8–23).

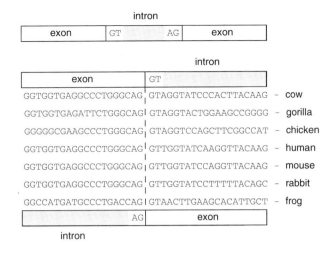

Figure 8–23 The relationship of DNA sequences from the β-globin genes in different species to exons and introns (Problem 8–27).

If you cannot decide which side is the intron, you will have to cut your weekend short and return to the city. In desperation, you consider the problem from an evolutionary perspective. You know that introns evolve faster (suffering more nucleotide changes) than exons because they are not constrained by function. Does this perspective allow you to identify the intron, or will you have to pack your bags?

*8–28 You are studying the transcriptional control of actin synthesis in nematode worms. Nematodes have four genes that encode actin mRNAs: three are clustered on chromosome 5 (genes 1 and 3 are identical in sequence) and one is located on the X chromosome (Figure 8–24). To identify the start site of transcription, you employ two techniques: S1 mapping and primer extension (outlined in Figure 8–25).

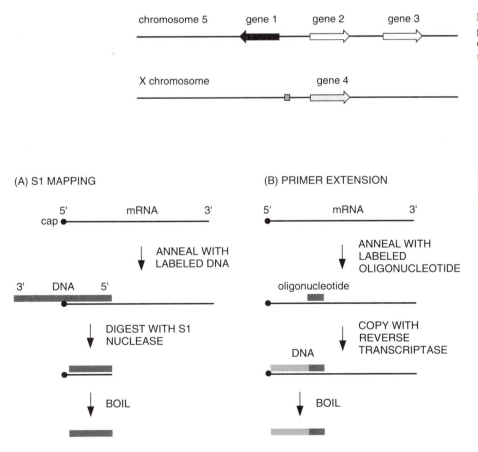

Figure 8–24 Location of the four actin genes in the nematode (Problem 8–28). Genes are indicated by arrows that show their direction of transcription.

Figure 8–25 The use of (A) S1 mapping and (B) primer extension to locate the 5′ ends of the actin mRNAs (Problem 8–28).

ACTIN GENE 1

DNA: 5' TATTATCAATTTAATTTTTCAGGTACATTAAAAACTAATCAAA<u>ATG</u>

RNA: 5' <u>XGUUUAAUUACCCAAGUUUG</u>AGGUACAUUAAAAACUAAUCAAAAUG

ACTIN GENE 2

DNA: 5' ATAATTCATAATTATTTTGTAGGCTAAGTTCCTCCTAATCTAATAAATC<u>ATG</u>

RNA: 5' <u>XGUUUAAUUACCCAAGUUUG</u>AGGCUAAGUUCCUCCUAAUCUAAUAAAUCAUG

ACTIN GENE 3

DNA: 5' TATTATCAATTTAATTTTTCAGGTACATTAAAAACTAATCAAA<u>ATG</u>

RNA: 5' <u>XGUUUAAUUACCCAAGUUUG</u>AGGUACAUUAAAAACUAAUCAAAAUG

LEADER RNA GENE

DNA: 5' GGTTTAATTACCCAAGTTTGAGGTAAACATTCAAACTGA

RNA: 5' <u>XGUUUAAUUACCCAAGUUUG</u>AGGUAAACAUUCAAACUGA

Figure 8–26 RNA and DNA sequences of actin genes and leader RNA genes (Problem 8–28). The start site for translation of the actin genes (ATG) is underlined in the DNA sequence. The leader RNA segment that is present at the 5′ ends of the actin genes and the leader RNA genes is underlined in the RNA sequences. The 5′ nucleotide on the RNAs (X) cannot be determined by primer extension.

To locate the 5′ end of the mRNAs by S1 mapping, you anneal a radioactive single-stranded segment from the 5′ end of the gene to the corresponding mRNA, digest the hybrid with S1 nuclease to remove all single strands, and analyze the protected fragment of radioactive DNA on a sequencing gel to determine its length (Figure 8–25A). For primer extension you hybridize specific oligonucleotides to each mRNA (genes 1 and 3 are identical), extend them to the 5′ end of the mRNA and analyze the resulting DNA segments on sequencing gels (Figure 8–25B).

For gene 4 the two techniques agree, which is the usual case. However, for genes 1, 2, and 3, the mRNAs appear to be 20 nucleotides longer when assayed by primer extension. When you compare the mRNA sequences (which can be determined by primer extension) with the sequences for the gene, you find that each of these mRNAs has an identical 20-nucleotide segment at its 5′ end that does not match the sequence of the gene (Figure 8–26).

Using an oligonucleotide complementary to this leader RNA segment, you discover that the corresponding DNA is repeated about 100 times in a cluster on chromosome 5, but it is a long way from the actin genes on chromosome 5. This leader gene encodes an RNA about 100 nucleotides long. The 5′ end of the leader RNA is identical to the segment found at the 5′ ends of the actin mRNAs (Figure 8–26).

A. Assuming that the leader RNA and the actin RNAs are joined by splicing according to the usual rules, indicate on Figure 8–26 the most likely point at which the RNAs are joined.

B. Since the leader gene and actin genes 1, 2, and 3 are all on the same chromosome, why is it *not* possible that transcription begins at a leader gene and extends through the actin genes to give a precursor RNA that is subsequently spliced to form the actin mRNAs?

C. How would you explain the formation of the actin mRNAs with the common leader segment?

8–29   Many higher eukaryotic genes contain a large number of exons. Correct splicing of such genes requires that neighboring exons be ligated to each other; if they are not, exons will be left out. Since all 5′ splice sites look alike, as do all 3′ splice sites, it is remarkable that skipping an exon occurs so rarely during splicing. Some mechanism must keep track of neighboring exons and ensure that they are brought together.

One proposal for maintaining exon order during splicing suggests that the splicing machinery binds to a splice site at one end of an intron and scans through the intron searching for a second splice site at the other end. Such a scanning mechanism would guarantee that an exon is never skipped. You are intrigued by this hypothesis and decide to test it. You construct two minigenes: one with a duplicated 5′ splice site and the other with a duplicated 3′ splice site (Figure 8–27). You transfect these minigenes into cells and analyze their RNA products to see which 5′ and 3′ splice sites are selected during splicing.

A. Draw a diagram of the products you expect from each minigene if the splicing machinery binds to a 5′ splice site and scans toward a 3′ splice site. Diagram the expected products if the splicing machinery scans in the opposite direction.

B. When you analyze the RNA produced from your transfected minigenes, you find that a mixture of products is generated from each minigene. Are neighboring exons brought together by intron scanning?

The Organization and Evolution of the Nuclear Genome
(MBOC 385–395)

8–30 Fill in the blanks in the following statements.

A. Tandem repeats of simple sequence are called _____ because the first DNAs of this type to be discovered had an unusual ratio of nucleotides that made it possible to separate them from the bulk of the cell's DNA as a minor component.

B. Some _____ move from place to place within chromosomes directly as DNA, while many others move via an RNA intermediate.

C. The nearly simultaneous transposition of several types of transposable elements, called _____, can produce cataclysmic changes in the genome.

D. Two transposable DNA sequences seem to have overrun the human genome: the longer _____, which accounts for about 4% of our DNA, and the shorter _____, which accounts for about 5% of our DNA.

8–31 Indicate whether the following statements are true or false. If a statement is false, explain why.

__ A. In any given species the functions of most genes have probably already been optimized with respect to variation by point mutation.

__ B. Tandemly repeated functional genes tend to remain the same due to unequal crossing over and gene conversion; however, the sequences of the nonfunctional spacer DNA between such genes tend to drift apart rapidly.

__ C. The separation of duplicated genes with distinct functions probably helps to stabilize them by protecting them from the homogenizing processes that act on closely linked genes of similar DNA sequence.

__ D. Long introns between exons provide an increased opportunity for recombination to duplicate exons or link exons from different genes.

__ E. The absence of introns in procaryotic genes indicates that introns were added to the eucaryotic line sometime after the evolutionary separation of procaryotes and eucaryotes.

__ F. Satellite DNA sequences are generally not transcribed and are most often located in the heterochromatin associated with the centromeric regions of chromosomes.

__ G. Although transposable elements are common in the genomes of higher eucaryotes, they move so rarely that they contribute very little to the variability of a species.

__ H. The organization of higher eucaryotic genomes—long noncoding segments and short coding segments—and the regulation of transcription from great distances mean that movements of transposable elements will usually affect gene expression rather than disrupt coding sequences.

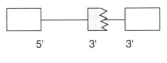
(A) MINIGENE 1

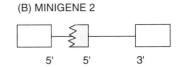
(B) MINIGENE 2

Figure 8–27 Minigene designed to test for intron scanning during RNA splicing (Problem 8–29). Minigene 1 (A) has two 3′ splice sites; minigene 2 (B) has two 5′ splice sites. Boxes represent complete (*open*) or partial (*shaded*) exons; 5′ and 3′ splice junctions are indicated.

Figure 8–28 Restriction map of a bacteriophage lambda clone carrying two *U2* genes (Problem 8–32).

___ I. By simultaneously changing several properties of an organism, transposition bursts increase the probability that two new traits that are useful in combination will appear in a single individual in a population.

___ J. Since *Alu* sequences are transcribed by RNA polymerase II, their continued movement in the genome depends on their insertion in the vicinity of a polymerase II promoter.

*8–32 You are interested in the genes that encode the human U2 small nuclear RNA (U2 snRNA), which is present at thousands of copies per nucleus and plays an important role in mRNA processing. Using radioactive U2 snRNA, you have isolated a bacteriophage lambda clone that carries two copies of the *U2* gene, which are located 6 kb apart. The restriction map of this clone is shown in Figure 8–28. When you cut human genomic DNA to completion with HindIII, HincII (H2), or KpnI (K) and analyze the restriction digest by blot hybridization against the *U2* gene, you detect a single intense band at 6 kb (Figure 8–29, lanes 9 to 11). If you cut genomic DNA with BglII (B), EcoRI (R), or XbaI (X), which do not cut the cloned genes (Figure 8–28), you also detect a single, intense band, but of a size greater than 50 kb (Figure 8–29, lanes 1 to 3). If you incubate the genomic DNA with HindIII and remove samples at various times, you see a ladder of bands (lanes 4 to 9). If you cut 2 ng of the cloned DNA with KpnI and run it alongside 10 mg of the genomic KpnI digest, two bands are visible—each of equal intensity to the 6-kb band from the genomic digest (compare lanes 11 and 12).

A. Explain how the restriction digests define the organization of the *U2* genes in the human genome.

B. Why are two bands visible in the digest of the cloned DNA (lane 12), whereas only one is visible in the digest of genomic DNA (lane 11)?

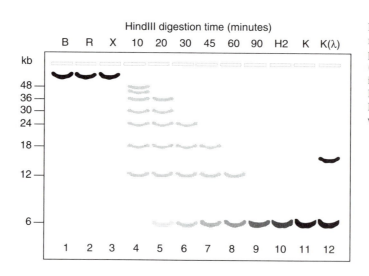

Figure 8–29 Autoradiograph of various restriction digests of human genomic DNA probed with a radiolabeled *U2* gene (Problem 8–32). Numbers under HindIII indicate time of digestion in minutes. B = BglII; R = EcoRI; X = XbaI; H2 = HincII; K = KpnI. K(λ) indicates cloned DNA that was digested with KpnI.

C. Given that 2 ng of cloned DNA produces a band of equal intensity to that from 10 µg of genomic DNA (lanes 11 and 12), calculate how many copies of the *U2* gene there are in the human genome. (The bacteriophage lambda clone is 43 kb, and the human genome is 3 million kb.)

*8–33   Color vision in humans is mediated by three different visual pigments that absorb light in the red, green, and blue part of the visible spectrum. Loss of any one of these pigments causes color blindness. Surprisingly, about 8% of all males have X-linked color vision defects that involve the red or green pigments. Blue color blindness, which is autosomal, is extremely rare.

The human genes for the visual pigments were found by searching for homologues of a cloned bovine rhodopsin gene, which encodes the visual pigment that mediates black and white vision in retinal rod cells. Four types of genes were identified: the rhodopsin gene and three others—one autosomal and two X-linked—that encode proteins that are structurally very similar to rhodopsin. The two genes on the X chromosome, which presumably encode the red and green pigments, are 98% identical throughout most of their length, in both exons and introns. Restriction maps of the two genes (A-type and B-type) are shown in Figure 8–30. The genes can be distinguished by restriction fragment length polymorphisms (RFLPs), one of which (generated by digestion with RsaI) is shown below the genes in Figure 8–30. To test whether these genes encode the red and green pigments, several normal, red-blind, and green-blind males were screened using a hybridization probe specific for the RsaI RFLP (Figure 8–31).

The surprising variability in the apparent number of A-type genes was investigated by digesting the DNA from selected individuals with NotI (which cleaves at very rare sites), separating the long restriction fragments by pulsed-field gel electrophoresis, and hybridizing with a probe that recognizes both genes (Figure 8–32, next page).

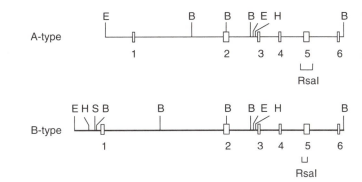

Figure 8–30 Restriction maps of A-type and B-type visual pigment genes (Problem 8–33). Exons are shown as small open boxes. RFLPs generated by RsaI digestion are indicated below the genes. E = EcoRI, B = BamHI, H = HindIII, and S = SalI.

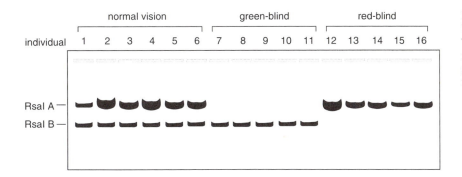

Figure 8–31 RsaI RFLPs in normal, green-blind, and red-blind males (Problem 8–33). The RFLP specific for the A-type gene is indicated as RsaI A; the RFLP for the B-type gene is RsaI B. Individual subjects are identified by a number.

A. Which gene encodes the red visual pigment, and which encodes the green visual pigment?

B. Genetic studies indicate that the genes encoding the red and green visual pigments are close together on the X chromosome. How do the above experiments prove that these genes are physically linked?

C. A probe that is specific for unique sequences just upstream from the 5′ end of the B-type gene hybridizes to a 32-kb restriction fragment generated by cutting with SfiI. One end of this SfiI fragment is generated by cleavage *within* the gene and the other by cleavage outside the gene. This SfiI fragment is also cleaved by NotI, which was used in the analysis shown in Figure 8–32. From this information decide which gene is at the 5′ end of the cluster and which gene is at the 3′ end. (The 5′ end of the cluster is defined by convention as the end nearer the 5′ end of the first gene.)

D. What is the basis for the variability in the number of A-type genes in males with normal color vision? How might your explanation for variability in gene number in normal males account for the high frequency of color blindness?

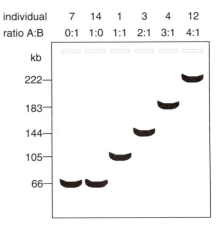

Figure 8–32 NotI digests of DNA from selected normal and color-blind individuals (Problem 8–33). Numbers for individuals correspond to the numbers in Figure 8–31. The ratio of A-type to B-type is estimated from the intensity of hybridization in Figure 8–31. The sizes of the NotI fragments are indicated in kb.

8–34 Almost two centuries ago Thomas Young advanced the hypothesis that humans perceive color by three independent light-sensitive mechanisms. These mechanisms are now known to be embodied in three classes of cone photoreceptor cells in the retina. Each class contains a different visual pigment—red, blue, or green—that determines the spectral properties of all the cones of that class. Two sorts of pigment abnormalities can lead to color blindness. Individuals who lack one of the visual pigments are missing one class of cones; they are called dichromats. Individuals who make a visual pigment with an anomalous absorption spectrum have all three classes of cones, but one of them is abnormal; they are called anomalous trichromats.

You have analyzed the structure of the genes for red and green pigments in 25 males with red-green color deficiencies. Examples of gene structures for red-blind (G^+R^-) and green-blind (G^-R^+) dichromats and red-anomalous (G^+R') and green-anomalous ($G'R^+$) trichromats are shown in Figure 8–33 along with three different structures for normal (G^+R^+) males (trichromats).

A. For each of the color-deficient males show how recombination between two normal (trichromat) gene arrays could give rise to the observed abnormal gene structure. (The green-blind dichromat—G^-R^+—is more difficult than the others.)

B. How do you think each of the four hybrid genes in the color-deficient males will be expressed? Will they be expressed like red genes or like green genes?

C. Based on the genetic structures in Figure 8–33, offer an explanation for each color deficiency.

*8–35 The Ty elements of the yeast *Saccharomyces cerevisiae* move to new locations in the genome by transposition through an RNA intermediate. Normally, the Ty-encoded reverse transcriptase is expressed at such a low level that transposition is very rare. To study the transposition process, you engineer a cloned version of the Ty element so that the gene for reverse transcriptase is linked to the galactose control elements. You also "mark" the element with a segment of bacterial DNA so that you can detect it specifically and thus distinguish it from other Ty elements in the genome. As a target gene to detect transposition, you use a defective histidine gene whose expression is dependent on the insertion of a Ty element near its 5′ end. You show that yeast cells carrying a plasmid with your modified Ty element generate *HIS*+ colonies at a frequency of 5×10^{-8} when grown on glucose. When the same cells are grown on galactose, however, the frequency of *HIS*+ colonies is 10^{-6}: an increase of twentyfold.

You notice that cultures of cells with the Ty-bearing plasmid grow normally on glucose but very slowly on galactose. To investigate this phenomenon, you isolate individual colonies that arise under three different conditions: *his*⁻

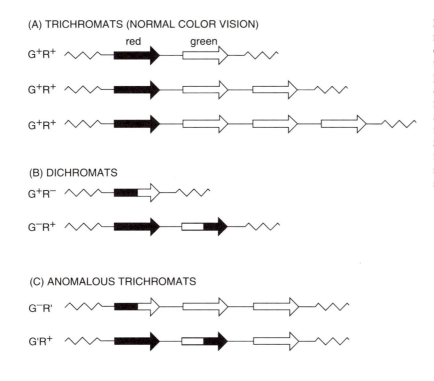

(A) TRICHROMATS (NORMAL COLOR VISION)

(B) DICHROMATS

(C) ANOMALOUS TRICHROMATS

Figure 8–33 Genetic structures for (A) normal trichromats, (B) color-blind dichromats, and (C) color-anomalous trichromats (Problem 8–34). Genes (or parts of genes) for red pigment are shown as black arrows; genes (or parts of genes) for green pigments are shown as white arrows. The base of an arrow represents the 5′ end of a gene; the tip represents the 3′ end. Thin lines indicate homologous intergenic regions; jagged lines indicate single-copy flanking chromosomal sequences.

colonies grown in the presence of glucose, *his⁻* colonies grown in the presence of galactose, and *HIS⁺* colonies grown in the presence of galactose. You "cure" each colony (eliminate the plasmid by growth under special conditions), isolate DNA from each culture, and analyze it by gel electrophesis and blot hybridization using the bacterial marker DNA as a probe. Your results are shown in Figure 8–34.

A. Why does transposition occur so much more frequently in cells grown on galactose than it does in cells grown on glucose?
B. As shown in Figure 8–34, cells grown on galactose in the presence of histidine (*his⁻*) have about the same number of marked Ty elements in their chromosomes as cells that were grown in the absence of histidine (*HIS⁺*). If transposition is independent of histidine selection, why is the frequency of Ty-induced *HIS⁺* colonies so low (10^{-6})?
C. Why do you think it is that cells with the Ty-bearing plasmid grow so slowly on galactose?

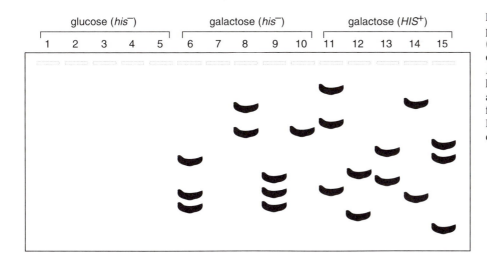

Figure 8–34 Analysis of cells that harbor a plasmid carrying a modified Ty element (Problem 8–35). Cells were initially grown on glucose or galactose as a carbon source. *His⁻* cells were grown in the presence of histidine; *HIS⁺* cells were grown in its absence. Bands indicate restriction fragments that hybridize to the marker DNA originally present in the Ty element carried by the plasmid.

Alu repeat
(300 nucleotides)

```
               TTAAATAGGCCGGG----------AAAAAAAAAAAAATTAAATA
              TGTGTGGGGATCAGG----------AAAAAAAAAAAAAATCTGTGGG
               TCTTCTTAGGCTGGG----------GAAAAAAAAAAAAATCTTCTTA
      ATAATAGTATCTGTCGGCTGGG----------AGAAAAAAAAAAATAAATAGTATCTGTC
             GGATGTTGTGGGGCCGGG----------AAAAAAAAAAAAAGGATGTTGTGG
             AGAACTAAAAGGGCTAGG----------AAAAAAGAGAAGAAGAACCGAAAG
```

Figure 8–35 Nucleotide sequences of the six *Alu* inserts in the human albumin-gene family (Problem 8–36). Dashed lines indicate nucleotides in the internal part of the *Alu* sequences.

8–36 *Alu* sequences are present at six sites in the introns of the human serum-albumin and α-fetoprotein genes. (These genes are evolutionary relatives that are located side by side in mammalian genomes.) The same pair of genes in the rat contains no *Alu* sequences. The lineages of rats and humans diverged more than 85 million years ago at the time of the mammalian radiation. Does the presence of *Alu* sequences in the human genes and their absence in the corresponding rat genes mean that *Alu* sequences invaded the human genes only recently, or does it mean that the *Alu* sequences have been removed in some way from the rat genes?

To examine this question, you have sequenced all six of the *Alu* sequences in the human albumin-gene family. The sequences around the ends of the inserted *Alu* elements are shown in Figure 8–35.

A. Mark the left and right boundaries of the inserted *Alu* sequences and underline the nucleotides in the flanking chromosomal DNA that have been altered by mutation. (Remember that *Alu* sequences create target-site duplications when they insert.)

B. The rate of nucleotide substitution in introns has been measured at about 3×10^{-3} mutations per million years at each site. Assuming the same rate of substitution into the intron sequences that flank these *Alu* sequences, calculate how long ago the *Alu* sequences inserted into these genes. (Lump all the *Alu* sequences together to make this calculation; that is, treat them as if they inserted at about the same time.)

C. Why are these particular flanking sequences used in the calculation? Why were larger segments of the intron not included? Why were the mutations in the *Alu* sequences themselves not used?

D. Did these *Alu* sequences invade the human genes recently (after the time of the mammalian radiation), or have they been removed from the rat genes?

Control of Gene Expression

An Overview of Gene Control
(MBOC 401–404)

9–1   Fill in the blanks in the following statements.

  A. _____ control determines when and how often a given gene is transcribed.
  B. _____ control determines how the primary RNA transcript is spliced or otherwise processed.
  C. _____ control determines which completed mRNAs in the cell nucleus are exported to the cytoplasm.
  D. _____ control determines which mRNAs in the cytoplasm are translated by ribosomes.
  E. _____ control selectively destabilizes certain mRNA molecules in the cytoplasm.
  F. _____ control activates, inactivates, or compartmentalizes specific protein molecules after they have been made.

9–2   Indicate whether the following statements are true or false. If a statement is false, explain why.

__ A. When the nucleus of a fully differentiated carrot cell is injected into a frog egg whose nucleus has been removed, the injected donor nucleus is capable of programming the recipient egg to produce a normal carrot.
__ B. It is likely that only a relatively small number of protein differences (perhaps several hundred) suffice to create very large differences in the morphology and behavior of the different types of cells in an organism.
__ C. One general feature of cell specialization is that different cell types often respond in different ways to the same extracellular signal.

DNA-binding Motifs in Gene Regulatory Proteins
(MBOC 404–417)

9–3   Fill in the blanks in the following statements.

  A. _____ proteins turn specific sets of genes on or off.
  B. The _____ motif has been found in hundreds of DNA-binding proteins from both eucaryotes and procaryotes; it is constructed from two α helices connected by a short extended chain of amino acids.

C. Homeotic selector genes all contain an almost identical stretch of 60 amino acids that is termed the _____.

D. The _____ DNA-binding motif utilizes one or more metal ions as a structural component.

E. The _____ motif was so named because two α helices, one from each monomer, are joined together to form a coiled-coil.

F. The ability of leucine zipper proteins to form heterodimers greatly expands the repertoire of DNA-binding specificities; this is an example of _____ control, in which combinations of proteins, rather than individual proteins, control a cellular process.

G. The _____ motif consists of a short α helix connected by a loop to a second, longer α helix.

H. Once the DNA recognition sequence for a gene regulatory protein has been determined, _____ can be used to purify the regulatory protein.

9–4   Indicate whether the following statements are true or false. If a statement is false, explain why.

__ A. The lambda repressor shuts off the viral genes that code for the protein components of new virus particles and thereby enables the viral genome to remain a silent passenger in the bacterial chromosome.

__ B. Each of the four possible nucleotide pairs (A-T, T-A, G-C, C-G) can be uniquely recognized by the specific arrangement of the atoms that protrude into the major groove of the DNA helix.

__ C. DNA has the same structure throughout, with 36° of helical twist between its adjacent nucleotide pairs and a uniform geometry.

__ D. Nucleotide sequences, typically less than 20 nucleotide pairs in length, function as fundamental components of genetic switches by serving as recognition sites for the binding of specific gene regulatory proteins.

__ E. Because the individual contacts are weak, the interactions between regulatory proteins and DNA are among the weakest interactions known in biology.

__ F. Helix-turn-helix proteins bind as symmetric dimers to DNA sequences that are composed of two very similar half-sites, which are also arranged symmetrically.

__ G. Like the helix-turn-helix motif of bacterial gene regulatory proteins, the helix-turn-helix motif of homeodomains is often embedded in different structural contexts.

__ H. A particular advantage of the zinc finger motif is that the strength and specificity of the DNA-protein interaction could be adjusted during evolution by changes in the number of zinc finger repeats.

__ I. Only α helices are involved in DNA-protein interactions.

__ J. The many types of leucine zipper proteins can all form heterodimers with one another, adding enormously to the combinatorial control of gene expression.

__ K. Naturally truncated HLH proteins can form heterodimers with full-length HLH proteins, enabling a cell to inactivate specific gene regulatory proteins.

__ L. Particular amino acid side chains always are used to recognize specific base pairs, forming a relatively simple amino acid–base pair recognition code.

__ M. The gel-mobility shift assay allows even trace amounts of a sequence-specific DNA-binding protein to be readily detected by its ability to retard the electrophoretic migration of a DNA fragment that contains the recognition sequence.

__ N. DNA affinity chromatography allows purification of unlimited amounts of DNA-binding proteins.

9–5   The binding of a protein to a DNA sequence can cause the DNA to bend in order to make appropriate contacts with chemical groups on the surface of the protein. Such protein-induced DNA bending can be readily detected by the way the protein-DNA complexes migrate through polyacrylamide gels. The rate of migration of bent DNA depends on the average distance between its ends as it gyrates in solution: the more bent the DNA, the closer together the ends are on average and the more slowly it migrates. If there are two sites of bending in the

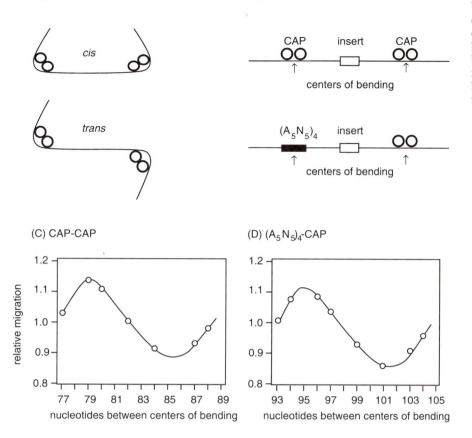

Figure 9–1 Bending of DNA by CAP binding (Problem 9–5). (A) *Cis* and *trans* configurations of a pair of bends. (B) Two constructs used to investigate DNA bending by CAP binding. Relationship between relative migration and number of nucleotides between the centers of bending in the CAP-CAP construct (C) and the $(A_5N_5)_4$-CAP construct (D).

DNA, the end-to-end distance depends on whether the bends are in the same (*cis*) or opposite (*trans*) direction (Figure 9–1A).

You have shown that the catabolite activator protein (CAP) causes DNA to bend by more than 90° when it binds to its regulatory site. You wish to know the details of the bent structure. Specifically, is the DNA at the center of the CAP-binding site bent so that the minor groove of the DNA helix is on the inside, or, alternatively, is the DNA bent so that the major groove is on the inside? To answer this, you prepare two kinds of constructs, as shown in Figure 9–1B. In one, you place two CAP-binding sequences on either side of a central site into which you insert a series of DNA segments that vary from 10 to 20 nucleotides in length. In the other, you flank the insertion site with one CAP-binding sequence and one $(A_5N_5)_4$ sequence, which is known to bend with the major groove on the inside. You now measure the migration of the constructs relative to the corresponding DNA with no insert and plot the relative migration versus the number of nucleotides between the centers of bending (Figure 9–1C and D).

A. Assuming that there are 10.6 nucleotides per turn of the DNA helix, estimate the number of turns that separate the centers of bending of the two CAP-binding sites at the point of minimum relative migration. How many helical turns separate the centers of bending at the point of maximum relative migration?
B. Is the relationship between the relative migration and the separation of the centers of bending of the CAP sites what you would expect if the *cis* configuration migrates slowest and the *trans* configuration migrates fastest? Explain why it is or is not.
C. How many helical turns separate the centers of bending at the point of minimum migration of the construct with one CAP site and one $(A_5N_5)_4$ site?
D. Which groove of the helix faces the inside of the bend at the center of bending of the CAP site?

*9–6 You have cloned four partial cDNAs for a transcription factor into an expression vector to test whether the encoded portions of the factor will bind to the DNA sequence that the complete protein recognizes. The partial cDNA clones extend for different distances toward the 5′ end of the gene (Figure 9–2). You transcribe and then translate these cDNA clones *in vitro* and then mix the translation products with highly radioactive DNA containing the DNA recognition sequence. When the mixtures are analyzed by polyacrylamide gel electrophoresis, some of the proteins encoded by the cDNA clones bind to the DNA fragment, causing a retardation in its migration (Figure 9–2, lanes 3, 4, and 5). When cDNA clones 3 and 4 are mixed together before transcription and translation, three bands appear in the gel retardation assay (Figure 9–2, lane 6).

A. Why are the retarded bands at different positions on the gel?
B. Where in the transcription factor is the binding domain for the DNA recognition sequence located?
C. Why are there three retarded bands when cDNA clones 3 and 4 are mixed together? What does that tell you about the structure of the transcription factor?

(A) MAP OF THE CLONES

(B) GEL RETARDATION ASSAY

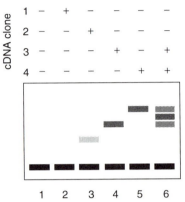

Figure 9–2 Structure of partial cDNAs encoding a transcription factor (A) and gel retardation assays (B) of the encoded proteins (Problem 9–6). Radioactivity of the bands is due to DNA. The band at the bottom of all lanes is the labeled DNA fragment that contains the binding site for the transcription factor.

How Genetic Switches Work
(MBOC 417–433)

9–7 Fill in the blanks in the following statements.

A. The _____ for the tryptophan biosynthetic genes is a short region of regulatory DNA that is recognized by a regulatory protein called the _____.
B. If the active, DNA-binding form of a gene regulatory protein turns genes off, the mode of gene regulation is called _____.
C. If the active, DNA-binding form of a gene regulatory protein turns genes on, the mode of gene regulation is called _____.
D. Eucaryotic RNA polymerase requires several additional proteins called _____ to initiate transcription.
E. _____ sequences, which can be thousands of nucleotide pairs away from a eucaryotic promoter, bind gene regulatory proteins and activate transcription.
F. A eucaryotic _____ region consists of the _____, where the general transcription factors and RNA polymerase assemble, plus all of the _____ sequences to which gene regulatory proteins bind.
G. Most gene regulatory proteins that activate transcription—that is, most _____ proteins—have two domains: one that binds a gene regulatory sequence and another that interacts with the transcription machinery.
H. Many eucaryotic gene regulatory proteins act as _____ proteins to suppress transcription.

9–8 Indicate whether the following statements are true or false. If a statement is false, explain why.

___ A. When the tryptophan repressor has bound two molecules of the amino acid tryptophan, its helix-turn-helix motif is distorted so that it can no longer bind to its operator DNA.
___ B. If a signal molecule binds to a gene activator protein and increases its affinity for its operator, the gene will be turned on, provided that any other requirements for transcription are met.
___ C. The *lac* operon is controlled positively by CAP and negatively by the lac repressor.
___ D. Most gene regulatory proteins in eucaryotes can act even when they are bound to DNA thousands of nucleotide pairs away from the promoter that they influence.

*Problems with an asterisk (*) are answered in the Instructor's Manual.

_ E. In order to bind to a promoter and initiate transcription, all three RNA polymerases in eucaryotes require a TATA sequence in the DNA and several general transcription factors.

_ F. The DNA between the enhancer and the promoter loops out to allow the proteins bound to the enhancer to interact directly either with one of the general transcription factors or with RNA polymerase itself.

_ G. General transcription factors and gene regulatory proteins are abundant proteins in the cell.

_ H. Acidic activators work by lowering the pH in the region of the promoter, which accelerates the assembly of the transcription initiation complex.

_ I. Like bacterial repressors, eucaryotic gene repressor proteins act by directly competing with the polymerase for access to the DNA.

_ J. Protein-protein interactions that are too weak to cause proteins to assemble in solution can cause the proteins to assemble on DNA.

_ K. The *Drosophila eve* gene has a control region that is composed of several regulatory modules, each of which responds to a particular mixture of the gene regulatory proteins that are unevenly distributed in the embryo.

_ L. Seven combinations of gene regulatory proteins—one combination for each stripe—activate *eve* expression, while many other combinations (all those found in the interstripe regions) keep the *eve* gene silent.

_ M. As much as 5% of the coding capacity of a mammalian genome may be devoted to the synthesis of proteins that serve as regulators of gene transcription.

_ N. Inactive gene regulatory proteins in mammalian cells are almost always activated by ligand binding, as they are in bacterial cells.

_ O. Unlike eucaryotes, which use three different RNA polymerases, procaryotes use only one type of core RNA polymerase molecule but modify it with different sigma subunits.

_ P. It seems likely that the close-packed arrangement of bacterial genetic switches developed from more extended forms of switches in response to the evolutionary pressure to maintain a small genome.

*9–9 In the absence of glucose *E. coli* can metabolize and grow on arabinose, a pentose sugar, using an inducible set of genes that are arranged in three groups on the chromosome (Figure 9–3). In one of these the *araA, araB,* and *araD* genes encode enzymes for the metabolism of arabinose while the *araC* gene encodes a regulatory protein that binds adjacent to arabinose promoters and coordinates the expression of the genes in the arabinose operon. (The other two groups of genes encode proteins involved in arabinose transport.)

To understand the regulatory properties of the araC protein, you isolate a mutant bacterium with a deletion of the *araC* gene. As shown in Table 9–1, the mutant strain does not induce expression of the *araA* gene when arabinose is added to the medium.

A. Do the results in Table 9–1 suggest that the araC protein is a positive regulator or a negative regulator of the arabinose operon?

B. What would the data in Table 9–1 have looked like if the araC protein were the opposite kind of gene regulatory protein?

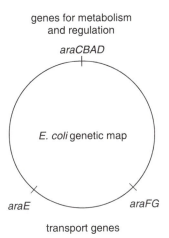

Figure 9–3 Chromosomal locations of the genes involved in arabinose metabolism (Problem 9–9).

Table 9–1 Response of Normal and Mutant Bacteria to the Presence and Absence of Arabinose (Problem 9–9)

Genotype	araA Gene Product	
	Minus Arabinose	Plus Arabinose
araC+	1	1000
araC-	1	1

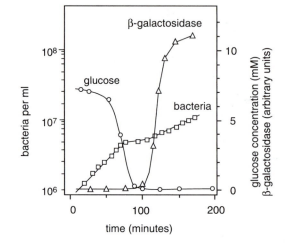

Figure 9–4 Growth of *E. coli* on a mixture of glucose and lactose (Problem 9–10).

9–10 *E. coli* grows faster on the monosaccharide glucose than it does on the disaccharide lactose for two reasons: (1) lactose is taken up more slowly than glucose and (2) lactose must be hydrolyzed to glucose and galactose (by β-galactosidase) before it can be further metabolized.

When *E. coli* is grown on a medium containing a mixture of glucose and lactose, it shows complex growth kinetics (Figure 9–4, squares). The bacteria grow faster at the beginning than at the end, and there is a lag between these two growth phases when they virtually stop growing. Assays of the concentrations of the two sugars in the medium show that glucose falls to very low levels after a few cell doublings (Figure 9–4, circles), but lactose remains high until near the end of the experimental time course. Although the concentration of lactose is high throughout the experiment, β-galactosidase, which is regulated as part of the *lac* operon, is not induced until more than 100 minutes have passed (Figure 9–4, triangles).

A. Explain the kinetics of bacterial growth during the experiment. Account for the rapid rate of initial growth, the slower rate of final growth, and the delay in growth in the middle of the experiment.

B. Explain why the *lac* operon is not induced by lactose during the rapid initial phase of bacterial growth.

*9–11 Transcription of the bacterial gene encoding the enzyme glutamine synthetase is regulated by the availability of nitrogen in the cell. The key transcriptional regulator is the ntrC protein, which stimulates transcription only when it is phosphorylated. Phosphorylation of the ntrC protein is controlled by the ntrB protein, which is both a protein kinase and a protein phosphatase. The balance between its kinase and phosphatase activities—hence the level of phosphorylation of ntrC and transcription of glutamine synthetase—is determined by other proteins that respond to the ratio of α-ketoglutarate and glutamine. (This ratio is a sensitive indicator of nitrogen availability because two nitrogens—as ammonia—must be added to α-ketoglutarate to make glutamine.)

Transcription of the gene for glutamine synthetase can be achieved *in vitro* by adding RNA polymerase, a special sigma factor, and phosphorylated ntrC protein to a linear DNA template containing the gene and its upstream regulatory region. DNA footprinting assays show that the ntrC protein binds to five sites upstream of the promoter. Although binding of the ntrC protein is only slightly increased by phosphorylation, transcription is absolutely dependent on phosphorylation. However, RNA polymerase binds strongly to the promoter even in the absence of the ntrC protein.

Activation of transcription by ntrC was further explored using three different templates: the normal gene with intact regulatory sequences, a gene with all of the ntrC-binding sites deleted, and a gene with only three ntrC-binding sites, all

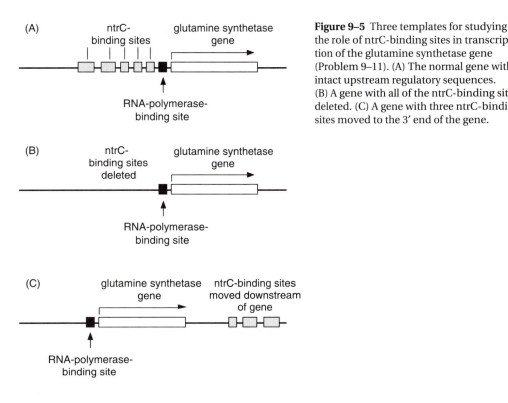

(A) ntrC-binding sites

glutamine synthetase gene

RNA-polymerase-binding site

(B) ntrC-binding sites deleted

glutamine synthetase gene

RNA-polymerase-binding site

(C) glutamine synthetase gene

ntrC-binding sites moved downstream of gene

RNA-polymerase-binding site

Figure 9–5 Three templates for studying the role of ntrC-binding sites in transcription of the glutamine synthetase gene (Problem 9–11). (A) The normal gene with intact upstream regulatory sequences. (B) A gene with all of the ntrC-binding sites deleted. (C) A gene with three ntrC-binding sites moved to the 3′ end of the gene.

at the 3′ end of the gene (Figure 9–5A, B, and C, respectively). In the absence of phosphorylated ntrC protein, no transcription occurred from any of the templates. In the presence of 100 nM phosphorylated ntrC protein, all three templates supported maximal transcription. However, the three templates differed significantly in the concentration of ntrC protein required for half-maximal rates of transcription: the normal gene (A) required 5 nM ntrC protein, the gene with 3′-binding sites (C) required 10 nM ntrC protein, and the gene without ntrC-binding sites (B) required 50 nM ntrC protein.

A. If RNA polymerase can bind to the promoter of the glutamine synthetase gene in the absence of the ntrC protein, why is the ntrC protein needed to activate transcription?

B. If the ntrC protein can bind to its binding sites regardless of its state of phosphorylation, why is phosphorylation necessary for transcription?

C. If the ntrC protein can activate transcription even when its binding sites are absent, what role do the binding sites play?

9–12 Regulation of arabinose metabolism in *E. coli* is fairly complex. Not only are the genes scattered around the chromosome (see Figure 9–3), but the regulatory protein, araC, acts as both a positive and a negative regulator. For example, in the regulation of the *araBAD* cluster of genes (Figure 9–6A), araC binding at site 1 in the presence of arabinose (and the absence of glucose) increases transcription roughly 100-fold over the basal level measured in the absence of araC protein. Binding of araC at site 2 in the absence of arabinose represses transcription of the *araBAD* genes about 10-fold below the basal level measured in the absence of araC protein. The combined effects of negative regulation at site 2 (in the absence of arabinose) and positive regulation at site 1 (in the presence of arabinose) means that addition of arabinose causes a 1000-fold increase in transcription of the *araBAD* genes.

Positive regulation by binding at site 1 seems straightforward since that site lies adjacent to the promoter and presumably facilitates RNA polymerase binding or stimulates open complex formation. You are more puzzled, however, by the results of binding at site 2. Site 2 lies 270 nucleotides upstream from the start site of transcription. Regulatory effects over such distances seem more

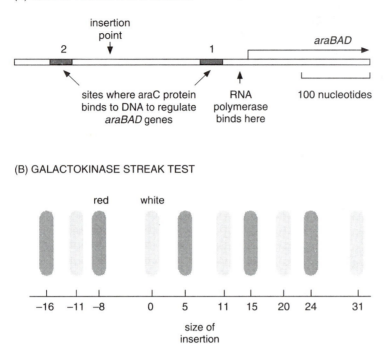

(A) *araBAD* REGULATORY REGION

insertion
point

2

sites where araC protein
binds to DNA to regulate
araBAD genes

1

RNA
polymerase
binds here

araBAD

100 nucleotides

(B) GALACTOKINASE STREAK TEST

red   white

−16 −11 −8   0   5   11   15   20   24   31

size of
insertion

Figure 9–6 Arrangement (A) of araC-binding sites in the *araBAD* operon and results (B) of altering the spacing between the araC-binding sites (Problem 9–12). In (B) the *araBAD* genes have been replaced by the *galK* gene and various numbers of nucleotides have been inserted or deleted at the insertion point, as indicated on the scale at the bottom. Colonies that do not express galactokinase are white; those that do are red.

reminiscent of enhancers in eucaryotic cells. To study the mechanism of repression at site 2 more easily, you move the entire regulatory region so that it is in front of the *galK* gene, whose encoded enzyme, galactokinase, is simpler to assay than are the enzymes encoded by the *araBAD* genes.

To determine the importance of the spacing between the two araC-binding sites, you insert or delete nucleotides at the insertion point indicated in Figure 9–6A. The activity of the promoter in the absence of arabinose is then assayed by growing the bacteria on special indicator plates, on which the bacterial colonies are white if the promoter is fully repressed and red if galactokinase is produced. You streak out bacteria containing the altered spacings against a scale that shows how many nucleotides were deleted or inserted (Figure 9–6B). Much to your surprise, red and white streaks are interspersed.

When you show your results to your advisor, she is very pleased and tells you that these experiments distinguish among three potential mechanisms of repression from a distance. (1) An alteration in the structure of the DNA could propagate from the repression site to the transcription site, making the promoter an unfavorable site for RNA polymerase binding. (2) The protein could bind cooperatively (oligomerize) at the repression site in such a way that additional subunits continue to be added until the growing chain of subunits extends to and covers the promoter, blocking transcription. (3) The DNA could form a loop so that the protein bound at the distant repression site could interact with proteins (or DNA) at the transcription start site.

Which of these general mechanisms do your data support, and how do you account for the patterns of red and white streaks?

9–13 The purification of specific transcription factors has caused you no end of trouble because the assays are slow and the factors tend to be unstable. By the time you identify the right fraction, the factor is often inactive. One day you have a brilliant idea for speeding up the assay. You realize that you can make a DNA sequence that contains no C nucleotides. If this sequence is placed next to a promoter and incubated in the appropriate reaction mix, the promoter should direct the synthesis of a transcript that contains no G nucleotides. Moreover, if GTP is omitted, the only long RNA transcript should be made from

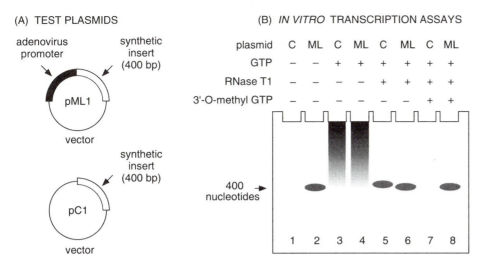

(A) TEST PLASMIDS

(B) *IN VITRO* TRANSCRIPTION ASSAYS

plasmid	C	ML	C	ML	C	ML	C	ML
GTP	–	–	+	+	+	+	+	+
RNase T1	–	–	–	–	+	+	+	+
3'-O-methyl GTP	–	–	–	–	–	–	+	+

400 nucleotides →

1 2 3 4 5 6 7 8

Figure 9–7 Structure of test plasmids (A) and results of transcription assays (B) under various conditions (Problem 9–13). All reactions contain RNA polymerase II, transcription factors, and ^{32}P-CTP. Other components are listed above each lane: (+) means the component is present in the reaction mixture; (–) means the component is absent.

the synthetic DNA sequence because all other transcripts would terminate when a G was required. If you can demonstrate this, then you can rapidly assay transcription of specific sequences simply by measuring incorporation of a radioactive nucleotide!

To test your idea, you construct two plasmids carrying the synthetic sequence: one with a promoter from adenovirus (pML1), the other without (pC1). You mix each of these two plasmids with pure RNA polymerase II, your best preparations of transcription factors, and ^{32}P-CTP. In addition, you add various combinations of GTP, RNase T1 (which cleaves RNA adjacent to each G nucleotide), and 3' O-methyl GTP (which terminates transcription whenever it is incorporated into a growing chain). You measure the products by gel electrophoresis with the result shown in Figure 9–7B.

A. Why is the 400-nucleotide transcript absent in lane 4 but present in lanes 2, 6, and 8?
B. Can you guess the source of the synthesis in lane 3 when the promoterless pC1 plasmid is used?
C. Why is a transcript of about 400 nucleotides present in lane 5 but not in lane 7?
D. Your goal in developing this ingenious assay was to aid the purification of transcription factors. One of your colleagues points out that purification will begin with crude cell extracts, which will contain GTP. Can you assay specific transcription in crude extracts? How?

*9–14 Using your rapid assay for specific transcription (see Problem 9–13), you establish that transcripts accumulate linearly for about an hour and then reach a plateau. Your assay conditions use a 25 µl reaction volume containing 16 mg/ml of DNA template (the pML1 plasmid, which is 3.5 kb in length) with all other components in excess. From the specific activity of the ^{32}P-CTP and the total radioactivity in transcripts, you calculate that at the plateau 2.4 pmol of CMP were incorporated. Each transcript is 400 nucleotides long and has an overall composition of C_2AU. (A nucleotide pair weighs 660 daltons.)

A. How many transcripts are produced per reaction?
B. How many templates are present in each reaction?
C. How many transcripts are made per template in the reaction?

9–15 You have developed an *in vitro* transcription system using a defined segment of DNA that is transcribed under the control of a viral promoter. Transcription of this DNA occurs when you add purified RNA polymerase II, TFIID (the TATA binding factor), and TFIIB and TFIIE (which bind to RNA polymerase). However, the low efficiency of the *in vitro* system suggests that there may be an additional regulatory sequence that binds a transcription factor that is not

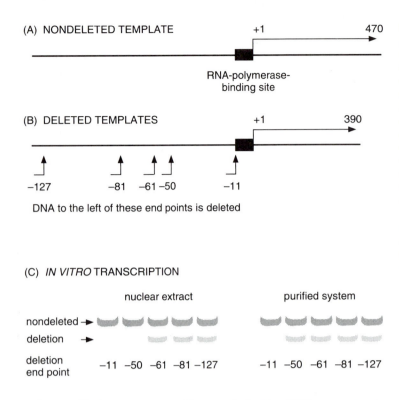

(A) NONDELETED TEMPLATE

+1   470

RNA-polymerase-
binding site

(B) DELETED TEMPLATES

+1   390

−127   −81 −61 −50   −11

DNA to the left of these end points is deleted

(C) *IN VITRO* TRANSCRIPTION

nuclear extract   purified system

nondeleted →

deletion →

deletion
end point   −11 −50 −61 −81 −127   −11 −50 −61 −81 −127

Figure 9–8 Transcription from a viral promoter (Problem 9–15). (A) Nondeleted template. (B) Deleted templates. The nondeleted template gives rise to a transcript that is 80 nucleotides longer than the transcript from the templates that carry deletions in the upstream regulatory region. The deletions remove DNA to the left of the indicated end points. Nucleotides are numbered from the start site of transcription (+1); negative numbers indicate nucleotides in front of the start site of transcription. (C) Results of transcription of a mixture of nondeleted and deleted templates in a crude extract (*left*) and using purified components (*right*). Negative numbers identify the particular deletion template that was included in each mixture.

present among the purified components. To search for the DNA sequence to which this putative regulatory factor binds, you make a series of deletions upstream of the start site for transcription (Figure 9–8B) and compare their transcriptional activity in the purified system and in crude extracts. As an internal control, you mix each of the deletions with a nondeleted template that encodes a slightly longer transcript (Figure 9–8A). The results of these assays are shown in Figure 9–8C. Deletions up to −61 have no effect on transcription and the −11 deletion inactivates transcription in both the purified system and the crude extract. Surprisingly, the −50 deletion is transcribed as efficiently as the nondeleted template in the purified system, but not in the crude system.

You purify the protein that is responsible for this effect and show that it stimulates transcription approximately tenfold from the nondeleted template and from the −61 deletion template, but does not stimulate transcription from the −50 deletion template. Footprinting analysis shows that the factor binds to a specific, short sequence upstream of the TATA site. Furthermore, although the factor binds very transiently to its site in the absence of TFIID (a 20-second half-life), in the presence of TFIID it binds stably (with a half-life greater than 5 hours).

A. Where is the binding site for the stimulatory factor located?
B. How is that a *stimulatory* factor, when added to the other purified transcription components, causes transcripts from the −50 deletion template to be absent from the gel?
C. Why do you think there is such a marked difference in stability of binding of the stimulatory factor in the presence and absence of TFIID?

9–16   The large subunit of eucaryotic RNA polymerase II has a unique C-terminal domain (the CTD), which in yeast comprises 27 near-perfect repeats of the sequence YSPTSPS. It appears that RNA polymerase with a nonphosphorylated CTD forms a preinitiation complex at the promoter and subsequent phosphorylation of the CTD disengages the polymerase from the rest of the preinitiation complex, allowing RNA synthesis to begin.

You are using a genetic approach to identify proteins that interact with the CTD. When you replace the normal RNA polymerase II gene with one in which the CTD is trimmed to 11 YSPTSPS repeats, the yeast is viable at 30°C but unable

(A) TEMPLATES

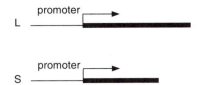

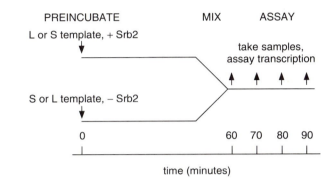

(B) DESIGN OF THE EXPERIMENT

(C) EXPERIMENTAL RESULTS

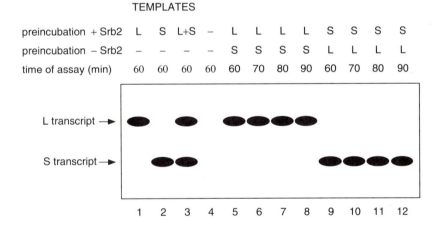

to grow at 12°C. (Cold-sensitive mutants commonly occur in proteins involved in multisubunit assemblies.) Cold sensitivity allows you to select revertants that can grow at 12°C. Upon analysis, some of these revertants prove to be mutations in the RNA polymerase gene that have elongated their CTDs, while others are dominant mutations in previously unknown genes. You choose one of the latter, called *SRB2*, to analyze its role in transcription.

Nuclear extracts prepared from yeast deleted for the *SRB2* gene (which are viable, but cold sensitive) cannot transcribe added DNA templates, but such extracts can be activated by addition of Srb2 protein expressed in *E. coli*. To test the role of Srb2 in transcription, you prepare plasmid DNA containing either a short or a long G-free sequence downstream of a TATA-box containing promoter (Figure 9–9A). You incubate these templates separately in a nuclear extract prepared from yeast deleted for the *SRB2* gene, in the presence or absence of a limiting amount of recombinant Srb2 protein. This preincubation is carried out in the absence of added NTPs so that transcription cannot begin.

Then you mix the preincubations and start transcription at various times later by adding NTPs (Figure 9–9B). You allow synthesis to proceed for just 7 minutes to preclude reinitiation of transcription, and display the products of transcription on a gel. Whichever template is preincubated with Srb2 in addition to the nuclear extract is the one that is transcribed in the assay (Figure 9–9C). In contrast to these results, if an excess of Srb2 protein is mixed with one template at the beginning of the preincubation, transcription is observed from both templates after mixing and addition of NTPs.

A. Does Srb2 show a preference for either template where the preincubation is carried out with the individual templates or a mixture of the templates (Figure 9–9C, lanes 1 to 3)?
B. Do your results indicate that the Srb2 protein acts stoichiometrically or catalytically? How so?
C. Does the Srb2 protein form part of the preinitiation complex or does it act after transcription has begun?
D. What do you think happens during the preincubation that so strongly favors transcription from the template that was included in the preincubation?
E. Do these results indicate that the Srb2 protein binds to the CTD of RNA polymerase II?

*9–17 Hormone receptors for glucocorticoids alter their conformation upon hormone binding to become DNA-binding proteins that activate a specific set of responsive genes. Genetic and molecular studies indicate that the DNA- and hormone-binding sites occupy distinct regions of the C-terminal half of the glucocorticoid receptor. Hormone binding could generate a functional DNA-binding protein in either of two ways: by altering receptor conformation to create a DNA-binding domain or by altering the conformation to uncover a preexisting DNA-binding domain.

These possibilities have been investigated by comparing the activities of a series of carboxy-terminal deletions (Figure 9–10). Fragments of the cDNA, which correspond to the N-terminal portion of the receptor, were inserted into a vector so they would be expressed upon transfection into appropriate cells. The capacity of the receptor fragments to activate responsive genes was tested in transient co-transfections with a reporter plasmid carrying a glucocorticoid response element linked to the chloramphenicol acetyltransferase *(CAT)* gene. As shown in Figure 9–10, cells co-transfected with a cDNA for the full-length receptor responded as expected: in the absence of glucocorticoid no *CAT* activity was detected; in the presence of glucocorticoid (dexamethasone) *CAT* activity was readily detected. Six mutant receptors, lacking 27, 101, 123, 180, 287, and 331 carboxy-terminal amino acids, failed to activate *CAT* expression in the presence or absence of dexamethasone. In contrast, four mutant receptors,

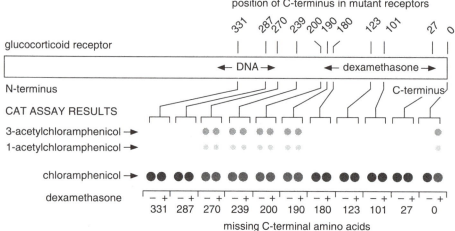

Figure 9–10 Effect of C-terminal deletions on the activity of the glucocorticoid receptor (Problem 9–17). The schematic diagram at the top illustrates the positions of the DNA-binding site and the glucocorticoid-binding site in the receptor, as well as the positions of the C-terminal deletions. The lower diagram shows the results of a standard *CAT* assay obtained by mixing cell extracts with ^{14}C-chloramphenicol: the lowest spot is unreacted chloramphenicol; the upper spots show the attachment of either one or two acetyl groups to chloramphenicol. The presence (+) or absence (–) of dexamethasone is indicated below appropriate lanes.

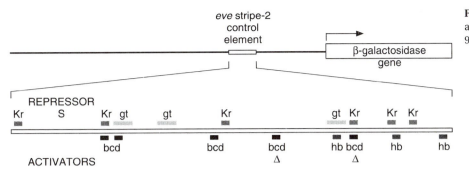

Figure 9–11 Binding sites for repressors and activators of *eve* stripe 2 (Problem 9–18).

lacking 190, 200, 239, and 270 carboxy-terminal amino acids, activated *CAT* expression in the presence and absence of dexamethasone. Separate experiments indicated that the mutant receptors were synthesized equally.

How do these experiments distinguish between the proposed models for hormone-dependent conversion of the normal receptor to a DNA-binding form? Does hormone binding create a DNA-binding site or does it uncover a preexisting DNA-binding site?

9–18　The protein encoded by the *even-skipped (eve)* gene of *Drosophila* is a transcriptional regulator required for controlling segmentation in the middle of the body. It first appears about 2 hours after fertilization at a uniform level in all the embryonic nuclei, but not long after it forms a pattern of 7 stripes. Each stripe is under the control of a separate module in the promoter that provides binding sites for both repressors and activators of *eve* transcription. In the case of stripe 2, two activators, hunchback (hb) and bicoid (bcd), and two repressors, giant (gt) and Krüppel (Kr), are required to give the normal pattern. The binding sites for these proteins have been mapped onto the 670-nucleotide segment shown in Figure 9–11; deletion of this upstream sequence abolishes *eve* expression in stripe 2. The patterns of expression of hunchback, bicoid, giant, and Krüppel in the embryo are shown in Figure 9–12. It seems that *eve* expression in stripe 2 occurs only where the two activators are present and the two repressors are absent, a simple enough rule.

To check if this rule is correct, you construct a β-galactosidase reporter gene driven by a 5-kb upstream segment from the *eve* promoter, which also includes the controlling elements for stripes 3 and 7. In addition to the normal upstream element, you make three mutant versions in which various of the binding sites in the *eve* stripe-2 control segment have been deleted. (Note, however, that because many of the binding sites overlap it is not possible to delete all of one kind of site without affecting some of the other sites.)

Construct 1.　Deletion of all the Krüppel-binding sites
Construct 2.　Deletion of all the giant-binding sites
Construct 3.　Deletion of two bicoid-binding sites (indicated by Δs in Figure 9–11)

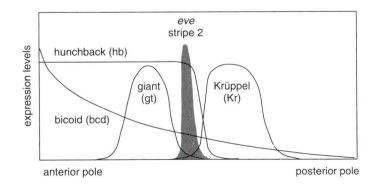

Figure 9–12 Expression of repressors and activators of *eve* stripe 2 in the *Drosophila* embryo (Problem 9–18).

You make flies containing these novel genetic constructs integrated into their chromosomes and determine the patterns of β-galactosidase expression in their embryos, which are shown in Figure 9–13.

A. Match the mutant embryos to the mutant constructs.
B. You began these experiments to test the simple rule that *eve* expression in stripe 2 occurs in the embryo where the two activators are present and the two repressors are absent. Do the results with the mutant embryos confirm this rule?
C. Offer a plausible explanation for why there is no expression of β-galactosidase at the anterior pole of mutant embryo D in Figure 9–13.
D. In the *eve* stripe-2 control segment the binding sites for the two activators do not overlap, nor do the binding sites for the two repressors; however, it is often the case that binding sites for activators overlap binding sites for repressors. What does this overlap suggest about the mode of genetic control of *eve* in stripe 2, and what might be the consequences of this on stripe morphology?

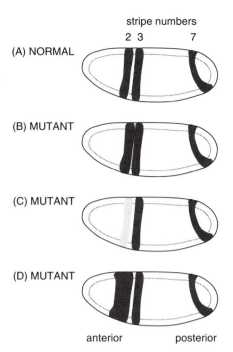

Figure 9–13 Embryonic expression of β-galactosidase from constructs with normal or mutated *eve* stripe-2 control elements (Problem 9–18).

Chromatin Structure and the Control of Gene Expression

(MBOC 433–439)

9–19 Fill in the blank in the following statement.

A. The silencing of the *ADE2* gene when it is relocated to the end of a chromosome by a genetic rearrangement is an example of a _____.

9–20 Indicate whether the following statements are true or false. If a statement is false, explain why.

___ A. Although some gene regulatory proteins can bind to DNA and displace nucleosomes, the general transcription factors seem unable to assemble onto a promoter that is packaged into a nucleosome.
___ B. Inactive forms of chromatin probably contain special proteins that make the DNA unusually inaccessible.
___ C. Mammalian globin genes appear to be activated in two steps: in the first step a set of gene regulatory proteins binds to the globin promoter, and in the second step the entire globin locus opens up for transcription.
___ D. Since most of the DNA in eucaryotic cells is not under tension, topological effects are not thought to play any role in eucaryotic gene expression.

9–21 How does the packing of DNA into chromatin affect transcription in eucaryotes? You have decided to tackle this issue head-on using the C-minus transcription unit you developed in Problem 9–13. This template is transcribed very well in the presence of RNA polymerase II and four transcription factors—TFIIA, TFIIB, TFIID, and TFIIE.

To test the effect of chromatin on transcription, you first assemble the template into nucleosomes (using an extract from frog oocytes), purify the chromatin template, and then add the transcription components. There is no transcription (Figure 9–14, lane 2). You then try a different order of steps. You first incubate the template with the transcription components (in the absence of NTPs), then assemble the template into nucleosomes and purify the chromatin template. Now when you add the transcription components (in the presence of NTPs), transcription proceeds just as well as it does on the naked DNA template (compare lanes 1 and 3). You conclude that one or more of the transcription components must bind to the template and keep the promoter accessible.

To investigate this phenomenon in more detail, you carry out two additional kinds of experiments. In one you leave out individual transcription components during the preincubation (lanes 4 to 8). In the second you leave out individual transcription components during the transcription assay (lanes 9 to 13).

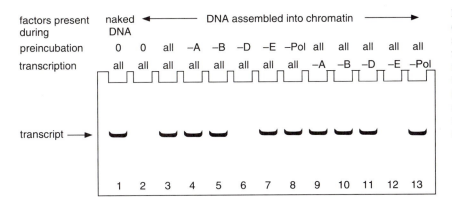

Figure 9–14 Effects of chromatin assembly on transcription (Problem 9–21). Transcription templates were preincubated with none, all, or all minus one of the transcription components (for example, –A means that TFIIA was left out and –Pol means that RNA polymerase II was left out). The transcription assay was carried out in the presence of all or all minus one of the transcription components.

A. Which of the transcription components must be present during the preincubation to keep the template active during chromatin assembly?
B. Which of the transcription components form a complex with the template that is stable to chromatin formation and subsequent purification?
C. Which of the transcription components must be added during the assay in order to produce a transcript?

The Molecular Genetic Mechanisms That Create Specialized Cell Types

(MBOC 439–453)

9–22 Fill in the blanks in the following statements.

A. *Salmonella* bacteria can synthesize two types of flagellin by a reversible DNA rearrangement called _____.
B. Haploid *S. cerevisiae* cells can fuse in a process known as _____.
C. The mating type of haploid yeast cells is determined by the _____ locus.
D. At the heart of the complex regulation of bacteriophage lambda are two gene regulatory proteins, the _____ protein and the _____ protein, which repress each other's synthesis.
E. The pattern of _____ on the parental DNA strand acts as a template for the modification of the daughter DNA strand, causing this pattern to be inherited directly following DNA replication.
F. If expression of a gene depends on whether it was inherited from the mother or the father, the gene is said to be subject to _____.
G. CG sequences are underrepresented in the mammalian genome as a whole, but they are present at 10 to 20 times their average density in selected regions called

_____.

9–23 Indicate whether the following statements are true or false. If a statement is false, explain why.

___ A. Reversible genetic rearrangements are a common way of regulating gene expression in procaryotes and mammalian cells.
___ B. Mating-type switching in yeast is irreversible because the original mating-type gene at the MAT locus is discarded when it is replaced by the other mating-type gene.
___ C. The lambda repressor protein and the cro protein can repress each other's synthesis, creating a two-state molecular switch that specifies lysogeny when the repressor dominates and lysis when the cro protein dominates.
___ D. The fibroblasts and other cell types that are converted to muscle cells by myogenic proteins probably have already accumulated a number of gene

regulatory proteins that can cooperate with the myogenic proteins to switch on muscle-specific genes.

___ E. In the combinatorial control of gene expression, gene regulatory proteins serve many purposes that overlap with those of other gene regulatory proteins.

___ F. Because the condensed X chromosome is reactivated in the formation of the germ cells in the female, no permanent change can have occurred in its DNA.

___ G. Gene control mechanisms that rely entirely on the action of diffusible gene regulatory proteins are sufficient to explain how the inactive X chromosome is kept inactive through successive cell divisions.

___ H. The maintenance methylase perpetuates the preexisting pattern of CG methylation; the establishment methylase initially sets up the pattern of CG methylation in the egg.

___ I. Transgenic mice lacking the maintenance methylase die as young embryos, suggesting that it may be important to reinforce developmental decisions by methylation.

___ J. The CG islands in the promoters of housekeeping genes have been preserved during evolution because sequence-specific DNA-binding proteins protect the 5-methyl C nucleotides in the island from accidental deamination and conversion to T nucleotides.

*9–24   It is relatively common for pathogenic organisms to change their coats periodically in order to evade the immune surveillance of their host. *Salmonella* (a bacterium that can cause food poisoning) can exist in two antigenically distinguishable forms, or phases as they were called by their dicoverer in 1922. Bacteria in the two different phases synthesize different kinds of flagellin, which is the protein that makes up the flagellum. Phase 1 bacteria switch to phase 2 and vice versa about once per thousand cell divisions. Two kinds of explanation were originally considered for the switch mechanism: a DNA rearrangement, such as insertion or inversion, and a DNA modification, such as methylation.

The two flagellins responsible for phase variation are encoded by the unlinked genes, *H1* and *H2*, each of which encodes a completely functional flagellin. The genetic element that enables the bacteria to switch phases is very closely linked to the *H2* gene. To distinguish between the mechanisms of switching, a segment of DNA containing the *H2* gene was cloned. When this segment was introduced into *E. coli*, which have no flagella, most bacteria that picked up the plasmid became able to swim, indicating that they were synthesizing the flagellin encoded by the *H2* gene. A few colonies of *E. coli*, however, were nonmotile even though they carried the plasmid. When DNA was prepared from cultures grown from these nonmotile colonies and introduced into a fresh culture of *E. coli*, some of the transformed bacteria were able to swim, indicating that H2 flagellin synthesis had been switched on.

Plasmid DNA was prepared from these switching cultures, digested with a restriction enzyme, heated to separate the DNA strands, and then slowly cooled to allow DNA strands to reanneal. The DNA molecules were then examined by electron microscopy. About 5% of the molecules contained a bubble, formed by two equal-length single-stranded DNA segments, at a unique position near one end. Two examples are shown in Figure 9–15.

A. All *Salmonella* can swim no matter which type of flagellin they are synthesizing. In contrast, *E. coli* switch between a form that is able to swim and one that is immotile. Why?

B. Explain how these results distinguish between a mechanism of switching that involves a DNA rearrangement and one that involves a DNA modification.

C. How do these results distinguish among site-specific DNA rearrangements that involve deletion of DNA, addition of DNA, or inversion of DNA?

9–25   One of the key regulatory proteins produced by the yeast mating-type locus is a repressor protein known as α2 (see MBOC, Figure 9–58). In haploid cells of the

Figure 9–15 Reannealed DNA fragments from cultures of switching *E. coli* (Problem 9–24). Arrows indicate single-stranded bubbles.

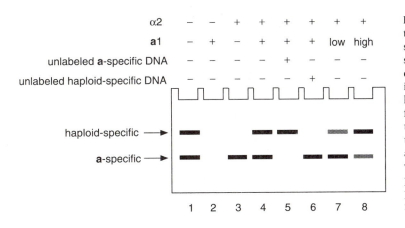

Figure 9–16 Binding of regulatory proteins to fragments of DNA containing the **a**-specific or haploid-specific regulatory sequences (Problem 9–25). Various combinations of regulatory proteins were incubated with a mixture of **a**-specific and haploid-specific radioactive DNA fragments (shown in lane 1). At the end of the incubation the samples were precipitated with antibody against the proteins and the DNA fragments in the precipitate were run on the gel. The gel was then placed against x-ray film to visualize the positions of the radioactive DNA fragments.

α mating type, α2 is essential for turning off a set of genes that are specific for the **a** mating type. In **a**/α diploid cells the α2 repressor collaborates with the product of the **a**1 gene to turn off a set of haploid-specific genes in addition to the **a**-specific genes. Two distinct but related types of conserved DNA sequences are found upstream of these two sets of regulated genes: one in front of the **a**-specific genes and the other in front of the haploid-specific genes. Given the relatedness of these upstream sequences, it is most likely that α2 binds to both; however, its binding properties must be modified in some way by the **a**1 protein before it can recognize the haploid-specific sequence. You wish to understand the nature of this modification. Does **a**1 catalyze covalent modification of α2, or does it modify α2 by binding to it stoichiometrically?

To study these questions, you perform three types of experiments. In the first, you measure the binding of **a**1 and α2, alone and together, to the two kinds of upstream regulatory DNA sites. As shown in Figure 9–16, **a**1 alone does not bind DNA fragments that contain either regulatory site (lane 2), whereas α2 binds to **a**-specific fragments but not to haploid-specific fragments (lane 3). The mixture of **a**1 and α2 binds to **a**-specific *and* haploid-specific fragments (lane 4).

In the second series of experiments you add a vast excess of unlabeled DNA containing the **a**-specific sequence to the reaction along with the mixture of **a**1 and α2 proteins. Under these conditions the haploid-specific fragment is still bound (Figure 9–16, lane 5). Similarly, if you add an excess of unlabeled haploid-specific DNA to the reaction mixture, the **a**-specific fragment is still bound (lane 6).

In the third set of experiments you vary the ratio of **a**1 relative to α2. When α2 is in excess, binding to the haploid-specific fragment is decreased (Figure 9–16, lane 7); when **a**1 is in excess, binding to the **a**-specific fragment is decreased (lane 8).

A. In the presence of **a**1, is α2 present in two forms with different binding specificities or in one form that can bind to both regulatory sequences? How do your experiments distinguish between these alternatives?

B. An α2 repressor with a small deletion in its DNA-binding domain does not bind to DNA fragments containing the haploid-specific sequence. If this mutant protein is expressed in a diploid cell along with normal α2 and **a**1 proteins, however, the haploid-specific genes are turned on. (These genes are normally off in a diploid—see MBOC, Figure 9–58.) In the light of this result and your other experiments, do you consider it more likely that **a**1 catalyzes a covalent modification of α2, or that **a**1 modifies α2 by binding to it stoichiometrically to form an **a**1 α2 complex?

9–26   You have discovered a new strain of yeast with a novel mating system. There are two haploid mating types, called M and F. Cells of opposite mating type can mate to form M/F diploid cells. These diploids can undergo meiosis and spor-

Table 9–2 Phenotypes of Mutants That Affect Mating in a New Strain of Yeast (Problem 9–26)

	Mutant	Mating Phenotype	Genes Expressed
Haploid cells	wild-type M	M	Msg
	M1⁻	nonmating	Msg, Fsg
	M2⁻	M	Msg
	M1⁻, M2⁻	nonmating	Msg, Fsg
	wild-type F	F	Fsg
	F1⁻	F	Fsg
	F2⁻	nonmating	Fsg, Msg
	F1⁻, F2⁻	nonmating	Fsg, Msg
Diploid cells	wild-type M/F	nonmating	Ssg
	M1⁻/F1⁻	F	Fsg
	M1⁻/F2⁻	nonmating	Msg, Fsg, Ssg
	M2⁻/F1⁻	nonmating	none
	M2⁻/F2⁻	M	Msg

Msg = M-specific genes; Fsg = F-specific genes; Ssg = sporulation-specific genes

ulate, but they cannot mate with each other or with either haploid mating type.

Your genetic analysis of the strains shows there are four genes that control mating type. When the genes *M1* and *M2* are at the mating-type locus, the cells are mating-type M; when the genes *F1* and *F2* are at the mating-type locus, the cells are mating-type F. You have also identified three sets of regulated genes: one that is expressed specifically in M-type haploids (Msg), one in F-type haploids (Fsg), and one in sporulating cells (Ssg). You obtain viable mutants (M1⁻, M2⁻, F1⁻, F2⁻) that are defective in each of the mating-type genes and study their effects on the mating phenotype and on expression of the different sets of specific genes they express. Your results with haploid and diploid cells containing different combinations of mutants are shown in Table 9–2.

Suggest a regulatory scheme to explain how the *M1, M2, F1,* and *F2* gene products control the expression of the M-specific, F-specific, and sporulation-specific sets of genes. Indicate which gene products are activators and which are repressors of transcription, and decide whether the gene products act alone and/or in combination.

9–27 You are interested to know whether a transcriptional complex can remain bound to the DNA during DNA replication. If it could, it might serve as a sort of biological memory that would allow daughter cells to inherit the parental pattern of gene expression. You have just the system to test this idea. You can assemble an active transcription complex on the *Xenopus* 5S RNA gene carried on a plasmid, induce its replication, and then test for transcription from the replicated genes. You are able to carry out all these steps *in vitro.*

To distinguish between replicated and unreplicated templates, you take advantage of restriction enzymes (DpnI, MboI, and Sau3A) that are sensitive in different ways to the methylation state of their recognition sequence GATC (Figure 9–17). This sequence is present once at the beginning of the 5S RNA gene and, if the DNA is cleaved at this site, no transcription occurs. If the template is grown in wild-type *E. coli*, GATC will be methylated at the A on both strands by the bacterial *dam* methylase. Replication of fully methylated DNA *in vitro* generates daughter duplexes that are methylated only on one strand (hemimethylated) in the first round and unmethylated DNA in subsequent rounds. Your idea is to start with fully methylated DNA and induce its replication *in vitro.* You can then assay transcription from the replicated DNA

(A) CUTTING PATTERNS

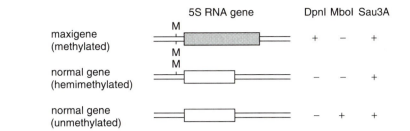

Figure 9–17 Sensitivity of 5S RNA genes in different methylation states (A) and transcriptional activity with and without replication (B) (Problem 9–27). The 5S RNA maxigene is shaded and the normal gene is white. M indicates that a strand is methylated. Sensitivity to cleavage by a restriction enzyme is indicated by (+); insensitivity is indicated by (–). In (B) the positions of the RNA transcripts from the maxigene and the normal 5S RNA gene are indicated by arrows.

(B) TRANSCRIPTIONAL ACTIVITY

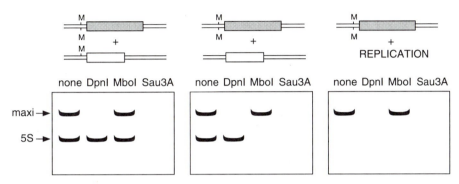

by treating the DNA with DpnI, which cuts unreplicated DNA (which is fully methylated) thus preventing its transcription, but does not cut replicated DNA (which is hemimethylated or unmethylated).

To test whether the analytical part of your scheme will work, you construct a slightly longer than normal 5S RNA gene (maxigene), whose RNA transcript can be distinguished from that of the normal 5S RNA gene (Figure 9–17A). You then prepare mixtures of the fully methylated maxigene with either the hemimethylated or unmethylated normal gene and test their transcription before and after digestion with DpnI, MboI, and Sau3A. The specificity of the restriction enzymes is shown in Figure 9–17A and the results of the transcription experiments are shown in Figure 9–17B.

To test the effect of replication on transcription, you assemble transcription complexes on the fully methylated maxigene, induce replication, and assay transcriptional activity before and after cleavage with restriction enzymes. The results are shown in Figure 9–17B.

A. Does the methylation status of the 5S RNA gene affect its transcription? Explain your answer.

B. Does the pattern of transcription after cleavage with the various restriction enzymes match your expectations? Explain your answer.

C. In your experiment about half of the DNA molecules were replicated. Does the pattern of transcription after replication and cleavage indicate that the transcription complex remains bound to the 5S RNA gene during replication?

D. In these experiments you were careful to show that greater than 90% of the molecules were assembled into active transcription complexes and that 50% of the molecules were replicated. How would the pattern have changed if only 50% of the molecules were assembled into active transcription complexes? Would your conclusions have changed?

*9–28 You are studying the role of DNA methylation in the control of gene expression using the human γ-globin gene as a test system. Globin mRNA can be detected when this gene is incorporated into the genome of mouse fibroblasts, even though it is expressed at much lower levels than it is in red cells. If the gene is

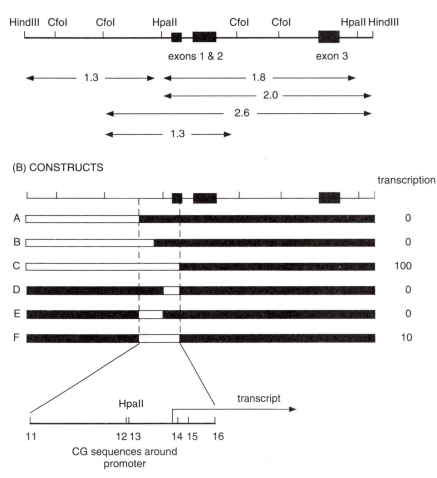

(A) RESTRICTION MAP

HindIII CfoI   CfoI   HpaII   CfoI   CfoI   HpaII HindIII

exons 1 & 2   exon 3

1.3
1.8
2.0
2.6
1.3

(B) CONSTRUCTS

transcription

A   0
B   0
C   100
D   0
E   0
F   10

HpaII   transcript

11   12 13   14 15   16
CG sequences around
promoter

Figure 9–18 Effects of methylation on transcription of the γ-globin gene (Problem 9–28). (A) HindIII fragment containing the γ-globin gene. Sites of cleavage of the methylation-sensitive restriction enzymes, CfoI and HpaII, are indicated along with the sizes of the larger fragments that are observed on the gel in Figure 9–19. (B) Methylated constructs of the γ-globin gene. The methylated segments of the gene are shown in black. The six CG sites around the promoter are shown in more detail below the constructs. The level of expression of γ-globin RNA from each construct is shown on the right as a percentage of the expression from a fully unmethylated construct.

first methylated at all 27 CG sites, however, its expression is blocked completely. You are using this system to decide whether a single critical methylation site is sufficient to determine globin expression.

You use a combination of site-directed mutagenesis and primed synthesis in the presence of 5-methyl dCTP to create several different γ-globin constructs that are unmethylated in various regions of the gene. These constructs are illustrated in Figure 9–18, with the methylated regions shown in black. The arrangement of six methylation sites around the promoter is shown below the constructs. Sites 11, 12, and 13 are unmethylated in construct E, sites 14, 15, and 16 are unmethylated in construct D, and all six sites are unmethylated in construct F. You incorporate these constructs into mouse fibroblasts, grow cell lines containing individual constructs, and measure γ-globin RNA synthesis relative to cell lines containing the fully unmethylated construct (Figure 9–18B).

To check whether the methylation patterns were correctly inherited, you isolate DNA samples from cell lines containing constructs B, C, and F and digest them with HindIII plus CfoI or HpaII. CfoI and HpaII do not cleave if their recognition sites are methylated. The cleavage sites for these enzymes are shown in Figure 9–18A along with the sizes of relevant restriction fragments larger than 1 kb. You separate the cleaved DNA samples on a gel and visualize them by hybridization to the radiolabeled HindIII fragment (Figure 9–19).

A. To create some of the methylated DNA substrates you used a single-stranded version of the gene as a template and primed synthesis of the second strand using 5-methyl dCTP instead of dCTP in the reaction. This method creates a

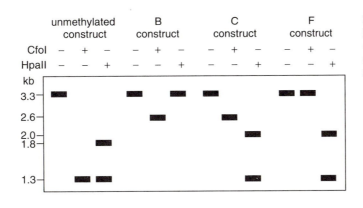

Figure 9–19 Restriction patterns from cell lines containing constructs B, C, and F (Problem 9–28). HindIII was included in all digests. Fragments less than 1 kb are not shown.

DNA molecule containing 5-methyl C in place of C in one strand. After isolation of colonies containing such constructs, you find they have 5-methyl C on both strands but only at CG sequences. Explain the retention of 5-methyl C in CG sequences and their loss elsewhere.

B. Do the restriction patterns of the constructs in the isolated cell lines (Figure 9–19) indicate that the CG sequences that were methylated during creation of the constructs (Figure 9–18) were maintained in the cell?

C. Does the γ-globin RNA synthesis associated with the cell lines containing the various constructs (Figure 9–18) indicate that a single critical site of methylation determines whether the gene is expressed?

9–29   The embryonic mouse fibroblast cell line, 10T½ (10 T and a half), is a very stable line with cells that look and behave like fibroblasts. If these cells are exposed to a medium containing 5-azacytidine (5-aza C) for 24 hours, however, they will then differentiate into cartilage, fat, or muscle cells when they grow to a high cell density. (Treatment with 5-aza C reduces the general level of DNA methylation, allowing some previously inactive genes to become active.) If the cells are grown at low cell density after the treatment, they retain their original fibroblastlike shape and behavior, but even after many generations of growth they still differentiate when they reach a high cell density. 10T½ cells that have not been exposed to 5-aza C do not differentiate no matter what the cell density.

When the treated cells differentiate, about 25% turn into muscle cells (myoblasts). The high frequency of myoblast formation leads your advisor to hypothesize that a single master regulatory gene, which is normally repressed by methylation, may trigger the entire transformation. Accordingly, he persuades you to undertake a high-risk, high-payoff project: clone the gene! You assume the gene is off before treatment with 5-aza C and on in the induced myoblasts. If this assumption is valid, you should be able to find sequences corresponding to the gene among the cDNA copies of mRNAs that are synthesized after 5-aza C treatment.

Your strategy is to screen an existing cDNA library from normal myoblasts (which according to your assumption will also express the gene) using a set of radioactive probes to identify likely cDNA clones. You prepare three radioactive probes.

Probe 1.   You isolate RNA from 5-aza-C-induced myoblasts and prepare radioactive cDNA copies.

Probe 2.   You hybridize the radioactive cDNA from the induced myoblasts with RNA from untreated 10T½ cells and discard all the RNA:DNA hybrids.

Probe 3.   You isolate RNA from normal myoblasts and prepare radioactive cDNA copies, which you then hybridize to RNA from untreated 10T½ cells; you discard the RNA:DNA hybrids.

The Molecular Genetic Mechanisms That Create Specialized Cell Types

Table 9–3 Patterns of Myoblast cDNA Hybridization with Radioactive Probes (Problem 9–29)

Class	Probe 1	Probe 2	Probe 3
A	+	−	−
B	+	−	+
C	+	+	−
D	+	+	+

The first probe hybridizes to a large number of clones from the cDNA library, but only about 1% of those clones hybridize to probes 2 and 3. Overall, you find four distinct patterns of hybridization (Table 9–3).

A. What is the purpose of hybridizing the radioactive cDNA from the two kinds of myoblasts to the RNA from untreated 10T½ cells? In other words, why are probes 2 and 3 useful?

B. What general kinds of genes would you expect to find in each of the four classes of cDNA clones (A, B, C, and D in Table 9–3)? Which class of cDNA clone is most likely to contain sequences corresponding to the putative muscle regulatory gene you are seeking?

Posttranscriptional Controls
(MBOC 453–468)

9–30 Fill in the blanks in the following statements.

A. Although controls on the initiation of gene transcription are the predominant form of gene regulation, _____ controls can act later in the pathway from RNA to protein to modulate the amount of gene product that is made.

B. The premature termination of transcription of an RNA molecule as a means for controlling gene expression is known as _____.

C. Many genes in higher eucaryotes produce several different spliced mRNAs from a single primary transcript by means of _____.

D. A _____ is any DNA sequence that is transcribed as a single unit and encodes one or a set of closely related polypeptide chains.

E. In trypanosomes all mRNAs possess a common 5′ capped leader sequence that is transcribed separately and added to the RNA transcript by _____.

F. In some instances the actual sequence of nucleotides in a primary RNA transcript is altered in a process known as _____.

G. In eucaryotes changes in the rate of protein synthesis in response to various situations is thought to be regulated by the initiation factor _____.

H. The translation of ferritin mRNA molecules is blocked by proteins that bind to their 5′ ends in the absence of iron; this is an example of _____ control.

I. Usually, the completion of the synthesis of a protein is automatic once its synthesis has begun, but a process called _____ can alter the final protein that is made.

9–31 Indicate whether the following statements are true or false. If a statement is false, explain why.

___ A. In adenovirus and HIV, the proteins that assemble at the promoter seem to determine whether or not the polymerase will be able to pass through specific sites of attenuation downstream.

___ B. Although alternative splicing can generate several different versions of the protein encoded by a gene, the different versions are always made in different cell types.

__ C. Sex determination in *Drosophila* depends on a cascade of regulated RNA splicing events that involves three crucial genes.

__ D. A change in the site of RNA transcript cleavage and poly-A addition can change the carboxyl terminus of a protein only by adding amino acids to it or removing amino acids from it.

__ E. The modern definition of a gene is any DNA sequence that is transcribed as a single unit and encodes one polypeptide chain or a set of closely related ones.

__ F. The requirement that RNA exported through nuclear pores possess a nucleotide cap at the 5′ end of the RNA and a poly-A tail at the 3′ end prevents excised introns from reaching the cytosol.

__ G. Certain mRNA molecules are directed to specific intracellular locations by signals in the mRNA sequence, which are typically located in the 3′ untranslated region.

__ H. In the RNA transcripts that code for proteins in the mitochondria of trypanosomes, one or more U nucleotides are either added or removed from selected regions of a transcript, thereby altering the meaning of the message.

__ I. Leaky scanning allows some genes to produce the same protein with and without a signal sequence at its amino terminus so that the protein is directed to two different compartments in the cell.

__ J. Hydrolysis of GTP to GDP by initiation factor eIF-2 is used to drive formation of the first peptide bond in the nascent protein.

__ K. Transferrin receptor mRNA stability and ferritin mRNA translatability are mediated by the same iron-sensitive RNA-binding protein; in both cases binding of the protein to the mRNA increases the level of the encoded protein.

__ L. The regulation of histone mRNA stability depends on a short 3′ stem-and-loop structure that replaces the poly-A tail that is present at the 3′ ends of other mRNAs.

__ M. The observation that maturing oocytes cannot translate mRNAs with short poly-A tails, but can translate them once the poly-A tail has been lengthened, suggests that some critical interaction between proteins at the 5′ and 3′ ends of the mRNA must occur for initiation of protein synthesis.

__ N. Because selenocysteine is incorporated at UGA codons, many proteins—other than the ones that require selenocysteine—have a selenocysteine residue at their carboxyl terminus.

__ O. Many of the RNA-catalyzed reactions in present-day cells may represent molecular fossils—descendants of the complex network of RNA-mediated reactions that are presumed to have dominated cellular metabolism in the beginning.

9–32 The segmentation of eucaryotic genes into exons and introns presents an opportunity for the production of multiple gene products from a single gene by alternative RNA processing. Developmental programs often use differential splicing or differential polyadenylation to produce tissue-specific variants from a single transcription unit.

The gene encoding the small peptide hormone calcitonin is one such differentially utilized gene. The calcitonin gene contains six exons. In thyroid cells an mRNA that encodes calcitonin is produced; it contains exons 1, 2, 3, and 4 and uses a polyadenylation site at the end of exon 4. In neuronal cells no calcitonin is produced from this gene. Instead, calcitonin gene-related peptide (CGRP) is produced; its mRNA consists of exons 1, 2, 3, 5, and 6. The gene and its tissue-specific pattern of processing are diagrammed in Figure 9–20. In both cell types transcription begins in the same place and extends beyond exon 6.

The mechanism of differential processing of the calcitonin/CGRP transcript is not understood. Because different poly-A sites and different splice sites are utilized in the two processing pathways, the tissue-specific factors that regulate calcitonin and CGRP expression could be involved either in poly-adenylation or in splicing. There are two popular models. One is that thyroid cells produce calcitonin because they contain a specific factor that recognizes the poly-A site

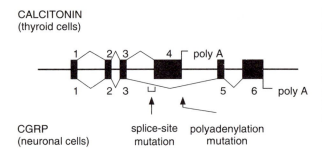

CALCITONIN
(thyroid cells)

CGRP
(neuronal cells)

splice-site
mutation

polyadenylation
mutation

Figure 9–20 Structure and tissue-specific splicing of the gene encoding calcitonin and CGRP (Problem 9–32). Black boxes indicate exons. The splicing/polyadenylation choices used to produce calcitonin are diagrammed above the line; those used to produce CGRP, below the line. Arrows mark the positions of the splice-site and polyadenylation mutations.

in exon 4 with high efficiency and causes cleavage of the precursor RNA before splicing of exon 3 to exon 5 can occur. Neuronal cells lack this factor with the result that splicing of exon 3 to exon 5 predominates, leading to CGRP mRNA production. A second model is that splice-site selection determines which RNA is produced. Thyroid cells produce calcitonin because they splice exon 3 to exon 4; neuronal cells produce CGRP because they splice exon 3 to exon 5. Presumably, one or both cell types produce a factor that favors one splice over the other.

To test these hypotheses, the splicing and polyadenylation signals at the ends of exon 4 were altered by mutation (Figure 9–20). The altered genes were transfected into a lymphocyte cell line, which produces only calcitonin from the wild-type gene. The mutant lacking the exon-4 polyadenylation site produced no mRNA at all; the mutant lacking the exon-4 splice site produced only CGRP mRNA.

A. Does the lymphocyte cell line contain the splicing and polyadenylation factors necessary to produce both calcitonin and CGRP mRNAs?

B. If differential processing results from polyadenylation-site selection, which mutant would you expect to produce CGRP mRNA when transfected into the lymphocyte cell line?

C. If differential processing results from splice-site selection, which mutant would you expect to produce CGRP mRNA when transfected into the lymphocyte cell line?

D. Which model for differential processing best explains the ability of the lymphocyte cell line to produce calcitonin mRNA but not CGRP mRNA?

9–33 Ferritin is the protein that stores iron in many tissues. The synthesis of ferritin increases up to twofold in the presence of iron. You wish to define the molecular mechanism for this induction of ferritin synthesis. In particular, you wish to know whether regulation is transcriptional or posttranscriptional.

You carry out a series of experiments in which rats are given an injection of iron (as a ferric salt solution) or an injection of saline, each with or without simultaneous administration of actinomycin D (which is a powerful inhibitor of RNA synthesis). Three hours later the rats are sacrificed, their livers are homogenized, and polysomal and supernatant fractions are prepared. RNA is then extracted from each fraction and translated in a cell-free system in the presence of radioactive amino acids. You measure total protein synthesis by incorporation of label into protein, and you measure the synthesis of ferritin by precipitation using ferritin-specific antibodies (Table 9–4).

A. For each sample calculate the percentage of total protein synthesis that is due to synthesis of ferritin. Given that 85% of the bulk mRNA is bound to ribosomes in the polysomal fraction and 15% is free (in the supernatant fraction), determine the percentage of total ferritin mRNA in the polysomal and supernatant fractions under each condition of treatment. (Assume that the percentages of polysomal and supernatant mRNA are not changed by treatment with actinomycin D.)

B. How do your results distinguish between transcriptional and posttranscriptional control of ferritin synthesis by iron?

Table 9–4 Synthesis of Ferritin in the Rat After Various Treatments (Problem 9–33)

Injection	Actinomycin D	Fraction	Total Protein Synthesis	Ferritin Synthesis
Saline	absent	polysomes	750,000	700
		supernatant	225,000	1400
Iron	absent	polysomes	500,000	900
		supernatant	400,000	500
Saline	present	polysomes	800,000	800
		supernatant	600,000	3000
Iron	present	polysomes	780,000	1380
		supernatant	550,000	700

Numbers show radioactivity (cpm) incorporated into total proteins or into ferritin.

C. How would you account for the twofold increase in ferritin synthesis in the presence of iron?

9–34 A very active cell-free protein synthesis system can be prepared from reticulocytes, which are immature red blood cells that have already lost their nucleus but still contain ribosomes. A reticulocyte lysate can translate each globin mRNA many times provided heme is added (Figure 9–21). Heme serves two functions: it is required for assembly into hemoglobin, and, surprisingly, it is required to maintain a high rate of protein synthesis. If heme is omitted, protein synthesis stops after a brief lag (Figure 9–21).

The first clue to the molecular basis for the effect of heme on globin synthesis came from a simple experiment. A reticulocyte lysate was incubated for several hours in the absence of heme. When 5 μl of this preincubated lysate were then added to 100 μl of a fresh lysate in the presence of heme, protein synthesis in this fresh lysate was rapidly inhibited (Figure 9–21). When further characterized, the inhibitor (termed heme-controlled repressor, or HCR) was shown to be a protein with a molecular weight of 180,000. Pure preparations of HCR at a concentration of 1 μg/ml completely inhibit protein synthesis in a fresh, heme-supplemented lysate.

A. Calculate the ratio of HCR molecules to ribosomes and globin mRNA at the concentration of HCR that inhibits protein synthesis. Reticulocyte lysates contain 1 mg/ml ribosomes (molecular weight, 4 million), and the average polysome contains 4 ribosomes per globin mRNA.

B. Do the results of this calculation favor a catalytic or a stoichiometric mechanism for HCR inhibition of protein synthesis in a reticulocyte lysate?

*9–35 The level of tubulin gene expression is established in cells by an unusual regulatory pathway in which the intracellular concentration of free tubulin dimers (composed of one α-tubulin and one β-tubulin subunit) regulates the rate of new tubulin synthesis. The initial evidence for such an autoregulatory pathway came from studies with drugs that cause assembly or disassembly of all cellular tubulin. For example, when cells are treated with colchicine, which causes microtubule depolymerization into tubulin subunits, there is a tenfold repression of tubulin synthesis. This autoregulation of tubulin synthesis by tubulin dimers occurs not at the level of transcription but, rather, at the level of tubulin mRNA stability. The first 12 nucleotides of the coding portion of the mRNA were found to contain the site responsible for this autoregulatory control.

Since the critical segment of the mRNA involves a coding region, it is not clear whether the regulation of mRNA stability results from an interaction of free tubulin dimers with the RNA or with the nascent protein. Either interaction

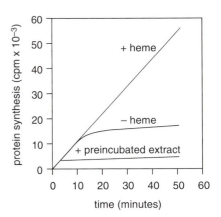

Figure 9–21 Protein synthesis in a reticulocyte lysate (Problem 9–34). The lysate to which preincubated extract was added also contained added heme.

might plausibly trigger a nuclease that would destroy the mRNA.

These two possibilities were tested by mutagenizing the regulatory region on a cloned version of the gene. The mutant genes were then transfected into cells and the stability of their mRNAs was assayed when the intracellular concentration of free tubulin dimers was increased. The results from a dozen mutants that affect a short region of the mRNA are shown in Figure 9–22.

Does the regulation of tubulin mRNA stability result from an interaction with the RNA or from an interaction with the encoded protein? Explain the reasoning behind your answer.

*9–36 Two closely related forms of apolipoprotein B (apo-B) are found in blood as constituents of the plasma lipoproteins. In humans, apo-B100 is synthesized in the liver and is necessary for the assembly of very low density lipoproteins (VLDL) and for the transport of endogenously synthesized triglycerides. Apo-B48 is synthesized by the intestine and is essential for chylomicron formation and for absorption and transport of dietary cholesterol and triglycerides. Several studies indicate that apo-B48 represents the amino-terminal half of apo-B100. What is the relationship between these two gene products? Are they produced from different genes, or are they produced from the same gene by tissue-specific processing of the RNA or tissue-specific cleavage of the protein?

Using recombinant DNA techniques, cDNA copies of the mRNA in human liver and human intestinal cells were cloned. Characterization of several clones from each tissue revealed a single difference: cDNAs from intestinal cells had a T, as part of a stop codon, at a point where the cDNAs from liver cells had a C, as part of a glutamine codon (Figure 9–23). To test for these presumptive differences in the mRNA more directly and to search for corresponding differences in the genome, the PCR (polymerase chain reaction) technique was used to amplify the region that contains the alteration. RNA and DNA were isolated from intestinal cells and from liver cells and then subjected to PCR amplification using oligonucleotides that flank the region of interest. The resulting amplified DNA segments from each of the four samples were tested for the alteration directly by hybridization to oligonucleotides containing either the liver cDNA sequence (oligo-Q) or the intestinal cDNA sequence (oligo-STOP). The results are shown in Table 9–5.

How do these results distinguish among the possibilities for the tissue-specific production of apo-B100 in liver cells and apo-B48 in intestinal cells? Are the two forms of apo-B produced by transcriptional control from two different genes, by a processing control of the RNA transcript from a single gene, or by differential cleavage of the protein product from a single gene?

9–37 The *c-fos* gene is the cellular homolog of the oncogene carried by the FBJ murine osteosarcoma virus. Its function is unknown. Activation of *c-fos* is one of the earliest transcriptional responses to growth factors: the *c-fos* transcription rate in mouse cells increases markedly within 5 minutes, reaches a maximum by 10 to 15 minutes, and abruptly decreases thereafter.

```
          M   R   E   I    regulation
--ATGAGGGAAATC--                +
-----T---------                 -
-----C---------                 +
------C--------                 -
-------A-------                 +
--------T------                 -
---------G-----                 -
---------G-----                 +
-----TA--------                 -
-----C-C-------                 +
-----G--A------                 -
-----G---T-----                 -
-----C----G----                 +
```

Figure 9–22 Effects of mutations on the regulation of tubulin mRNA stability (Problem 9–35). The wild-type sequence for the first 12 nucleotides of the coding portion of the gene is shown at the top, and the first 4 encoded amino acids beginning with methionine (M) are indicated above the codons. The nucleotide changes in the 12 mutants are shown below; only the altered nucleotides are indicated. Regulation of mRNA stability is shown on the right: (+) indicates wild-type response to changes in intracellular tubulin concentration and (–) indicates no response to changes.

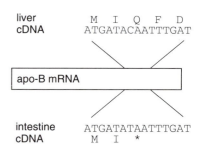

Figure 9–23 The location of the sequence differences in cDNA clones from apo-B RNA isolated from liver and intestine (Problem 9–36). The encoded amino acid sequences indicated are above the cDNA sequences.

Table 9–5 Hybridization of Specific Oligonucleotides to the Amplified Segments from Liver and Intestine RNA and DNA (Problem 9–36)

	RNA		DNA	
	Liver	Intestine	Liver	Intestine
oligo-Q	+	–	+	+
oligo-STOP	–	+	–	–

The oligonucleotide complementary to the sequence derived from liver cDNA is oligo-Q: the oligonucleotide complementary to the intestinal cDNA sequence is oligo-STOP. Hybridizations were carried out under very stringent conditions so that a single nucleotide mismatch was sufficient to prevent hybridization.

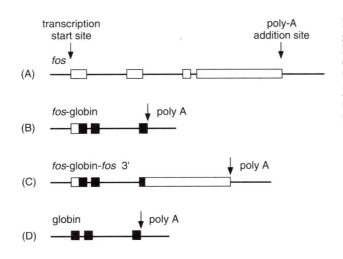

Figure 9–24 Structures of the human *c-fos* gene and the hybrid genes of *c-fos* with the human β-globin gene (Problem 9–37). Open boxes denote *c-fos* exons; black boxes denote β-globin exons. The junctions between *fos* and globin sequences in the hybrid genes are located in exons.

To study the mechanism of transient induction of the human *c-fos* gene, you clone a DNA fragment that contains the complete transcription unit starting 750 nucleotides upstream of the transcription start site and ending 1.5 kb downstream of the poly-A addition site (Figure 9–24A). When you transfer this cloned segment into mouse cells (so you can distinguish the human *c-fos* mRNA from the endogenous cell product) and stimulate cell growth one day later by adding serum (a rich source of growth factors), you observe the same sort of transient induction (Figure 9–25A), although the timescale is slightly longer than for the mouse gene.

You show by analyzing a series of deletion mutants that an enhancerlike element (SRE, serum response element) 300 nucleotides upstream of the transcription start site is necessary for increased transcription in response to serum. You also investigate the stability of the *c-fos* mRNA by creating various hybrid genes containing portions of the human β-globin gene, which encodes a very stable mRNA. The structures of the globin gene and two of the hybrid genes are shown in Figure 9–24B, C, and D. When the hybrid genes are transfected into mouse cells in low serum, they respond to serum added 24 hours later as indicated in Figure 9–25B and C.

A. Which portion of the *c-fos* gene confers instability on *c-fos* mRNA?

B. The *fos*-globin hybrid gene carries all the normal *c-fos* regulatory elements, including the SRE, and yet the mRNA is present at time zero (24 hours after transfection but before serum addition) and does not increase appreciably after serum addition. Can you account for this behavior in terms of mRNA stability?

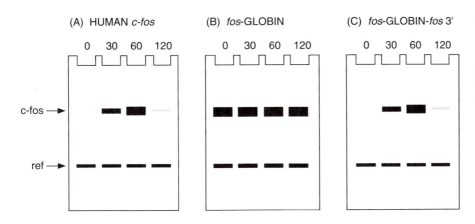

Figure 9–25 Responses of *c-fos* and hybrid genes to addition of serum (Problem 9–37). The upper band in each gel is the transcript containing *c-fos* sequences. The lower band is a reference transcript (to control for recoveries of RNA) from a gene that does not respond to serum addition. Sampling times (in minutes) are indicated above each lane.

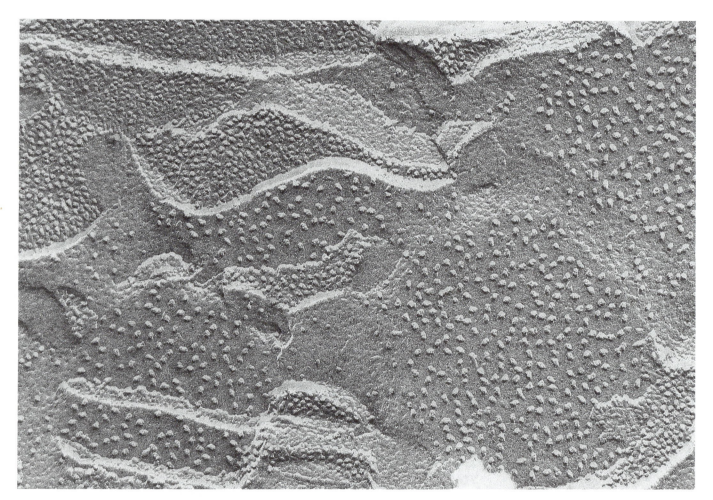

Freeze-fracture electron micrograph of the thylakoid membranes from a chloroplast.
(Courtesy of L.A. Staehelin.)

Membrane Structure

10

- The Lipid Bilayer
- Membrane Proteins

The Lipid Bilayer
(MBOC 478–485)

10–1 Fill in the blanks in the following statements.

A. Lipid molecules in biological membranes are arranged as a continuous double layer called the _____, which is about 5 nm thick.

B. The most abundant lipids in cell membranes are _____.

C. All the lipids found in membranes are said to be _____ because they have one hydrophilic end and one hydrophobic end.

D. The hydrophilic end of a phospholipid consists of a _____ head group, and the hydrophobic end contains two _____ tails.

E. When amphipathic molecules are placed in an aqueous environment, they tend to aggregate so as to bury their hydrophobic ends and expose their hydrophilic ends to water, giving rise to two different kinds of structures, either spherical _____ or planar _____, with the hydrophobic tails sandwiched between the hydrophilic head groups.

F. Synthetic bilayers containing defined lipids or mixtures of lipids can be made either as spherical vesicles, called _____, or as planar bilayers, called _____ membranes.

G. Special ER-membrane-bound enzymes called _____ catalyze the rapid flip-flop of specific phospholipids from the cytosolic monolayer where they are made to the opposite monolayer.

H. Artificial lipid bilayers made from a single type of phospholipid change from a liquid state to a rigid crystalline state or *vice versa* at a characteristic temperature; this change of state is called a _____.

I. Eucaryotic plasma membranes contain especially large amounts of _____, which enhances the mechanical stability of the lipid bilayer.

J. Sugar-containing lipids called _____ are found only in the outer half of the bilayer and their sugar groups are exposed at the cell surface.

K. Glycolipids that contain sialic acid are called _____. One example is _____, to which cholera toxin binds.

10–2 Indicate whether the following statements are true or false. If a statement is false, explain why.

__ A. A lipid bilayer is the fundamental structural component of all cell membranes.

__ B. Maintenance of the lipid bilayer in the plasma membrane requires special enzymes and the hydrolysis of ATP.

__ C. Although lipid molecules are free to diffuse in the plane of the bilayer, they cannot flip-flop across the bilayer unless enzyme catalysts called phospholipid translocators are present in the membrane.

_ D. The temperature at which a eucaryotic membrane "freezes" is determined solely by how much cholesterol it contains.

_ E. The phospholipid head groups on the outside of the cell carry a net positive charge because the choline head groups of phosphatidylcholine—$(CH_3)_3N^+CH_2CH_2OH$—are located predominantly in the outer monolayer.

_ F. The cytoplasmic face of the red cell membrane carries a net negative charge because of the relative excess of phosphatidylserine present on this side of the bilayer.

_ G. Glycolipids are never found on the cytoplasmic face of membranes in living cells.

Table 10–1 Results of Treatment of Red Blood Cells with Enzymes Isolated from Snake Venom (Problem 10–3)

Purified Enzyme	Hemolysis
Protease	no
Neuraminidase	no
Phospholipase	yes

*10–3 A friend of yours has just returned from a nearly disastrous African safari. While crossing the Limpopo River, he was bitten by a poisonous water snake and nearly died from extensive hemolysis. A true biologist at heart, your friend captured the snake before he passed out; he has asked you to analyze the venom to discover the basis of its hemolytic activity. You find that the venom contains a protease (which breaks peptide bonds in proteins), a neuraminidase (which removes sialic acid residues from gangliosides), and a phospholipase (which cleaves bonds in phospholipids). Treatment of isolated red blood cells with these purified activities gave the results shown in Table 10–1. Analysis of the products of hemolysis produced by phospholipase treatment showed an enormous increase in free phosphorylcholine (choline with a phosphate group attached) and diacylglycerol (glycerol with two fatty acid chains attached).

A. What is the substrate for the phospholipase, and where is it cleaved?

B. In light of what you know of the structure of the plasma membrane, explain why the phospholipase causes lysis of the red blood cells, but the protease and neuraminidase do not.

10–4 You wish to determine the distribution of the phospholipids in the plasma membrane of the human red blood cell. Phospholipids make up 60% of the lipids in the red cell bilayer, with cholesterol (23%) and glycolipids (3%) accounting for most of the rest. The phospholipids comprise phosphatidylcholine (17%), phosphatidylethanolamine (18%), sphingomyelin (18%), phosphatidylserine (7%), with the remaining few percent distributed among a variety of phospholipids. To measure the distribution of the individual phospholipids, you react intact red cells and red cell ghosts (1) with two different phospholipases and (2) with a membrane-impermeant fluorescent reagent, abbreviated SITS, which specifically labels primary amine groups.

Treatment with sphingomyelinase degrades up to 85% of the sphingomyelin in intact red cells (without causing lysis) and slightly more in red cell ghosts. The phospholipases in sea snake venom release only phosphatidylcholine breakdown products from intact cells (without causing lysis), but in red cell ghosts they degrade phosphatidylserine and phosphatidylethanolamine as well. SITS labels phosphatidylethanolamine and phosphatidylserine almost to completion in red cell ghosts but reacts less than 1% as well with intact red blood cells.

A. From these results, which are summarized in Table 10–2, deduce the distribution of the four principal phospholipids in red cell membranes. Which of the phospholipids, if any, are located in both monolayers of the membrane?

B. Why did you use red blood cells for these experiments?

*10–5 The behavior of lipids in the two monolayers of a membrane can be studied conveniently by labeling individual molecules with nitroxide groups, which are stable organic free radicals (Figure 10–1). Such spin-labeled lipids can be detected by electron spin-resonance (ESR) spectroscopy, a technique that does not disturb living cells. To introduce spin-labeled lipids into cell mem-

*Problems with an asterisk (*) are answered in the Instructor's Manual.

Table 10–2 Sensitivity of Phospholipids in Human Red Cells and Red Cell Ghosts to Phospholipases and a Membrane-impermeant Label (Problem 10–4)

Phospholipid	Sphingomyelinase		Sea Snake Venom		SITS Fluorescence	
	Red Cells	Ghosts	Red Cells	Ghosts	Red Cells	Ghosts
Phosphatidylcholine	−	−	+	+	−	−
Phosphatidylethanolamine	−	−	−	+	−	+
Sphingomyelin	+	+	−	−	−	−
Phosphatidylserine	−	−	−	+	−	+

nitroxide
radical

spin-labeled
phospholipid 1

spin-labeled
phospholipid 2

Figure 10–1 Structures of two nitroxide-labeled lipids (Problem 10–5). The nitroxide radical is shown at the top, and its position of attachment to the phospholipids is illustrated schematically below.

branes, one first sonicates a mixture of labeled and unlabeled phospholipids to prepare small lipid vesicles. When these vesicles are added to intact cells under appropriate conditions, they fuse with the cell, thereby transferring the labeled lipids into the plasma membrane.

The two spin-labeled phospholipids shown in Figure 10–1 were incorporated into intact human red cell membranes in this way. To determine whether they were introduced equally into the two monolayers of the bilayer, ascorbic acid (vitamin C), which is a water-soluble reducing agent that does not cross membranes, was added to the medium to destroy any nitroxide radicals exposed on the outside of the cell. The ESR signal was followed as a function of time in the presence and absence of ascorbic acid as indicated in Figure 10–2.

A. Ignoring for the moment the difference in extent of loss of ESR signal, offer an explanation for why phospholipid 1 (Figure 10–2A) reacts faster with ascorbate than does phospholipid 2 (Figure 10–2B). Note that phospholipid 1 reaches a plateau in about 15 minutes, whereas phospholipid 2 takes almost an hour.

B. To investigate the difference in extent of loss of ESR signal with the two phospholipids, the experiments were repeated using red cell ghosts that had been resealed to make them impermeable to ascorbate. In these experiments the loss of ESR signal for both phospholipids was negligible in the absence of ascorbate and reached a plateau at 50% in the presence of ascorbate. Offer an explanation for the difference in extent of loss of ESR signal in these experiments with red cell ghosts and the experiments shown in Figure 10–2, which used intact red cells.

C. Were the spin-labeled phospholipids introduced equally into the two monolayers of the red cell membrane?

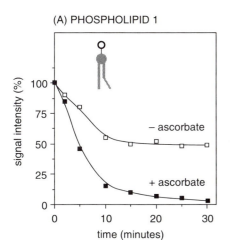

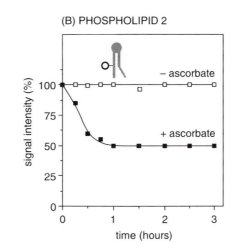

Figure 10–2 Decrease in ESR signal intensity as a function of time in intact red cells in the presence and absence of ascorbate (Problem 10–5).
(A) Phospholipid 1. (B) Phospholipid 2.

10–6   The asymmetric distribution of phospholipids in the two monolayers of the plasma membrane implies that very little spontaneous flip-flop occurs or, alternatively, that any spontaneous flip-flop is rapidly corrected by phospholipid translocators that return phospholipids to their appropriate monolayer. The rate of phospholipid flip-flop in the plasma membrane of intact red blood cells has been measured to decide between these alternatives.

One experimental measurement used the same two spin-labeled phospholipids described in Problem 10–5 (Figure 10–1). To measure the rate of flip-flop from the inner to the outer monolayer, red cells with spin-labeled phospholipids exclusively in the inner monolayer were incubated for various times in the presence of ascorbate and the loss of ESR signal was followed. To measure the rate of flip-flop from the outer to the inner monolayer, red cells with spin-labeled phospholipids exclusively in the outer monolayer were incubated for various times in the absence of ascorbate and the loss of ESR signal was followed. The results of these experiments are illustrated in Figure 10–3.

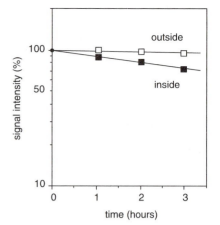

Figure 10–3 Decrease in ESR signal intensity of red cells containing spin-labeled phospholipids in the outer monolayer (outside) and inner monolayer (inside) of the plasma membrane (Problem 10–6).

A. From the results in Figure 10–3, estimate the rate of flip-flop from the inner to the outer monolayer and from the outer to the inner monolayer. A convenient way to express such rates is as the half-time of flip-flop—that is, the time it takes for half the phospholipids to flip-flop from one monolayer to the other.

B. From what you learned about the behavior of the two spin-labeled phospholipids in Problem 10–5, deduce which one was used to label the inner monolayer of the intact red blood cells, and which one was used to label the outer monolayer.

C. Propose a method to generate intact red cells that contain spin-labeled phospholipids exclusively in the inner monolayer, and a method to generate cells spin-labeled exclusively in the outer monolayer.

Membrane Proteins
(MBOC 485–504)

10–7   Fill in the blanks in the following statements.

A. Proteins that extend across the bilayer and are exposed to an aqueous environment on both sides of the membrane are called _____ proteins.

B. While still in the ER, the transmembrane segment of some single-pass transmembrane proteins is cleaved off and a _____ is added, leaving the protein bound to the noncytoplasmic surface of the membrane solely by this anchor.

C. _____ proteins can be released from membranes by gentle procedures, such as extraction by a salt solution, whereas _____ proteins can be removed only by totally disrupting the bilayer with detergents or organic solvents.

D. In _____ proteins, the polypeptide crosses the membrane only once, whereas in _____ proteins, the polypeptide chain crosses multiple times.

E. The outer membrane of *E. coli* is penetrated by various _____, which allow hydrophilic solutes of up to 600 daltons to diffuse across the outer lipid bilayer.

F. The most useful agents for disrupting hydrophobic associations and destroying the bilayer are _____, which are small amphipathic molecules that tend to form micelles in water.

G. The technique of using a membrane-impermeant radioactive or fluorescent labeling reagent to determine the sidedness of membrane proteins is termed _____.

H. _____ is a long fibrous molecule composed of a complex of two very large polypeptide chains, arranged in a filamentous meshwork on the cytoplasmic surface of the red blood cell membrane.

I. _____ is the anion channel in red cells that is responsible for the exchange of HCO_3^- for Cl^- when CO_2 from the tissues is delivered to the lungs.

J. In _____ electron microscopy, cells are frozen in liquid nitrogen and the resulting block of ice is fractured with a knife, splitting the lipid bilayer into its two monolayers.

K. The purple membrane of the bacterium *Halobacterium halobium* is a specialized patch in the plasma membrane containing a single species of protein molecule, _____, which converts light energy into a proton and voltage gradient that in turn drives production of ATP.

L. Like membrane lipids, membrane proteins are able to rotate about an axis perpendicular to the plane of the bilayer (_____ diffusion), many are able to move laterally in the membrane (_____ diffusion), and they do not tumble (_____) across the bilayer.

M. Direct evidence that some plasma membrane proteins are mobile in the plane of the membrane was provided in 1970 by an elegant experiment using hybrid cells called _____, which were produced artificially by fusing mouse and human cells.

N. The lateral diffusion rates of membrane proteins that contain a chromophore or bind a fluorescent ligand can be quantitated using a technique called _____.

O. The carbohydrate-rich zone at the surface of most eucaryotic cells is known as the _____ or _____.

P. Proteins that recognize specific sugar residues are called _____.

Q. _____ are a family of cell-cell adhesion molecules that contain a carbohydrate-binding domain exposed at the cell surface.

10–8   Indicate whether the following statements are true or false. If a statement is false, explain why.

__ A. The basic structure of biological membranes is determined by the lipid bilayer, but their specific functions are carried out largely by proteins.

__ B. Membrane proteins form an extended monolayer on both surfaces of the lipid bilayer.

__ C. Proteins that span a lipid bilayer twice are likely to have their transmembrane segments arranged as β sheets.

__ D. Hydropathy plots are useful for identifying hydrophobic polypeptide segments that are long enough to span a membrane as an α helix.

__ E. Intrachain (and interchain) disulfide (S—S) bonds form readily between cysteine residues on the cytosolic side of membranes, but not on the noncytosolic side.

__ F. In SDS polyacrylamide-gel electrophoresis, individual proteins migrate in the electrical field at rates determined by their molecular weight: the larger the protein, the more it is retarded by the complex meshwork of polyacrylamide molecules that constitutes the gel and, therefore, the more slowly it migrates.

__ G. Human red blood cells contain no internal membranes other than the nuclear membrane.

__ H. One can determine if a membrane protein is exposed on the external side of the plasma membrane by covalent attachment of a labeling reagent or by protease digestion only if the membrane is intact.

__ I. Spectrin, ankyrin, band 3, band 4.1, adducin, and actin are linked together noncovalently on the cytoplasmic surface of the red cell membrane to form a filamentous network, which is thought to be involved in maintaining the biconcave shape of the red cell.

__ J. Each molecule of bacteriorhodopsin contains a single chromophore called retinal, which, when activated by a photon of light, causes a conformational change in the protein that results in the transfer of protons from the inside to the outside of the cell.

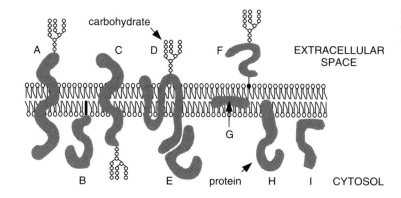

Figure 10–4 A variety of possible associations of proteins with a membrane (Problem 10–9).

_ K. The mobility of membrane proteins can be restricted by interactions with structures outside the cell or inside the cell.

_ L. Although membrane domains with different protein compositions are well known, there are at present no examples of membrane domains that differ in lipid composition.

_ M. Although most proteoglycans are extracellular matrix molecules, integral membrane proteoglycans also exist.

_ N. Whereas all the carbohydrate in the plasma membrane faces outward on the external surface of the cell, all the carbohydrate on internal membranes faces inward toward the cytosol.

_ O. The carbohydrate that makes up the glycocalyx is always attached to glycoproteins and proteoglycans that are integral membrane proteins.

10–9 Which of the arrangements of membrane-associated proteins indicated in Figure 10–4 have been found in biological membranes?

*10–10 Proteins that span a membrane have a characteristic structure in the region of the bilayer. Which, if any, of the three 20-amino-acid sequences listed below is the most likely candidate for such a transmembrane segment? Explain the reasons for your choice.

    A. ITLIYFGVMAGVIGTI LLIS
    B. ITPIYFGPMAGVIGTP LLIS
    C. ITEIYFGRMAGVIGTDLLIS

10–11 Enzymatic digestion of sealed right-side-out red cell ghosts was originally used to determine the sidedness of the major membrane-associated proteins: spectrin, band 3, and glycophorin. These experiments made use of sialidase, which removes sialic acid residues from protein, and pronase, which cleaves peptide bonds. The proteins from normal ghosts and enzyme-treated ghosts were separated by SDS polyacrylamide-gel electrophoresis and then stained for protein and carbohydrate (Figure 10–5).

  A. How does the information in Figure 10–5 allow you to decide whether the carbohydrate of glycophorin is on the cytoplasmic or external surface, and how does it allow you to decide which of the red cell proteins are exposed on the external side of the cell?

  B. When you show your deductions to a colleague, she challenges your conclusion that some proteins are not exposed on the external surface and suggests instead that these proteins may be resistant to pronase digestion. What control experiment can you propose to test this possibility?

  C. How would you modify this enzymatic approach in order to determine which red cell proteins span the plasma membrane?

*10–12 Estimates of the number of membrane-associated proteins per cell and the fraction of the plasma membrane occupied by such proteins provide a useful quantitative basis for understanding the structure of the plasma membrane. These calculations are straightforward for proteins in the plasma membrane

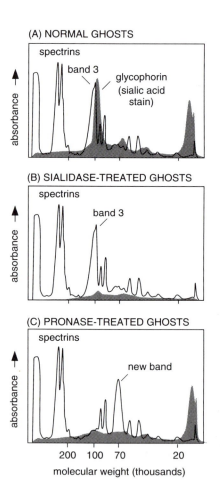

Figure 10–5 Analysis of proteins associated with red cell ghosts before and after digestion with sialidase and pronase (Problem 10–11). Membrane-associated proteins were separated by SDS polyacrylamide-gel electrophoresis and then stained for protein and for carbohydrate. Lines indicate the distribution of proteins, and shaded regions indicate the distribution of carbohydrate.

Table 10–3 Proportion of Stain Associated with Three Membrane-associated Proteins (Problem 10–12)

Protein	Molecular Weight	Percent of Stain
Spectrin	250,000	25
Band 3	100,000	30
Glycophorin	30,000	2.3

of a red blood cell because red cells are readily isolated from blood and they contain no internal membranes to confuse the issue. Plasma membranes are prepared, the membrane-associated proteins are separated by SDS polyacrylamide-gel electrophoresis, and then they are stained with a dye (Coomassie blue). Because the intensity of color is roughly proportional to the mass of protein present in a band, quantitative estimates can be made as shown in Table 10–3.

A. From the information in Table 10–3, calculate the number of molecules of spectrin, band 3, and glycophorin in an individual red blood cell. Assume that 1 ml of red cell ghosts contains 10^{10} cells and 5 mg of total membrane protein.

B. Calculate the fraction of the plasma membrane that is occupied by band 3. Assume that band 3 is a cylinder 3 nm in radius and 10 nm in height and is oriented in the membrane as shown in Figure 10–6. The total surface area of a red cell is 10^8 nm^2.

10–13 One difficult problem in molecular biology is to define the associations between different proteins in complex assemblies. The associations involving spectrin, ankyrin, band 3, and actin, which generate the filamentous meshwork on the cytoplasmic surface of the red cell plasma membrane, have been investigated in several ways. One general method is to use antibodies that are specific for individual proteins. A mixture of two proteins is incubated together, and then an antibody specific for one of them is added. The resulting antibody-protein complexes are then precipitated and analyzed. This technique, when applied to pairwise mixtures of spectrin, ankyrin, band 3, and actin, yields the results summarized in Table 10–4. From the information in the table, deduce the associations between these proteins.

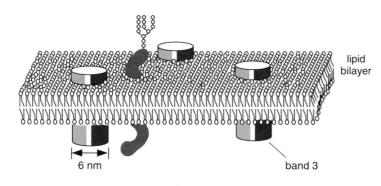

Figure 10–6 Schematic diagram of band 3, represented as a cylinder, in the plasma membrane (Problem 10–12).

Table 10–4 Precipitation of Red Cell Plasma Membrane Proteins by Antibodies Specific for Individual Proteins (Problem 10–13)

Protein Mixture	Antibody Specificity	Proteins in Pellet
1. Band 3 + actin	actin	actin
2. Band 3 + spectrin	spectrin	spectrin
3. Band 3 + ankyrin	ankyrin	band 3 + ankyrin
4. Actin + spectrin	spectrin	actin + spectrin
5. Actin + ankyrin	ankyrin	ankyrin
6. Spectrin + ankyrin	spectrin	spectrin + ankyrin

*10–14 You are studying the binding of proteins to the cytoplasmic face of cultured neuroblastoma cells and have found a method that gives a good yield of inside-out vesicles from the plasma membrane. Unfortunately, your preparations of inside-out vesicles are contaminated with variable amounts of right-side-out vesicles. Nothing you have tried avoids this variable contamination. A friend suggests that you pass your vesicles over an affinity column made of lectin coupled to solid beads. What is the point of your friend's suggestion?

*10–15 In the epithelial cells that line the tubules of the kidney, tight junctions confine certain plasma membrane proteins to the apical surface and others to the basolateral surface. Two structures for tight junctions have been proposed. In one ingenious model, which is based on observations from freeze-fracture electron microscopy, each sealing strand in a tight junction results from a membrane fusion that forms cylinders of lipid at the points of fusion (Figure 10–7A). The other model proposes that each strand of a tight junction is formed by a chain of transmembrane proteins whose extracellular domains bind to one another to seal the epithelial sheet (Figure 10–7B).

As you compare these lipid and protein models for tight junctions, you realize that they might be distinguishable on the basis of lipid diffusion between the apical and basolateral surfaces. Both models predict that lipids in the cytoplasmic leaflet will be able to diffuse freely between the apical and the basolateral surfaces. However, the models suggest different fates for lipids in the external leaflet. In the lipid model, lipids in the outer leaflet will be confined to either the apical surface or the basolateral surface, since the cylinder of lipids interrupts the external leaflet, preventing diffusion through it. By contrast, in the protein model the apical and basolateral surfaces appear to be connected by a continuous external leaflet, suggesting that lipids in the external leaflet should be able to diffuse freely between the two surfaces.

You have exactly the experimental tools to resolve this issue! You have been working with a line of dog kidney cells that forms an exceptionally tight epithelium with well-defined apical and basolateral surfaces. In addition, after infection with influenza virus, the cells express a fusogenic protein only on their apical surface. This feature allows you to fuse liposomes specifically to the apical surface of infected cells very efficiently by brief exposure to low pH, which activates the fusogenic protein. Thus, you can add fluorescently labeled lipids to the apical surface and detect their migration to the basolateral surface using fluorescence microscopy.

For the experiment you prepare two sets of labeled liposomes: one with a fluorescent lipid only in the outer leaflet, the other with the fluorescent lipid equally distributed between the inner and outer leaflets. You fuse these two sets of liposomes to epithelia in which about half the cells were infected with virus. By adjusting the focal plane of the microscope, you examine the apical

Figure 10–7 Models for tight-junction structure (Problem 10–15). (A) Three views of the cylinder of lipids that is proposed to form a tight junction according to the lipid model. (B) Schematic representation of the protein model for tight-junction structure.

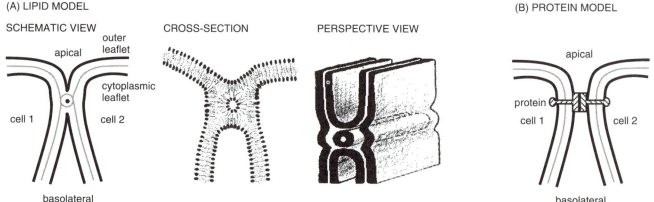

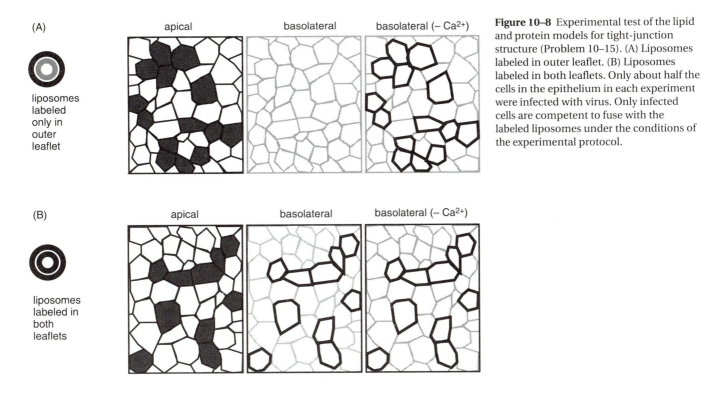

Figure 10–8 Experimental test of the lipid and protein models for tight-junction structure (Problem 10–15). (A) Liposomes labeled in outer leaflet. (B) Liposomes labeled in both leaflets. Only about half the cells in the epithelium in each experiment were infected with virus. Only infected cells are competent to fuse with the labeled liposomes under the conditions of the experimental protocol.

(A) liposomes labeled only in outer leaflet

apical basolateral basolateral (– Ca²⁺)

(B) liposomes labeled in both leaflets

apical basolateral basolateral (– Ca²⁺)

and basolateral surfaces for fluorescence. As a control, you remove Ca^{2+} from the medium—a treatment that disrupts tight junctions—and reexamine the basolateral surface. The results are shown in Figure 10–8.

You are delighted! These results show clearly that the lipids in the external leaflet are confined to the apical surface, whereas lipids in the cytoplasmic leaflet diffuse freely between the apical and basolateral surfaces. Triumphantly, you show these results to your advisor as proof that the lipid model for tight-junction structure is correct. He examines your results carefully, shakes his head knowingly, gives you that penetrating look of his, and tells you that, although the experiments are exquisitely well done, you have drawn exactly the wrong conclusion. These results prove that the lipid model is incorrect.

What has your advisor seen in the data that you have overlooked? How do your results disprove the lipid model? If the protein model is correct, why do you think it is that the fluorescent lipids are confined to the apical surface?

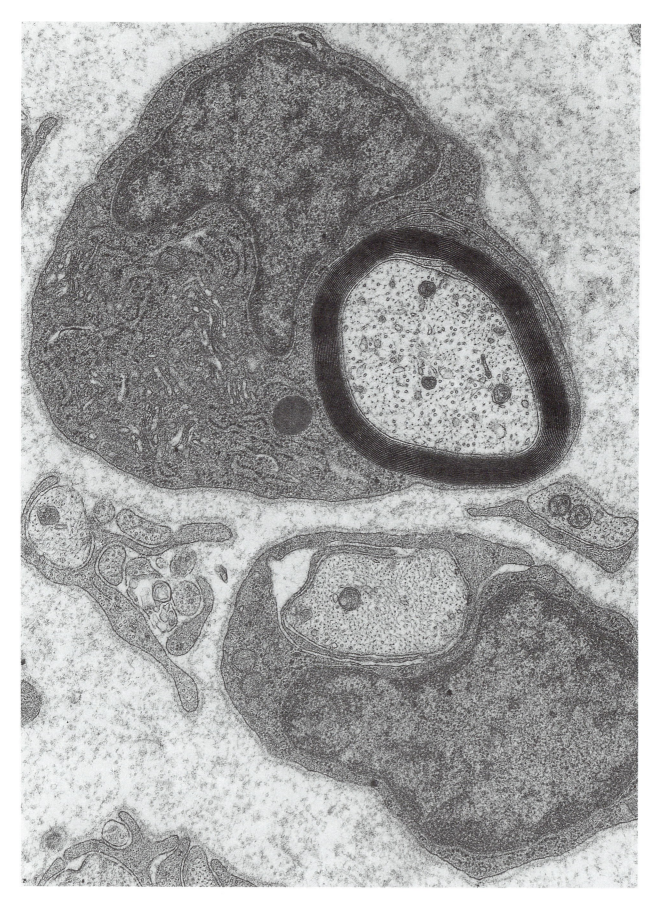

How can one cell locally block membrane transport in another? Electron micrograph of a section from a nerve in the leg of a young rat. See Figure 11–24 on page 532 of MBOC, 3rd Edition, for details. (From C. Raine, in Myelin [P. Morell, ed.]. New York: Plenum, 1976.)

Membrane Transport of Small Molecules and the Ionic Basis of Membrane Excitability

Principles of Membrane Transport
(MBOC 508–512)

11–1 Fill in the blanks in the following statements.

A. Specific proteins called _____ proteins must be present in order for cell membranes to be permeable to small polar molecules such as ions, sugars, and amino acids.

B. There are two major classes of membrane transport proteins: _____ proteins, which bind specific solutes and change conformation to transfer the solute across the membrane; and _____ proteins, which form water-filled pores that allow specific solutes to cross the membrane down their electrochemical gradients.

C. Two general transport processes control the entry of solutes into cells: _____ transport requires no energy input by the cell, whereas _____ transport pumps specific solutes across a membrane against an electrochemical gradient.

D. The concentration gradient for a charged solute and the membrane potential constitute the _____ gradient for that solute.

E. Small hydrophobic molecules that make membranes more permeable to specific inorganic ions are called _____. There are two kinds, known as _____ carriers and _____ formers.

11–2 Indicate whether the following statements are true or false. If a statement is false, explain why.

___ A. The plasma membrane is highly impermeable to all charged molecules.

___ B. All membrane transport proteins so far known traverse the lipid bilayer, and their polypeptide backbones generally extend back and forth across the membrane a number of times.

___ C. Carrier proteins transport their ligands like a revolving door, thereby maintaining a sealed lipid bilayer.

___ D. Studies using recombinant DNA have revealed that membrane transport is mediated by a surprisingly large number of protein families, each of which contains a small number of isoforms.

___ E. Ionophores operate by shielding the charge of the transported ion so that it can penetrate the hydrophobic interior of the lipid bilayer.

11–3 Cytochalasin B, which is often used as an inhibitor of actin-based motility systems, is also a very potent competitive inhibitor of D-glucose uptake into mammalian cells. When red blood cell ghosts are incubated with ^{3}H-cytochalasin B and then irradiated with ultraviolet light, the cytochalasin becomes cross-linked to the glucose transporter. Cytochalasin is not cross-linked to the transporter if an excess of D-glucose is present during the labeling reaction; however, an excess of L-glucose (which is not transported) does not interfere with labeling.

If membrane proteins from labeled ghosts are separated by SDS polyacrylamide-gel electrophoresis, the transporter appears as a fuzzy radioactive band extending from 45,000 to 70,000 daltons. If labeled ghosts are treated with an enzyme that removes attached sugars before electrophoresis, the fuzzy band disappears and a much sharper band at 46,000 daltons takes its place.

A. Why does D-glucose, but not L-glucose, prevent cross-linking of cytochalasin to the glucose transporter?

B. Why does the glucose transporter appear as a fuzzy band on SDS polyacrylamide gels?

*11–4 Glutamate is a major excitatory neurotransmitter in the central nervous system. Glutamate binds to a variety of receptors, some of which control intracellular signaling via G proteins and others that control cation channels. One of the latter, the NMDA receptor, opens a Ca^{2+} channel in response to N-methyl-D-aspartate (NMDA), which is an analogue of glutamate.

The NMDA receptor was cloned from mRNA isolated from rat forebrain. The mRNA was first size fractionated on a sucrose gradient and individual fractions were then injected into *Xenopus* oocytes. A 3–5 kb fraction conferred NMDA responsiveness, as assessed by electrophysiological measurements, on the injected oocytes, which normally do not respond to NMDA. A cDNA library was prepared from the active mRNA fraction and then subdivided into 10 aliquots containing 1000 clones each. RNA was made from each aliquot by *in vitro* transcription and then injected into ooctyes. Aliquots whose RNA conferred NMDA responsiveness were further divided into aliquots containing fewer cDNA clones and retested for the ability of the transcribed RNA to confer NMDA responsiveness to injected ooctyes. Repeated subdivision of NMDA-responsive cDNA aliquots allowed the cDNA for the NMDA receptor to be cloned as outlined in Figure 11–1.

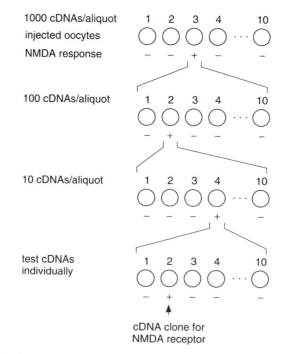

Figure 11–1 Pooling strategy for isolating cDNA for NMDA receptor (Problem 11–4).

*Problems with an asterisk (*) are answered in the Instructor's Manual.

112   Chapter 11 : Membrane Transport

Table 11–1 Pharmacology of the Cloned NMDA Receptor (Problem 11–4)

Agonist		Antagonist		
For NMDA Receptor	**For Other Glu Receptors**	**For NMDA Receptor**	**For Other Glu Receptors**	**Response**
NMDA (100 µM)				100%
Ibotenate (100 µM)				71%
L-HCA (100 µM)				85%
	Kainate (500 µM)			< 5%
	AMPA (50 µM)			< 5%
NMDA (100 µM)		D-APV (10 µM)		27%
NMDA (100 µM)		7-Cl-KYNA (50 µM)		2%
NMDA (100 µM)			GAMS (100 µM)	97%
NMDA (100 µM)			JSTX (100 nM)	83%

Oocytes injected with the putative cDNA for the cloned NMDA receptor were tested with a variety of pharmacologic agents whose receptor specificities and effective concentrations were already defined from previous work.

HCA = homocysteate; AMPA = α-amino-3-hydroxy-5-methyl-4-isoxazolepropionate; APV = 2-amino-5-phosphonovalerate; 7-Cl-KYNA = 7-chlorokynurenate; GAMS = γ-D-glutamylaminomethyl sulphonate; JSTX = Joro spider toxin.

The identity of the NMDA receptor was confirmed using a variety of pharmacologic agents that are known agonists (activators) or antagonists (inhibitors) of the NMDA receptor or other glutamate receptors (Table 11–1).

A. The pooling strategy outlined in Figure 11–1 is an efficient way to find one particular cDNA in a complex mixture. How many oocyte injections and analyses for NMDA responsiveness would have been required if the original cDNA library were tested one cDNA at a time? How many were required using the pooling strategy?

B. How do the data in Table 11–1 confirm that the cDNA for the NMDA receptor was cloned?

Carrier Proteins and Active Membrane Transport
(MBOC 512–522)

11–5 Fill in the blanks in the following statements.

A. Most animal cells take up glucose from the extracellular fluid by passive transport through glucose carriers that operate as _____.

B. The transport of sugars into intestinal cells occurs by inward _____ of Na^+ ions along with the sugar molecules.

C. The band 3 protein of the human red blood cell, which is an _____ carrier, couples transport of Cl^- in one direction with HCO_3^- transport in the other and is thus a good example of an _____ transporter.

D. The _____ is inhibited by ouabain.

E. The network of tubular sacs in muscle cells that stores Ca^{2+} by means of an ATP-driven Ca^{2+} pump is called the _____.

F. Ion-driven carriers are said to mediate _____ active transport, whereas transport ATPases are said to mediate _____ active transport.

G. In the epithelial cells of the gut Na^+-linked symporters in the apical (absorptive) domain of the plasma membrane and Na^+-independent transport proteins in the basolateral domain bring about the _____ of absorbed solutes.

H. Thin fingerlike projections called _____ on the apical surfaces of kidney and intestinal epithelial cells increase their absorptive area by as much as 25 times.

I. The transport ATPases in the bacterial plasma membrane belong to a diverse family of transport proteins known as the _____ because each member contains a highly conserved ATP-binding cassette.

J. Overexpression of the _____ in human cancer cells can make these cells simultaneously resistant to a variety of chemically unrelated cytotoxic drugs that are widely used in cancer chemotherapy.

11–6 Indicate whether the following statements are true or false. If a statement is false, explain why.

___ A. The Na^+-K^+ pump consumes a third of the total ATP supply of a typical animal cell and is responsible for maintaining the high concentration of K^+ inside cells, for controlling cell volume, and for driving the uptake of sugars and amino acids in the intestine and kidney.

___ B. ATP supplies the energy for the Na^+-K^+ pump by phosphorylating an aspartate residue if, and only if, Na^+ is bound; this aspartylphosphate becomes dephosphorylated if, and only if, K^+ is bound. The conformational changes associated with this phosphorylation-dephosphorylation cycle drive the pump.

___ C. Since the Na^+-K^+ ATPase exchanges equal numbers of Na^+ and K^+ ions in each pumping cycle, it is electrically neutral.

___ D. When an action potential depolarizes the muscle cell membrane, the Ca^{2+} pump is responsible for pumping Ca^{2+} from the sarcoplasmic reticulum into the cytosol to initiate muscle contraction.

___ E. The light-activated proton pump of *Halobacterium* synthesizes ATP from ADP and P_i when H^+ ions are pumped across the membrane out of the cell.

___ F. Most cells have one or more types of Na^+-driven antiporters in their plasma membrane that regulate intracellular pH, keeping it at about 7.2.

___ G. The defective gene responsible for cystic fibrosis encodes a Cl^- channel that is a member of the ABC transporter superfamily.

*11–7 Insulin is a small protein hormone that binds to a receptor in the plasma membrane of fat cells. This binding dramatically increases the rate of uptake of glucose into the cells. The increase occurs within minutes and is not blocked by inhibitors of protein synthesis or glycosylation. Therefore, insulin must increase the activity of the glucose transporter in the plasma membrane without increasing the total number of transporters in the cell.

The two experiments described below suggest a possible mechanism for the insulin effect. In the first experiment, the initial rate of glucose uptake in control and insulin-treated cells was measured, with the results shown in Figure 11–2. In the second experiment, the concentration of glucose transporter in fractionated membranes from control and insulin-treated cells was measured, using the binding of radioactive cytochalasin B as the assay (see Problem 11–3), as shown in Table 11–2.

A. Deduce the mechanism by which glucose transport is increased in insulin-treated cells.

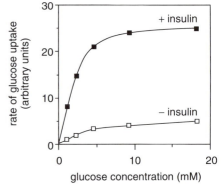

Figure 11–2 Rate of glucose uptake into cells in the presence and absence of insulin (Problem 11–7).

Table 11–2 Amount of Glucose Transporter Associated with the Plasma Membrane and Internal Membranes in the Presence and Absence of Insulin (Problem 11–7)

Membrane Fraction	Bound ^{3}H-Cytochalasin B (cpm/mg vesicle protein)	
	Untreated Cells (– Insulin)	Treated Cells (+ Insulin)
Plasma membrane	890	4480
Internal membranes	4070	480

B. Transport proteins, like enzymes, can be characterized by the kinetic parameters K_M (the concentration of substrate at which the rate of transport is half-maximal) and V_{max} (the rate of transport achieved at saturating substrate concentration). Does insulin stimulation alter either of these kinetic properties of the glucose transporter? How can you tell from these data?

11–8   How much energy does it take to pump substances across membranes? Or, to put it another way, since active transport is usually driven directly or indirectly by ATP, how steep a gradient can ATP hydrolysis maintain for a particular solute? For transport into the cell the free-energy change (ΔG_{in}) per mole of solute moved across the plasma membrane is

$$\Delta G_{in} = -2.3RT \log_{10} \frac{C_o}{C_i} + zFV$$

Where R = the gas constant, 1.98×10^{-3} kcal/°K mole
T = the absolute temperature in °K (37°C = 310°K)
C_o = solute concentration outside the cell
C_i = solute concentration inside the cell
z = the valence (charge) on the solute
F = Faraday's constant, 23 kcal/V mole
V = the membrane potential in volts (V)

Since $\Delta G_{in} = -\Delta G_{out}$, the free-energy change for transport out of the cell is

$$\Delta G_{out} = 2.3RT \log_{10} \frac{C_o}{C_i} - zFV$$

At equilibrium, where $\Delta G = 0$, the equations can be rearranged to the more familiar form known as the Nernst equation.

$$V = 2.3 \frac{RT}{zF} \log_{10} \frac{C_o}{C_i}$$

For the questions below, assume that hydrolysis of ATP to ADP and P_i proceeds with a ΔG of –12 kcal/mole; that is, ATP hydrolysis can drive active transport with a ΔG of +12 kcal/mole. Assume that V is –60 mV.

A. What is the maximum concentration gradient that can be achieved by the ATP-driven active transport into the cell of an uncharged molecule such as glucose, assuming that 1 ATP is hydrolyzed for each solute molecule that is transported?

B. What is the maximum concentration gradient that can be achieved by active transport of Ca^{2+} from the inside to the outside of the cell? How does this maximum compare with the actual concentration gradient observed in mammalian cells (see MBOC Table 11–1)?

C. Calculate how much energy it takes to drive the Na^+-K^+ pump. This remarkable molecular device transports five ions for every molecule of ATP that is hydrolyzed: 3 Na^+ out of the cell and 2 K^+ into the cell. The pump typically maintains internal Na^+ at 10 mM, external Na^+ at 145 mM, internal K^+ at 140 mM, and external K^+ at 5 mM. As shown in Figure 11–3, Na^+ is transported against the membrane potential, whereas K^+ is transported with it. (The ΔG for the overall reaction is equal to the sum of the ΔG values for transport of the individual ions.)

D. How efficient is the Na^+-K^+ pump? That is, what fraction of the energy available from ATP hydrolysis is used to drive transport?

*11–9   A principal function of the plasma membrane is to control the entry of nutrients into the cell. This function is especially important for the epithelial cells that line the gut because they are responsible for absorbing virtually all the nutrients that enter the body. As befits this important role, their plasma membranes are specialized so that the surface facing the gut is folded into

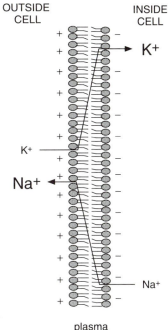

OUTSIDE CELL   INSIDE CELL

K+

K+

Na+

Na+

plasma membrane

Figure 11–3 Na^+ and K^+ gradients and direction of pumping across the plasma membrane (Problem 11–8). Large letters symbolize high concentrations and small letters symbolize low concentrations. Both Na^+ and K^+ are pumped against chemical concentration gradients; but Na^+ is pumped up the electrical gradient, whereas K^+ runs down the electrical gradient.

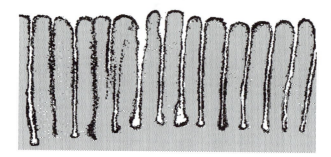

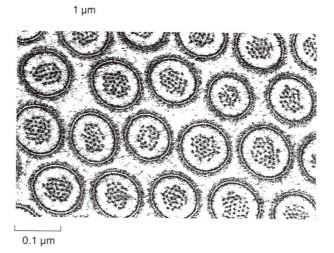

1 µm

0.1 µm

Figure 11–4 Microvilli of intestinal epithelial cells in profile and cross-section (Problem 11–9). (Courtesy of Dr. Ito and Dr. Ichikawa.)

numerous fingerlike projections, termed microvilli. Microvilli increase the surface area of the intestinal cells, providing for more efficient absorption of nutrients. Microvilli are shown in profile and cross-section in Figure 11–4. From the dimensions given in the figure, estimate the increase in surface area that microvilli provide (for the portion of the plasma membrane in contact with the lumen of the gut) relative to the corresponding surface of a cell with a "flat" plasma membrane.

Ion Channels and Electrical Properties of Membranes
(MBOC 523–547)

11–10 Fill in the blanks in the following statements.

A. _____ proteins form hydrophilic pores across membranes; almost all such proteins in eucaryotic plasma membranes are concerned with inorganic ion transport and are therefore referred to as _____.

B. Three kinds of perturbation that can cause gated ion channels to open or close are _____, _____, and _____.

C. The uneven distribution of ions on either side of the plasma membrane gives rise to a voltage across the membrane known as the _____. This voltage depends crucially on the existence of _____ channels, which make most animal cells much more permeable to K^+ than to Na^+.

D. The equilibrium condition, in which there is no net flow of ions across the plasma membrane, defines the _____, which is expressed quantitatively by the _____ equation.

E. Every neuron consists of a _____ (containing the nucleus), usually one long _____ to conduct signals toward distant targets, and several shorter _____, which extend like antennae to receive signals from other nerve cells.

F. An _____, or nerve impulse, is a traveling wave of electrical excitation that can carry a message without attenuation from one end of a neuron to the other.

G. In nerve and skeletal muscle cells a depolarizing stimulus causes _____ to open, allowing a small amount of _____ to enter the cell down its electrochemical gradient.

H. In many nerve cells _____ help bring the activated plasma membrane back to its original negative potential by allowing an efflux of _____.

I. The axons of many vertebrate neurons are insulated by a _____, which greatly increases the rate at which an axon can conduct an action potential.

J. A revolutionary technique that has made it possible to study the behavior of single channels in cell membranes is called _____ recording.

K. Transmission of neuronal signals across a _____ is indirect: a small signal molecule known as a _____ is released from the nerve terminal and diffuses to the target cell where it provokes an electrical change by binding to _____ ion channels.

L. _____ such as acetylcholine open Na$^+$ channels, which causes depolarization of the postsynaptic membrane, whereas _____ such as GABA open Cl$^-$ channels, which suppresses depolarization.

M. A _____ is the specialized chemical synapse between a motor neuron and a skeletal muscle cell.

N. _____ and _____ summation of individual excitatory and inhibitory postsynaptic potentials (PSPs) provide the means by which the rates of firing of many presynaptic neurons jointly control the _____ PSP in the body of a single postsynaptic cell.

O. Action potentials are initiated at the _____, a unique region of each neuron where voltage-gated Na$^+$ channels are plentiful.

P. _____ channels repolarize the membrane after each action potential to prepare the cell to fire again; _____ channels ensure that the firing rate is proportional to the strength of the depolarizing stimulus over a very broad range.

Q. _____ channels help to decrease the response of the cell to an unchanging, prolonged stimulation: a process called _____.

R. Whereas occasional single action potentials leave no lasting trace, a short burst of repetitive firing causes _____, such that subsequent single action potentials in the presynaptic cells evoke a greatly enhanced response in the postsynaptic cells.

S. The _____ channels are doubly gated, opening only when glutamate is bound to the receptor and the membrane is strongly depolarized.

11–11 Indicate whether the following statements are true or false. If a statement is false, explain why.

___ A. Channel proteins (such as the voltage-gated Na$^+$ channel) transport ions much faster than carrier proteins (such as the Na$^+$-K$^+$ pump), but they cannot be coupled to an energy source; therefore, transport mediated by channels is always passive.

___ B. Carrier proteins saturate at high concentrations of the transported molecule when all its binding sites are occupied; channel proteins, on the other hand, do not bind the ions they transport and thus the flux of ions through a channel does not saturate.

___ C. The resting membrane potential of a typical animal cell arises predominantly through the action of the Na$^+$-K$^+$ pump, which in each cycle transfers 3 Na$^+$ ions out of the cell and 2 K$^+$ into the cell, leaving an excess of negative charges inside the cell.

___ D. The membrane potential arises from movements of charge that leave ion concentrations practically unaffected and result in only a very slight discrepancy in the number of positive and negative ions on the two sides of the membrane.

___ E. Upon stimulation of a nerve cell, two processes limit the entry of Na^+ ions: (1) the membrane potential reaches the Na^+ equilibrium potential, which stops further net entry of Na^+, and (2) the Na^+ channels are inactivated and cannot reopen until the original resting potential has been restored.

___ F. The voltage gradient across the lipid bilayer changes dramatically during the passage of a nerve impulse, causing conformational changes in voltage-gated channel proteins.

___ G. The aggregate current crossing the membrane of an entire cell indicates the degree to which individual channels are open.

___ H. Inactivation of voltage-gated K^+ channels is carried out by the amino terminus of each K^+ channel subunit, which acts like a tethered ball to occlude the cytoplasmic end of the pore soon after it opens.

___ I. Transmitter-gated ion channels open in response to specific neurotransmitters in their environment but are insensitive to the membrane potential; therefore, they cannot by themselves generate an action potential.

___ J. Because the concentration of Cl^- is much higher inside the cell than outside, opening Cl^- channels depolarizes the membrane.

___ K. Unlike voltage-gated cation channels, which are selective for specific inorganic cations, the acetylcholine receptor is not very selective, allowing passage of Na^+, K^+, and Ca^{2+}.

___ L. When a nerve impulse stimulates a muscle cell to contract, five different sets of ion channels are sequentially activated, all within a few milliseconds.

___ M. A motor neuron will fire an action potential along its axon if the number of excitatory PSPs is greater than the number of inhibitory PSPs.

___ N. Adaptation allows a neuron to react sensitively to change, even against a high background of steady stimulation.

___ O. The underlying rule in the hippocampus seems to be that long-term potentiation occurs on any occasion where a presynaptic cell fires at a time when the postsynaptic membrane is strongly depolarized.

11–12   The squid giant axon occupies a unique position in the history of our understanding of cell membrane potentials and action potentials. Its large size (0.2–1.0 mm in diameter and 5–10 cm in length) allowed electrodes, large by modern standards, to be inserted so that intracellular voltages could be measured. When an electrode is stuck into an intact giant axon, a membrane potential of –70 mV is registered. When the axon, suspended in a bath of seawater, is stimulated to conduct a nerve impulse, the membrane potential rises transiently from –70 mV to +40 mV.

The Nernst equation relates equilibrium ionic concentrations to the membrane potential.

$$V = 2.3 \, \frac{RT}{zF} \log_{10} \frac{C_o}{C_i}$$

For univalent ions and 20°C (293°K),

$$V = 58 \text{ mV} \times \log_{10} \frac{C_o}{C_i}$$

A. Using this equation, calculate the potential across the resting membrane (1) assuming that it is due solely to K^+ and (2) assuming that it is due solely to Na^+. (The Na^+ and K^+ concentrations in axon cytoplasm and in seawater are given in Table 11–3.) Which calculation is closer to the measured resting potential? Which calculation is closer to the measured action potential? Explain why these assumptions approximate the measured resting and action potentials.

B. If the solution bathing the squid giant axon is changed from seawater to an artificial seawater in which NaCl is replaced with choline chloride, there is no

Table 11–3 Ionic Composition of Seawater and of Cytoplasm from the Squid Giant Axon (Problem 11–12)

Ion	Cytoplasm	Seawater
Na^+	65 mM	430 mM
K^+	344 mM	9 mM

effect on the resting potential, but the nerve no longer generates an action potential upon stimulation. What would you predict would happen to the magnitude of the action potential if the concentration of Na^+ in the external medium were reduced to a half or a quarter of its normal value, using choline chloride to maintain osmotic balance?

*11–13   The number of Na^+ ions entering the squid giant axon during an action potential can be calculated from theory. Because the cell membrane separates positive and negative charges, it behaves like a capacitor. From the known capacitance of biological membranes, the number of ions that enter during an action potential can be calculated. Starting from a resting potential of –70 mV, it can be shown that 1.1×10^{-12} moles of Na^+ must enter the cell per cm^2 of membranes during an action potential.

To determine experimentally the number of entering Na^+ during an action potential, a squid giant axon (1 mm in diameter and 5 cm in length) was suspended in a solution containing radioactive Na^+ (specific activity $= 2 \times 10^{14}$ cpm/mole) and a single action potential was propagated down its length. When the cytoplasm was analyzed for radioactivity, a total of 340 cpm were found to have entered the axon.

A. How well does the experimental measurement match the theoretical calculation?

B. How many moles of K^+ must cross the membrane of the axon, and in which direction, to reestablish the resting potential after the action potential is over?

C. Given that the concentration of Na^+ inside the axon is 65 mM, calculate the fractional increase in internal Na^+ concentration that results from the passage of a single action potential down the axon.

D. At the other end of the spectrum of nerve sizes are small dendrites about 0.1 μm in diameter. Assuming the same length (5 cm), the same internal Na^+ concentration (65 mM), and the same resting and action potentials as for the squid giant axon, calculate the fractional increase in internal Na^+ concentration that results from the passage of a single action potential down a dendrite.

E. Is the Na^+-K^+ pump more important for the continuing performance of a giant axon or a dendrite?

11–14   Intracellular changes in ion concentration often trigger dramatic cellular events. For example, when a clam sperm contacts a clam egg, it triggers ionic changes that result in the breakdown of the egg nuclear envelope, condensation of chromosomes, and initiation of meiosis. Two observations confirm that ionic changes initiate these cellular events: (1) suspending clam eggs in seawater containing 60 mM KCl triggers the same intracellular changes as do sperm; (2) suspending eggs in artificial seawater lacking calcium prevents activation by 60 mM KCl.

A. How does 60 mM KCl affect the resting potential of eggs? The intracellular K^+ concentration is 344 mM and that of normal seawater is 9 mM. Remember from Problem 11–12 that

$$V = 58\,\text{mV} \times \log_{10} \frac{C_o}{C_i}$$

B. What does the lack of activation by 60 mM KCl in calcium-free seawater suggest about the mechanism of KCl activation?

C. What would you expect to happen if the calcium ionophore, A23187, was added to a suspension of eggs (in the absence of sperm) in (1) regular seawater and (2) calcium-free seawater?

*11–15   One important parameter for understanding any particular membrane transport process is to know the number of copies of the specific transport protein present in the cell membrane. You wish to measure the number of voltage-gated Na^+ channels in the rabbit vagus nerve. You have found a potent toxin in spider venom that specifically inactivates voltage-gated Na⁺ channels in these nerve cells; moreover, you have shown that the toxin can be labeled with ^{125}I without affecting its toxic properties. Assuming that each channel binds one toxin molecule, the number of Na⁺ channels in a segment of vagus nerve will be equal to the maximum number of bound toxin molecules.

To measure the number of Na⁺ channels, you incubate identical segments of nerve for 8 hours with increasing amounts of labeled toxin. You then wash the segments to remove unbound toxin and measure the radioactivity associated with the nerve segments to establish a titration curve for binding, which is shown in the upper curve in Figure 11–5. You are puzzled because you expected to see binding reach a maximum (saturate) at high concentrations of toxin; however, no distinct end point was reached. Indeed, binding continued to increase with the same slope at even higher concentrations of toxin than those shown in the upper curve in the figure. After careful thought, you design a control experiment in which the binding of labeled toxin is measured in the presence of a large molar excess of unlabeled toxin. The results of this experiment, which are shown in the lower curve in Figure 11–5, make everything clear and allow you to calculate the number of Na⁺ channels in the membrane of the vagus nerve axon.

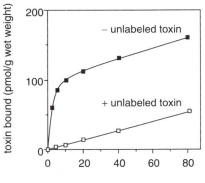

Figure 11–5 Toxin-binding curves in the presence and absence of tetrodotoxin (Problem 11–15).

A. Why does binding of the labeled toxin not saturate? What is the point of the control experiment, and how does it work?
B. Given that 1 g of vagus nerve has an axonal membrane area of 6000 cm² and assuming that the Na⁺ channel is a cylinder with a diameter of 6 nm, calculate the number of Na+ channels per square micrometer of axonal membrane and the fraction of the cell surface occupied by the channel. (Use 100 pmol as the amount of toxin specifically bound to the receptor.)

11–16 To make antibodies against the acetylcholine receptor from electric eel electric organ, you inject the purified receptor into mice. You note an interesting correlation: mice with high levels of antibodies against the receptor appear weak and sluggish; those with low levels are lively. You suspect that the antibodies against the eel acetylcholine receptors are cross-reacting with the mouse acetylcholine receptors, causing many of the receptors to be destroyed. Since a reduction in the number of acetylcholine receptors is also the basis for the human autoimmune disease myasthenia gravis, you wonder whether an injection of the drug neostigmine might give a temporary restoration of strength, as it does for myasthenic patients. Sure enough, when you inject your mice with neostigmine, they immediately stand up and become very lively. Propose an explanation for how neostigmine might restore temporary function to a neuromuscular synapse with a reduced number of acetylcholine receptors.

*11–17 Acetylcholine-gated cation channels at the neuromuscular junction open in response to acetylcholine released by the nerve terminal and allow Na⁺ ions to enter the muscle cell, which causes membrane depolarization and ultimately leads to muscle contraction (see MBOC Figure 11–34). Patch-clamp measurements in young rat muscle show that there are two classes of channel—a 4-pA channel and a 6-pA channel—that respond to acetylcholine (Figure 11–6).

Calculate the number of ions that enter through each type of channel in one millisecond. (One Amp is a current of one Coulomb per second. An ion with a single charge such as Na⁺ carries a charge of 1.6×10^{-19} Coulomb.)

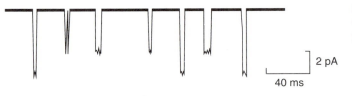

2 pA

40 ms

Figure 11–6 Patch-clamp measurements of acetylcholine-gated cation channels in young rat muscle (Problem 11–17).

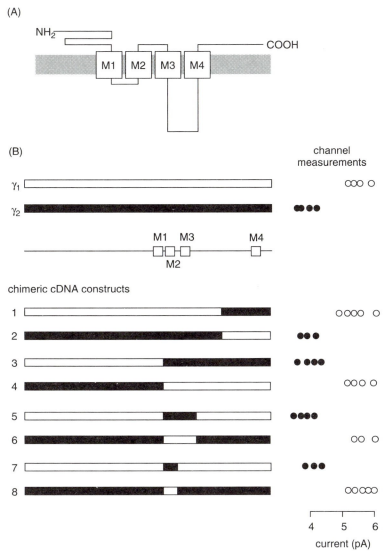

(A)

(B)

channel measurements

Figure 11–7 Arrangement of a γ subunit in the membrane (A) and the current through channels made with the γ_1 subunit, the γ_2 subunit, or chimeric subunits (B) (Problem 11–18). The current through the different channels is indicated on the right. Open circles indicate channels with conductance like those made from the γ_1 subunit; filled circles indicate channels with conductance like those made from the γ_2 subunit. M1, M2, M3, and M4 indicate the four trans-membrane segments in the γ subunits.

11–18 The acetylcholine-gated cation channel is a pentamer of homologous proteins containing two copies of the α subunit and one copy each of the β, γ, and δ subunits (see MBOC Figure 11–32). The two distinct acetylcholine-gated channels in young rat muscle (see Problem 11–17) differ in their γ subunits. The 6-pA channel contains the γ_1 subunit, whereas the 4-pA channel contains the γ_2 subunit.

To pinpoint the portions of the γ_1 and γ_2 subunits responsible for the conductance differences of the two channels, you prepare a series of chimeric cDNAs in which different portions of the γ_1 and γ_2 cDNAs have been swapped (Figure 11–7). Since it seems likely that channel conductance is determined by the transmembrane segments, you have designed the chimeric cDNAs to test different combinations of the four transmembrane segments (M1 to M4 in Figure 11–7). You mix mRNA made from each of the chimeric cDNAs with mRNA from the α, β, and δ subunits, inject the mRNAs into *Xenopus* oocytes, and measure the current through individual channels by patch clamping (Figure 11–7B).

A. Which transmembrane segment is responsible for the differences in conductance through the two types of acetylcholine-gated cation channels?

B. Why do you suppose that the difference in channel conductance is all due to one transmembrane segment?

C. Assume that a threonine residue in the critical transmembrane segment contributes to the constriction that forms the ion-selectivity filter for the channel. How would the current through the channel change if a glycine were substituted at that position? If a leucine were substituted at that position? How might you account for the different currents through channels constructed with γ_1 and γ_2 subunits?

Intracellular Compartments and Protein Sorting

- The Compartmentalization of Higher Cells
- The Transport of Molecules into and out of the Nucleus
- The Transport of Proteins into Mitochondria and Chloroplasts
- Peroxisomes
- The Endoplasmic Reticulum

The Compartmentalization of Higher Cells
(MBOC 551–560)

12–1　Fill in the blanks in the following statements.

A. A eucaryotic cell is elaborately subdivided into functionally distinct, membrane-bounded compartments called _____.

B. The interior of each intracellular compartment, called its _____, is like a separate subcellular reaction vessel endowed with specialized functions.

C. The fate of proteins synthesized in the cytosol depends on their amino acid sequence, which can contain _____ that direct their delivery to locations outside the cytosol.

D. For some sorting steps, the sorting signal resides in a continuous stretch of amino acids called the _____; this signal is often removed from the finished protein by a specialized _____ once the sorting process has been completed.

E. For some sorting steps, the sorting signal results from a particular three-dimensional arrangement of amino acids on the protein's surface; these _____ generally remain in the finished protein.

12–2　Indicate whether the following statements are true or false. If a statement is false, explain why.

___ A. Internal membranes partition the cell into functionally distinct compartments, each with boundaries established by impermeable membranes.

___ B. In terms of both area and mass, the plasma membrane is only a minor membrane in most eucaryotic cells.

___ C. The plasma membrane in a eucaryotic cell cannot provide enough surface area or house enough membrane-bound enzyme molecules to support all the vital functions that membranes must sustain in such a large cell.

___ D. The interior of the nucleus and the lumen of the ER are both topologically equivalent to the outside of the cell.

___ E. The lumenal spaces of each of the organelles that communicate by means of transport vesicles are topologically equivalent to one another and to the outside of the cell.

___ F. Depending on the individual protein, a signal peptide or a signal patch may direct the protein to the ER, mitochondria, chloroplasts, or nucleus.

___ G. The signal peptides attached to proteins that share the same destination are functionally interchangeable, even though their amino acid sequences can vary greatly.

___ H. If a membrane-bounded organelle, such as the ER or Golgi, were removed from a eucaryotic cell, the cell could regenerate the organelle from the information present in the DNA.

12–3   The rough endoplasmic reticulum (ER) is the site of synthesis of many different classes of membrane proteins. Some of these proteins remain in the ER, whereas others are sorted to compartments, such as the Golgi apparatus, lysosomes, and the plasma membrane. One measure of the difficulty of the sorting problem is the degree of "purification" that must be achieved during transport from the ER to the other compartments. For example, if membrane proteins bound for the plasma membrane represented 90% of all proteins in the ER, then only a small degree of purification would be needed (and the sorting problem would appear relatively easy). On the other hand, if plasma membrane proteins represented only 0.01% of the proteins in the ER, a very large degree of purification would be required (and the sorting problem would appear correspondingly more difficult).

   What is the magnitude of the sorting problem? What fraction of the membrane proteins in the ER are destined for other compartments? A few simple considerations allow one to estimate the answers to these questions. Assume that all proteins on their way to other compartments remain in the ER 30 minutes on average before exiting, and that the ratio of proteins to lipids in the membranes of all compartments is the same.

   A. In a typical growing cell that is dividing once every 24 hours the equivalent of one new plasma membrane must transit the ER every day. If the ER membrane is 20 times the area of a plasma membrane, what is the ratio of plasma membrane proteins to other membrane proteins in the ER?
   B. If in the same cell the Golgi membrane is three times the area of the plasma membrane, what is the ratio of Golgi membrane proteins to other membrane proteins in the ER?
   C. If the membranes of all other compartments (lysosomes, endosomes, inner nuclear membrane, and secretory vesicles) that receive membrane proteins from the ER are equal in total area to the area of the plasma membrane, what fraction of the membrane proteins in the ER of this cell are permanent residents of the ER membrane?

The Transport of Molecules into and out of the Nucleus
(MBOC 561–568)

12–4   Fill in the blanks in the following statements.

   A. The _____ encloses the DNA and defines the nuclear compartment.
   B. The nucleus is bounded by two concentric membranes: the _____ membrane, which contains specific proteins that act as binding sites for the feltlike _____ that supports it, and the _____ membrane, which is continuous with the ER membrane.
   C. The nucleus is perforated by _____, each of which is embedded in a large, elaborate structure known as the _____.
   D. The selectivity of nuclear transport resides in _____ signals, which are present only in nuclear proteins.
   E. The nuclear lamina is a meshwork of interconnected protein subunits called _____.

12–5   Indicate whether the following statements are true or false. If a statement is false, explain why.

   ___ A. The perinuclear space is continuous with the lumen of the ER.
   ___ B. At a nuclear pore, as elsewhere, the lipid bilayers of the inner and outer nuclear membranes are distinct from one another.

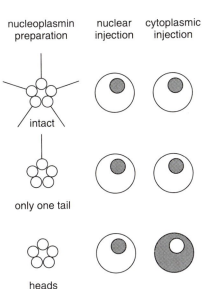

C. Nuclear localization signals can be located almost anywhere in the amino acid sequence of a protein and generally consist of a short sequence that is rich in lysine and arginine and usually contains proline.

D. Although import of proteins into the nucleus requires specific signals and uses the energy of ATP hydrolysis, export of RNA and ribosomal subunits from the nucleus is unlikely to require either signals or energy.

E. Because nuclear protein molecules need to be imported repeatedly, their signals are not cleaved off after transport into the nucleus.

F. Some proteins are kept out of the nucleus, until needed, by inactivating the nuclear localization signal by phosphorylation.

12–6 In principle, proteins might accumulate in the nucleus in two ways. (1) Proteins might diffuse into the nucleus passively and accumulate there by binding to a resident of the nucleus, such as a chromosome. (2) Proteins might be actively transported into the nucleus and accumulate there re-gardless of their affinity for nuclear components. Passive diffusion and ac-tive transport are particularly difficult to distinguish between since most, if not all, proteins in the nucleus are linked directly or indirectly to nuclear residents. Although the physical size of the nuclear pore (apparent diameter 9 nm) often is cited in favor of active transport, the arguments are not satisfying because the shapes of most nuclear proteins are unknown.

One straightforward experiment to address this problem used several forms of radioactive nucleoplasmin, which is a large pentameric protein involved in chromatin assembly. In addition to the intact protein, nucleoplasmin heads, tails, and heads with a single tail were injected into the cytoplasm or into the nucleus of frog oocytes (Figure 12–1). All forms of nucleoplasmin, except heads, accumulated in the nucleus when injected in the cytoplasm, and all forms were retained in the nucleus when injected there.

A. What portion of the nucleoplasmin molecule is responsible for localization in the nucleus?

B. How do these experiments distinguish between active transport, in which a nuclear localization signal triggers transport by the nuclear pore complex, and passive diffusion, in which a binding site for a nuclear component allows accumulation in the nucleus?

Figure 12–1 Cellular location of injected nucleoplasmin and nucleoplasmin components (Problem 12–6). Schematic diagrams of autoradiographs of cells show the cytoplasm and nucleus with the location of the nucleoplasmin indicated by the shaded areas.

*12–7 You have just joined a laboratory that is engaged in defining the nuclear transport machinery in yeast. Your advisor, who is known for her extraordi-narily clever ideas, has given you a project with enormous potential. In principle, it would allow a genetic selection for conditional-lethal mutants in the nuclear transport apparatus.

She gave you the two plasmids shown in Figure 12–2. Each plasmid consists of a hybrid gene under the control of a regulatable promoter. The hybrid gene is a fusion between a gene whose product is normally imported into the nucleus and the gene for the restriction enzyme EcoRI. The plasmid pNL⁺ contains a functional nuclear localization signal; the plasmid pNL⁻ contains a nonfunctional signal. The promoter, which is from the yeast *GAL1* gene, allows transcription of the hybrid gene only when the sugar galactose is present in the growth medium.

Following her instructions, you introduce the plasmids into yeast (in the absence of galactose) and then assay the transformed yeast in medium containing glucose and in medium containing galactose. Your results are shown in Table 12–1 (p. 126). You don't remember what your advisor told you to expect, but you know you will be expected to explain these results at the weekly lab meeting. Why do yeasts with the pNL⁺ plasmid grow in the presence of glucose but die in the presence of galactose?

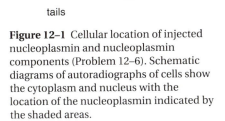

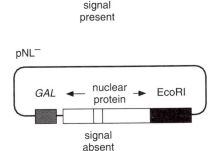

Figure 12–2 Two plasmids for investigat-ing nuclear localization in yeast (Problem 12–7).

---

*Problems with an asterisk (*) are answered in the Instructor's Manual.

12–8   Now that you understand why plasmid pNL⁺ (Figure 12–2) kills cells in galactose-containing medium, you begin to understand how your advisor intends to exploit its properties to select mutants in the nuclear transport machinery. You also understand why she emphasized that the desired mutants would have to be *conditionally* lethal. Since nuclear import is essential to the cell, a fully defective mutant could never be grown and thus would not be available for study. By contrast, conditionally lethal mutants can be grown perfectly well under one set of conditions (permissive conditions); under a different set of conditions (restrictive conditions), however, the cells exhibit the defect, which can then be studied.

Table 12–1 Results of Growth Experiments with Yeast Carrying the Plasmids pNL⁺ or pNL⁻ (Problem 12–7)

Plasmid	Glucose Medium	Galactose Medium
pNL⁺	growth	death
pNL⁻	growth	growth

With this overall strategy in mind, you design a selection scheme for temperature-sensitive (ts) mutants in the nuclear translocation machinery. You want to find mutants that grow well at low temperature (the permissive condition) but are defective at high temperature (the restrictive condition). You plan to mutagenize cells containing the pNL⁺ plasmid at low temperature and then shift them to high temperature in the presence of galactose. You reason that at the restrictive temperature the nuclear transport mutants will not take up the killer protein encoded by pNL⁺ and, therefore, will not be killed. Normal cells, however, will transport the killer protein into the nucleus and die. After one or two hours at the high temperature to allow selection against normal cells, you intend to lower the temperature and put the surviving cells on nutrient agar plates containing glucose. You anticipate that the nuclear translocation mutants will recover at the low temperature and form colonies.

When you show your advisor your scheme, she is very pleased at your initiative. However, she sees a critical flaw in your experimental design that would prevent isolation of nuclear translocation mutants, but she won't tell you what it is—but she hints that it has to do with the killer protein made during the high-temperature phase of the experiment.

A. What is the critical flaw in your original experimental protocol?

B. How might you modify your protocol to correct this flaw?

C. Assuming for the moment that your original protocol would work, can you think of any other types of mutants (not involved in nuclear transport) that would survive your selection scheme?

12–9   The gene for the κ-light chain of immunoglobulins contains a DNA sequence element (an enhancer) in one of its introns that regulates its expression. A protein called NF-κB binds to this enhancer and can be found in nuclear extracts of B cells, which actively synthesize immunoglobulins. By contrast, NF-κB activity is absent from precursor B cells (pre-B cells), which do not express immunoglobulins. You are interested in the strategy by which NF-κB regulates expression of κ-light-chain genes.

If pre-B cells are treated with phorbol ester (which activates protein kinase C, causing the phosphorylation of some cellular proteins), NF-κB activity appears within a few minutes in parallel with the transcriptional activation of the κ-light chain gene. This activation of NF-κB is not blocked by cycloheximide, which is an inhibitor of protein synthesis.

You discover by chance that cytoplasmic extracts of unstimulated pre-B cells contain an inactive form of NF-κB that can be activated by treatment with mild denaturants. (Denaturants disrupt noncovalent bonds but do not break covalent bonds.) To understand the relationship between NF-κB activation and transcriptional regulation of κ-light-chain genes, you isolate nuclear (N) and cytoplasmic (C) fractions from pre-B cells before and after stimulation by phorbol ester. You then measure NF-κB activity, before and after treatment with the mild denaturants, by its ability to form a complex with labeled DNA fragments that carry the enhancer. The results are shown in Figure 12–3.

A. How does the subcellular localization of NF-κB change in pre-B cells in response to phorbol ester treatment?

B. Do you think that the activation of NF-κB in response to phorbol ester occurs because NF-κB is phosphorylated by protein kinase C?

C. Outline a molecular mechanism to account for activation of NF-κB by treatment with phorbol ester.

The Transport of Proteins into Mitochondria and Chloroplasts

(MBOC 568–574)

12–10 Fill in the blanks in the following statements.

A. From inside to outside the four subcompartments of a mitochondrion are the _____, the _____, the _____, and the _____.

B. Cytosolic proteins that are destined for import into mitochondria are called _____ proteins.

C. The _____ at which the inner and outer mitochondrial membranes appear to be joined are thought to be the sites at which import into the matrix occurs.

D. Chloroplasts have an extra membrane-bounded compartment, which is called the _____.

E. The matrix space of the chloroplast is termed the _____.

12–11 Indicate whether the following statements are true or false. If a statement is false, explain why.

__ A. The relatively small number of proteins encoded in the genomes of mitochondria and chloroplasts are located mostly in the inner membranes of both organelles.

__ B. Mitochondrial signal peptides are amphipathic α helical structures with positively charged amino acids on one side and negatively charged amino acids on the other side.

__ C. The electrochemical proton gradient apparently drives insertion of the mitochondrial signal peptide into the outer mitochondrial membrane during the initial penetration of the precursor protein into the mitochondria.

__ D. Since a matrix protease can remove the N-terminus of an imported protein while the C-terminus is accessible to externally added proteases, precursor proteins must penetrate both the inner and outer membranes at the same time.

___ E. Removal of cytosolic and mitochondrial chaperonins from translocated proteins as they move from the cytosol to the matrix accounts for part of the ATP dependence of the import of mitochondrial proteins.

___ F. The ATP-dependent release of cytosolic hsp70 is thought to push mitochondrial proteins across the double membrane.

___ G. Proteins destined for the inner mitochondrial membrane apparently pass through the outer and inner membranes into the matrix and are then subsequently reinserted into the inner membrane.

___ H. Since the outer mitochondrial membrane contains very large pores, it does not present a permeability barrier to proteins.

___ I. Both chloroplasts and mitochondria exploit the electrochemical proton gradient to help drive the transport of precursor proteins across their outer and inner membranes.

12–12   To aid your studies of protein import into mitochondria, you treat yeast cells with cycloheximide, which blocks ribosome movement along mRNA. When you examine these cells in the electron microscope, you are surprised to find cytosolic ribosomes attached to the outside of the mitochondria. You have never seen attached ribosomes in the absence of cycloheximide. To investigate this phenomenon further, you prepare mitochondria from cycloheximide-treated cells and extract the mRNA that is bound to the mitochondria-associated ribosomes. You translate this mRNA *in vitro* and compare the protein products with similarly translated mRNA from the cytosol. The results are clear-cut: the mitochondria-associated ribosomes are translating mRNAs that encode mitochondrial proteins.

You are astounded! Here, clearly visible in the electron micrographs, seems to be proof that protein import into mitochondria occurs during translation. How can you rationalize this result with the prevailing view that mitochondrial proteins are imported after they have been synthesized and released from ribosomes?

*12–13   Amphipathic helices are one of the key features of signal peptides used for protein import into mitochondria. A helix is amphipathic if one side is hydrophilic and the other side is hydrophobic. A simple way to decide whether a sequence of amino acids might form an amphipathic helix is to arrange the amino acids around what is known as a "helix-wheel projection" (Figure 12–4A). This representation shows the positions of the amino acids around an α helix as viewed from the top of the helix. If hydrophobic and hydrophilic amino acids are intermixed in such a diagram, the helix is not amphipathic; however, if the hydrophobic and hydrophilic amino acids are segregated on opposite sides, the helix is amphipathic.

Using the helix-wheel projection, decide which of the three peptides in Figure 12–4B can form an amphipathic helix and could serve as a mitochondrial import signal.

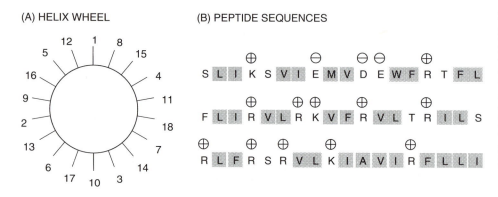

Figure 12–4 Potential signal peptides for mitochondrial import (Problem 12–13). (A) Helix-wheel projection of an α helix. Numbers show the positions of the first 18 amino acids of an α helix; amino acid 19 would occupy the same position as amino acid 1. (B) Amino acid sequences of three peptides. The N-termini are shown at the left; hydrophobic amino acids are shaded; the charge on charged amino acids is indicated in a circle; and uncharged hydrophilic amino acids are unmarked.

12–14 Chloroplasts contain six compartments—outer membrane, intermembrane space, inner membrane, stroma, thylakoid membrane, and thylakoid lumen (Figure 12–5)—each of which is populated by specific sets of proteins. Many of these proteins are encoded by nuclear genes, translated in the cytosol, and then posttranslationally directed to the appropriate chloroplast compartment. To investigate the import of proteins into chloroplasts, you have cloned the cDNAs for ferredoxin (FD), which is located in the stroma, and plastocyanin (PC), which is located in the thylakoid lumen. Furthermore, using recombinant DNA techniques, you have constructed two hybrid genes: ferredoxin with the plastocyanin signal peptide (PCFD) and plastocyanin with the ferredoxin signal peptide (FDPC). You translate mRNAs from these four genes *in vitro*, mix the translation products with isolated chloroplasts for a few minutes, reisolate the chloroplasts after protease treatment, and fractionate them to find which compartments the proteins have entered (Figure 12–6A). The status of the normal and hybrid proteins at each stage of the experiment are shown in Figure 12–6B: each lane in the gels corresponds to a stage of the experiment as indicated alongside the experimental protocol in Figure 12–6A.

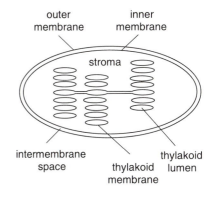

Figure 12–5 The six compartments of chloroplasts (Problem 12–14).

A. How efficient is chloroplast uptake of ferredoxin and plastocyanin in your *in vitro* system? How can you tell?
B. Are ferredoxin and plastocyanin localized to their appropriate chloroplast compartments in these experiments? How can you tell?
C. Are the hybrid proteins imported as you would expect if the N-terminal signal peptides determined their final location? Comment on any significant differences.
D. Why are there three bands in experiments with plastocyanin and PCFD but only two bands in experiments with ferredoxin and FDPC? To the extent you can, identify the bands and their relationship to each other.
E. Based on your experiments, propose a model for the import of proteins into the stroma and thylakoid lumen.

Figure 12–6 Import of ferredoxin and plastocyanin into chloroplast compartments (Problem 12–14). Samples were treated as indicated in (A) and then analyzed on gels as shown in (B). Each lane in (B) corresponds to a particular experimental treatment in (A). Ferredoxin gene segments are black and labeled FD; plastocyanin gene segments are white and labeled PC. In all four genes the signal peptide is encoded by the short segment at the left end. The normal ferredoxin gene is FDFD; the normal plastocyanin gene is PCPC.

(A) EXPERIMENTAL PROTOCOL

TRANSLATE mRNA *IN VITRO* — lane 1

↓

ADD CHLOROPLASTS — lane 2

↓

TREAT WITH PROTEASE, REISOLATE CHLOROPLASTS — lane 3

↓

FRACTIONATE CHLOROPLASTS

↓

inner and outer membranes — lane 4

stroma — lane 5

thylakoids — lane 6

thylakoids plus protease — lane 7

(B) GEL ANALYSIS

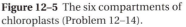

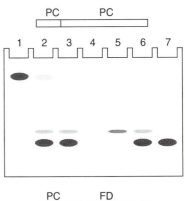

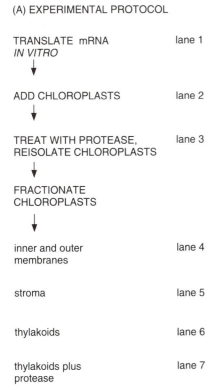

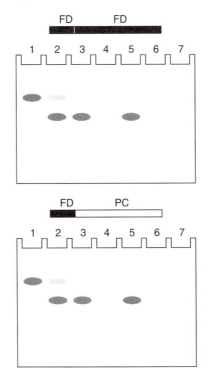

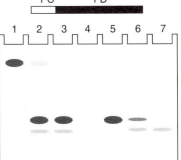

Peroxisomes
(MBOC 574–577)

12–15 Fill in the blanks in the following statements.

    A. _____ resemble the ER in being a self-replicating membrane-bounded organelle that exists without a genome of its own.

    B. Plants, but not animals, can convert fatty acids to sugars by a series of reactions known as the _____ , which is why the peroxisomes that carry out these reactions are also called _____.

12–16 Indicate whether the following statements are true or false. If a statement is false, explain why.

  __ A. Peroxisomes are confined to a few types of eucaryotic cells.

  __ B. Peroxisomal oxidation reactions are particularly important in liver and kidney cells, whose peroxisomes detoxify various toxic molecules that enter the bloodstream.

  __ C. The membrane "shell" of the peroxisome forms by budding from the ER, whereas the "content" is imported from the cytosol.

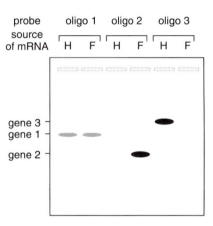

Figure 12–7 Hybridization of specific oligonucleotide probes to mRNA isolated from trypanosomes from humans (H) and from tsetse flies (F) (Problem 12–17). The intensity of the bands on the autoradiograph reflects the concentrations of the mRNAs.

*12–17 Trypanosomes are single-celled parasites that cause sleeping sickness when they infect humans. Trypanosomes from humans carry the enzymes for a portion of the glycolytic pathway in a peroxisomelike organelle, termed the glycosome. By contrast, trypanosomes from the tsetse fly—the intermediate host—carry out glycolysis entirely in the cytosol. This intriguing difference has alerted the interest of the pharmaceutical company that employs you. Your company wishes to exploit this difference to control the disease.

    You decide to study the enzyme phosphoglycerate kinase (PGK) because it is in the affected portion of the glycolytic pathway. Trypanosomes from the tsetse fly express PGK entirely in the cytosol, whereas trypanosomes from humans express 90% of the total PGK activity in glycosomes and only 10% in the cytosol. When you clone PGK genes from trypanosomes, you find three forms that differ slightly from one another. Exploiting these small differences, you design three oligonucleotides that hybridize specifically to the mRNAs from each gene. Using these oligonucleotides as probes, you determine which genes are expressed by trypanosomes from humans and from tsetse flies. The results are shown in Figure 12–7.

    A. Which PGK genes are expressed in trypanosomes from humans? Which are expressed in trypanosomes from tsetse flies?

    B. Which PGK gene probably encodes the glycosomal form of PGK?

    C. Do you think that the minor cytosolic PGK activity in trypanosomes from humans in due to inaccurate sorting into glycosomes? Explain your answer.

12–18 A number of serious human diseases are known in which the patients' cells lack functional peroxisomes. Patients with Zellweger syndrome—the best known of these diseases—show multiple developmental and psychomotor abnormalities and usually survive no more than a few months after birth. To understand these peroxisome-deficiency diseases, more must be known about the biogenesis of peroxisomes.

    You feel it is time to take a genetic approach to this problem and decide to see if Chinese hamster ovary (CHO) cells can be mutagenized to give cells that mimic those from Zellweger patients. (CHO cells are especially useful for isolating recessive mutations because they behave as if they were haploid for much of their genome.) You develop an assay for peroxisomal function in which mutagenized cell colonies are incubated with a soluble radioactive precursor that is converted into a readily detectable insoluble product by a peroxisomal enzyme. A laborious screen of 25,000 colonies is finally rewarded by the discovery of two colonies that do not have the insoluble radioactive

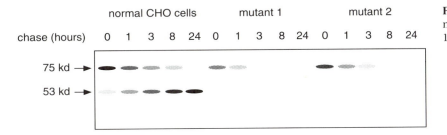

Figure 12–8 Pulse-chase experiments with normal and mutant CHO cells (Problem 12–18).

product. Sure enough, these mutant cells lack typical peroxisomes, as judged by electron microscopy.

To confirm their peroxisomal deficiency, you test directly for two peroxisomal enzymes: catalase and acyl CoA oxidase. The levels of catalase in the mutants is virtually the same as in normal CHO cells, except that it is dispersed in the cytosol instead of localized in peroxisomes. By contrast, acyl CoA oxidase activity is absent in both mutant cell lines.

To investigate the acyl CoA oxidase deficiency, you perform a pulse-chase experiment: you grow cells for 1 hour in medium containing ^{35}S-methionine, then transfer them to unlabeled medium and immunoprecipitate acyl CoA oxidase at various times after transfer (Figure 12–8). To clarify the relationship between the 75-kd and 53-kd forms of the oxidase, you isolate mRNA from wild-type and mutant cells, translate it *in vitro*, and immunoprecipitate acyl CoA oxidase: all three sources of mRNA give similar levels of the 75-kd form, but none of the 53-kd form.

You are most curious as to whether the two mutant cell lines have defects in the same gene or different genes. To answer this, you fuse wild-type and mutant cells in various combinations and examine the fused cells (heterocaryons) for the location of catalase and the presence of acyl CoA activity. As shown in Table 12–2, when two mutant cells were fused, the heterycaryons had normal peroxisomes as judged by catalase localization and acyl CoA oxidase activity.

A. As shown in Figure 12–8, acyl CoA oxidase exists in two forms in normal CHO cells. How do you think the two forms are related? Which one, if either, do you suppose is the active enzyme?

B. Why do the mutant cells have only the 75-kd form of acyl CoA oxidase, and why do you think it disappears during the chase in the pulse-chase experiment? How does this differ from the case of catalase?

Table 12–2 Analysis of Peroxisomal Functions in Heterocaryons (Problem 12–18)

		Heterocaryon Analysis	
Fusion	Cells Fused	Catalase Localization	Acyl CoA Oxidase Activity
1.	Normal × Normal	peroxisomal	present
2.	Normal × Mutant 1	peroxisomal	present
3.	Normal × Mutant 2	peroxisomal	present
4.	Mutant 1 × Mutant 1	cytosolic	absent
5.	Mutant 2 × Mutant 2	cytosolic	absent
6.	Mutant 1 × Mutant 2	peroxisomal	present
7.	Mutant 2 × Mutant 1	peroxisomal	present

Normal CHO cells and the mutant cells, as isolated, are HPRT$^+$ ouabain^s. In preparation for fusion, modified versions of each cell line were selected to be defective for the HPRT gene and resistant to the poison ouabain (HPRT$^-$ ouabain^r). Neither of these mutations affects the peroxisomal phenotype. In each fusion one of the cell lines is HPRT$^+$ ouabain^s, while the other is HPRT$^-$ ouabain^r. To select specifically for heterocaryons, the fused cells are grown in the presence of ouabain, which kills ouabain^s cells, and under conditions (HAT medium) that demand a functional HPRT gene for survival. Under these conditions, any unfused cells and cells arising from fusions of like cells will die: only true heterocaryons will survive and grow.

C. Do the mutations in the two mutant cell lines affect the same gene or different genes? How can you tell?

D. Are the mutations in the mutant cells recessive or dominant? How can you tell?

The Endoplasmic Reticulum
(MBOC 577–594)

12–19 Fill in the blanks in the following statements.

A. The membrane of the _____ typically constitutes more than half of the total membrane of the cell, and the internal space, called the _____, often occupies more than 10% of the total cell volume.

B. In mammalian cells the import of proteins into the ER begins before the polypeptide chain is completed, that is, it occurs _____, whereas import of proteins into mitochondria is _____.

C. _____, which are synthesizing proteins that are being concurrently translocated into the ER, coat the surface of the ER, and create regions termed _____.

D. When a ribosome happens to be making a protein with an _____, the signal directs the ribosome to the ER membrane.

E. When many ribosomes bind to a single mRNA, a _____ is formed.

F. When cells are disrupted by homogenization, the ER is fragmented into many small closed vesicles called _____.

G. The _____ hypothesis postulated that the amino-terminal leader serves as a signal peptide that directs the secreted protein to the ER membrane.

H. The signal peptide is guided to the membrane of the ER by at least two components: a _____, which binds to the signal peptide in the cytosol, and the _____, which is located in the ER membrane.

I. The amino-terminal ER signal peptide of a soluble protein itself has two signaling functions: in addition to directing the protein to the ER membrane, it is thought to serve as a _____.

J. _____ proteins can be inserted into the ER in three ways: in the simplest case an amino-terminal signal peptide initiates translocation, but an additional hydrophobic segment, a _____, blocks the translocation process and anchors the protein in the membrane.

K. In _____ proteins the polypeptide passes back and forth repeatedly across the lipid bilayer.

L. The ER lumen contains a chaperone protein known as _____ that recognizes incorrectly folded proteins, probably by binding to exposed hydrophobic amino acids, which would normally be buried in the interior of the protein.

M. Most of the soluble and membrane-bound proteins sequestered in the lumen of the ER are _____, which carry covalently attached sugars.

N. During protein _____, a preformed precursor oligosaccharide is transferred *en bloc* from a special lipid molecule, dolichol, to an asparagine residue on the target protein, in a reaction catalyzed by an _____.

O. The carboxyl terminus of some membrane proteins destined for the plasma membrane is covalently attached to a sugar residue of a glycolipid, which provides a _____ consisting of two fatty acids.

P. The rapid flip-flop of phospholipids across the ER membrane is thought to be mediated by _____ that are head-group specific.

Q. Water-soluble carrier proteins, called _____ proteins, are thought to be important for the transfer of phospholipids from the ER to mitochondria, plastids, and peroxisomes.

12–20   Indicate whether the following statements are true or false. If a statement is false, explain why.

___ A. In mammalian cells the import of proteins into the ER begins before the polypeptide chain is completely synthesized—that is, it occurs co-translationally.

___ B. Free ribosomes and membrane-bound ribosomes are identical.

___ C. Detoxification by the cytochrome P450 family of enzymes involves cleavage of harmful drugs or metabolites into units small enough to be excreted in the urine.

___ D. Rough microsomes can be separated readily from smooth microsomes because their higher concentration of proteins makes them more dense.

___ E. Although the smooth ER and the rough ER are continuous with one another, the rough ER contains several proteins that are not present in the smooth ER.

___ F. If a protein that is normally secreted by the cell is synthesized *in vitro* in the presence of microsomes, it will be protected from degradation by added protease.

___ G. The signal peptide, when it emerges from the ribosome, binds to a hydrophobic site on the ribosome causing a pause in protein synthesis, which resumes when the signal recognition particle binds to the signal peptide.

___ H. Ribosomes of the rough ER use the energy released during protein synthesis to drive their growing polypeptide chains through the ER membrane.

___ I. Newly synthesized polypeptide chains are transferred across the ER membrane through a pore in a protein translocator.

___ J. Regardless of the topology of the membrane protein, the amino terminus of a cleaved signal sequence is never exposed to the lumen of the ER.

___ K. In a protein with a cleavable signal peptide and multiple hydrophobic membrane-spanning segments, the odd-numbered segments (counting from the N-terminus) act as start-transfer peptides and the even-numbered segments act as stop-transfer peptides.

___ L. It seems likely that, by binding to an unfolded chain, BiP helps to keep improperly folded proteins in the ER and thus out of the Golgi.

___ M. The ER lumen contains a mixture of thiol-containing reducing agents that prevent the formation of S—S linkages (disulfide bonds) by maintaining the cysteine residues of lumenal proteins in reduced (—SH) form.

___ N. *N*-linked oligosaccharides are much more common on glycoproteins than are *O*-linked oligosaccharides.

___ O. Some membrane proteins are attached to the cytoplasmic surface of the plasma membrane through a C-terminal linkage to a glycosylphosphatidylinositol anchor.

___ P. The initial formation of phosphatidic acid and its subsequent modifications to form other phospholipid molecules all take place in the cytosolic half of the ER lipid bilayer.

___ Q. Phospholipids are added to the ER and subsequently transported to all the other membrane-bounded compartments of the cell by transport vesicles.

12–21   Translocation of proteins across the membrane of the ER is usually studied using microsomes. Microsomes from the rough ER carry ribosomes attached to their outer surface. Translocation of proteins across microsomal membranes can be assessed by several experimental criteria: (1) the newly synthesized protein is protected from added proteases, but not when detergents are present to solubilize the protecting lipid bilayer; (2) the newly synthesized proteins are glycosylated by oligosaccharide transferases, which are localized exclusively to the lumen of the ER; (3) the signal peptides are cleaved by signal peptidase, which is also active only on the lumenal side of the ER membrane.

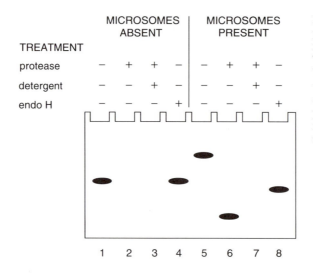

	MICROSOMES ABSENT				MICROSOMES PRESENT			
TREATMENT								
protease	−	+	+	−	−	+	+	−
detergent	−	−	+	−	−	−	+	−
endo H	−	−	−	+	−	−	−	+

1 2 3 4 5 6 7 8

Figure 12–9 Results of translation of a pure mRNA in the presence and absence of microsomal membranes (Problem 12–21). Treatments of the products of translation before electrophoresis are indicated at the top of each lane. Electrophoresis was on an SDS polyacrylamide gel, which separates proteins on the basis of size with lower molecular weight proteins migrating farther down the gel.

You want to use these criteria to decide whether the protein synthesized from a purified mRNA is translocated across microsomal membranes. Therefore, you translate the mRNA into protein in a cell-free system in the absence or presence of microsomes. You then prepare samples from these translation reactions in four different ways: (1) no treatment, (2) add a protease, (3) add a protease and detergent, and (4) disrupt microsomes and add endo-glycosi-dase H (endo H), which removes *N*-linked sugars that are added in the ER. An electrophoretic analysis of these samples is shown in Figure 12–9.

A. Explain the experimental results that are seen in the absence of microsomes (Figure 12–9, lanes 1 to 4).

B. Using the three criteria outlined in the problem, decide whether the experimental results in the presence of microsomes (lanes 5 to 8) indicate that the protein is translocated across microsomal membranes. Explain the migration of the proteins in lanes 5, 6, and 8.

C. Is the protein anchored in the membrane, or is it translocated all the way through the membrane?

*12–22 The segregation of secretory proteins and membrane proteins into the lumen of the ER is normally coupled tightly to protein synthesis. The co-translational nature of translocation in eucaryotes provides a sensitive and specific assay for the early steps in the biosynthesis of these proteins. However, it also poses a serious obstacle to elucidating the mechanism of translocation. For example, in such a coupled system it is difficult to determine whether the ribosome "pushes" the protein across the membrane or the translocation machinery "pulls" the protein across.

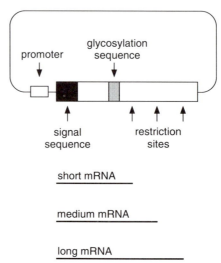

Figure 12–10 A cloned gene for testing the coupling of translation and translocation (Problem 12–22). Protein coding sequences start and stop at the ends of the large rectangle; promoter sequences are located in the small rectangle. Cleavage sites for restriction enzymes used to truncate the transcription template are indicated with arrows. The three different mRNA products of transcription from the truncated templates are indicated below the map of the gene.

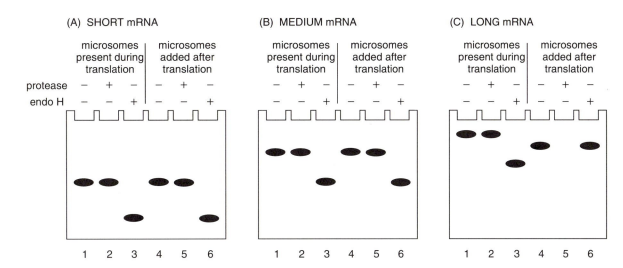

(A) SHORT mRNA

	microsomes present during translation			microsomes added after translation		
protease	−	+	−	−	+	−
endo H	−	−	+	−	−	+
	1	2	3	4	5	6

(B) MEDIUM mRNA

	microsomes present during translation			microsomes added after translation		
protease	−	+	−	−	+	−
endo H	−	−	+	−	−	+
	1	2	3	4	5	6

(C) LONG mRNA

	microsomes present during translation			microsomes added after translation		
protease	−	+	−	−	+	−
endo H	−	−	+	−	−	+
	1	2	3	4	5	6

In an attempt to uncouple translation from translocation you have cloned a gene onto a plasmid adjacent to a promoter for a bacteriophage RNA polymerase (Figure 12–10). This arrangement allows you to transcribe the gene *in vitro* by adding the phage RNA polymerase. In addition, by cutting the plasmid at different positions within the gene, you can create shorter mRNAs than normal (Figure 12–10). You prepare the three mRNAs indicated in Figure 12–10 and translate them *in vitro*. In one experiment you add microsomes before translation begins. In a second experiment you add microsomes after translation is completed (along with cycloheximide to inhibit any additional protein synthesis). To assess translocation, you treat some samples with protease or disrupt the microsomes and treat with endoglycosidase H (endo H), and then display the products by SDS-gel electrophoresis (Figure 12–11).

A. Are the proteins from each of the mRNAs translocated into microsomes when the microsomes are present during translation? How can you tell?
B. Have you managed to uncouple translocation from translation in any of the experiments? Explain your answer.
C. Do your experiments support the idea that ribosomes push proteins across the membrane of the ER, or the idea that the translocation machinery pulls proteins across the membrane? How so?
D. Why do you think that translocation of the shorter proteins can be uncoupled from translation, whereas translocation of the longer proteins cannot?

12–23 Four membrane proteins are represented schematically in Figure 12–12. The boxes represent membrane-spanning segments and the arrows represent sites for cleavage of the signal peptides. Using the rules for co-translational

Figure 12–11 Results of experiments to test the coupling of translation and translocation (Problem 12–22). The results with short (A), medium (B), and long (C) mRNAs are shown. Treatments of samples before electrophoresis are indicated above the gels. Endo H removes sugars of the type added in the ER.

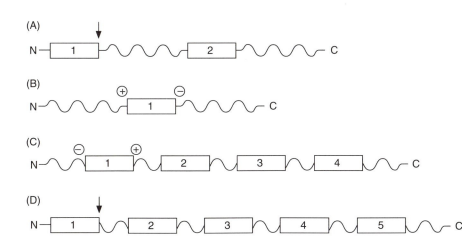

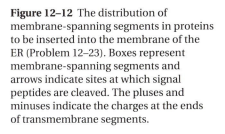

Figure 12–12 The distribution of membrane-spanning segments in proteins to be inserted into the membrane of the ER (Problem 12–23). Boxes represent membrane-spanning segments and arrows indicate sites at which signal peptides are cleaved. The pluses and minuses indicate the charges at the ends of transmembrane segments.

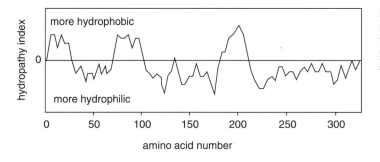

Figure 12–13 Hydropathy plot of a membrane protein (Problem 12–24). The three hydrophobic peaks indicate the positions of three potential membrane-spanning segments.

insertion described in MBOC Figures 12–44 and 12–45, predict how each of the mature proteins will be arranged across the membrane of the ER. Indicate clearly the N and C-termini relative to the cytosol and the lumen of the ER, and label each box as a start-transfer or stop-transfer peptide.

*12–24 You want to know the organization of a bacterial protein in the membrane of *E. coli*. The gene for this protein has been cloned onto a plasmid and its sequence is known. The hydropathy plot of the protein in Figure 12–13 indicates three potential membrane-spanning segments. You plan to check these assignments and determine their arrangement in the membrane by making hybrid fusion proteins of various different lengths with the membrane protein at the N-terminus and alkaline phosphatase at the C-terminus. Alkaline phosphatase is easy to assay in whole cells and has no significant hydrophobic stretches. Moreover, when it is on the cytoplasmic side of the membrane, its activity is low, and when it is on the external side of the membrane (in the periplasmic space), its activity is high.

You isolate six fusions of the membrane protein with alkaline phosphatase. The structures of the hybrid proteins (with the C-terminal amino acid of the membrane protein numbered) and the assayed levels of alkaline phosphatase activity (HIGH or LOW) are indicated in Figure 12–14. You also use two conveniently located restriction sites in the gene for the membrane protein to create deletions that remove codons for amino acids 68 to 103. The results from these modified plasmids are also shown in Figure 12–14.

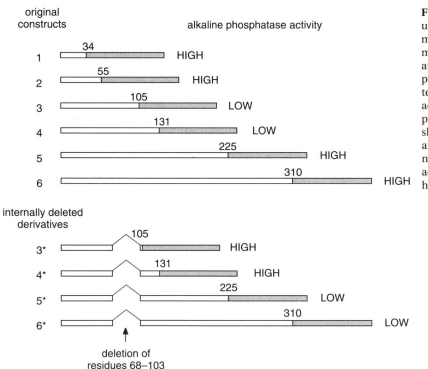

Figure 12–14 Structures of hybrid proteins used to determine the organization of a membrane protein (Problem 12–24). The membrane protein (unshaded segment) is at the N-terminus and alkaline phosphatase (shaded segment) is at the C-terminus of hybrid protein. The amino acids deleted from the modified hybrid proteins are indicated by the inverted V-shaped segment. The most C-terminal amino acid of the membrane protein is numbered in each hybrid protein. The activity of alkaline phosphatase in each hybrid protein is shown on the right.

A. How is the protein organized in the membrane? Explain how the results with the fusion proteins indicate this arrangement.

B. How is the organization of the membrane protein altered by the deletion? Are your measurements of alkaline phosphatase activity in the internally deleted plasmids consistent with the altered arrangement?

C. Are the N-terminus and the C-terminus of the mature membrane protein (the normal, nonhybrid protein) on the same side of the membrane?

12–25 Mitochondria and peroxisomes, as opposed to most other cellular membranes, acquire new phospholipids in soluble form from phospholipid exchange proteins. One such protein, PC exchange protein, specifically transfers phosphatidylcholine (PC) between membranes. Its activity is measured by mixing red blood cell ghosts (intact plasma membranes with cytoplasm removed) with synthetic phospholipid vesicles containing radioactively labeled PC in both monolayers of the vesicle bilayer. After incubation at 37°C, the mixture is centrifuged briefly so that ghosts form a pellet, whereas the vesicles stay in the supernatant. The amount of exchange is determined by measuring the radioactivity in the pellet.

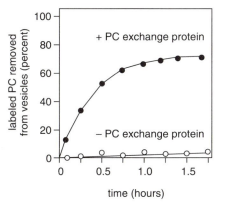

Figure 12–15 Transfer of labeled PC from donor vesicles to red cell membranes by PC exchange protein (Problem 12–25).

Figure 12–15 shows the result of an experiment along these lines, using labeled (donor) vesicles with an outer radius of 10.5 nm and a bilayer 4.0 nm in thickness. No transfer occurred in the absence of the exchange protein, but in its presence up to 70% of the labeled PC in the vesicles could be transferred to the red cell membranes.

Several control experiments were performed to explore the reason why only 70% of the label in donor vesicles was transferred.

1. Five times as many membranes from red cell ghosts were included in the incubation: the transfer still stopped at the same point.

2. Fresh exchange protein was added after 1 hour: it caused no further transfer.

3. The labeled lipids remaining in donor vesicles at the end of the reaction were extracted and made into fresh vesicles: 70% of the label in these vesicles was exchangeable.

When the red cell ghosts that were labeled in this experiment were used as donor membranes in the reverse experiment (that is, transfer of PC from red cell membranes to synthetic vesicles), 96% of the label could be transferred to the acceptor vesicles.

A. What possible explanations for the 70% limit do each of the three control experiments eliminate?

B. What do you think is the explanation for the 70% limit? (HINT: the area of the outer surface of these small donor vesicles is about 2.5 times larger than the area of the inner surface.)

C. Why do you think that almost 100% of the label in the red cell membrane can be transferred back to the vesicle?

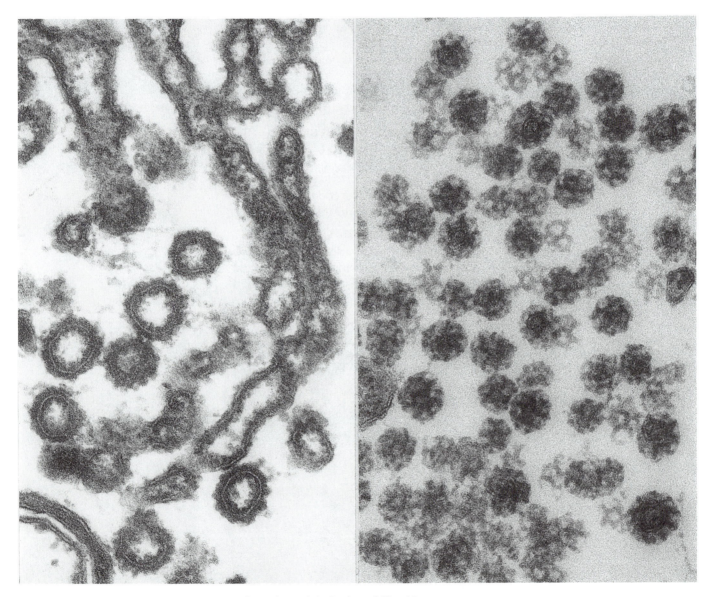

Two kinds of coated vesicles: (*left*) a vesicle from the Golgi of cultured fibroblasts; (*right*) one from bovine brain. Which (if any) have clathrin coats? See Figure 13–47 on page 635 of MBOC, 3rd Edition, for details. (Courtesy of Lelio Orci. From L. Orci, B. Glick, and J. Rothman, *Cell* 46:171–184, 1986. © Cell Press.)

Vesicular Traffic in the Secretory and Endocytic Pathways

Transport from the ER Through the Golgi Apparatus
(MBOC 600–610)

13–1 Fill in the blanks in the following statements.

A. The _____ is a major site of carbohydrate synthesis as well as a sorting and dispatching station for the products of the ER.

B. Each _____ is a collection of four to six flattened, membrane-bounded cisternae and has two distinct faces: a _____ (or entry face) and a _____ (or exit face).

C. Proteins and lipids enter a network of interconnected tubular and cisternal structures, called the _____ network, in transport vesicles from the ER and exit from a similar network, the _____ network, in transport vesicles destined for the cell surface or another compartment.

D. Vesicles destined for the Golgi apparatus bud from a specialized region of the ER called the_____, whose membrane lacks bound ribosomes and is often located between the rough ER and the Golgi apparatus.

E. In cells treated with _____ the Golgi apparatus largely disappears and the Golgi proteins end up in the ER, where they intermix with ER proteins.

F. A single species of _____ oligosaccharide is attached *en bloc* to many proteins in the ER, and this oligosaccharide is then trimmed while the protein is still in the ER.

G. _____ oligosaccharides have no new sugars added to them in the Golgi apparatus, whereas _____ oligosaccharides contain a variable number of additional sugars.

H. Proteins exported from the ER are processed in the Golgi; they enter the _____, which is thought to be continuous with the *cis* Golgi network, then move to the _____, consisting of the central cisternae of the stack, and finally to the _____, where glycosylation is completed.

I. In the Golgi some proteins have sugars added to the OH groups of selected serine or threonine side chains, a process called _____ glycosylation.

J. The Golgi apparatus confers the heaviest glycosylation of all on special proteins, which it modifies to produce _____.

13–2 Indicate whether the following statements are true or false. If a statement is false, explain why.

___ A. Plant and animal cells typically have a single Golgi stack.

___ B. There is one strict requirement for the exit of a protein from the ER: it must be correctly folded.

___ C. It is thought that brefeldin A accelerates the return transport from the Golgi to the ER, ultimately causing the Golgi apparatus to empty into the ER.

___ D. All glycosylation of membrane and exported proteins occurs in the Golgi apparatus.

___ E. The transport of proteins between the different cisternae of the Golgi stack is thought to be mediated by transport vesicles, which bud from one cisterna and fuse with the next.

___ F. Proteoglycan core proteins are converted in the Golgi apparatus into proteoglycans by the addition of one or more *O*-linked glycosaminoglycan chains.

___ G. All of the glycoproteins and glycolipids in intracellular membranes have their oligosaccharides facing the luminal side, whereas those in the plasma membrane have their oligosaccharides facing outside the cell.

___ H. *N*-linked oligosaccharides aid in the transport of proteins through the ER and Golgi. Drugs that block glycosylation generally do not interfere with transport through the ER and Golgi.

13–3   Until recently, there were two competing models for how material progresses through the Golgi apparatus (Figure 13–1). In the cisternal-progression model, new cisternae form continuously as vesicles from the ER coalesce at the *cis* face of the Golgi apparatus. Each newly formed cisterna moves through the stack (with appropriate modifications occurring to their contents) and finally breaks up into transport vesicles at the *trans* face. In the vesicle transport model, the cisternae remain fixed and the maturing glycoproteins move from the *cis* to the *trans* cisternae inside transport vesicles.

One test of these two models made use of mutant cells that are defective in the addition of galactose, which occurs in the *trans* compartment of the Golgi. The mutant cells were infected with vesicular stomatitis virus (VSV) to provide a convenient marker protein, the viral G protein. At an appropriate point in the infection an inhibitor of protein synthesis was added to stop further synthesis of G protein. The infected cells were then incubated briefly with a radioactive precursor of GlcNAc, which in the absence of protein synthesis is added only in the medial compartment of the Golgi. Next, the infected mutant cells were fused with uninfected wild-type cells to form a common cytoplasm containing both wild-type and mutant Golgi stacks. After a few minutes the cells were dissolved with detergent and all the G protein was precipitated using G-specific antibodies. After separation from the antibodies the G proteins carrying galactose were precipitated with a lectin that binds galactose. The radioactivity in the precipitate and in the supernatant was measured. The results of this experiment along with control experiments (which used mutant cells only or wild-type cells only) are shown in Table 13–1.

A. The movement of proteins between which two compartments of the Golgi apparatus is being tested in this experiment? Explain your answer.

B. If proteins moved through the Golgi apparatus by cisternal progression, what would you predict for the results of this experiment? If proteins moved

(A) CISTERNAL PROGRESSION MODEL

(B) VESICLE TRANSPORT MODEL

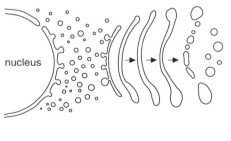

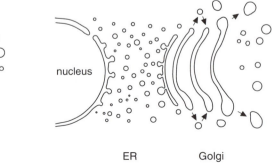

ER   Golgi   ER   Golgi

Figure 13–1 The cisternal progression model (A) and the vesicle transport model (B) for movement of material through the Golgi apparatus (Problem 13–3).

Table 13–1 Addition of Galactose to G Protein After Fusion of VSV-infected Cells to Uninfected Cells (Problem 13–3)

	Precipitate	Supernatant
1. Infected mutant cells fused with uninfected wild-type cells	45%	55%
2. Infected mutant cells fused with uninfected mutant cells	5%	95%
3. Infected wild-type cells fused with uninfected wild-type cells	85%	15%

Radioactive G protein was purified and then reacted with a lectin that precipitates galactose-containing proteins. The percentage of total radioactivity in the precipitate (G protein with galactose) and the supernatant (G protein without galactose) is shown.

through the Golgi via transport vesicles, what would you predict for the results of this experiment?

C. Which model is supported by the results in Table 13–1?

*13–4   You have isolated several mutant cell lines that are defective in their ability to add carbohydrate to exported proteins. Using an easily purified protein that carries only N-linked complex oligosaccharides, you have analyzed the sugars in the N-linked oligosaccharides that are added in the different mutant cells. Each mutant is unique in the kinds and numbers of different sugars contained in its oligosaccharides (Table 13–2).

A. Arrange the mutants in the order that corresponds to the steps in the pathway for processing N-linked oligosaccharides.

B. Which of these mutants are defective in processing events that occur in the ER? Which mutants are defective in processing steps that occur in the Golgi?

C. Which of the mutants are likely to be defective in a processing enzyme that is directly responsible for modifying N-linked oligosaccharides? Which mutants might not be defective in a processing enzyme but, rather, in another enzyme that affects oligosaccharide processing indirectly?

13–5   The vesicular stomatitis virus (VSV) G protein is a typical membrane glycoprotein. In addition to its signal peptide, which is removed after import into the ER, the G protein contains a single membrane-spanning segment that anchors the protein in the plasma membrane so that a small C-terminal domain is exposed to the cytoplasm and a much larger N-terminal domain is outside

Table 13–2 Analysis of the Sugars Present in the N-linked Oligosaccharides from Wild-Type Cells and from Mutant Cell Lines Defective in Oligosaccharide Processing (Problem 13–4)

Cell Line	Man	GlcNAc	Gal	NANA	Glc
Wild-type	3	4	2	2	0
Mutant A	3	4	0	0	0
Mutant B	5	3	0	0	0
Mutant C	9	2	0	0	3
Mutant D	9	2	0	0	0
Mutant E	5	2	0	0	0
Mutant F	3	3	0	0	0
Mutant G	8	2	0	0	0
Mutant H	9	2	0	0	2
Mutant I	3	4	2	0	0

Abbreviations: Man = mannose; GlcNAc = N-acetylglucosamine; Gal = galactose; NANA = N-acetylneuraminic acid, or sialic acid; Glc = glucose.

Numbers indicate the number of sugar monomers in the oligosaccharide.

*Problems with an asterisk (*) are answered in the Instructor's Manual.

		← membrane-spanning segment →	
pMS20	**K**	S S I A S F F F I I G L I I G L F L V L	**R**
pMS18	**K**	S S I A S F F F I I G L - - G L F L V L	**R**
pMS16	**K**	S S I A S F F F I I G - - - - L F L V L	**R**
pMS14	**K**	S S I A S F F F I I - - - - - - F L V L	**R**
pMS12	**K**	S S I A S F F F I - - - - - - - - L V L	**R**
pMS8	**K**	S S I A - - - - - - - - - - - - F L V L	**R**
pMS0	**K**	- - - - - - - - - - - - - - - - - - - -	**R**

Figure 13–2 The membrane-spanning domains of normal and mutant VSV G proteins (Problem 13–5). Plasmid numbers indicate the number of amino acids in the membrane-spanning segment; for example, pMS20 contains the wild-type, 20-amino-acid membrane-spanning segment. Dashed lines indicate amino acids that are missing in the other plasmids. Boxed letters indicate the basic amino acids that flank the membrane-spanning segment.

the cell. The membrane-spanning segment consists of 20 uncharged and mostly hydrophobic amino acids that are flanked by basic amino acids (Figure 13–2). Twenty amino acids arranged in an α helix is just sufficient to span the 3-nm thickness of the lipid bilayer of the membrane.

To test the length requirements for membrane-spanning segments, you modify a cloned version of the G protein to generate a series of mutants in which the membrane-spanning segment is shorter, as indicated in Figure 13–2. When you introduce the modified plasmids into cultured cells, roughly the same amount of G protein is synthesized from each mutant as from wild-type cells. You analyze the cellular distribution of the altered G proteins in several ways.

1. You examine the cellular location of the modified G proteins by immunofluorescence microscopy, using G-specific antibodies tagged with fluorescent markers.
2. You characterize the attached oligosaccharide chains by digesting the G proteins with endoglycosidase H (endo H), which removes *N*-linked oligosaccharides up to a certain stage in their processing in the Golgi (see Figure 13–11 in MBOC).
3. You determine whether the altered G proteins retain the small C-terminal cytoplasmic domain (which characterizes the normal G protein) by treating isolated microsomes with a protease. In the normal G protein this domain is sensitive to protease treatment and removed.

The results of these experiments are summarized in Table 13–3.

A. To the extent these data allow, deduce the intracellular location of each altered G protein that fails to reach the plasma membrane.

Table 13–3 Results of Experiments Characterizing the Cellular Distribution of G Proteins from Normal and Mutant Cells (Problem 13–5)

Plasmid	Cellular Location	Endo H Treatment	Protease Treatment
pMS20	plasma membrane	resistant	sensitive
pMS18	plasma membrane	resistant	sensitive
pMS16	plasma membrane	resistant	sensitive
pMS14	plasma membrane	resistant	sensitive
pMS12	intracellular	+ / – resistant	sensitive
pMS8	intracellular	sensitive	sensitive
pMS0	intracellular	sensitive	resistant

+ / – in the endo H column indicates that only about 30% of the G protein was sensitive to endo H; other modified G proteins were either totally resistant or sensitive. In the protease column "sensitive" indicates that the small C-terminal domain was removed by protease treatment; "resistant" indicates that the molecular weight of the G protein was unchanged by the protease treatment.

B. For the VSV G protein, what is the minimum length of the membrane-spanning segment that is sufficient to anchor the protein in the membrane?

C. What is the minimum length of the membrane-spanning segment that is consistent with proper sorting of the G protein?

Transport from the *Trans* Golgi Network to Lysosomes
(MBOC 610–618)

13–6   Fill in the blanks in the following statements.

A. _____ are membranous bags of hydrolytic enzymes used for the controlled intracellular digestion of macromolecules; all the hydrolytic enzymes are _____ that require a low-pH environment for optimal activity.

B. Most plant and fungal cells contain one or several very large, fluid-filled vesicles called _____.

C. _____ is a pathway to degradation in lysosomes that is used in all cell types for disposal of obsolete parts of the cell itself.

D. Lysosomal hydrolases carry a unique marker in the form of _____ groups, which are recognized by transmembrane _____ proteins in the *trans* Golgi network.

E. _____ diseases are caused by genetic defects that affect one or more of the lysosomal hydrolases and result in accumulation of their undigested substrates in lysosomes.

13–7   Indicate whether the following statements are true or false. If a statement is false, explain why.

__ A. Lysosomal membranes contain a proton pump that utilizes the energy of ATP hydrolysis to pump protons out of the lysosome, thereby maintaining the lumen at a low pH.

__ B. Lysosomes are heterogeneous organelles that are found in all nucleated eucaryotic cells.

__ C. Vacuoles in plant cells are related to lysosomes of animal cells, but their functions are remarkably diverse.

__ D. Late endosomes are converted to mature lysosomes by the loss of distinct endosomal membrane proteins and a further decrease in their internal pH.

__ E. Proteins that contain KFERQ sequences on their surface are specifically exported from lysosomes to the cytosol to prevent their degradation.

__ F. Lysosomal hydrolases are marked for delivery to lysosomes by having mannose 6-phosphate (M6P) groups added to their *N*-linked oligosaccharides in the lumen of the Golgi.

__ G. If cells were treated with a weak base such as ammonia or chloroquine, which raises the pH of organelles toward neutrality, M6P receptors would be expected to accumulate in the Golgi because they cannot bind to the lysosomal enzymes.

__ H. Since all glycoproteins arrive in the Golgi with identical *N*-linked oligosaccharides, the signal for adding the M6P units to oligosaccharides must reside somewhere in the polypeptide chain of each hydrolase.

__ I. In I-cell disease the lysosomes in some cell types contain a normal complement of lysosomal enzymes, implying that there is a second pathway for sorting hydrolases to lysosomes that is used in some cells but not in others.

*13–8   The principal pathway for transport of lysosomal hydrolases from the *trans* Golgi network (pH7) to the late endosomes (pH6) and the recycling of M6P receptors back to the Golgi depends on the pH difference between those two compartments. From what you know about M6P receptor binding and recycling and the pathways for delivery of material to lysosomes, describe the consequences of changing the pH in those two compartments.

A. What do you think would happen if the pH in late endosomes were raised to pH7?

B. What do you think would happen if the pH in the *trans* Golgi network were lowered to pH6?

13–9   Most lysosomal hydrolases are tagged with several oligosaccharide chains that can acquire multiple M6P groups. The multiplicity of M6P groups substantially increases the affinity of these hydrolases for M6P receptors and markedly improves the efficiency of sorting to lysosomes. The reasons for the increase in affinity are twofold: one relatively straightforward and the other, more subtle. Both can be appreciated at a conceptual level (without resorting to detailed kinetic analysis or heavy mathematics) by considering the simple case of a hydrolase with one M6P compared with the same hydrolase with four M6P groups.

The binding of a hydrolase (H) to the M6P receptor (R) can be represented as

$$\text{Hydrolase} + \text{Receptor} \; \underset{k_{OFF}}{\overset{k_{ON}}{\rightleftharpoons}} \; \text{Hydrolase} - \text{Receptor complex}$$

$$H + R \; \underset{k_{OFF}}{\overset{k_{ON}}{\rightleftharpoons}} \; HR$$

At equilibrium the rate of association of hydrolase with the receptor ($k_{ON}[H][R]$) equals the rate of dissociation ($k_{OFF}[HR]$)

$$k_{ON}[H][R] = k_{OFF}[HR]$$

$$\frac{k_{ON}}{k_{OFF}} = \frac{[HR]}{[H][R]} = K$$

where K is the equilibrium constant. Because the equilibrium constant is a measure of the strength of binding between two molecules, it is sometimes called the affinity constant: the larger the value of the affinity constant, the stronger the binding.

A. Consider first the hypothetical situation in which the hydrolase and the M6P receptor are both soluble—that is, the receptor is not in a membrane. Assume that the M6P receptor has a single binding site for M6P groups, and think about the interaction between a single receptor and a hydrolase molecule. How do you think the rate of association will change if the hydrolase has one M6P group or four M6P groups? How will the rate at which the hydrolase dissociates from a single receptor change if the hydrolase has one M6P group or four? Given the effect on association and dissociation, how will the affinity constants differ for a hydrolase with one M6P group versus a hydrolase with four M6P groups?

B. Consider the situation in which a hydrolase with four M6P groups has bound to one receptor already. Assuming that the first receptor is locked in place, how will the binding to the first receptor influence the affinity constant for binding to a second receptor? (For simplicity, assume that the binding to the first receptor does not interfere with the ability of other M6P groups on the hydrolase to bind to a second receptor.)

C. In the real situation, as it occurs during sorting, the M6P receptors are in the Golgi membrane, whereas the hydrolases are initially free in the lumen of the Golgi. Consider the situation in which a hydrolase with four M6P groups has bound to one receptor already. In this case how do you think the binding of the hydrolase to the first receptor will influence the affinity constant for binding to a second receptor? (Think about this question from the point of view of how the binding changes the distribution of the hydrolase with respect to the lumen and the membrane.)

13–10 Patients with Hunter's syndrome or Hurler's syndrome rarely live beyond their teens. Analysis indicates that patients accumulate glycosaminoglycans in lysosomes due to the lack of specific lysosomal enzymes necessary for their degradation. When cells from patients with the two syndromes are fused, glycosaminoglycans are degraded properly, indicating that the cells are missing different degradative enzymes. Even if the cells are just cultured together, they still correct each others' defects. Most surprising of all, the medium from a culture of Hurler's cells corrects the defect in Hunter's cells (and vice versa). The corrective factors in the media are inactivated by treatment with proteases, by treatment with periodate, which destroys carbohydrate, and by treatment with alkaline phosphatase, which removes phosphates.

A. What do you think the corrective factors are, and how do you think they correct the lysosomal defects?

B. Why do you think the treatments with protease, periodate, and alkaline phosphatase inactivate the corrective factors?

C. Would you expect a similar sort of correction scheme to work for mutant or missing cytosolic enzymes?

*13–11 Children with I-cell disease synthesize perfectly good lysosomal enzymes but secrete them outside the cell instead of sorting them to lysosomes. The mistake occurs because the cells lack GlcNAc-P-transferase, which is required to create the M6P marker that is essential for proper sorting. In principle, I-cell disease could also be caused by deficiencies in two other proteins: the phosphoglycosidase that removes GlcNAc to expose M6P and the M6P receptor itself.

These three potential kinds of I-cell disease could be distinguished by the ability of various culture supernatants to correct defects in mutant cells. Imagine that you have cell lines from three hypothetical I-cell patients (A, B, and C) that give the results below:

1. The supernatant from normal cells corrects the defects in B and C but not the defect in A.

2. The supernatant from A corrects the defect in Hurler's cells, but the supernatants from B and C do not.

3. If the supernatants from the mutant cells are first treated with the phosphoglycosidase that removes GlcNAc, then the supernatants from A and C correct the defect in Hurler's cells, but the supernatant from B does not.

From these results deduce the nature of the defect in each of the mutant cell lines.

13–12 The thyroid hormone thyroxine is composed of two linked, iodinated tyrosine residues (Figure 13–3). It is stored in the thyroid gland in a structure called a follicle, as part of a much larger protein called thyroglobulin. The follicle consists of a cellular epithelium surrounding an extracellular space, or lumen. When the thyroid is stimulated by TSH (thyroid-stimulating hormone), thyroxine is digested out of thyroglobulin by proteases and released into the bloodstream.

The actual pathway for thyroxine release was difficult to identify. When the thyroid is stimulated by TSH, "colloid droplets" appear in the cytoplasm of the follicle cells. The similarity of the material in these droplets to the material in

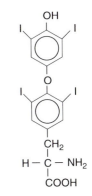

Figure 13–3 Structure of thyroxine (Problem 13–12).

the lumen of the follicle sparked an intense debate: Do the droplets represent material on the way out of the cell to the lumen (to replenish the supply), or do they represent lumenal material engulfed by the cell (to generate thyroxine)? This question has been resolved by a series of experimental observations.

1. If the colloid in the lumen is prelabeled with ^{131}I and the follicles are then stimulated with TSH under conditions that block further incorporation of iodine, the intracellular droplets are labeled.
2. Intracellular droplets form about 4 minutes after exposure to TSH and are seen first in the apical cell processes, which abut the lumen of the follicle, then in the apical region of the cell, and finally in the basal region.
3. Thyroglobulin in the lumen of the follicle carries M6P, which normally targets proteins for delivery to lysosomes.

Given these observations, propose a pathway for thyroxine production and release from follicle cells.

Transport from the Plasma Membrane via Endosomes: Endocytosis
(MBOC 618–626)

13–13   Fill in the blanks in the following statements.

A. The routes that lead from the cell surface inward to lysosomes start with the process of _____, by which cells take up macromolecules, particulate substances and, in specialized cases, even other cells.
B. _____ is a specialized form of endocytosis in which large particles such as microorganisms and cell debris are ingested.
C. In mammals there are two classes of white blood cells that act as professional phagocytes: _____ and _____.
D. Whereas the endocytic vesicles involved in pinocytosis are small and uniform, _____ have diameters that are determined by the size of the ingested particle.
E. Virtually all eucaryotic cells continually ingest bits of their plasma membrane in the form of small endocytic vesicles in the process known as _____.
F. Ingestion of plasma membrane by endocytosis and addition of membrane to the plasma membrane by the converse process are linked processes that can be considered to constitute an _____.
G. Pinocytosis begins at specialized regions of the plasma membrane called _____, which invaginate into the cell and pinch off to form _____.
H. Since extracellular fluid is trapped in pinocytotic vesicles, substances dissolved in the extracellular fluid are internalized, a process called _____.
I. In most animal cells clathrin-coated pits and vesicles provide an efficient pathway for taking up specific macromolecules from the extracellular fluid in a process called _____.
J. Most cholesterol is transported in the blood bound to protein in the form of particles known as _____.
K. The _____ compartment is composed of two distinguishable components: _____, just beneath the plasma membrane, and _____, close to the Golgi apparatus and near the nucleus.
L. The protein that carries iron in the blood is known as _____.
M. Some receptors on the surface of polarized epithelial cells transfer specific macromolecules from one extracellular space to another by a process called _____.

13–14 Indicate whether the following statements are true or false. If a statement is false, explain why.

__ A. Any particle that is bound to the surface of a phagocyte will be ingested by phagocytosis.

__ B. The endocytic cycle begins at specialized regions of the plasma membrane known as coated pits, at which clathrin and associated proteins drive the invagination of the membrane.

__ C. Receptor-mediated endocytosis provides a selective concentrating mechanism that increases the efficiency of internalization of particular ligands more than 1000-fold.

__ D. Like the LDL receptor, most of the more than 35 different receptors known to participate in receptor-mediated endocytosis enter coated pits only after they have bound their specific ligands.

__ E. All the molecules that enter early endosomes ultimately reach late endosomes where they become mixed with newly synthesized acid hydrolases and end up in lysosomes.

__ F. In the acidic environment of the early endosome, many internalized receptor proteins change their conformation and release their ligand.

__ G. Although the relationship between early and late endosomes is unclear, they do, in fact, differ in their protein composition.

__ H. During transcytosis, vesicles that form from coated pits on the apical surface fuse with the plasma membrane on the basolateral surface and in that way transport molecules across the epithelium.

__ I. In polarized epithelial cells endocytosis from either the apical or basolateral domains ultimately delivers molecules to be degraded to a common late-endosomal compartment.

13–15 Cells take up extracellular molecules by receptor-mediated endocytosis and by fluid-phase endocytosis. The efficiencies of these two pathways were compared by incubating human epithelial carcinoma cells with epidermal growth factor (EGF), to measure receptor-mediated endocytosis, and with horseradish peroxidase (HRP), to measure fluid-phase endocytosis. A quantitative comparison of uptake was made by incubating cells with 40-nM ferritin-labeled EGF and 20-μM HRP (500-fold higher concentration than EGF) for various times, after which the cells were fixed and stained for HRP activity and examined for the presence of ferritin in vesicles. Both EGF and HRP were present in small vesicles with an internal radius of 20 nm, but EGF was very common, whereas HRP was rare.

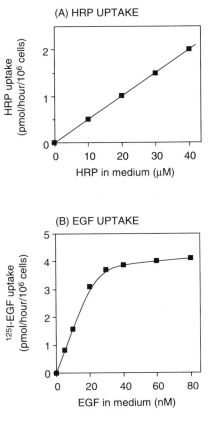

Figure 13–4 Uptake of HRP and EGF as a function of their concentration in the medium (Problem 13–15).

The rates of uptake of EGF and HRP were compared as shown in Figure 13–4. Uptake of HRP was linear with respect to both time and concentration: cells took up HRP at a rate of 1 pmol/hour at 20 μM HRP (Figure 13–4). EGF uptake showed an initial linear phase but reached a plateau at higher concentrations of EGF in the medium (Figure 13–4).

A. Explain why the shapes of the curves in Figure 13–4 are different for HRP and EGF.

B. Calculate the number of EGF receptors on the surface of each cell.

C. Calculate how many HRP molecules get taken up by each endocytic vesicle (radius 20 nm) when the medium contains 1 mg/ml HRP (molecular weight 40,000). (The volume of a sphere is $^4/_3\pi r^3$.)

D. The scientists who did these experiments said at the time, "These calculations clearly illustrate how cells can internalize EGF by endocytosis while excluding all but insignificant quantities of extracellular fluid." Explain what they meant.

13–16 Cholesterol is an essential component of the plasma membrane, but people who have very high levels of cholesterol in their blood tend to have heart attacks. Cholesterol in the blood is carried in the form of cholesterol esters in low-density lipoprotein (LDL) particles. LDL binds to a high-affinity receptor

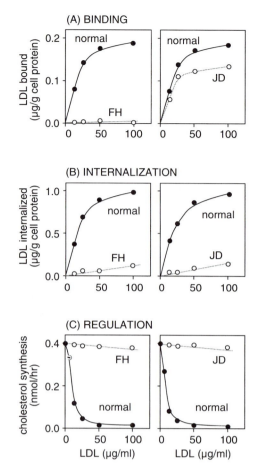

Figure 13–5 LDL metabolism in normal cells and in cells from patients with severe familial hypercholesterolemia (Problem 13–16). (A) High-affinity surface binding of LDL. (B) Internalization of LDL. (C) Regulation of cholesterol synthesis by LDL. Binding and uptake of LDL can be followed by labeling LDL either with ferritin particles, which can be visualized by electron microscopy, or with radioactive iodine, which can be measured in a gamma counter. Surface binding can be reversed by washing with negatively charged polymers, but internalized label cannot be washed away.

on the cell surface, enters the cell via a coated pit, and ends up in lysosomes. There its protein coat is degraded, and cholesterol esters are released and hydrolyzed to cholesterol. The released cholesterol enters the cytosol and inhibits the enzyme HMG CoA reductase, which controls the committed step in cholesterol biosynthesis. Patients with severe hypercholesterolemia cannot remove LDL from the blood. As a result, their cells do not turn off normal cholesterol synthesis, which makes the problem worse.

LDL metabolism can be conveniently divided into three stages experimentally: binding of LDL to the cell surface, internalization of LDL, and regulation of cellular synthesis by LDL. Skin cells from a normal person and two patients suffering from severe familial hypercholesterolemia were grown in culture and tested for LDL binding, LDL internalization, and LDL regulation of cholesterol synthesis. The results are shown in Figure 13–5.

A. In Figure 13–5A the surface binding of LDL by normal cells is compared with LDL binding by cells from patients FH and JD. Why does binding by normal cells and by JD's cells reach a plateau? What explanation can you suggest for the lack of LDL binding by FH's cells?

B. In Figure 13–5B internalization of LDL by normal cells increases as the external LDL concentration is increased, reaching a plateau fivefold higher than the amount of externally bound LDL. Why does LDL not enter cells from patients FH or JD?

C. In Figure 13–5C the regulation of cholesterol synthesis by LDL in normal cells is compared with cells from FH and JD. Why does increasing the external LDL concentration inhibit cholesterol synthesis in normal cells but not affect it in cells from FH or JD?

D. How would you expect the rate of cholesterol synthesis to be affected if normal cells and cells from FH or JD were incubated with cholesterol itself? (Free cholesterol crosses the plasma membrane by diffusion.)

*13–17 Refer to the data in Problem 13–16. What is wrong with JD's metabolism of LDL? JD's cells bind LDL with the same affinity as normal and in almost the same amounts as normal, but the binding does not lead to internalization of LDL. Two classes of explanation could account for JD's problem:

1. JD's LDL receptors are defective in some internal portion so that the receptors cannot enter the cell, even though the LDL-binding domains on the cell surface are perfectly normal.
2. JD's LDL receptors are normal, but there is a mutation in the cellular internalization machinery such that the loaded LDL receptor cannot be brought in.

To distinguish between these explanations, JD's parents were studied. Since the gene encoding the LDL receptor is autosomal, each parent must have donated one of their two genes to JD. JD's mother suffered from mildly elevated blood cholesterol. At 4°C her cells bound only half as much LDL as normal, but when her cells were warmed to 37°C, the bound LDL was internalized at the same rate as normal. JD's father also had mild hypercholesterolemia, but his cells bound 50% more LDL than normal at 4°C. When his cells were warmed to 37°C, about half of the label was internalized normally, but the other half was not internalized at all.

The association of this family's LDL receptors with coated pits was studied by electron microscopy, using LDL that was labeled with ferritin. The results are shown in Table 13–4.

A. Why does JD's mother have mild hypercholesterolemia? What kind of LDL-receptor gene might she have passed on to JD?
B. Why does JD's father have mild hypercholesterolemia? Can you make an argument to distinguish between explanations 1 and 2 (above), based on the inability of cells from JD's father to internalize all the LDL they bind?
C. Can you account for JD's hypercholesterolemia from the behavior of the LDL receptors in his parents?
D. What do the electron microscopic studies suggest is wrong with JD's LDL uptake?

13–18 The recycling of the membrane receptors for transferrin has been studied by labeling transferrin receptors on the cell surface and following their fate at 0°C and 37°C. Intact cells at 0°C were reacted with radioactive iodine under conditions that label cell-surface proteins. If the cells were kept on ice and treated with trypsin, which destroys the receptors without damaging the integrity of the cell, the radioactive transferrin receptors were completely degraded. (The transferrin receptors were detected as a spot after separation

Table 13–4 Distribution of LDL Receptors on the Surface of Cells from JD and His Parents as Compared with Normal Individuals (Problem 13–17)

| | Number of LDL Receptors | |
	In Pits	Outside Pits
Normal male	186	195
Normal female	186	165
JD	10	342
JD's father	112	444
JD's mother	91	87

of cell proteins by two-dimensional polyacrylamide-gel electrophoresis.) If the cells were first warmed to 37°C for 1 hour and then treated with trypsin on ice, about 70% of the initial radioactivity was present in the spot. At both temperatures, however, most of the receptors, as visualized by a protein stain, remained intact.

A second sample of cells that had been surface-labeled at 0°C and incubated at 37°C for 1 hour was analyzed with transferrin-specific antibodies. If intact cells were reacted with antibody, 0.54% of the labeled proteins were bound by antibody. If the cells were first dissolved in detergent, 1.76% of the labeled proteins were bound by antibody.

A. Why does trypsin treatment destroy the labeled transferrin receptors, but not the majority of the receptors, when the cells are kept on ice? Why do the labeled receptors become resistant to trypsin when the cells are incubated at 37°C?

B. What fraction of the total transferrin receptor is on the cell surface after a 1-hour incubation at 37°C? Do the two experimental approaches agree?

*13–19   The average time for transferrin receptors to cycle from the cell surface through the endosomal compartment and back again to the cell surface has been determined by labeling cell-surface receptors with radioactive iodine at 0°C and then following their fate at 37°C. At various times after shifting labeled cells to 37°C, samples were diluted into ice-cold medium that contained trypsin. The amount of radioactivity in trypsin-resistant transferrin receptors was measured after separation from other membrane components by two-dimensional polyacrylamide-gel electrophoresis. The results are shown in Figure 13–6.

A. The initial rate of internalization of labeled transferrin receptors is indicated by the dashed line. Why does the rate of internalization of labeled receptors decline with time?

B. Using the initial rate of internalization, estimate the fraction of surface receptors that are internalized each minute.

C. What fraction of the *total* receptor population is internalized each minute?

D. At the rate determined in part C, how many minutes would it take for the equivalent of the entire population of receptors to be internalized? Explain why this time equals the average time for a receptor to cycle from the cell surface through the endosomal compartment and back to the cell surface.

E. On average, how long does each transferrin receptor spend on the cell surface?

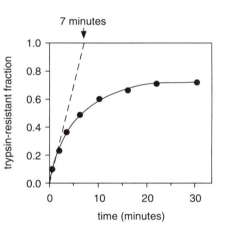

Figure 13–6 Fraction of labeled transferrin receptor that is trypsin resistant as a function of time after labeling (Problem 13–19). The dashed line indicates the initial rate of internalization.

Transport from the *Trans* Golgi Network to the Cell Surface: Exocytosis

(MBOC 626–633)

13–20   Fill in the blanks in the following statements.

A. The fusion of transport vesicles from the *trans* Golgi network with the plasma membrane is called _____.

B. While all cells use the _____ secretory pathway, specialized secretory cells also use a _____ secretory pathway.

C. Cells that are specialized for secreting some of their products rapidly on demand concentrate and store these products in _____.

D. Histamine is a small molecule secreted by _____ cells in response to specific ligands that bind to receptors on the cell surface.

E. Nerve cells make use of a specialized class of tiny secretory vesicles called _____, which store neurotransmitter molecules.

13–21   Indicate whether the following statements are true or false. If a statement is false, explain why.

___ A. In a cell capable of regulated secretion, at least three classes of proteins must be separated before they leave the *trans* Golgi network—those destined for lysosomes, those destined for secretory vesicles, and those destined for immediate delivery to the cell surface.

___ B. When a gene encoding a secretory protein is transferred to a secretory cell that normally does not make the protein, the foreign protein is not packaged into secretory vesicles.

___ C. Proteolytic processing is common in the secretory pathway to delay activation of hydrolytic enzymes and to generate active peptides that are too short to be co-translationally transported into the ER lumen.

___ D. Once a secretory vesicle is properly positioned beneath the plasma membrane, it will fuse with the membrane and release its contents to the cell exterior.

___ E. When a mast cell is stimulated at one local region of its cell surface, exocytosis occurs all over the cell surface.

___ F. Although exocytosis should greatly increase the surface area of the plasma membrane, it does so only transiently because membrane components are removed from the surface by endocytosis almost as fast as they are added by exocytosis.

___ G. Unlike other secretory vesicles, synaptic vesicles are believed to be generated not from the Golgi membrane but by local recycling from the plasma membrane via endosomes.

___ H. In polarized cells proteins from the ER that are destined for different domains travel together until they reach the *trans* Golgi network, where they are separated and dispatched in secretory or transport vesicles to the appropriate plasma membrane domain.

13–22   You are interested in exocytosis and endocytosis in a line of cultured liver cells that secrete albumin and take up transferrin. To distinguish between these events, you add transferrin tagged with colloidal gold to the medium, and then after a few minutes you fix the cells, prepare thin sections, and react them with ferritin-labeled antibodies against albumin. Colloidal gold and ferritin are both electron dense and therefore readily visible when viewed by electron microscopy; moreover, they can be easily distinguished from one another on the basis of size and density.

   A. Will this experiment allow you to identify vesicles in the exocytic and endocytic pathways? How?

   B. Not all the gold-labeled vesicles are clathrin coated. Why?

13–23   Liver cells secrete a broad spectrum of proteins into the blood via the constitutive pathway. You are interested in how long it takes for different proteins to be secreted. Accordingly, you add ^{35}S-methionine to cultured liver cells to label proteins as they are synthesized. You then sample the medium at various times to measure the appearance of individual labeled proteins. As shown in Figure 13–7, albumin appears after 20 minutes, transferrin appears after 50 minutes, and retinol-binding protein appears after 90 minutes. You are surprised at the variability in secretion rates, which bear no obvious relationship to the size, function, or quantity of the individual proteins.

   Why do transferrin and the retinol-binding protein take so much longer than albumin to be secreted? You suspect that the slow step in constitutive secretion of these proteins occurs either in the ER or in the Golgi apparatus. To determine which, you label cells for 4 hours, which is a long enough labeling period so that the labeled proteins in the cells reach the normal steady-state distribution of unlabeled proteins. (At a steady state the influx

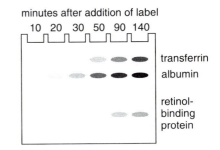

Figure 13–7 Time of appearance of secreted proteins in the medium (Problem 13–23). At various times after labeling, proteins were immunoprecipitated with specific antibodies, separated by gel electrophoresis, and subjected to autoradiography.

into a pathway exactly equals efflux from the pathway.) You then homogenize the cells to break the ER and Golgi into vesicles and separate the vesicles by density on a sucrose gradient. You measure the amount of labeled albumin and transferrin that are associated with the two types of vesicles (Figure 13–8).

Does the slow step in the constitutive secretion of transferrin occur in the ER or in the Golgi? Where does the slow step in the constitutive secretion of albumin occur? How do these experiments allow you to decide?

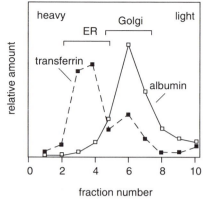

Figure 13–8 Distribution of albumin and transferrin in vesicles derived from the ER and Golgi (Problem 13–23). Labeled albumin and transferrin were assayed by immunoprecipitation, electrophoresis, and autoradiography.

*13–24 Polarized epithelial cells must make an extra sorting decision since their plasma membranes are divided into apical and basolateral domains, which are populated by distinctive sets of proteins. Proteins destined for the apical or basolateral domain of the plasma membrane seem to travel there directly from the *trans* Golgi network. One way to sort proteins to these domains would be to use a specific sorting signal for one class of proteins, which would then be actively recognized and directed to one domain, and to allow the other class to travel via a default pathway to the other domain.

Consider the following experiment to identify the default pathway. The cloned genes for several foreign proteins were engineered by recombinant DNA techniques so that they could be expressed in the polarized epithelial cell line MDCK. These proteins are secreted in other types of cells but are not normally expressed in these polarized cells. The cloned genes were introduced into the polarized MDCK cells, and their sites of secretion were assayed. Although the cells remained polarized, the foreign proteins were delivered in roughly equal amounts to the apical and basolateral domains.

A. What is the expected result of this experiment, based on the hypothesis that targeting to one domain of the plasma membrane is actively signaled and targeting to the other domain is via a default pathway?

B. Do these results support the concept of a default pathway as outlined above?

13–25 Synaptic vesicles are concentrated in the presynaptic cells at nerve synapses. Their distribution in a culture of neurons can be visualized in the following way. The culture is first exposed for 1 hour to a fluorescently tagged antibody specific for the luminal domain of synaptotagmin, a transmembrane protein that resides exclusively in the membrane of synaptic vesicles. The culture is then washed thoroughly to remove all of the synaptotagmin antibody. Next the cells are fixed with formaldehyde and treated with detergent to make them permeable to antibodies, and then they are exposed to a second fluorescently tagged antibody specific for a microtubule-associated protein found only in dendrites—the portion of the neuron with which the presynaptic cells form synapses. The results give the striking picture shown on the front cover of this book. The dendrites are beautifully outlined, and the nerve terminals are marked by dots of color—the rest of the presynaptic cell remains invisible in this procedure.

If antibodies do not cross intact membranes, how do the synaptic vesicles get labeled? If the procedure is repeated using an antibody specific for the cytoplasmic domain of synaptotagmin, the nerve terminals do not get labeled. Explain the results with the two different antibodies for synaptotagmin.

The Molecular Mechanisms of Vesicular Transport and the Maintenance of Compartmental Diversity
(MBOC 634–647)

13–26 Fill in the blanks in the following statements.

A. Most transport vesicles form from specialized regions of membrane and bud off as _____ with a distinctive cage of proteins covering the surface facing the cytosol.

B. The major protein component of _____ vesicles is _____, a protein complex that consists of three large and three small polypeptide chains that together form a three-legged structure called a triskelion.

C. _____ is a multisubunit complex that is required both to bind the clathrin coat to the membrane and to trap various transmembrane receptor proteins.

D. _____ vesicles are thought to mediate the nonselective vesicular transport of the default pathway, which includes transport from the ER to the Golgi, from one Golgi cisterna to another, and from the Golgi to the plasma membrane.

E. Vesicles in the default pathway have a coat that consists in part of a large protein complex called _____, comprising seven individual coat protein subunits.

F. Both the assembly and disassembly of the coatomer coat depend on a protein called _____, which is also thought to play a role in the assembly of clathrin coats.

G. The selective delivery of transport vesicles to their target membranes is thought to be mediated by proteins called _____, which exist as complementary sets: one on the vesicle membrane and its complement on the target membrane.

H. The crucial recognition step in the docking of a transport vesicle to its membrane is, according to one view, controlled by members of a family of monomeric GTPases called _____.

13–27 Indicate whether the following statements are true or false. If a statement is false, explain why.

__ A. In principle, the difference in the concentration of a particular membrane protein in two compartments can be maintained by using free energy to transfer the protein actively in one direction, against its concentration gradient.

__ B. Coatomer-coated vesicles mediate the outward, vesicular transport from the ER and Golgi cisternae, while clathrin-coated vesicles mediate the balanced counterflow of membrane in the opposite direction.

__ C. The formation of a clathrin-coated bud is believed to be driven by forces generated by the assembly of the coat proteins on the cytosolic surface of the membrane.

__ D. Clathrin-coated vesicles are all alike.

__ E. In contrast to clathrin coats, coatomer coats do not self-assemble but require ATP to drive their formation, and instead of disassembling as soon as the vesicle has pinched off from the donor membrane, the coatomer coat is retained until the vesicle docks with its target membrane.

__ F. GTP-binding proteins operate in a cycle that typically depends on two auxiliary components: a guanine nucleotide releasing protein (GNRP) to catalyze exchange of GDP for GTP and a GTPase-activating protein (GAP) to trigger the hydrolysis of the bound GTP.

__ G. ARF facilitates transfer of coatomer-coated vesicles in one direction only because GDP/GTP exchange is stimulated by a GNRP in the donor membrane and GTP hydrolysis is stimulated by a GAP in the target membrane.

__ H. Transport vesicles, whether or not they are selective in the way they pick up cargo from the donor compartment, have to be highly selective as to the target membrane with which they fuse.

__ I. Complementary Rab proteins on transport vesicles and their target membranes allow transport vesicles to bind selectively to their appropriate target membranes.

__ J. Docking of a transport vesicle on its target membrane and fusion of the two membranes are two distinct and separable processes.

Table 13–5 Relative Densities of G Protein in Golgi and Vesicle Lumena and Membranes (Problem 13–28)

Parameter Measured	Mean Density
Surface density over whole Golgi from *uninfected* cells	5/μm^2
Surface density over whole Golgi from *infected* cells	271/μm^2
Surface density over buds and vesicles of Golgi from *infected* cells	233/μm^2
Linear density over cisternal membranes of Golgi from *infected* cells	6/μm
Linear density over buds and vesicles of Golgi from *infected* cells	4/μm

___ K. The fusion protein of influenza virus undergoes a large conformational change at low pH, exposing a previously buried hydrophobic region that can interact with the lipid bilayer of a target membrane to promote fusion of the target membrane with the membrane surrounding the virus.

13–28 Vesicle transport between cisternae in Golgi stacks and onward to the plasma membrane via the default pathway occurs in coatomer-coated vesicles. This transport is sometimes referred to as bulk transport because the Golgi substituents do not become concentrated in the vesicles. Given that transport in clathrin-coated vesicles is so highly concentrating, you are skeptical that no concentration occurs in the default pathway for secretion.

To determine whether vesicles in the default pathway concentrate their contents, you infect cells with vesicular stomatitis virus (VSV) and follow the viral G protein. Your idea, an ambitious one, is to compare the concentration of G protein in the lumen of the Golgi stacks with that in the associated coatomer-coated transport vesicles. You intend to measure G-protein concentration by preparing thin sections of VSV-infected cells and incubating them with G-specific antibodies tagged with gold particles. Since the gold particles are visible in electron micrographs as small black dots, it is relatively straightforward to count dots in the lumena of transport vesicles (fully formed and just budding) and of the Golgi apparatus. You make two estimates of G-protein concentration: (1) the number of gold particles per cross-sectional area and (2) the number of gold particles per linear length of membrane. Your results are shown in Table 13–5.

Do the vesicles involved in bulk transport concentrate their contents or not?

*13–29 When the fungal metabolite brefeldin A is added to cells, the Golgi apparatus largely disappears and the Golgi proteins end up in the ER, where they mix with ER proteins. Brefeldin-A treatment also causes the rapid dissociation of some Golgi-associated peripheral membrane proteins, including one of the subunits of the coatomer coat. This implies that brefeldin A prevents transport involving coatomer-coated vesicles by blocking the assembly of coats and thus the budding of transport vesicles. In principle, brefeldin A could block formation of coatomer-coated vesicles at any point in the normal scheme for assembly, which is shown in Figure 13–9. The following observations identify the point of action of brefeldin A.

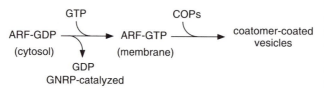

Figure 13–9 Normal pathway for formation of coatomer-coated vesicles (Problem 13–29). The monomeric GTPase ARF carries a bound GDP in its cytosolic form. In response to a guanine-nucleotide-releasing protein (GNRP), ARF releases GDP and binds GTP. Binding of GTP causes a conformational change that exposes a fatty acid tail on ARF, which promotes binding of ARF-GTP to the membrane. The subunits of the coatomer coat (COPs) bind to ARF-GTP on the membrane and form coatomer-coated vesicles.

1. ARF with bound GTPγS (a nonhydrolyzable analogue of GTP) will cause coatomer-coated vesicles to form when added to Golgi membranes. Formation of vesicles in this way is not affected by brefeldin A.
2. ARF with bound GDP exchanges GDP for GTP when added to Golgi membranes. Trypsin-treated Golgi membranes do not stimulate GTP-for-GDP exchange. The exchange reaction with normal Golgi membranes does not occur in the presence of brefeldin A.
3. ARF with bound GTP can be made to exchange GDP for GTP in the absence of Golgi membranes by treatment with phosphatidyl choline and cholate. This artificial exchange is not affected by brefeldin A.

A. Given these experimental observations, how do you think brefeldin A blocks formation of coatomer-coated vesicles?
B. In brefeldin-A treated cells why do you think the Golgi apparatus disappears and the Golgi proteins end up in the ER?

13–30 A variety of monomeric GTPases participate in the mechanics and control of vesicle formation and their delivery to specific target membranes. The *SEC4* gene of budding yeast encodes a 24-kd GTP-binding protein that plays an essential role in the secretion pathway leading to the construction of the daughter bud. Normally, about 80% of the Sec4 protein is found on the cytosolic surface of transport vesicles and 20% is free in the cytosol. When temperature-sensitive *sec4* mutants of yeast are incubated at 37°C (the restrictive temperature), growth ceases, and there is a notable accumulation of small vesicles in the daughter bud.

To define the role of the Sec4 protein in secretion, you decide to try a genetic approach that draws on the wealth of knowledge about monomeric GTPases gained through studies of the *ras* oncogene. Like the Ras protein, Sec4 has two cysteine residues at its C-terminus that, in Ras, represent sites for covalent attachment of a lipid, which is required for membrane binding. You make a *sec4* mutant, *sec4-ccΔ*, which lacks these two cysteines, and express it in yeast. The mutant protein binds GTP, but it does not bind to vesicles, does not allow the *sec4^ts* mutant to grow at high temperature, and does not inhibit the growth of wild-type yeast even when massively overexpressed.

Next you make a *sec4* mutant, *sec4-Ile133*, that encodes a Sec4 protein with an isoleucine residue in place of the normal asparagine residue at position 133 in the amino acid sequence. By analogy with similar *ras* mutants, this mutation should lock the protein into its active state, even though it cannot bind GTP or GDP. Unlike the *sec4-ccΔ* gene, introduction of this mutant protein markedly slows the growth of wild-type yeast when it is expressed at normal levels and completely inhibits growth when expressed at a level about five times higher than the normal Sec4 protein. The inhibited yeast are packed with small vesicles. If you now further mutate this gene to remove the two C-terminal cysteines, it proves to be completely harmless, and like the protein encoded by *sec4-ccΔ*, it is located exclusively in the cytosol.

A. Do you think the Sec4 protein is required for formation of vesicles, for vesicle fusion with target membranes, or for both? Based on its function, would you guess it was analogous to mammalian ARF or mammalian Rab proteins?
B. Outline how you think the normal Sec4 protein functions in vesicle formation and fusion. Why is there normally some Sec4 protein free in the cytosol of wild-type cells, and how does removal of the C-terminal cysteines prevent the Sec4-ccΔ protein from carrying out its normal function? In cells that express normal amounts of the Sec4-Ile133 protein, do you expect there to be more than, less than, or the same amount of mutant Sec4 protein in the cytosol as there is normal Sec4 protein in the cytosol of wild-type cells?
C. Why do you think expression of the Sec4-Ile133 protein in the presence of roughly the same amount of wild-type Sec4 protein inhibits growth of the yeast?

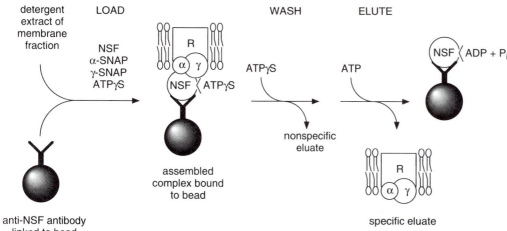

Figure 13–10 Procedure used to purify SNAP receptors (Problem 13–31). R stands for SNAP receptor.

detergent extract of membrane fraction

LOAD

NSF
α-SNAP
γ-SNAP
ATPγS

WASH

ELUTE

anti-NSF antibody linked to bead

assembled complex bound to bead

nonspecific eluate

specific eluate

13–31 Fusion between a vesicle and its target membrane requires the assembly of a complex membrane-fusion machine. The components of the fusion machinery are thought to include NSF, two SNAPs, and specific SNAP receptors (SNAREs) on the vesicle and its target membrane. During assembly of the machine the SNAPs first bind to their receptors, which can then bind NSF. Binding by NSF requires ATP, which is bound by NSF. Once the machine is properly assembled, NSF hydrolyzes its bound ATP and is released from the complex in preparation for membrane fusion.

You want to exploit the role of NSF in the assembly of the fusion machine in order to purify SNAP receptors. As a source of SNAP receptors, you use a crude detergent extract of membranes from bovine brain. You already have available a large supply of recombinant NSF, α-SNAP, and γ-SNAP. Your purification protocol is outlined in Figure 13–10. You first mix an excess of α- and γ-SNAPs with the crude membrane extract and then add NSF in the presence of ATPγS (a nonhydrolyzable analogue of ATP). NSF and any assembled complexes are then bound to beads through a bead-attached antibody specific for NSF. The beads are then placed in a column and washed extensively in the presence of ATPγS. Finally, you add buffer containing ATP to the column, collect the proteins that come off the column, and display them by gel electrophoresis (Figure 13–11). Microsequencing of the four non-SNAP bands

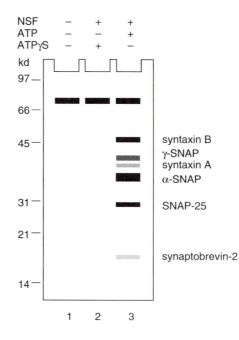

Figure 13–11 Proteins released from column upon addition of ATP (Problem 13–31).

NSF	–	+	+
ATP	–	–	+
ATPγS	–	+	–

reveals that they all had been previously identified as components of synaptic vesicles or presynaptic membranes.

A. Why did you use ATPγS initially and then collect candidate SNAP receptor proteins after addition of ATP?

B. Why did you use an excess of recombinant SNAPs instead of depending on the SNAPs that were likely to be present in the membrane extract?

C. On a molar basis the sum of all the putative SNAP receptors is approximately equal to half the sum of α- and γ-SNAPs. Is that what you expected?

D. Syntaxins A and B are part of the presynaptic plasma membrane, synaptobrevin-2 is a component of the synaptic vesicle membrane, and SNAP-25 (coincidentally named as synaptosome-associated protein of 25 kd) may also be a component of the synaptic vesicle membrane. Based on their location, which of these proteins are likely to be v-SNAREs and which ones are likely to be t-SNAREs?

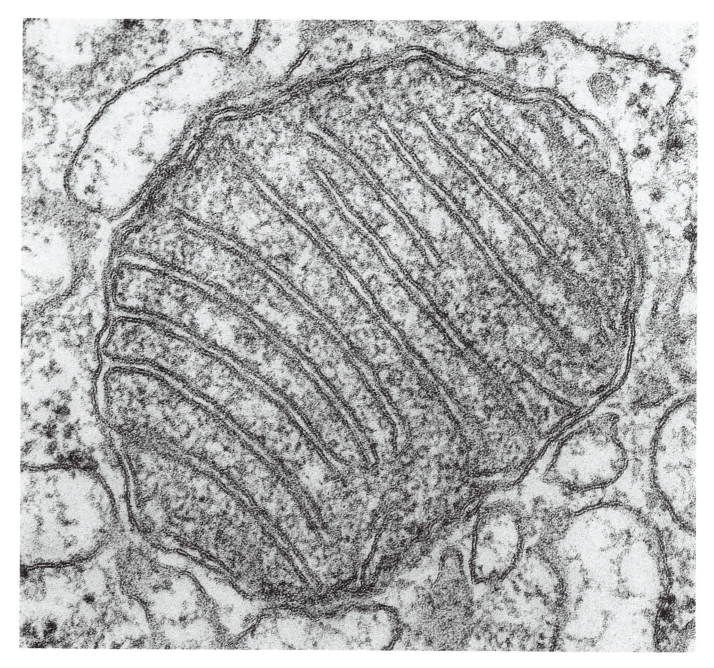

Electron micrograph of a mitochondrion. (Courtesy of Daniel S. Friend.)

Energy Conversion: Mitochondria and Chloroplasts

The Mitochondrion
(MBOC 655–672)

14–1 Fill in the blanks in the following statements.

A. The mitochondrial inner and outer membranes create two separate mitochondrial compartments: the internal _____ and a much narrower _____.

B. The mitochondrial _____ membrane resembles a sieve that is permeable to all molecules, including small proteins of 10,000 daltons or less.

C. The enzymes of the _____ are embedded in the _____ mitochondrial membrane; they are essential to the process of oxidative phosphorylation, which generates most of the animal cell's ATP.

D. The inner membrane is usually highly convoluted, forming a series of infoldings, known as _____, which greatly increase the area of the inner membrane.

E. Pyruvate and fatty acids are selectively transported from the cytosol into the mitochondrial matrix, where they are broken down into the two-carbon group on _____.

F. _____, which are composed of three molecules of fatty acid held in ester linkage to glycerol, have no charge and are virtually insoluble in water, coalescing into droplets in the cytosol.

G. The large, branched polymer of glucose that is contained in granules in the cell cytoplasm is known as _____.

H. The _____ accounts for about two-thirds of the total oxidation of carbon compounds in most cells, and its major end products are CO_2 and NADH.

I. The transfer of electrons from NADH and $FADH_2$ to oxygen releases a large amount of energy that is harnessed to drive the conversion of ADP + P_i to ATP in a process known as _____.

J. The energy released by the passage of electrons along the respiratory chain is stored as an _____ across the inner mitochondrial membrane.

K. The flow of electrons across the inner membrane generates a pH gradient and a membrane potential, which together exert a _____ force.

L. _____ synthesizes ATP from ADP and P_i in the mitochondrial matrix in a reaction that is coupled to the inward flow of protons.

M. The _____ for a reaction can be written as the sum of two parts: the first, called the _____, depends on the intrinsic character of the reacting molecules; the second depends on their concentrations.

14–2 Indicate whether the following statements are true or false. If a statement is false, explain why.

__ A. Due to the many specialized transport proteins in the inner and outer membranes, the intermembrane space and the matrix space are chemically equivalent to the cytosol with respect to small molecules.

__ B. The number of cristae is threefold greater in the mitochondria of a cardiac muscle cell than in the mitochondria of a liver cell, presumably reflecting the greater demand for ATP in heart cells.

__ C. To ensure a continuous supply of energy from oxidative metabolism, animal cells store fuel in the form of fatty acids and glucose.

__ D. The most important contribution of the citric acid cycle to metabolism is the extraction of high-energy electrons during the oxidation of the two acetyl carbon atoms to CO_2.

__ E. The energy released during transport down the respiratory chain in the mitochondrial inner membrane is used to pump protons across the inner membrane from the intermembrane space into the matrix.

__ F. Each respiratory enzyme complex in the electron-transport chain has a greater affinity for electrons than its predecessors, so that electrons pass sequentially from one complex to another until they are finally transferred to oxygen, which has the greatest affinity of all for electrons.

__ G. In a typical cell the membrane potential accounts for nearly three-quarters of the total proton-motive force across the inner membrane of a respiring mitochondrion.

__ H. The orientation of ATP synthase in the inner mitochondrial membrane is such that ATP is generated in the intermembrane space, allowing it to diffuse into the cytosol through the pores in the outer membrane.

__ I. The net change of disorder in the universe due to a reaction is reflected in the change in free energy associated with the reaction: the larger the increase in free energy (so that ΔG is very positive), the more favored the reaction.

__ J. The remarkable efficiency of cellular respiration is due primarily to the many intermediates in the oxidation pathways, which allow the huge amount of free energy released by oxidation to be parceled out into small packages.

*14–3 In 1904 Franz Knoop performed what was probably the first successful use of a labeling experiment to study metabolic pathways. He fed fatty acids labeled with a terminal benzene ring to dogs and analyzed their urine for excreted benzene derivatives. Whenever the fatty acid had an even number of carbon atoms, phenylacetic acid was excreted (Figure 14–1A). However, whenever the

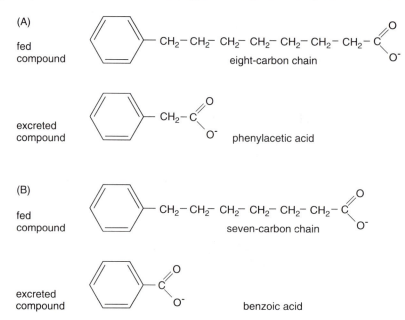

Figure 14–1 Fed and excreted derivatives of an even-chain (A) and an odd-chain (B) fatty acid (Problem 14–3).

*Problems with an asterisk (*) are answered in the Instructor's Manual.

160   **Chapter 14 : Energy Conversion: Mitochondria and Chloroplasts**

fatty acid had an odd number of carbon atoms, benzoic acid was excreted (Figure 14–1B). (Actually in both experiments the excreted compounds were found esterified to a sugar, which helped solubilize them; these modifications are irrelevant to the metabolism of the fatty acids.)

From these experiments Knoop deduced that oxidation of fatty acids to CO_2 and H_2O involved removal of two-carbon fragments from the carboxylic acid end of the chain. Can you explain the reasoning that led him to conclude that two-carbon fragments, as opposed to any other number, were removed, and that degradation was from the carboxylic acid end, as opposed to the other end?

14–4   In 1937 Hans Krebs deduced the operation of the citric acid cycle from careful observations on the oxidation of carbon compounds in minced preparations of pigeon flight muscle. (Pigeon breast is a rich source of mitochondria, but the function of mitochondria was unknown at the time.) The consumption of O_2 and the production of CO_2 were monitored with a manometer, which measures changes in volume of a closed system at constant pressure and temperature. Standard chemical methods were used to determine the concentrations of key metabolites. (Remember, radioactive isotopes were not available then.)

In one set of experiments Krebs measured the rate of consumption of O_2 during the oxidation of endogenous carbohydrates in the presence or absence of citrate. As shown in Table 14–1, addition of a small amount of citrate resulted in a large increase in the consumption of oxygen. Szent-Gyorgyi (1925) and Stare and Baumann (1936) had previously shown that fumarate, oxaloacetate, and succinate also stimulated respiration in extracts of pigeon breast muscle.

When metabolic poisons, such as arsenite or malonate (whose modes of action were undefined), were added to the minced muscles, the results were much different. In the presence of arsenite, 5.5 mmol of citrate were converted into about 5 mmol of α-ketoglutarate. In the presence of malonate an equivalent conversion of citrate into succinate occurred. Furthermore, in the presence of malonate roughly 5 mmol of oxygen were consumed (above background levels in the absence of citrate), which was twice as much as in the presence of arsenite.

Finally, Krebs showed that the minced muscles were actually capable of synthesizing citrate if oxaloacetate was added and all traces of oxygen were excluded. None of the other intermediates in the cycle led to a net synthesis of citrate in the absence of oxygen.

A. If citrate ($C_6H_8O_7$) were completely oxidized to CO_2 and H_2O, how many molecules of O_2 would be consumed per molecule of citrate? What is it about the results in Table 14–1 that caught Krebs's attention?

B. Why is the consumption of oxygen so low in the presence of arsenite and malonate? If citrate were oxidized to α-ketoglutarate ($C_5H_6O_5$), how much

Table 14–1 Respiration in Minced Pigeon Breast in the Presence and Absence of Citrate (Problem 14–4)

Time (minutes)	Oxygen Consumption (mmol)		
	No Citrate	3 mmol Citrate	Difference
30	29	31	2
60	47	68	21
90	51	87	36
150	53	93	40

oxygen would be consumed per molecule of citrate? If citrate were oxidized to succinate ($C_4H_6O_4$), how much oxygen would be consumed per molecule of citrate? Does the observed stoichiometry agree with the expectations based on these calculations?

C. Why in the absence of oxygen does oxaloacetate alone cause an accumulation of citrate? Would any of the other intermediates in the cycle cause an accumulation of citrate in the presence of oxygen?

D. Toward the end of the paper Krebs states, "While the citric acid cycle thus seems to occur generally in animal tissues, it does not exist in yeast or in *E. coli*, for yeast and *E. coli* do not oxidize citric acid at an appreciable rate." Why do you suppose Krebs got this point wrong?

14–5   The relationship of free-energy change (ΔG) to the concentrations of reactants and products is important because it predicts the direction of spontaneous chemical reactions. Familiarity with this relationship is essential for understanding energy conversions in cells. Consider, for example, the hydrolysis of ATP to ADP and inorganic phosphate (P_i).

$$ATP + H_2O \rightarrow ADP + P_i$$

The free-energy change due to ATP hydrolysis is

$$\Delta G = \Delta G° + RT \ \ln \frac{[ADP][P_i]}{[ATP]}$$
$$= \Delta G° + 2.3RT \log_{10} \frac{[ADP][P_i]}{[ATP]}$$

where the concentrations are expressed as molarities (by convention, the concentration of water is not included in the expression). R is the gas constant (1.98×10^{-3} kcal/°K mole), T is temperature (assume 37°C, which is 310°K), and $\Delta G°$ is the standard free-energy change (–7.3 kcal/mole for ATP hydrolysis to ADP and P_i).

A. Calculate the ΔG for ATP hydrolysis when the concentrations of ATP, ADP, and P_i are all equal to 1 M. What is the ΔG when the concentrations of ATP, ADP, and P_i are all equal to 1 mM?

B. In a resting muscle, the concentrations of ATP, ADP, and P_i are approximately 5 mM, 1 mM, and 10 mM, respectively. What is ΔG for ATP hydrolysis in resting muscle?

C. What will ΔG equal when the hydrolysis reaction reaches equilibrium? At $[P_i]$ = 10 mM, what will be the ratio of [ATP] to [ADP] at equilibrium?

D. Show that, at constant $[P_i]$, ΔG decreases by 1.4 kcal/mole for every tenfold increase in the ratio of [ATP] to [ADP], regardless of the value of $\Delta G°$. (For example, ΔG decreases by 2.8 kcal/mole for a 100-fold change, by 4.2 kcal/mole for a 1000-fold change, and so on.)

*14–6   One of the two ATP generating steps in glycolysis is outlined in Figure 14–2. This sequence of reactions yields ATP and produces pyruvate, which is subsequently converted to acetyl CoA and oxidized to CO_2 in the citric acid cycle. Under anaerobic conditions ATP production from phosphoenolpyruvate accounts for half of a cell's ATP supply. These "substrate-level" phosphorylation events (so named to distinguish them from oxidative phosphorylation in the mitochondrion) were the first to be understood. Consider the conversion of 3-phosphoglycerate to phosphoenolpyruvate, which constitutes the first two reactions in Figure 14–2.

The equation relating $\Delta G°$ to the equilibrium constant is

$$\Delta G° = -RT \ln K = -2.3 \ RT \log_{10} K$$

where K is the equilibrium ratio of the products over the reactants.

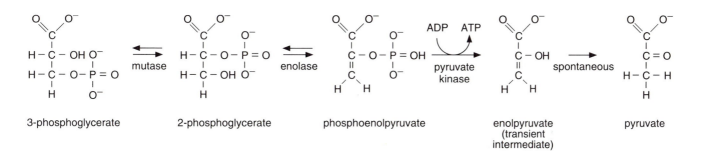

Figure 14–2 Conversion of 3-phospho-glycerate to pyruvate during glycolysis (Problem 14–6).

A. If 10 mM 3-phosphoglycerate is mixed with phosphoglycerate mutase, which catalyzes its conversion to 2-phosphoglycerate (Figure 14–2), the equilibrium concentrations at 37°C are 8.3 mM 3-phosphoglycerate and 1.7 mM 2-phosphoglycerate. How would the ratio of equilibrium concentrations change if 1 M 3-phosphoglycerate had been added initially? What is the equilibrium constant (K) for the conversion of 3-phosphoglycerate into 2-phosphoglycerate, and what is the $\Delta G°$ for the reaction?

B. If 10 mM phosphoenolpyruvate is mixed with enolase, which catalyzes its conversion to 2-phosphoglycerate, the equilibrium concentrations at 37°C are 2.9 mM 2-phosphoglycerate and 7.1 mM phosphoenolpyruvate. What is the $\Delta G°$ for conversion of phosphoenolpyruvate to 2-phosphoglycerate? What is the $\Delta G°$ for the reverse reaction?

C. What is the $\Delta G°$ for the conversion of 3-phosphoglycerate to phosphoenolpyruvate? ($\Delta G°$ for the overall reaction is the sum of the $\Delta G°$ values for the linked reactions.)

14–7   The study of substrate-level phosphorylation events such as that in Figure 14–2 led to the concept of the "high-energy" phosphate bond. The term "high-energy" bond is somewhat misleading since it refers not to the strength of the bond, but, rather, to the free-energy change (ΔG) upon hydrolysis. In the conversion of 3-phosphoglycerate to phosphoenolpyruvate (the first two reactions in Figure 14–2) a low-energy phosphate bond is turned into a high-energy phosphate bond. The standard free-energy change ($\Delta G°$) of hydrolysis of the phosphate group in 3-phosphoglycerate is about –3.3 kcal/mole, whereas hydrolysis of the phosphate group in phosphoenolpyruvate (converting it to pyruvate) has a $\Delta G°$ of –14.8 kcal/mole. How is it that moving the phosphate to the 2 position of glycerate and removing water, which has an overall $\Delta G°$ of 0.4 kcal/mole, can have such an enormous effect on the free energy for the subsequent hydrolysis of the phosphate bond?

Consider the set of reactions shown in Figure 14–3.

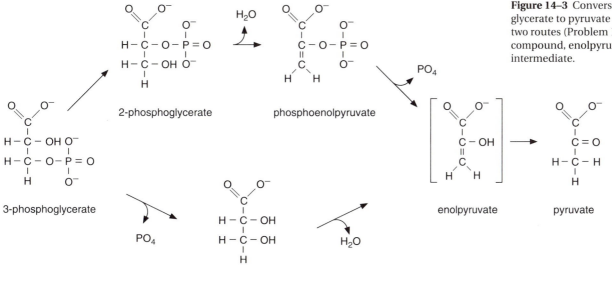

Figure 14–3 Conversion of 3-phospho-glycerate to pyruvate and phosphate by two routes (Problem 14–7). The bracketed compound, enolpyruvate, is a transient intermediate.

A. What is $\Delta G°$ for conversion of 3-phosphoglycerate to pyruvate by way of phosphoenolpyruvate (PEP)? (Remember that $\Delta G°$ for the overall reaction is the sum of the $\Delta G°$ values of the individual steps.)

B. What is $\Delta G°$ for conversion of 3-phosphoglycerate to pyruvate by way of glycerate? What is $\Delta G°$ for conversion of glycerate to pyruvate? (It will help to remember that thermodynamic quantities are state functions; that is, they describe differences between the initial and final states—they are independent of the pathway between the states.)

C. Propose an explanation for why the phosphate bond in phosphoenolpyruvate is a high-energy bond, whereas that in 3-phosphoglycerate is a low-energy bond. (Assume that removal of water from glycerate has a $\Delta G°$ of –0.5 kcal/mole.)

D. In cells the conversion of phosphoenolpyruvate to pyruvate is linked to the synthesis of ATP as shown in Figure 14–2. What is $\Delta G°$ for the linked reactions?

*14–8   The ADP-ATP antiporter in the mitochondrial inner membrane can exchange ATP for ATP, ADP for ADP, and ATP for ADP. Even though mitochondria can transport both ADP and ATP, there is a strong bias in favor of exchange of external ADP for internal ATP under phosphorylating conditions. You suspect that this bias is due to the conversion of ADP into ATP inside the mitochondrion. ATP synthesis would continually reduce the internal concentration of ADP and thereby create a favorable concentration gradient for import of ADP. The same process would increase the internal concentration of ATP, thereby creating favorable conditions for export of ATP.

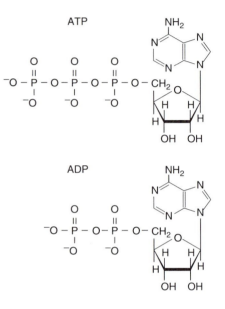

Figure 14–4 Structures of ATP and ADP (Problem 14–8).

To test your hypothesis, you conduct experiments on isolated mitochondria. In the absence of substrate (when the mitochondria are not respiring and the membrane is uncharged), you find that ADP and ATP are taken up at the same rate. When you add substrate, the mitochondria begin to respire, and ADP enters mitochondria at a much faster rate than ATP. As you expected, when you add an uncoupler (dinitrophenol, which collapses the pH gradient) along with the substrate, ADP and ATP enter at the same rate. However, when you add an inhibitor of ATP synthase (oligomycin) along with the substrate, ADP is taken up much faster than ATP. Your results are summarized in Table 14–2. You are puzzled by the results with oligomycin, since your hypothesis predicted that the rates of uptake would be equal.

When you show the results to your advisor, she compliments you on your fine experiments and agrees that they disprove the hypothesis. She suggests that you examine the structures of ATP and ADP (Figure 14–4) if you wish to understand the behavior of the antiporter. What is the correct explanation for the biased exchange by the ADP-ATP antiporter under some of the experimental conditions and an unbiased exchange under others?

Table 14–2 Entry of ADP and ATP into Isolated Mitochondria (Problem 14–8)

Experiment	Substrate	Inhibitor	Relative Rates of Entry
1	absent	none	ADP = ATP
2	present	none	ADP > ATP
3	present	dinitrophenol	ADP = ATP
4	present	oligomycin	ADP > ATP

Note: In all cases the initial rates of entry of ATP and ADP were measured.

The Respiratory Chain and ATP Synthase

(MBOC 672–684)

14–9   Fill in the blanks in the following statements.

A. The F_1ATPase is part of a larger transmembrane complex containing at least nine different polypeptide chains, which is now known as _____.

B. The _____ constitute a family of colored proteins that are related by the presence of a bound heme group, whose iron atom changes from the ferric to the ferrous state whenever it accepts an electron.

C. Proteins from a second major family of electron carriers have either two or four iron atoms bound to an equal number of sulfur atoms and to cysteine side chains, forming an _____ on the protein.

D. The simplest of the electron carriers is a small hydrophobic molecule known as ubiquinone, which, because it is a _____, can pick up or donate either one or two electrons at a time.

E. Mild ionic detergents, which solubilize selected components of the mitochondrial inner membrane in their native form, permitted identification and purification of the three major membrane-bound _____ in the pathway from NADH to oxygen.

F. The _____ accepts electrons from NADH and passes them through a flavin and at least five iron-sulfur proteins to ubiquinone.

G. The _____ accepts electrons from ubiquinone and passes them on to cytochrome c.

H. The _____ accepts electrons from cytochrome c and passes them to oxygen.

I. Pairs of compounds such as NADH and NAD^+ are called _____, since one compound is converted to the other by the addition of one or more electrons plus one or more protons.

J. A 50-50 mixture of NADH and NAD^+ maintains a defined "electron pressure," or _____, that is a measure of the electron carrier's affinity for electrons.

K. The direct inhibitory influence of the electrochemical proton gradient on the rate of electron transport is known as _____.

14–10   Indicate whether the following statements are true or false. If a statement is false, explain why.

__ A. The inside-out nature of submitochondrial particles was important for purification of the proteins responsible for oxidative phosphorylation because the particles can be readily provided with the membrane-impermeant metabolites that would normally be present in the matrix space.

__ B. If ATP synthase and bacteriorhodopsin, which is a light-driven proton pump, are incorporated into lipid vesicles, exposure of the vesicles to light will cause ATP to be made.

__ C. Although purified ATP synthase will hydrolyze ATP to ADP and P_i, the normal membrane-bound form in the mitochondrion is tightly regulated so that it will only synthesize ATP.

__ D. If the flow of protons through ATP synthase is blocked, the injection of a small amount of oxygen into an anaerobic preparation of submitochondrial particles will result in a burst of respiration that will cause the medium to become more basic.

__ E. The proteins that constitute the respiratory chain all use iron atoms as electron carriers.

__ F. The toxicity of the poisons cyanide and azide is due to their ability to bind tightly to the cytochrome oxidase complex and thereby block all electron transport.

___ G. The three respiratory enzyme complexes exist in structurally ordered arrays in the plane of the inner membrane to facilitate the correct transfer of electrons between appropriate complexes.

___ H. Since most cytochromes have a higher redox potential than iron-sulfur centers, the cytochromes tend to serve as electron carriers near the O_2 end of the respiratory chain.

___ I. The molecular mechanism by which electron transport is coupled to proton pumping is likely to be different for different respiratory enzyme complexes.

___ J. Lipophilic weak acids short-circuit the normal flow of protons across the inner membrane, thereby dissipating the proton-motive force, stopping ATP synthesis, and blocking the flow of electrons.

___ K. If a very large electrochemical gradient is imposed across the inner membrane, a reverse electron flow can be detected in some sections of the respiratory chain.

___ L. In brown fat cells mitochondrial respiration is partially uncoupled from ATP synthesis and much of the energy of oxidation is dissipated as heat.

___ M. Most bacteria, including strict anaerobes, maintain a proton-motive force across their plasma membrane, which is used to drive the flagellar motor and a variety of active transport processes.

14–11 The electrochemical proton gradient is undoubtedly the energy source for ATP synthesis during oxidative phosphorylation; however, the molecular mechanism by which the gradient is coupled to ATP synthesis remains to be determined. Is ATP synthesized directly from ADP and inorganic phosphate or is the phosphate transferred from an intermediate source, such as a phosphoenzyme or some other phosphorylated compound?

One elegant approach to this question analyzed the stereochemistry of the reaction mechanism. As is often done, the investigators studied the reverse reaction (ATP hydrolysis into ADP and phosphate) to gain an understanding of the forward reaction. (A fundamental principle of enzyme catalysis is that the forward and reverse reactions are precisely the reverse of one another.) All enzyme-catalyzed phosphate transfers occur with inversion of configuration about the phosphate atom; thus one-step mechanisms, in which the residue is transferred directly between substrates, result in inversion of the final product (Figure 14–5A).

To analyze the stereochemistry of ATP hydrolysis, the investigators first generated a version of ATP with three distinct atoms (S, ^{16}O, and ^{18}O) attached stereospecifically to the terminal phosphorus atom (Figure 14–5B). This compound was then hydrolyzed to ADP and phosphate by purified ATP synthase in the presence of H_2O that was enriched for ^{17}O. The resulting phosphate was analyzed by NMR to determine whether the configuration about the phosphorus atom had been inverted or retained (Figure 14–5B).

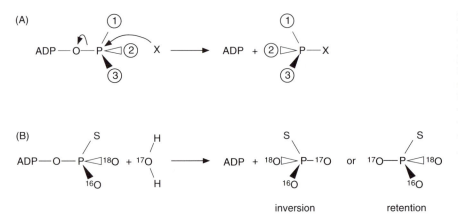

Figure 14–5 Stereochemistry of phosphate transfer reactions (Problem 14–11). (A) Inversion of configuration by a one-step phosphate transfer reaction. (B) Experimental setup for assaying stereochemistry of ATP synthesis by ATP synthase. Thin bonds are in the plane of the page; thick white bonds point behind the page; and thick black bonds project out of the page. Oxygen atoms are indicated by their atomic number.

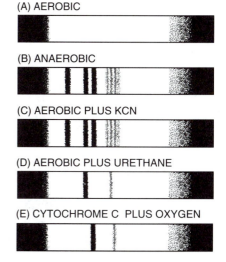

Figure 14–6 Cytochrome absorption bands (Problem 14–12). Numbers are the wavelenghts of light in nanometers.

700   600   500   450
red   blue

a   b c

cytochrome absorption bands

A. How does this experiment distinguish between synthesis of ATP directly from ADP and phosphate and synthesis of ATP through an intermediate phosphorylated substance?

B. The investigators' analysis showed that the configuration had been inverted. Does this result support direct synthesis of ATP or synthesis of ATP through a phosphorylated intermediate?

14–12   In 1925 David Keilin used a simple spectroscope to observe the characteristic absorption bands of the cytochromes that make up the electron-transport chain in mitochondria. A spectroscope passes a very bright light through the sample of interest and then through a prism to display the spectrum from red to blue. If there are molecules in the sample that absorb light of particular wavelengths, dark bands interrupt the colors of the rainbow. Keilin found that tissues from a wide variety of animals all showed the pattern in Figure 14–6. (This pattern had actually been observed several decades before by an Irish physician named MacMunn, but he thought all the bands were due to a single pigment. His work was all but forgotten by the 1920s.)

The different heat stabilities of the individual absorption bands and their different intensities in different tissues led Keilin to conclude that the absorption pattern was due to three components, which he labeled cytochromes a, b, and c (Figure 14–6). His key discovery was that the absorption bands disappeared when oxygen was introduced (Figure 14–7A) and then reappeared when the samples became anoxic (Figure 14–7B). He later confessed, "This visual perception of an intracellular respiratory process was one of the most impressive spectacles I have witnessed in the course of my work."

Keilin subsequently discovered that cyanide prevented the bands from disappearing when oxygen was introduced (Figure 14–7C). When urethane (a no longer used inhibitor of electron transport) was added, bands a and c disappeared in the presence of oxygen, but band b remained (Figure 14–7D). Finally, using cytochrome c extracted from dried yeast with water, he showed that the band due to cytochrome c remained when oxygen was present (Figure 14–7E).

A. Is it the reduced (electron-rich) or the oxidized (electron-poor) forms of the cytochromes that give rise to the bands Keilin observed?

B. From Keilin's observations, deduce the order in which the three cytochromes carry electrons from intracellular substrates to oxygen.

C. One of Keilin's early observations was that the presence of excess glucose prevented the disappearance of the absorption bands when oxygen was added. How do you think that rapid glucose oxidation to CO_2 might explain this observation?

*14–13   During operation of the respiratory chain, cytochrome c accepts electrons from the cytochrome b-c_1 complex and transfers them to the cytochrome oxidase complex. What is the relationship of cytochrome c to the two complexes it connects? Is cytochrome c linked simultaneously to both complexes like a wire, allowing electrons to flow through it from one complex to the other, or does cytochrome c move between the complexes like a ferry, picking up electrons from one and handing them over to the next? One source of information about this question is provided by the experiments described below.

(A) AEROBIC

(B) ANAEROBIC

(C) AEROBIC PLUS KCN

(D) AEROBIC PLUS URETHANE

(E) CYTOCHROME C PLUS OXYGEN

Figure 14–7 Cytochrome absorption bands under a variety of experimental conditions (Problem 14–12).

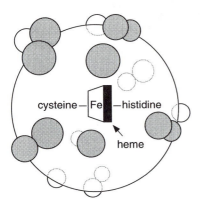

(A) FROM CYTOCHROME b-c₁ COMPLEX

lysines affecting transfer of electrons from the cytochrome b-c₁ complex to cytochrome c

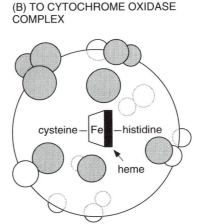

(B) TO CYTOCHROME OXIDASE COMPLEX

lysines affecting transfer of electrons from cytochrome c to the cytochrome b-c₁ complex

Figure 14–8 Positions of lysines that inhibit the transfer of electrons from (A) the cytochrome b-c₁ complex to cytochrome c and (B) from cytochrome c to the cytochrome oxidase complex (Problem 14–13). One edge of the heme group protrudes from the front surface of cytochrome c toward the reader. Circles show the positions of lysines. Solid circles are on the front surface of the molecule; faint circles are on the back surface. The larger the circle, the closer it is to the reader. Shaded circles indicate lysines that when modified inhibit electron transfer; open circles indicate lysines that when modified do not inhibit electron transfer.

Individual lysines on cytochrome c were modified to replace the positively charged amino group with a neutral group or with a negatively charged group. The effects of these modifications on electron transfer from the cytochrome b-c₁ complex to cytochrome c and from cytochrome c to the cytochrome oxidase complex were then measured. Some modifications had no effect (Figure 14–8, open circles), whereas others inhibited electron transfer from the cytochrome b-c₁ complex to cytochrome c (Figure 14–8A, shaded circles) or from cytochrome c to the cytochrome oxidase complex (Figure 14–8B, shaded circles). In an independent series of experiments, several lysines that inhibited transfer of electrons to and from cytochrome c were shown to be protected from acetylation when cytochrome c was bound either to the cytochrome b-c₁ complex or to the cytochrome oxidase complex.

How do these results distinguish the possibilities that cytochrome c connects the two complexes by binding to them simultaneously or by shuttling between the complexes?

14–14   How many molecules of ATP are formed from ADP + P$_i$ when a pair of electrons from NADH are passed down the electron-transport chain to oxygen? Is the number an integer or not? These deceptively simple questions are difficult to answer from purely theoretical considerations, but they can be measured directly with an oxygen electrode, as illustrated in Figure 14–9. At the indicated time, a suspension of mitochondria was added to a phosphate-buffered solution containing β-hydroxybutyrate, which can be oxidized by mitochondria to generate NADH + H⁺. After an initial rapid burst, oxygen consumption slowed to a background rate. When the rate of oxygen consumption stabilized, 500 nmol of ADP were added, causing a rapid increase in the rate of consumption until all the ADP had been converted into ATP, at which point the rate of oxygen consumption again slowed to the background rate.

A.   Why did the rate of oxygen consumption increase dramatically over the background rate when ADP was added; why did it return to the background rate when all the ADP had been converted to ATP?

B.   Why do you think it is that mitochondria consume oxygen at a slow background rate in the absence of added ADP?

C.   How many ATP molecules were synthesized per pair of electrons transferred down the electron-transport chain to oxygen (P/2e⁻ ratio)? How many ATP molecules were synthesized per oxygen atom consumed (P/O ratio)? (Remember that ½ O₂ + 2e⁻ → H₂O.)

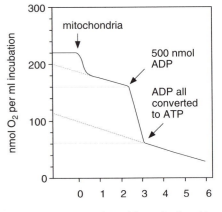

minutes after adding mitochondria

Figure 14–9 Consumption of oxygen by mitochondria under various experimental conditions (Problem 14–14).

D. In experiments like this one, what processes in addition to ATP production are driven by the electrochemical proton gradient?

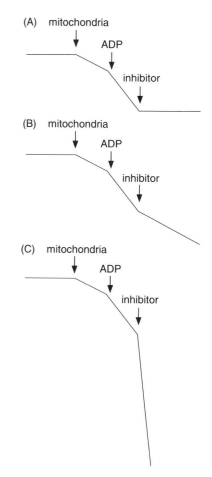

*14–15 Inhibitors of mitochondrial function have provided extremely useful tools for analyzing how mitochondria work. Figure 14–10 shows three distinct patterns of oxygen electrode traces obtained using a variety of inhibitors. In all experiments mitochondria were added to a phosphate-buffered solution containing succinate as the sole source of electrons for the respiratory chain. After a short interval ADP and then an inhibitor were added, as indicated in Figure 14–10. The rates of oxygen consumption at various times during the experiment are shown by downward sloping lines, with faster rates shown by steeper lines.

A. From the description of the inhibitors in the list that follows, assign each inhibitor to one of the oxygen traces in Figure 14–10. All these inhibitors stop ATP synthesis.

Inhibitor	Function
1. FCCP	Makes membranes permeable to protons
2. Malonate	Prevents oxidation of succinate
3. Cyanide	Inhibits cytochrome oxidase
4. Atractylate	Inhibits the ADP-ATP antiporter
5. Oligomycin	Inhibits ATP synthase
6. Butylmalonate	Blocks mitochondrial uptake of succinate

B. Using the same experimental protocol indicated in Figure 14–10, sketch the oxygen traces that you would expect for the sequential addition of the pairs of inhibitors in the list below.

1. FCCP followed by cyanide
2. FCCP followed by oligomycin
3. Oligomycin followed by FCCP

Figure 14–10 Oxygen traces showing three patterns of inhibitor effects on oxygen consumption by mitochondria (Problem 14–15).

*14–16 Methanogenic bacteria produce methane as the end product of electron transport. For example, *Methanosarcina barkeri*, when grown under hydrogen in the presence of methanol, transfers electrons from hydrogen to methanol, producing methane and water:

$$CH_3OH + H_2 \rightarrow CH_4 + H_2O$$

This reaction is analogous to those used by aerobic bacteria and mitochondria, which transfer electrons from carbon compounds to oxygen, producing water and carbon dioxide. However, given the peculiar biochemistry involved in methanogenesis, it was initially unclear whether methanogenic bacteria synthesized ATP by electron-transport-driven phosphorylation or by substrate-level phosphorylation.

In the experiments depicted in Figure 14–11 methanol is added to a culture of methanogenic bacteria grown under hydrogen, and the production of CH_4, the size of the electrochemical proton gradient (ΔG_{H+}), and the intracellular concentration of ATP are assayed. The effects of the addition of two types of inhibitor are measured: TCS, which dissipates the electrochemical proton gradient, and DCCD, which directly inhibits ATP synthase. In all the experiments, addition of methanol produces an increase in the electrochemical proton gradient and an increase in intracellular ATP. Addition of TCS (Figure 14–11A) dissipates the electrochemical proton gradient and stops ATP synthesis; however, it does not stop CH_4 production. Addition of DCCD (Figure 14–11B) stops ATP synthesis and inhibits CH_4 production, but it does not significantly affect the electrochemical proton gradient. Addition of TCS after ATP synthesis and CH_4 production have been stopped by DCCD (Figure 14–11C) dissipates the electrochemical proton gradient but stimulates production of CH_4.

A. Would you expect 2,4-dinitrophenol, which dissipates the electrochemical proton gradient in mitochondria, to have an effect on mitochondria analogous to the effect of TCS on methanogens, that is, stopping ATP production but not blocking production of CO_2? Why?

B. Would you expect oligomycin, which inhibits ATP synthase in mitochondria, to have an effect on mitochondria analogous to the effect of DCCD on methanogens, that is, blocking production of CO_2 without affecting the electrochemical proton gradient? Why?

C. Would you expect addition of 2,4-dinitrophenol to oligomycin-inihibited mitochondria to stimulate production of CO_2 in a manner analogous to the stimulation of CH_4 production by addition of TCS to DCCD-inhibited methanogens? Why?

D. Do methanogenic bacteria generate ATP by electron-transport-driven phosphorylation or by substrate-level phosphorylation?

14–17 The electrochemical proton gradient is responsible not only for ATP production in bacteria, mitochondria, and chloroplasts, but also for powering bacterial flagella. The flagellar motor is thought to be driven directly by the flux of protons through it. To test this idea, you analyze a motile strain of *Streptococcus*. These bacteria swim when glucose is available for oxidation, but they do not swim when glucose is absent (and no other substrate is available for oxidation). Using a series of ionophores that alter the pH gradient or the membrane potential (the two components of the electrochemical proton gradient), you make several observations.

1. Bacteria that are swimming in the presence of glucose stop swimming upon addition of the proton ionophore FCCP.
2. Bacteria that are swimming in the presence of glucose in a medium containing K^+ are unaffected by addition of the K^+ ionophore valinomycin.
3. Bacteria that are motionless in the absence of glucose in a medium containing K^+ remain motionless upon addition of valinomycin.
4. Bacteria that are motionless in the absence of glucose in a medium containing Na^+ swim briefly upon addition of valinomycin and then stop.

These observations are summarized in Table 14–3.

A. Explain how each of these observations is consistent with the idea that the flagellar motor is driven by a flux of protons. (The concentration of K^+ inside these bacteria is lower than the concentration of K^+ used in the medium.)

B. Wild-type bacteria can swim in the presence or absence of oxygen. However, mutant bacteria that are missing ATP synthase, which couples proton flow to ATP production, can swim only in the presence of oxygen. How are normal bacteria able to swim in the absence of oxygen when there is no electron flow? How do you think the loss of the ATP synthase prevents swimming in mutant bacteria?

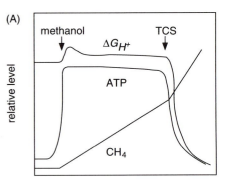

(A)

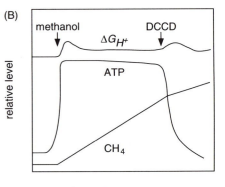

(B)

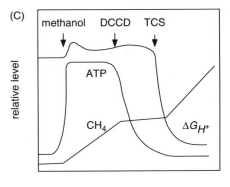

(C)

Figure 14–11 Time course of CH_4 production after addition of methanol to a culture of methanogenic bacteria growing under H_2 (Problem 14–16). The magnitude of the electrochemical proton gradient (ΔG_{H+}), the intracellular ATP concentration, and the production of CH_4 are expressed in arbitrary units.

Table 14–3 Effects of Ionophores on the Swimming of Normal Bacteria (Problem 14–17)

Observation	Glucose	Ionophore	Ion in Medium	Effect on Bacteria
1	present	FCCP (H^+)	—	stop swimming
2	present	valinomycin (K^+)	K^+	keep swimming
3	absent	valinomycin (K^+)	K^+	remain motionless
4	absent	valinomycin (K^+)	Na^+	swim briefly

Note: The specificity of each ionophore is shown in parentheses.

Chloroplasts and Photosynthesis

(MBOC 684–697)

14–18 Fill in the blanks in the following statements.

A. Chloroplasts are the most prominent member of the _____ family of organelles.

B. The inner chloroplast membrane surrounds a large central space called the _____, which is analogous to the mitochondrial matrix.

C. The photosynthetic light-absorbing system, the electron-transport chain, and an ATP synthase are all contained in a set of flattened disclike sacs, called _____.

D. The many reactions that occur in photosynthesis can be grouped into two broad categories: the _____ reactions and the _____ reactions.

E. The conversion of CO_2 into carbohydrate occurs by a cycle of reactions, which is called the _____ cycle.

F. The disaccharide _____ is the major form in which sugar is transported between plant cells, acting in plant cells as glucose acts in animal cells.

G. Like glycogen in animal cells, _____ is a large polymer of glucose that serves as a carbohydrate reserve.

H. Plants that pump CO_2 are called _____ plants; all others are called _____ plants.

I. The energy required to drive photosynthetic electron transport is derived from sunlight that is absorbed by _____ molecules.

J. A photosystem consists of two closely linked components: an _____, which is important for light harvesting, and a _____, which transfers excited electrons to a chain of electron acceptors.

K. Photosynthesis in plants and cyanobacteria produces both ATP and NADPH directly by a two-step process called _____.

L. The two electron-energizing steps catalyzed by photosystems I and II are linked together to form the _____ of photosynthesis.

M. During _____ chloroplasts switch photosystem I into a cyclic mode of operation in which its energy is directed into the synthesis of ATP instead of NADPH.

14–19 Indicate whether the following statements are true or false. If a statement is false, explain why.

___ A. In a general way, one might view the chloroplast as a greatly enlarged mitochondrion in which the cristae are condensed into a series of interconnected submitochondrial particles in the matrix space.

___ B. The conversion of CO_2 to carbohydrate requires light energy directly, whereas the formation of O_2 requires light energy only indirectly.

___ C. In the central reaction of carbon fixation, CO_2 from the atmosphere combines with the five-carbon compound ribulose 1,5-bisphosphate to give two molecules of the three-carbon compound 3-phosphoglycerate.

___ D. Both phosphate-bond energy (ATP) and reducing power (NADPH) are required for formation of organic molecules from CO_2 and H_2O.

___ E. To avoid the waste of photorespiration, many plants in hot, dry climates "pump" CO_2 into bundle-sheath cells to provide ribulose bisphosphate carboxylase with a high concentration of CO_2.

___ F. The process of energy conversion begins when a chlorophyll molecule is excited by a quantum of light and an electron is moved from one molecular orbital to another of higher energy.

___ G. When a molecule of chlorophyll in an antenna complex absorbs a photon, the excited electron is rapidly transferred from one molecule to another until it reaches the photochemical reaction center.

___ H. A photosystem enables light to activate a net electron transfer from a molecule such as a cytochrome, which is a weak electron donor, to a molecule such as quinone, which is a strong electron donor in its reduced form.

___ I. The purple baterium uses its photochemical reaction center to generate an electrochemical proton gradient across the plasma membrane, which is used to drive ATP synthesis and to drive a reverse electron flow to produce NADH.

___ J. The linking of two photosystems into the Z scheme of photosynthesis allows two electrons from H_2O to be energized sufficiently by two photons to reduce $NADP^+$ to NADPH.

___ K. The balance between noncyclic photophosphorylation, which generates ATP and NADPH, and cyclic photophosphorylation, which generates only NADPH, is regulated according to the need for ATP.

___ L. Intact thylakoid discs resemble submitochondrial particles in having a membrane whose electron-transport chain has its $NADP^+$-, ADP-, and phosphate-utilizing sites all freely accessible to the outside.

___ M. The export of glyceraldehyde 3-phosphate from the chloroplast provides not only the main source of fixed carbon to the rest of the cell, but also reducing power and ATP needed for other biosynthetic reactions in the cytosol.

14–20   Recalling Joseph Priestley's famous experiment in which a sprig of mint saved the life of a mouse in a sealed chamber, you decide to do an analogous experiment to see how a C_3 and a C_4 plant do when confined together in a sealed environment. You place a corn plant (C_4) and a geranium (C_3) in a sealed plastic chamber with normal air (300 parts per million CO_2) on a windowsill in your laboratory. What happens to the two plants? Do they compete or collaborate? If they compete, which one wins and why?

*14–21   How much energy is available in visible light? How much energy does sunlight deliver to the earth? How efficient are plants at converting light energy into chemical energy? The answers to these questions provide an important backdrop to the subject of photosynthesis

Each quantum or photon of light has an energy of hv, where h is Planck's constant (1.58×10^{-37} kcal sec/photon) and v is the frequency in sec^{-1}. The frequency of light is equal to c/λ, where c is the speed of light (3.0×10^{17} nm/ sec) and λ is the wavelength in nm. Thus, the energy ($\mathscr{E}$) of a photon is

$$\mathscr{E} = hv = hc/\lambda$$

A. Calculate the energy of a mole of photons (6×10^{23} photons/mole) at 400 nm (violet light), at 680 nm (red light), and at 800 nm (near infrared light).

B. Bright sunlight strikes the earth at the rate of about 0.3 kcal/sec per square meter. Assuming for the sake of calculation that sunlight consists of monochromatic light of wavelength 680 nm, how many seconds does it take for a mole of photons to strike a square meter?

C. Assuming that it takes 8 photons to fix 1 molecule of CO_2 as carbohydrate under optimal conditions (8 to 10 photons is the currently accepted value), calculate how long it would take a tomato plant with a leaf area of 1 square meter to make a mole of glucose from CO_2. Assume that photons strike the leaf at the rate calculated above and, furthermore, that all the photons are absorbed and used to fix CO_2.

D. If it takes 112 kcal/mole to fix a mole of CO_2 into carbohydrate, what is the efficiency of conversion of light energy into chemical energy after photon capture? Assume again that 8 photons of red light (680 nm) are required to fix 1 molecule of CO_2.

14–22   Recently, your boss has expanded the lab's interest in photosynthesis from algae to higher plants. You have been assigned to study photosynthetic carbon fixation in the cactus, but so far you have had no success: cactus plants do not seem of fix $^{14}CO_2$, even in direct sunlight. Under the same conditions your colleagues studying dandelions get excellent incorporation of $^{14}CO_2$ within seconds of adding it and are busily charting new biochemical pathways.

Depressed, you leave the lab one day without dismantling the labeling chamber. The following morning you remove the cactus and, much to your surprise, find it has incorporated a great deal of $^{14}CO_2$. Evidently, the cactus fixed carbon during the night. When you repeat your experiments at night in complete darkness, you find that cactus plants incorporate label splendidly.

Although you are forced to shift your work habits (and your spouse is suspicious), at least now you can make some progress. A brief exposure to $^{14}CO_2$ labels one compound—malate—almost exclusively. During the night labeled malate builds up to very high levels in specialized vacuoles inside chloroplast-containing cells. In addition, the starch in these same cells disappears. During the day the malate disappears, and labeled starch is formed in a process that requires light. Furthermore, you find that $^{14}CO_2$ reappears in these cells during the day.

These results remind you in some ways of CO_2 pumping in C_4 plants, but they are quite distinct in other ways.

A. Why is light required for starch formation in the cactus? Is it required for starch formation in C_4 plants?
B. Using the reactions of the CO_2 pump in C_4 plants as a starting point, sketch a brief outline of CO_2 fixation in the cactus. Which reactions occur during the day and which at night?
C. A cactus depleted of starch could not fix CO_2, but a C_4 plant could. Why is starch required for CO_2 fixation in the cactus but not in C_4 plants?
D. Can you offer an explanation for why this method of CO_2 fixation is advantageous for cactus plants?

*14–23 Careful experiments comparing absorption and action spectra of plants ultimately led to the notion of two cooperating photosystems in chloroplasts. The absorption spectrum is the amount of light captured by photosynthetic pigments at different wavelengths. The action spectrum is the rate of photosynthesis (for example, O_2 evolution or CO_2 fixation) resulting from the capture of photons.

The first measurement of an action spectrum was probably made in 1882 by T.W. Englemann, who used simple equipment and an ingenious experimental design. He placed a filamentous green alga into a test tube along with a suspension of oxygen-seeking bacteria. He allowed the bacteria to use up the available oxygen and then illuminated the alga with light that had been passed through a prism to form a spectrum. After a short time he observed the results shown in Figure 14–12. Sketch the action spectrum for this alga and explain how this experiment works.

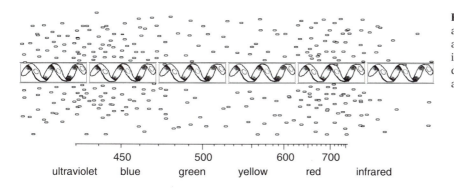

450   500   600   700

ultraviolet   blue   green   yellow   red   infrared

Figure 14–12 Experiment to measure the action spectrum of a filamentous green alga (Problem 14–23). Bacteria, which are indicated by the tiny rectangles, were distributed evenly throughout the test tube at the beginning of the experiment.

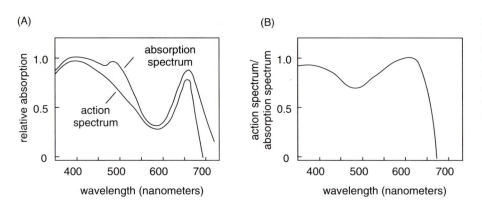

(A)

relative absorption

absorption spectrum

action spectrum

1.0

0.5

0

400   500   600   700

wavelength (nanometers)

(B)

action spectrum/ absorption spectrum

1.0

0.5

0

400   500   600   700

wavelength (nanometers)

Figure 14–13 Absorption and action spectra (A) and the ratio of action spectrum to absorption spectrum (B) for the alga *Chlorella* (Problem 14–24). The action spectrum shows the evolution of O_2 at different wavelengths. The ratio of the action spectrum to the absorption spectrum is shown on an arbitrary scale.

14–24 If all pigments captured energy and delivered it to the photosystem with equal efficiency, then the absorption spectrum and the action spectrum would have the same shape; however, the two spectra differ slightly (Figure 14–13A). When a ratio of the two spectra is displayed (Figure 14–13B), the most dramatic difference is the so-called "red drop" at long wavelengths. In 1957 Emerson found that if shorter wavelength light (650 nm) was mixed with the less effective longer wavelength light (700 nm), the rate of O_2 evolution was much enhanced over either wavelength given alone. This result, along with others, suggested that two photosystems (now called photosystem I and photosystem II) were cooperating with one another and led to the familiar Z scheme for photosynthesis.

One clue to the order in which the two photosystems are linked came from experiments in which illumination was switched between 650 nm and 700 nm. As shown in Figure 14–14, a shift from 700 nm to 650 nm was accompanied by a transient burst of O_2 evolution, whereas a shift from 650 nm to 700 nm was accompanied by a transient depression in O_2 evolution.

Using your knowledge of the Z scheme of photosynthesis, explain why these so-called chromatic transients occur, and deduce whether photosystem II, which accepts electrons from H_2O, is more sensitive to 650-nm light or to 700-nm light.

*14–25 The most compelling early evidence for the Z scheme of photosynthesis came from measuring the oxidation state of the cytochromes in algae under different regimes of illumination (Figure 14–15). Illumination with light at 680 nm caused oxidation of cytochromes (indicated by the upward trace); additional illumination with light at 562 nm caused reduction of the cytochromes (indicated by the downward trace); and both effects could be reversed by turning the lights off (Figure 14–15A). In the presence of the herbicide DCMU, which blocks electron transport through the cytochromes and stops O_2 evolution, the reduction with 562-nm light disappeared and was replaced by a small additional oxidation (Figure 14–15B).

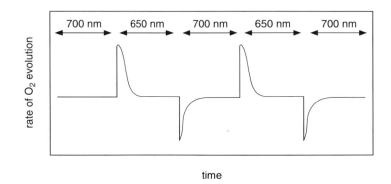

700 nm   650 nm   700 nm   650 nm   700 nm

rate of O_2 evolution

time

Figure 14–14 Chromatic transients observed upon switching between 650-nm light and 700-nm light (Problem 14–24). The intensities of light at the two wavelengths were adjusted beforehand so that alone each gave the same rate of O_2 evolution.

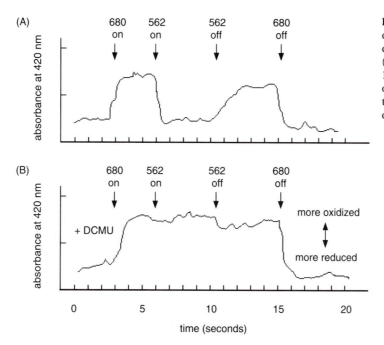

Figure 14–15 Oxidation state of cytochromes after illumination of algae with different wavelength light in the absence (A) and presence (B) of DCMU (Problem 14–25). An upward trace indicates oxidation of the cytochromes; a downward trace indicates reduction of the cytochromes.

A. In these algae, which wavelength stimulates photosystem I and which stimulates photosystem II?

B. How do these results support the Z scheme for photosynthesis; that is, how do they support the idea that there are two photosystems that are linked by cytochromes?

C. On which side of the cytochromes does DCMU block electron transport—on the side nearer photosystem I or the side nearer photosystem II?

14–26   Photosystem II accepts electrons from water, generating O_2, and donates them via the electron-transport chain to photosystem I. Each photon absorbed by photosystem II can effect the transfer of only a single electron, and yet four electrons must be removed from water to generate a molecule of O_2. Thus, four photons are required to evolve a molecule of O_2.

$$2H_2O + 4h\nu \rightarrow 4e^- + 4H^+ + O_2$$

How do four photons cooperate in the production of O_2? Is it necessary that four photons arrive at a single reaction center simultaneously? Can four activated reaction centers cooperate to evolve a molecule of O_2? Or is there some sort of "gear wheel" that collects the four electrons from H_2O and transfers them one at a time to a reaction center?

To investigate this problem, you expose dark-adapted spinach chloroplasts to a series of brief saturating flashes of light (2 μsec) separated by short periods of darkness (0.3 second) and measure the evolution of O_2 that results from each flash. Under this lighting regime most photosystems capture a photon during each flash. As shown in Figure 14–16, O_2 is evolved with a distinct periodicity: the first burst of O_2 occurs on the third flash, and subsequent peaks occur every fourth flash thereafter. If you first inhibit 97% of the photosystem II reaction centers with DCMU and then repeat the experiment, you observe the same periodicity of O_2 production, but the peaks are only 3% of the uninhibited values.

A. How do these results distinguish among the three possibilities posed at the outset (simultaneous action, cooperation among reaction centers, and a gear wheel)?

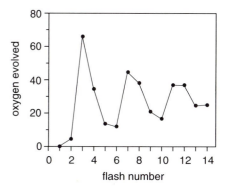

Figure 14–16 Oxygen evolution by spinach chloroplasts in response to saturating flashes of light (Problem 14–26). The chloroplasts were placed in the dark for 40 minutes prior to the experiment to allow them to come to the same "ground" state.

B. Why do you think it is that the first burst of O_2 occurs after the third flash, whereas additional peaks occur at four-flash intervals? (Consider what this observation implies about the dark-adapted state of the chloroplasts.)

C. Can you suggest a reason why the periodicity in O_2 evolution becomes less pronounced with increasing flash number?

*14–27 Chloroplasts synthesize ATP much like mitochondria; they couple electron transport to the pumping of protons and then harvest the energy in the resulting electrochemical proton gradient by directing the protons through an ATP synthase to make ATP. One of the earliest and most convincing tests of the chemiosmotic coupling of electron transport and ATP synthesis used thylakoid vesicles obtained from spinach chloroplasts.

In these experiments formation of ATP was assayed in a suspension of thylakoid vesicles (stromal surface facing outward, that is, right-side-out). These vesicles were first acidified to pH 4 and then made alkaline in the presence of ADP and $^{32}PO_4$. As shown in Figure 14–17, the yield of ATP was greater when the vesicles were acidified with succinic acid than it was when they were acidified with HCl. Furthermore, the yield of ATP increased with increasing concentration of succinic acid (even though pH 4 solutions were used in all experiments). As shown in Figure 14–18, the yield of ATP also depended on the pH of the alkaline stage of the experiment, with the yield of ATP increasing up to pH 8.5.

A. Why does acidification with succinic acid yield more ATP than acidification with HCl? And why does the yield of ATP increase with increasing concentrations of succinic acid? (Succinic acid has two carboxylic acid groups with pKs of 4.2 and 5.5.)

B. Why is the yield of ATP so critically dependent on the pH of the alkaline stage of the experiment?

C. Predict the effect of the following treatments during the alkaline incubation, and rationalize your prediction.

1. Addition of FCCP, which is an uncoupler of electron transport and ATP synthesis
2. Addition of DCMU, which blocks electron transport through the cytochromes
3. Illumination with bright light
4. Incubation in total darkness

14–28 *Thiobacillus ferrooxidans* is a bacterium that lives on slag heaps at pH 2. In the mining industry these bacteria are used to recover copper and uranium from low-grade ore by an acid leaching process. The bacteria oxidize Fe^{2+} to produce Fe^{3+}, which in turn oxidizes (and solubilizes) these minor components of the ore. It is remarkable that the bacterium can live in such an environment. It does so by exploiting the pH difference between the environment and its cytoplasm (pH 6.5) to drive synthesis of ATP and NADPH, which it can then use to fix CO_2 and nitrogen. In order to keep its cytoplasmic pH constant, *T. ferrooxidans* uses electrons from Fe^{2+} to reduce O_2 to water, thereby removing the protons.

$$4Fe^{2+} + O_2 + 4H^+ \rightarrow 4Fe^{3+} + 2H_2O$$

What are the energetics of these various processes? Is the flow of electrons from Fe2+ to O_2 energetically favorable? Is the electrochemical proton gradient across the membrane sufficient to permit the synthesis of ATP? How difficult is it to reduce NADP+ using electrons from Fe2+? These are key questions for understanding how *T. ferrooxidans* manages to thrive in such an unlikely niche. Addressing them requires an introduction to the energetics of redox chemistry.

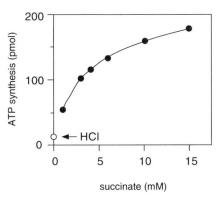

Figure 14–17 Yield of ATP with HCl and with increasing concentrations of succinic acid (Problem 14–27). The yield of ATP with HCl is shown on the Y axis, where the concentration of succinic acid is zero. In all cases the thylakoid suspension was treated with acid for 60 seconds, then treated with alkali for 15 seconds in the presence of ADP and $^{32}PO_4$, at which point the reaction was stopped and the yield of radioactive ATP was measured.

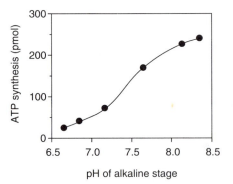

Figure 14–18 Yield of ATP as a function of the pH during the alkaline stage of the experiment (Problem 14–27).

In a redox reaction

$$aA_{ox} + bB_{red} \rightarrow cA_{red} + dB_{ox}$$

the tendency to donate electrons is given by ΔE_o, which is computed by subtracting E_o (the standard half-cell potential) for the half-cell that donates electrons (is oxidized) from the E_o for the half-cell that accepts electrons (is reduced).

$$\Delta E_o = E_o{}^A - E_o{}^B$$

When a redox reaction takes place under nonstandard conditions (standard conditions are 25°C or 298°K and all concentrations at 1 M), the tendency to donate electrons is

$$\Delta E = \Delta E_o - 2.3 \frac{RT}{nF} \log_{10} \frac{[A_{red}]^c [B_{ox}]^d}{[A_{ox}]^a [B_{red}]^b}$$

where

$R = 1.98 \times 10^{-3}$ kcal/°K mole
$T =$ temperature in °K
$n =$ the number of electrons transferred
$F = 23$ kcal/V mole

ΔG is related to ΔE by the equation

$$\Delta G = -nF \, \Delta E$$

Note that the signs of ΔG and ΔE are opposite; thus, a favorable redox reaction has a positive ΔE and a negative ΔG.

A. What is ΔE for the reduction of O_2 by Fe^{2+}, assuming that the reaction occurs under standard conditions? The half-cell potentials are

$$Fe^{3+}/Fe^{2+}, \ E_o = 0.77V$$

$$4H^+ + O_2/2H_2O, \ E_o = 0.82V$$

What is ΔG for this reaction?

B. Write a balanced equation for the reduction of $NADP^+ + H^+$ by Fe^{2+}. What is ΔE for this reaction under standard conditions?

$$NADP^+ + 2H^+/NADPH + H^+, \ E_o = -0.32 \text{ V}$$

What is ΔE if the concentrations of Fe^{3+} and Fe^{2+} are equal, the concentration of NADPH is tenfold greater than that of $NADP^+$ (ignore the protons), and the temperature is 310°K? What is ΔG under these two conditions?

C. The ΔG available from the transport of protons from the outside (pH 2) to the inside (pH 6.5) of *T. ferrooxidans* is given by the Nernst equation

$$\Delta G = 2.3 \, RT \log_{10} \frac{[H^+]_{in}}{[H^+]_{out}} + nFV$$

where V is the membrane potential. In *T. ferrooxidans* grown at pH 2, the membrane potential is zero. Assuming that T = 310°K and that $\Delta G = 11$ kcal/ mole for ATP synthesis under the prevailing intracellular conditions in *T. ferrooxidans*, how many protons (to the nearest integer) would have to enter the cell through the ATP synthase to drive ATP synthesis? In order for proton transport to be coupled to ATP synthesis, could the protons pass through the ATP synthase one at a time, or would they all have to pass through at once?

D. How many protons (to the nearest integer) would have to enter the cell to drive the reduction of NADP⁺ by Fe²⁺ under standard conditions? Comment on the coupling of proton transport to NADP⁺ reduction.

E. How many moles of Fe^{2+} does *T. ferrooxidans* have to oxidize to Fe^{3+} to fix 1 mole of CO_2 into carbohydrate via the Calvin cycle?

The Evolution of Electron-Transport Chains
(MBOC 697–703)

14–29 Fill in the blank in the following statement.

A. In the process of _____, ATP is made by a substrate-level phosphorylation event that harnesses the energy released from the partial oxidation of a hydrogen-rich organic molecule.

14–30 Indicate whether the following statements are true or false. If a statement is false, explain why.

___ A. The excreted end-products of fermentation differ in different organisms, but they tend to be organic acids, thereby accomplishing the excretion of protons.

___ B. A lowering of the pH of the local environment would favor survival of bacteria with transmembrane proton pumps to pump H⁺ out of the cell to prevent death from intracellular acidification.

___ C. The major evolutionary breakthrough in energy metabolism was the development of photochemical reaction centers that could produce molecules such as NADH from environmental molecules such as H_2S.

___ D. The evolution of organisms capable of using water to reduce CO_2 involved the cooperation of a photosystem I derived from green bacteria and a photosystem II derived from purple bacteria.

___ E. Although cyanobacteria arose about 3×10^9 years ago, the oxygen content of the atmosphere did not increase significantly until about 2×10^9 years ago due to the precipitation of large amounts of ferric oxides.

___ F. It is believed that mitochondria arose by the endocytosis of a bacterium that had lost the ability to survive on light energy alone and came to rely entirely on respiration.

The Genomes of Mitochondria and Chloroplasts
(MBOC 704–717)

14–31 Fill in the blanks in the following statements.

A. Mutations that are not inherited according to the Mendelian rules that govern the inheritance of nuclear genes are said to display _____ and are likely to be located in organelle genes.

B. In budding yeast where a limited number of mitochondria enter the bud during mitosis, a cell with a mixture of mutant and wild-type mitochondria can give rise to daughter cells with a single type of mitochondria in a process known as _____.

C. In higher animals, where mitochondria enter the zygote primarily through the egg cytoplasm, mitochondria are said to display _____ inheritance.

D. Yeast mutants with large deletions in their mitochondrial DNA form unusually small colonies when grown on low glucose; all mutants with such defective mitochondria are called _____ mutants.

E. The _____ is the central metabolic pathway in mammals for the disposal of cellular breakdown products that contain nitrogen.

F. According to the _____, eucaryotic cells started out in evolution as anaerobic organisms without mitochondria or chloroplasts and then established a stable symbiotic relationship with a bacterium.

14–32 Indicate whether the following statements are true or false. If a statement is false, explain why.

___ A. Individual energy-converting organelles replicate their DNA in synchrony with the nuclear DNA and divide when the cell divides, thereby maintaining a constant amount of organelle DNA.

___ B. Although it is not known how organelle DNA is packaged, it is more likely to resemble the structure of bacterial genomes rather than eucaryotic chromatin because there are no histones in organelles.

___ C. Although the protein synthetic machinery of chloroplasts is very similar to that in bacteria, it is much less similar to the machinery in mitochondria, which synthesize proteins more like the cytoplasm.

___ D. In higher plants, many of the chloroplast ribosomal proteins are actually encoded in the cell nucleus, but these nuclear genes have a clear bacterial ancestry.

___ E. The mitochondrial genetic code differs slightly from the nuclear code, but it is identical in mitochondria from all species that have been examined.

___ F. The simplicity of the mitochondrial genetic system, and its reduced fidelity, may contribute to the increased rate of nucleotide substitutions observed in mitochondrial genomes.

___ G. Plant mitochondrial genomes vary greatly in DNA content yet seem to encode only a few more proteins than the much smaller mitochondrial genomes from animal cells.

___ H. The presence of introns in organelle genes is not surprising since similar introns have been found in related genes from bacteria whose ancestors are thought to have given rise to mitochondria and chloroplasts.

___ I. Mutations that are inherited according to Mendelian rules affect nuclear genes; mutations whose inheritance violates Mendelian rules are likely to affect organelle genes.

___ J. The green and white patches in variegated leaves are caused by the mitotic segregation of normal and defective mitochondria.

___ K. Mitochondria, which divide by fission, can replicate indefinitely in the cytoplasm of proliferating eucaryotic cells even in the complete absence of a mitochondrial genome.

___ L. Mitochondria from different tissues of the same organism contain the same complement of nuclear and mitochondrial proteins.

___ M. Chloroplasts tend to make most of the lipids they require, whereas mitochondria import most of their lipids.

___ N. Both animal and plant mitochondria probably descended from a purple photosynthetic bacterium, which had previously lost its ability to carry out photosynthesis and was left with only a respiratory chain.

14–33 Mouse cells contain roughly 1000 mitochondrial DNA molecules. The replication of an individual mitochondrial genome takes only about an hour, which is about 5% of the cell-generation time (20 hours). Since the average number of mitochondrial genomes per cell is constant, they must replicate, on average, once per cell cycle. How is the replication of mitochondrial genomes coordinated? Does mitochondrial DNA replicate at random times throughout the cell cycle or is their replication, like nuclear DNA replication, confined to one particular stage, for example, S phase?

You have devised an elegant way to answer this question. In outline, your basic approach is to label mouse cell mitochondrial DNA with a short exposure (2 hours) to ^{3}H-thymidine, then to chase with nonradioactive thymidine for various times, and finally to label with 5-bromodeoxyuridine (BrdU) for 2 hours (Figure 14–19). Any DNA that was replicated during both exposures will be radioactive (due to the ^{3}H-thymidine) and have a higher density than normal mitochondrial DNA (due to the BrdU). Since mitochondrial DNA is circular, you can separate it cleanly from nuclear DNA and

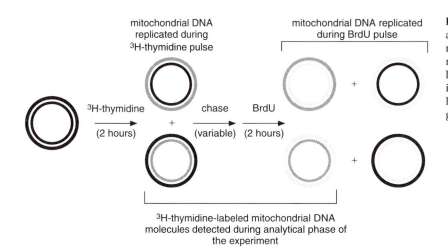

mitochondrial DNA
replicated during
³H-thymidine pulse

mitochondrial DNA replicated
during BrdU pulse

³H-thymidine
(2 hours)

chase
(variable)

BrdU
(2 hours)

³H-thymidine-labeled mitochondrial DNA
molecules detected during analytical phase of
the experiment

Figure 14–19 Experimental design to assess the timing of mitochondrial DNA replication (Problem 14–33). Unlabeled mitochondrial DNA is indicated with black lines; DNA labeled with ³H-thymidine is indicated with dark gray lines; DNA labeled with BrdU is indicated with light gray lines.

examine the two DNAs independently. Using density-gradient analysis, you separate heavy DNA (BrdU-labeled) from light DNA (non-BrdU-labeled) and measure the amount of radioactivity associated with each. You have just completed your analysis of the mitochondrial DNA and the results are shown in Figure 14–20. Your results show that a relatively constant fraction of ³H-labeled mitochondrial DNA was shifted to a higher density, regardless of the separation of the labeling periods (the chase times).

A. Do these results fit better with your expectations for replication during a specific part of the cell cycle or with replication at random times? Why?

B. One of your colleagues criticizes the design of these experiments. He suggests that you cannot distinguish between random or timed replication because the cells were growing asynchronously (that is, within the cell population all different stages of the cell cycle were represented). How does this concern affect your interpretation?

C. Sketch your expectations (on Figure 14–20) for the results of your impending analysis of the nuclear DNA (assume that the DNA synthesis phase—S phase—of the cell cycle is 5 hours long).

D. Another of your colleagues wants to know what your results would look like if mitochondrial DNA was replicated at all times during the cell cycle, but once an individual molecule was replicated, it had to wait exactly one cell cycle before it was replicated again.

*14–34 The majority of mRNAs, tRNAs, and rRNAs in human mitochondria are transcribed from one strand of the genome. These RNAs are all present initially on one very long transcript, which is 93% the length of the DNA strand. However, during mitochondrial protein synthesis these RNAs function as separate, independent species of RNA. The relationship of the individual RNAs to the primary transcript and many of the special features of the mitochondrial genetic system have been revealed by comparing the sequences of the RNAs with the nucleotide sequence of the genome. An overview of the map is shown in Figure 14–21.

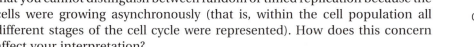

Figure 14–20 The fraction of mitochondrial DNA labeled with ³H-thymidine that is shifted to heavy density after various lengths of chase (Problem 14–33). The fraction of DNA that is density shifted is the radioactivity at the heavy density divided by the total radioactivity. The length of chase is the time from the end of the ³H-thymidine labeling to the beginning of the BrdU labeling.

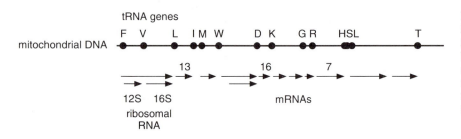

tRNA genes

F V   L   I M W   D K   G R   HSL   T

mitochondrial DNA

13   16   7

12S 16S
ribosomal
RNA

mRNAs

Figure 14–21 Transcription map of human mitochondrial DNA (Problem 14–34). Individual tRNA genes are indicated by black circles; the amino acids they carry are shown in the one-letter code. The three mRNAs whose detailed sequences are shown in Figure 14–22 are indicated by number.

```
tRNA^L                                          tRNA^I
TTCTTAACAACATACCCAT.........CTCAAACCTAAGAAATATG          DNA
         ACAUACCCAU.........CUCAAACCUAAAAAAAAAA          mRNA 13
              M  P                E  T  *               protein


tRNA^D                                          tRNA^K
TATATCTTAATGGCACATG.........CTCTAGAGCCCACTGTAAA          DNA
         AUGGCACAUG.........CUCUAGAGCCAAAAAAAAA          mRNA 16
           M  A  H                 S  *                 protein


tRNA^R                                          tRNA^H
ATTTACCAAATGCCCCTCA.........TTTTCCTCTTGTAAATATA          DNA
         AUGCCCCUCA.........UUUUCCUCUUAAAAAAAAA          mRNA 7
           M  P  L              F  S  S  *             protein
```

Figure 14–22 Arrangements of tRNA and mRNA sequences at three places on the human mitochondrial genome (Problem 14–34). Underlined sequences indicate tRNA genes. The sequences of the mRNAs are shown below the corresponding genes. The middle portions of the mRNAs and their genes are indicated by dots. The 5′ ends of the sequences are shown at the left. The 5′ ends of the mRNAs are unmodified and the 3′ ends have poly-A tails. The encoded protein sequences are indicated below the mRNAs, with the letter for the amino acid immediately under the center nucleotide of the codon. An asterisk (*) indicates a termination codon.

Three segments of the nucleotide sequence of the human mitochondrial genome are shown in Figure 14–22 along with the three mRNAs that are generated from those regions. The nucleotides that encode tRNA species are underlined; the amino acids encoded by the mRNAs are indicated below the center base of the codon.

A. In terms of codon usage and mRNA structure, in what two ways does initiation of protein synthesis in mitochondria differ from initiation in the cytosol?

B. In what two ways are the termination codons for protein synthesis in mitochondria unusual? (The termination codons are shown in Figure 14–22 as asterisks.)

C. Does the arrangement of tRNA and mRNA sequences in the genome suggest a possible mechanism for processing the primary transcript into individual RNA species?

*14–35 Chloroplast DNA from higher plants is remarkably constant in size (120–180 kb) and in sequence arrangement. By contrast, the corresponding plant mitochondrial DNAs are quite variable in sequence arrangement and in size, ranging from 218 kb for turnip to 2400 kb for muskmelon. Some portion of this variability may result from the transfer of DNA from chloroplasts to mitochondria.

One experiment to search for genetic transfers between organellar genomes used a defined restriction fragment from spinach chloroplasts, which carried information for the gene for the large subunit of ribulosebisphosphate carboxylase. This gene has no known mitochondrial counterpart. Mitochondrial and chloroplast DNAs were prepared from zucchini, corn, spinach, and pea. All these DNAs were digested with the same restriction enzyme, and the resulting fragments were separated by electrophoresis. The fragments were then transferred to a filter and hybridized to a radioactive preparation of the spinach probe fragment. A schematic representation of the autoradiograph is shown in Figure 14–23.

A. It is very difficult to prepare mitochondrial DNA that is not contaminated to some extent with chloroplast DNA. How do these experiments control for contamination of the mitochondrial DNA preparation by chloroplast DNA?

B. Which of these plant mitochondrial DNAs appear to have acquired chloroplast DNA?

14–36 A friend of yours has been studying a pair of mutants in the fungus *Neurospora*, which she has whimsically named *poky* and *puny*. Both mutants grow at about the same rate, but much more slowly than wild-type. Your friend has

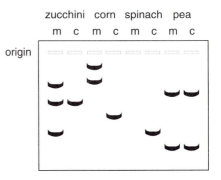

Figure 14–23 Patterns of hybridization of a probe from spinach chloroplast DNA to mitochondrial and chloroplast DNAs from zucchini, corn, spinach, and pea (Problem 14–35). Lanes labeled *m* contain mitochondrial DNA; lanes labeled *c* contain chloroplast DNA. Restriction fragments to which the probe hybridized are shown as dark bands.

Table 14-4 Genetic Analysis of *Neurospora* Mutants (Problem 14-36)

Cross	Protoperithecial Parent		Fertilizing Parent	Spore Counts	
				Fast Growth	Slow Growth
1	*poky*	×	wild	0	1749
2	wild	×	*poky*	1334	0
3	*puny*	×	wild	850	799
4	wild	×	*puny*	793	801
5	*poky*	×	*puny*	0	1831
6	*puny*	×	*poky*	754	710
7	wild	×	wild	1515	0
8	*poky*	×	*poky*	0	1389
9	*puny*	×	*puny*	0	1588

been unable to find any supplement that improves their growth rates. Her biochemical analysis shows that each mutant displays a different abnormal pattern of cytochrome absorption. To characterize the mutants genetically, she crossed them to wild-type and to each other and tested the growth rates of the progeny. She has come to you because she is puzzled by the results.

She explains that haploid nuclei from the two parents fuse during a *Neurospora* mating and then divide meiotically to produce four haploid spores, which can be readily tested for their growth rates. The parents contribute unequally to the diploid: one parent (the protoperithecial parent) donates a nucleus and the cytoplasm; the other (the fertilizing parent) contributes little more than a nucleus—much like egg and sperm in higher organisms. As shown in Table 14-4, the "order" of the crosses sometimes makes a difference: a result she has not seen before.

Can you help your friend understand these results?

14-37 Mutants of yeast that are defective in mitochondrial function grow on fermentable substrates such as glucose, but they fail to grow on nonfermentable substrates such as glycerol. Nuclear and mitochondrial mutations can be distinguished by genetic crosses. Nearly 200 different nuclear genes have been defined by complementation analysis of nuclear petite (*pet*) mutants. Surpris-

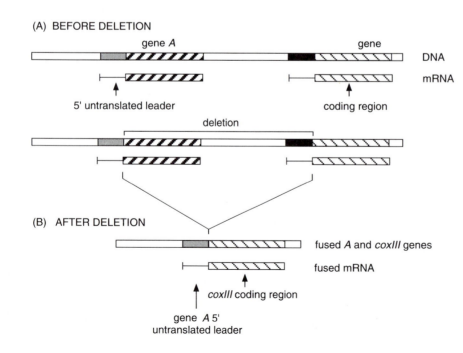

Figure 14-24 Schematic representation of the mitochondrial deletion mutations that suppress the nuclear mutation *pet494* (Problem 14-37).

ingly, mutations in about a quarter of these genes affect the expression of single, or restricted sets of, mitochondrial gene products without blocking overall gene expression.

For example, *pet494* mutants contain normal levels of all the known mitochondrial gene products except subunit III of cytochrome oxidase (coxIII). However, *pet494* mutants contain normal levels of coxIII mRNA, which appears normal in size. Thus, some posttranscriptional step in expression of the gene for coxIII appears to be regulated by the normal *PET494* gene product. It could be required to promote translation of coxIII mRNA or to stabilize coxIII during assembly of the cytochrome oxidase complex. To distinguish between these possibilities, mitochondrial mutations that suppress the respiratory defect of *pet494* mutations were selected and analyzed. All these mutations were deletions that fused the coxIII coding region to the 5' untranslated region of another gene, as illustrated in Figure 14–24.

A. How do these results distinguish between a requirement for translation of coxIII mRNA and a requirement for stablization of the coxIII protein?
B. Each of the mitochondrial deletions eliminates one or more genes that are essential for mitochondrial function, yet these yeast strains contain all the normal mitochondrial mRNAs and make normal colonies when grown on nonfermentable substrates. How can this be?

*14–38 At the cellular level evolutionary theories are particularly difficult to test since fossil evidence is lacking. The possible evolutionary origins of mitochondria and chloroplasts must be sought in living organisms. Fortunately, living forms resembling the ancestral types required by the endosymbiotic theory for the origin of mitochondria and chloroplasts can be found today. For example, the plasma membrane of the free-living aerobic bacterium *Paracoccus denitrificans* contains a respiratory chain that is nearly identical to the respiratory chain of mammalian mitochondria—both in the types of respiratory components present and in its sensitivity to respiratory inhibitors such as antimycin and rotenone. Indeed, no significant feature of the mammalian respiratory chain is absent from *Paracoccus. Paracoccus* effectively assembles in a single organism all those features of the mitochondrial inner membrane that are otherwise distributed at random among other aerobic bacteria.

Imagine that you are a protoeucaryotic cell looking out for your evolutionary future. You have been observing *proto-Paracoccus* and are amazed at its incredibly efficient use of oxygen in generating ATP. With such a source of energy your horizons would be unlimited. You plot to hijack a *proto-Paracoccus* and make it work for you and your descendants. You plan to take it into your cytoplasm, feed it any nutrients it needs, and harvest the ATP. Accordingly, one dark night you trap a lone *proto-Paracoccus*, surround it with your plasma membrane, and imprison it in a new cytoplasmic compartment. To your relief, the *proto-Paracoccus* seems to enjoy its new environment. After a day of waiting, however, you feel as sluggish as ever. What has gone wrong with your scheme?

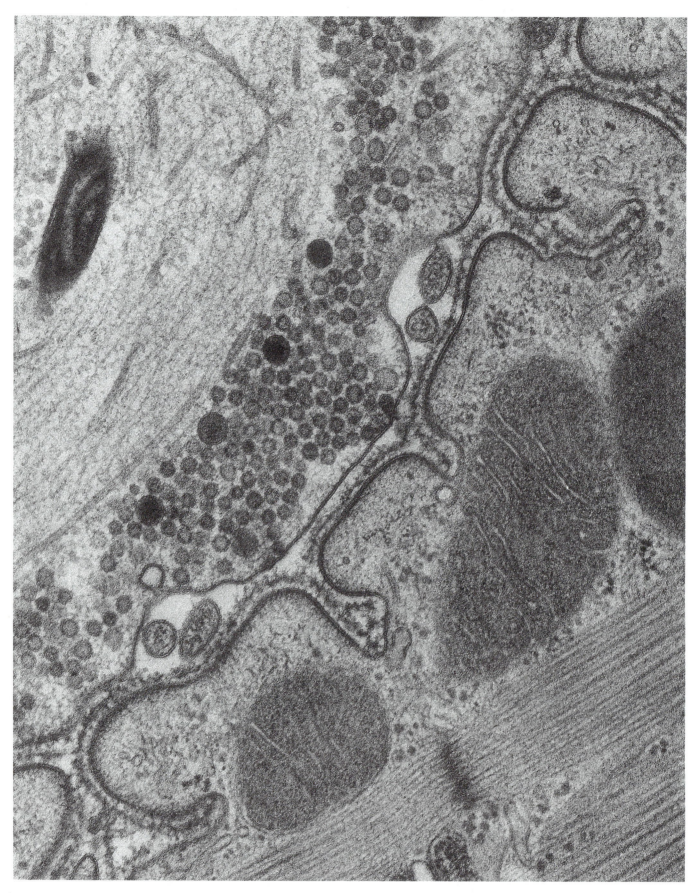

Cell signaling poised and ready for action: an electron micrograph of a section of a neuromuscular junction. How many of its parts can you recognize? (Courtesy of John Heuser.)

Cell Signaling

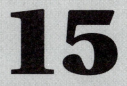

General Principles of Cell Signaling
(MBOC 721–734)

15–1   Fill in the blanks in the following statements.

A. Regardless of the nature of the signal, the target cell responds by means of a specific protein called a _____, which binds the signaling molecule and then initiates a response in the target cell.

B. Signaling molecules that a cell secretes may act as _____, affecting only cells in the immediate environment of the signaling cell, in a process called _____.

C. When a nerve impulse reaches the nerve terminals at the end of the axon, it stimulates the terminals to secrete a chemical signal called a _____, which is delivered rapidly to the postsynaptic target cell in a process termed _____.

D. _____ cells secrete their signaling molecules, called _____, into the bloodstream, which carries the signal to target cells throughout the body.

E. Cells can send signals to other cells of the same type and to themselves in a process called _____.

F. _____, which are derived from fatty acids (mainly arachidonic acid), are released to the cell exterior, where they influence cells in their immediate neighborhood.

G. _____ are specialized cell-cell junctions that can form between closely apposed plasma membranes, directly connecting the cytoplasms of the joined cells.

H. Acetylcholine indirectly causes smooth muscle cells in blood vessel walls to relax by inducing the epithelial cells to make and release the gas _____, which then signals the smooth muscle cells to relax.

I. Steroid hormones, thyroid hormones, and retinoids diffuse across the plasma membrane and bind to specific receptors that are structurally related and constitute the _____.

J. _____ hormones are all made from cholesterol.

K. The _____ hormones, which are made from the amino acid tyrosine, act to increase metabolism in a wide variety of cell types.

L. The _____, which are made from vitamin A, play important roles as local mediators in vertebrate development.

M. _____ receptors are involved in rapid synaptic signaling between electrically excitable cells.

N. _____ receptors act indirectly to regulate the activity of a separate plasma-membrane-bound target protein, which can be an enzyme or an ion channel.

O. _____ receptors, when activated, either function directly as enzymes or are associated with enzymes.

15–2 Indicate whether the following statements are true or false. If a statement is false, explain why.

___ A. There is no fundamental distinction between signaling molecules that bind to cell-surface receptors and those that bind to intracellular receptors.

___ B. The receptors involved in paracrine, synaptic, and endocrine signaling all have very high affinity for their respective signaling molecules.

___ C. Autocrine signaling is thought to be one possible mechanism underlying the "community effect" observed in early development, where a group of identical cells can respond to a differentiation-inducing signal but a single isolated cell of the same type cannot.

___ D. Gap junctions allow the free exchange of small molecules and proteins between linked cells.

___ E. When deprived of the appropriate set of signals for survival, a cell will activate a suicide program and kill itself—a process called programmed cell death.

___ F. A signaling molecule will elicit identical responses in different target cells if it binds to identical receptors.

___ G. The speed with which a cell responds to shutting off the signal depends on the rate of turnover of the molecules that the signal affects.

___ H. NO acts only locally because it has a short half-life—about 5 to 10 seconds—in the extracellular space before it is converted to nitrates and nitrites by oxygen and water.

___ I. Water-soluble signaling molecules, which have very short circulating lifetimes, usually mediate responses of short duration, whereas water-insoluble ones, which persist in the blood for hours to days, tend to mediate longer-lasting responses.

___ J. The bewildering diversity of known receptor proteins can be reduced to a much smaller number of large families of related proteins.

___ K. Although it is estimated that about 1% of our genes encode protein kinases, individual mammalian cells typically contain fewer than 10 distinct kinds of these enzymes.

*15–3 Succinylcholine, which is an acetylcholine analogue, is used by surgeons as a muscle relaxant. Care must be taken in its use because some individuals recover abnormally slowly from this paralysis, with life-threatening consequences. Such individuals are deficient in an enzyme called pseudocholinesterase, which is normally present in the blood.

If succinylcholine is an analogue of acetylcholine, why do you think it causes muscles to relax and not contract as acetylcholine does?

*15–4 Radioimmunoassay (RIA) is a powerful tool for quantifying virtually any substance of biological interest because it is sensitive, accurate, and fast. RIA technology arose from studies on adult onset diabetes, which demonstrated the existence of antibodies with high affinity for insulin and developed methods to distinguish free insulin from antibody-bound insulin.

How can high-affinity antibodies and separation techniques be exploited to measure low concentrations of insulin? When a small amount of anti-insulin antiserum is mixed with an equally small amount of very highly radioactive insulin, some binds and some remains free according to the equilibrium.

$$\text{Insulin} + \text{Antibody} \leftrightarrows \text{Insulin-Antibody Complex}$$

$$K = k\frac{[\text{Insulin-Antibody Complex}]}{[\text{Insulin}][\text{Antibody}]}$$

*Problems with an asterisk (*) are answered in the Instructor's Manual.

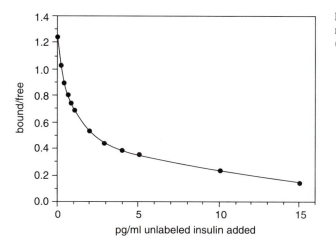

Figure 15–1 Calibration curve for radioimmunoassay of insulin (Problem 15–4).

When increasing amounts of unlabeled insulin are added to a fixed amount of labeled insulin and antiserum, the ratio of bound to free radioactive insulin decreases as expected from the equilibrium expression (Figure 15–1). If the concentration of the unlabeled insulin is known, then the resulting curve serves as a calibration against which other unknown samples can be compared.

You have three samples of insulin whose concentrations are unknown. When mixed with the same amount of radioactive insulin and anti-insulin antibody used in Figure 15–1 the three samples gave the following ratios of bound to free insulin:

Sample 1	0.67
Sample 2	0.31
Sample 3	0.46

A. What is the concentration of insulin in each of these unknown samples?
B. What portion of the standard curve is the most accurate, and why?
C. If the antibodies were raised against pig insulin, which is similar but not identical to human insulin, would the assay still be valid for measuring human insulin concentrations?

15–5 The optimal sensitivity of radioimmunoassay occurs when the concentrations of the unknown and the radioactive tracer are equal. Thus the sensitivity of radioimmunoassay is limited by the specific activity of the radioactive ligand. The more highly radioactive the ligand, the smaller the amount needed per assay and, therefore, the smaller amount of unknown that can be detected.

You wish to measure an unknown sample of insulin (molecular weight, 11,466) using an insulin tracer labeled with radioactive iodine, which has a half-life of 7 days. Assume that you can attach one atom of radioactive iodine per molecule of insulin, that iodine does not interfere with antibody binding, that each radioactive disintegration has a 50% probability of being detected as a "count," and that the limit of detection in the radioimmunoassay requires a total of at least 1000 counts per minute (cpm) distributed between the bound and free fractions.

A. Given these parameters, calculate how many picograms of radioactive insulin will be required for an assay.
B. At optimal sensitivity, how many picograms of unlabeled insulin will you be able to detect?

15–6 You are studying the expression of genes linked to a segment of the Moloney murine sarcoma virus, which is regulated by glucocorticoids (steroid hormones). You made a series of constructs with the viral segment in both orientations, upstream and downstream of a reporter gene (chloramphenicol

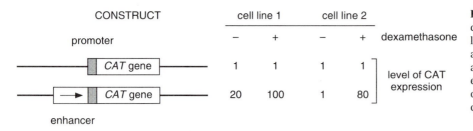

Figure 15–2 Results of transfecting two different constructs into two different cell lines (Problem 15–6). Presence (+) and absence (–) of the hormone (dexamethasone) is indicated. Numbers indicate expression of the *CAT* gene product; in all cases, the values are expressed relative to a construct without the viral segment.

acetyl transferase, *CAT*). You then transfected the constructs into two cell lines derived from different tissues and measured CAT activity in the presence and absence of a glucocorticoid (dexamethasone). The orientation and location of the viral segment made relatively little difference in the expression of the reporter gene, which is the expected result if the viral segment carries an enhancer of transcription. Results from the *CAT* gene alone and a construct with the viral segment in one orientation are shown in Figure 15–2. You are puzzled by the results with cell line 1 because the viral segment increased CAT expression twentyfold in the absence of dexamethasone.

A. Do both cell lines contain glucocorticoid receptors? How can you tell?
B. How does cell line 1 differ from cell line 2? Propose an explanation for the difference.
C. Based on your explanation, predict the outcome of an experiment in which a variety of shorter pieces of the viral segment are placed in front of the *CAT* gene and tested for CAT activity in the two cell lines.

*15–7   Glucocorticoids induce transcription of mouse mammary tumor virus (MMTV) genes. Using hybrid constructs containing the MMTV long terminal repeat (LTR) linked to an easily assayable gene, you have shown that the LTR contains regulatory elements that respond to glucocorticoids. To map the glucocorticoid response elements within the LTR, you generate a series of deletions that remove different extents of the LTR (Figure 15–3) and then measure the binding of the DNA segments to glucocorticoid receptors. To measure binding, you cut each of the mutant DNAs into fragments, purify the

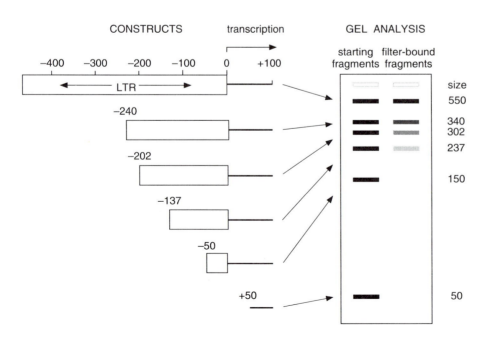

Figure 15–3 Schematic diagram of the LTR deletions and the electrophoretic patterns of the starting and bound mixtures of DNA fragments (Problem 15–7). Nucleotides are numbered relative to the start site for transcription, which is at +1. Negative numbers are in front of the start site, and positive numbers are after it. The electrophoretic pattern is an autoradiograph of the agarose gel.

1. CCAAGGAGGGGACAGTGGCTGGACTAATAG
2. GGACTAATAGAACATTATTCTCCAAAAACT
3. TCGTTTTAAGAACAGTTTGTAACCAAAAAC
4. AGGATGTGAGACAAGTGGTTTCCTGACTTG
5. AGGAAAATAGAACACTCAGAGCTCAGATCA
6. CAGAGCTCAGATCAGAACCTTTGATACCAA
7. CATGATTCAGCACAAAAAGAGCGTGTGCCA
8. CTGTTATTAGGACATCGTCCTTTCCAGGAC
9. CCTAGTGTAGATCAGTCAGATCAGATTAAA
10. GATCAGTCAGATCAGATTAAAAGCAAAAAG
11. TTCCAAATAGATCCTTTTTGCTTTTAATCT

Figure 15–4 DNA sequences that are bound by the glucocorticoid receptor (Problem 15–7).

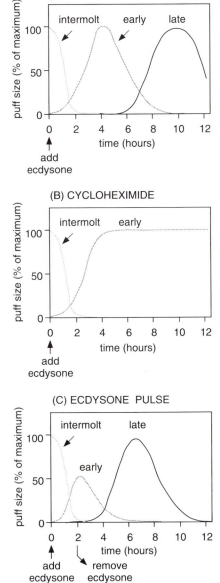

(A) NORMAL

(B) CYCLOHEXIMIDE

(C) ECDYSONE PULSE

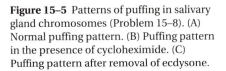

Figure 15–5 Patterns of puffing in salivary gland chromosomes (Problem 15–8). (A) Normal puffing pattern. (B) Puffing pattern in the presence of cycloheximide. (C) Puffing pattern after removal of ecdysone.

fragments containing LTR sequences, and label the ends with ³²P. You then incubate a mixture of the labeled fragments with the purified glucocorticoid receptor. You assess binding by passing the incubation mixture through a nitrocellulose filter, which binds protein (and any attached DNA) but not free DNA. You display the DNA fragments that were bound to the receptor by agarose gel electrophoresis. The electrophoretic patterns of the starting mixture of fragments and the bound fragments are shown in Figure 15–3.

A. Where in the LTR are the glucocorticoid response elements located?
B. A list of eleven MMTV sequences (several from the LTR) that are bound by the glucocorticoid receptor are shown in Figure 15–4. Can you find the consensus sequence to which the receptor binds?

15–8 *Drosophila* larvae molt in response to an increase in the concentration of the steroid hormone ecdysone. The polytene chromosomes of the *Drosophila* salivary glands are an excellent experimental system in which to study the pattern of gene activity initiated by the hormone because active genes enlarge into puffs that are visible in the light microscope. Furthermore, the size of a puff is proportional to the rate at which it is being transcribed. Prior to addition of ecdysone, a few puffs—termed intermolt puffs—are already active. Upon exposure of dissected salivary glands to ecdysone, these intermolt puffs regress, and two additional sets of puffs appear. One set of puffs (early puffs) arises within a few minutes after addition of ecdysone; the other set (late puffs) arises within 4 to 10 hours. The concentration of ecdysone does not change during this time period. The pattern of puff appearance and disappearance is illustrated for a typical puff in each category in Figure 15–5A.

Two critical experiments help to define the relationships between the different classes of puff. In the first, cycloheximide, which blocks protein synthesis, is added at the same time as ecdysone. As illustrated in Figure 15–5B, under these conditions the early puffs do not regress and the late puffs are not induced. In the second experiment, ecdysone is washed out after a 2-hour exposure. As illustrated in Figure 15–5C, this treatment causes an immediate regression of the early puffs and a *premature induction* of the late puffs.

A. Why do you think the early puffs do not regress and the late puffs are not induced in the presence of cycloheximide? Why do you think the intermolt puffs are unaffected?
B. Why do you think the early puffs regress immediately when ecdysone is removed? Why do you think the late puffs arise prematurely under these conditions?
C. Outline a model for ecdysone-mediated regulation of the puffing pattern.

Signaling via G-Protein-linked Cell-Surface Receptors

(MBOC 734–759)

15–9   Fill in the blanks in the following statements.

A. _____ functionally couple G-protein-linked receptors to their target enzymes or ion channels in the plasma membrane.

B. Most G-protein-linked receptors activate a chain of events that alters the concentration of one or more small intracellular signaling molecules, referred to as _____.

C. _____ is synthesized from ATP by a plasma-membrane-bound enzyme _____, and it is rapidly and continuously destroyed by one or more _____.

D. A trimeric G protein that activates adenylyl cyclase is called a _____.

E. _____, which mediate some of the actions of adrenaline and noradrenaline, are coupled by G proteins to the activation of adenylyl cyclase.

F. _____, which is responsible for the disease symptoms of cholera, is an enzyme that catalyzes the transfer of ADP ribose from intracellular NAD^+ to the α chain of a stimulatory G protein.

G. While the β-adrenergic receptors are functionally coupled to adenylyl cyclase by G_s, the α_2-adrenergic receptors are coupled to this enzyme by an _____.

H. _____, made by the bacterium that causes whooping cough, catalyzes the ADP ribosylation of α_i.

I. Cyclic AMP exerts its effects in animals cells mainly by activating the enzyme _____, which catalyzes the transfer of the terminal phosphate group from ATP to specific serines or threonines of selected proteins.

J. Genes that are activated by cyclic AMP contain a short DNA sequence, called the cyclic AMP response element, which is recognized by a specific gene regulatory protein called _____.

K. The dephosphorylation of phosphorylated serines and threonines is catalyzed by four groups of _____.

L. Some extracellular signaling molecules stimulate the incorporation of radioactive phosphate into _____, a minor phospholipid in cell membranes.

M. In the inositol signaling pathway an activated receptor stimulates a trimeric G protein called _____, which in turn activates _____, which then cleaves PIP_2 to generate two products: _____ and _____.

N. The initial transient increase in Ca^{2+} when the inositol phospholipid signaling pathway is activated is often followed by a series of Ca^{2+} "spikes," each lasting seconds or minutes; these _____ can persist for as long as receptors are activated on the cell surface.

O. The enzyme activated by diacylglycerol is called _____ because it is Ca^{2+}-dependent.

P. _____ functions as a multipurpose intracellular Ca^{2+} receptor, mediating many Ca^{2+}-regulated processes.

Q. Most of the effects of Ca^{2+} in cells are mediated by protein phosphorylations catalyzed by a family of _____.

R. _____ has a remarkable property: it can function as a molecular memory device, switching to an active state when exposed to Ca^{2+}/calmodulin and then phosphorylating itself so that it remains active even after the Ca^{2+} is withdrawn.

S. Olfactory cells in the lining of the nose recognize odorants by means of specific G-protein-linked _____.

T. Cyclic-nucleotide-gated ion channels are involved in signal transduction in vertebrate vision, where the crucial cyclic nucleotide is _____.

U. In _____ in the vertebrate retina the phototransduction apparatus is in the outer segment, which contains a stack of discs, each formed by a closed sac of membrane in which photosensitive _____ molecules are embedded.

V. In vision the activated photoreceptor binds to the trimeric G-protein transducin, causing the α subunit to dissociate and activate _____, which hydrolyzes cyclic GMP.

15–10   Indicate whether the following statements are true or false. If a statement is false, explain why.

__ A. G proteins and monomeric GTPases are active when GTP is bound and inactive when GDP is bound.

__ B. Extracellular signaling molecules that work by controlling cyclic AMP levels do so by altering the activity of cyclic AMP phosphodiesterase rather than the activity of adenylyl cyclase.

__ C. The ability to respond quickly to change is assured by the short lifetime of activated α_s: the GTPase activity of α_s is stimulated when α_s binds to adenylyl cyclase, so that the bound GTP is hydrolyzed to GDP, rendering both α_s and the adenylyl cyclase inactive.

__ D. The trimeric G proteins are remarkably versatile intracellular signaling molecules: the α subunit, or the βγ subunit, or both can serve as the active regulatory components.

__ E. A-kinase differs in different cell types, explaining why the effects of cyclic AMP vary depending on the target cell.

__ F. The activity of any protein regulated by phosphorylation depends on the balance at any instant between the activities of the kinases that phosphorylate it and the phosphatases that are constantly dephosphorylating it.

__ G. The resting concentration of Ca^{2+} in the cytosol is kept low by Ca^{2+} pumps in the plasma membrane, the ER membrane, and the inner mitochondrial membrane.

__ H. In the two well-defined pathways of Ca^{2+} signaling—the one specific for nerve cells and the ubiquitous pathway—the rise in cytosolic Ca^{2+} derives from extracellular Ca^{2+}.

__ I. In the inositol-phospholipid signaling pathway phosphoinositol is cleaved from phosphatidylinositol and then phosphorylated twice to generate the intracellular signaling molecule, inositol trisphosphate (IP_3).

__ J. Two mechanisms operate to terminate the IP_3-mediated Ca^{2+} response: some of the IP_3 is dephosphorylated, rendering it inactive, and the Ca^{2+} that enters the cytosol is rapidly pumped out of the cell.

__ K. Ca^{2+} oscillations in hormone-secreting pituitary cells may be a way to maximize secretory output while avoiding the toxic effects of a sustained rise in cytosolic Ca^{2+}.

__ L. The two branches of the inositol phospholipid signaling pathway provide independent routes for producing a cell response: a number of cell types, for example, can be stimulated to proliferate in culture when treated either with a Ca^{2+} ionophore or a C-kinase activator.

__ M. The allosteric activation of calmodulin by Ca^{2+} is analogous to the allosteric activation of A-kinase by cyclic AMP: when calmodulin binds Ca^{2+}, it is converted into an active protein kinase.

__ N. As in the case of cyclic AMP, the response of a target cell to an increase in free Ca^{2+} concentration in the cytosol depends on which CAM-kinase-regulated target proteins are present in the cell.

__ O. The cyclic AMP and Ca^{2+} intracellular signaling pathways interact at several levels in the hierarchy of control.

__ P. In some cases trimeric G proteins directly activate or inactivate ion channels in the plasma membrane of the target cell, thereby altering the ion permeability, and hence the excitability, of the membrane.

___ Q. In visual transduction, receptor activation is caused by light, and it leads to a rise in the intracellular concentration of cyclic GMP.

___ R. By contrast with the more direct signaling pathways used by intracellular receptors, catalytic cascades of intracellular mediators provide numerous opportunities for amplifying the responses to extracellular signals.

___ S. One mechanism for steepening a signaling response is to require that more than one intracellular effector molecule or complex bind to some target macromolecule in order to induce a response.

___ T. If the product of an enzyme allosterically activates the enzyme, the enzyme will be activated in a self-accelerating, runaway fashion.

*15–11 You are trying to purify adenylyl cyclase from brain. The assay is based on the conversion of α-^{32}P-ATP to cAMP. You can easily detect activity in crude brain homogenates stimulated by isoproteronol, which binds to β-adrenergic receptors, but the enzyme loses activity when low molecular weight cofactors are removed by dialysis. What single molecule do you think you could add back to the dialyzed homogenate to restore activity?

15–12 You wish to measure the number of β-adrenergic receptors on the membranes of frog erythrocytes. These receptors normally bind adrenaline and stimulate adenylyl cyclase activity; however, you have chosen to use a competitive inhibitor of adrenaline (alprenolol), which binds to the receptors 500 times more tightly. Your basic experimental protocol is to mix labeled alprenolol with erythrocyte membranes, leave them for 10 minutes at 37°C, pellet the membranes by centrifugation, and measure the radioactivity in the pellet. You perform the experiment in two ways. First, you measure the binding of increasing amounts of ^{3}H-alprenolol to a fixed amount of erythrocyte membranes in order to determine total binding. Second, you repeat the experiment in the presence of a vast excess of unlabeled alprenolol to measure nonspecific binding. Your results are shown in Figure 15–6.

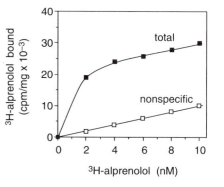

Figure 15–6 Binding of ^{3}H-alprenolol to frog erythrocyte membranes (Problem 15–12).

A. Sketch in the curve for specific binding of alprenolol to β-adrenergic receptors. Has alprenolol binding to the receptors reached saturation?

B. Assuming that one molecule of alprenolol binds per receptor, calculate the number of β-adrenergic receptors on the membrane of a frog erythrocyte. The specific activity of the labeled alprenolol is 1×10^{13} cpm/mmol, and there are 8×10^8 frog erythrocytes per milligram of membrane protein.

15–13 Does cyclic AMP work exclusively through A-kinase or do cyclic AMP-binding proteins have other critical roles, for example, as DNA-binding proteins as in bacteria? You are working with a hamster cell line that makes genetic analysis of this question possible. Because of chromosomal rearrangements, this cell line possesses only one functional copy of many genes, making isolation of recessive mutations much easier than in a fully diploid line. In addition, high intracellular levels of cyclic AMP stop their growth. By stimulating adenylyl cyclase with cholera toxin and by inhibiting cyclic AMP phosphodiesterase with theophylline, cyclic AMP can be artificially elevated. Under these conditions only cells that are resistant to the effects of cyclic AMP can grow.

In this way you isolate several resistant colonies that grow under the selective conditions and assay them for A-kinase activity: they are all defective. About 10% of the resistant lines are completely missing A-kinase activity. The remainder possess A-kinase activity, but a very high level of cyclic AMP is required for activation. To characterize the resistant lines further, you fuse them with the parental cells and test the hybrids for resistance to cholera toxin. Hybrids between parental cells and A-kinase negative cells are sensitive to cholera toxin, which indicates that these mutations are recessive. By contrast, hybrids between parental cells and resistant cells with altered A-kinase responsiveness are resistant to cholera toxin, which indicates that these mutations are dominant.

A. Is A-kinase an essential enzyme in these hamster cells?

B. A-kinase is a tetramer consisting of two catalytic (protein kinase) and two regulatory (cyclic AMP-binding) subunits. Propose an explanation for why the mutations in some cyclic AMP-resistant cell lines are recessive and why others are dominant.

C. Do these experiments support the notion that all cyclic AMP effects in hamster cells are mediated by A-kinase?

*15–14 A particularly graphic illustration of the kind of subtle, yet important, role of cyclic AMP in the whole organism comes from studies of the fruit fly *Drosophila melanogaster*. In search of the gene for cyclic AMP phosphodiesterase, one laboratory measured enzyme levels in flies with chromosomal duplications or deletions and found consistent alterations in flies with mutations involving bands 3D3 and 3D4 on the X chromosome. Duplications in this region have about 1.5 times the normal activity; deletions have about half the normal activity. Deletions in this region caused partial female sterility.

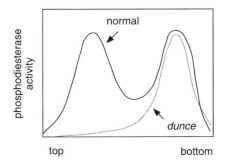

Figure 15–7 Sucrose gradient analysis of cyclic AMP phosphodiesterase activity in homogenates of normal and *dunce* flies (Problem 15–14).

An independent laboratory in the same institution was led to the same chromosomal region through work on behavioral mutants of fruit flies. The researchers had developed a learning test in which flies were presented with two metallic grids, one of which was electrified. If the electrified grid was painted with a strong-smelling chemical, normal flies quickly learned to avoid it even when it was longer electrified. The mutant flies, on the other hand, never learned to avoid the smelly grid; they were aptly called *dunce* mutants. The *dunce* mutation was mapped genetically to bands 3D3 and 3D4. Flies that had been selected for this kind of stupidity were also partially female sterile.

Is the learning defect really due to lack of cyclic AMP phosphodiesterase or are the responsible genes simply closely linked? Further experiments showed that the level of cyclic AMP in *dunce* flies was 1.6 times higher than in normal flies. Furthermore, sucrose gradient analysis of homogenates of *dunce* and normal flies revealed two cyclic AMP phosphodiesterase activities, one of which was missing in *dunce* flies (Figure 15–7).

A. Why do *dunce* flies have higher levels of cyclic AMP than normal flies?

B. Explain why homozygous (both chromosomes affected) duplications of the nonmutant *dunce* gene cause cyclic AMP phosphodiesterase levels to be elevated 1.5-fold and why homozygous deletions of the gene reduce enzyme activity to half the normal value?

C. What would you predict would be the effect of caffeine, a phosphodiesterase inhibitor, on the learning performance of normal flies?

D. Does the experimental evidence prove that the *dunce* gene is the structural gene for cyclic AMP phosphodiesterase? If not, how else might these results arise?

15–15 You are baffled. You are studying the control of cyclic AMP levels in brain slices, and have confirmed that signaling molecules such as isoproteronol that act through β-adrenergic receptors cause a modest increase in cyclic AMP, as expected from G-protein-mediated coupling between the receptor and adenylyl cyclase. The problem is that you find a puzzling synergy between isoproteronol and a number of pharmacological agents that by themselves have no effect on cyclic AMP levels or even reduce them. What is the explanation for this paradoxical augmentation of cyclic AMP levels?

A biochemist friend of yours has suggested a possible explanation: she has found in *in vitro* experiments that βγ subunits from inhibitory trimeric G proteins stimulate type II adenylyl cyclase, which is expressed in brain. To test this idea in cells, you plan to express the cDNAs for the component proteins in human kidney cells, which lack the receptors found in brain. In this way you hope to reconstruct the puzzling effects you observed in brain slices, but in a much simpler background.

Signaling via G-Protein-linked Cell-Surface Receptors

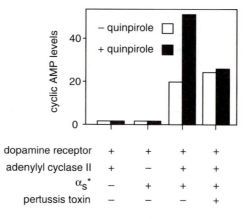

Figure 15–8 Measurements of cyclic AMP levels in cells transfected with cDNAs for the dopamine receptor, type II adenylyl cyclase, and α_s^* (Problem 15–15).

dopamine receptor	+	+	+	+
adenylyl cyclase II	+	−	+	+
α_s^*	−	+	+	+
pertussis toxin	−	−	−	+

You transfect the kidney cells with various combinations of cDNAs for type II adenylyl cyclase, the dopamine receptor (which interacts with an inhibitory G protein), and a mutated (constitutively active) α_s^* subunit. You measure the levels of cyclic AMP in the resulting cell lines in the absence or presence of quinpirole, which activates the dopamine receptor (Figure 15–8). You also measure the effects of pertussis toxin, which blocks the signal from G_i-coupled receptors by modifying the α_i subunit in such a way that it can no longer bind GTP or dissociate from its $\beta\gamma$ subunit (Figure 15–8).

A. Did you succeed in reproducing the paradoxical result you observed in brain slices in the transfected kidney cells? How so?
B. Explain the effects of pertussis toxin in your experiments.
C. What do your experiments indicate is required for maximal activation of type II adenylyl cyclase? Propose a molecular explanation for the augmented activation of type II adenylyl cyclase.
D. Predict the results of expressing the cDNA for the α subunit of transducin, which does not bind to adenylyl cyclase but binds tightly to free $\beta\gamma$ subunits.

*15–16 The mating behavior of yeast depends on signaling peptides termed pheromones that bind to G-protein-linked pheromone receptors. For example, when the α-factor pheromone binds to a yeast cell of the **a** mating type, it blocks cell-cycle progression, arresting growth until a mating partner is found. Genetic analysis of this system has been an extremely fruitful source of information on all aspects of signal transduction pathways, from receptors to gene activation. The focus in this problem, however, is on how the G protein transmits the signal that is initiated by α-factor binding to the α-factor pheronome receptor (Figure 15–9).

The α, β, and γ subunits of the yeast G protein show strong similarity to their mammalian counterparts, but whereas most of the work with mammalian G proteins is biochemical, with only modest clues from genetics, in yeast almost all the information comes from genetics with little biochemical corroboration.

In yeast nonmutant (wild-type) cells grow normally in the absence of the α-factor pheromone, but in its presence their growth is arrested and they undergo a normal mating response. Yeast mutants in which one or more of the genes for the α-factor pheromone receptor or the various components of the G protein have been deleted have characteristic phenotypes in the absence and in the presence of the α-factor pheromone (Table 15–1). Strains with any of these genes deleted cannot undergo the mating response and are therefore termed sterile.

A. Based on genetic analysis of the yeast mutants, decide which component of the G protein normally transmits the mating signal to the downstream effector molecules.
B. Predict the growth and mating phenotypes in the absence and presence of the α-factor pheromone of strains with the following mutant α subunits:

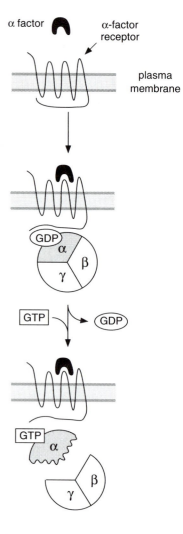

Figure 15–9 G-protein-linked α-factor pheromone receptor (Problem 15–16).

| | Phenotype | |
Mutation	Minus α Factor	Plus α Factor
None (wild-type)	normal growth	arrested growth, mating response
Receptor deleted	normal growth	normal growth, sterile
α subunit deleted	arrested growth	arrested growth, sterile
β subunit deleted	normal growth	normal growth, sterile
γ subunit deleted	normal growth	normal growth, sterile
Both α and β deleted	normal growth	normal growth, sterile
Both α and γ deleted	normal growth	normal growth, sterile

    1.   An α subunit that can bind GTP but cannot hydrolyze it.

    2.   An α subunit with an altered amino terminus that cannot be myristoylated. Addition of the fatty acid myristoylate is critical for localizing the α subunit to the membrane.

    3.   An α subunit that cannot bind to the activated pheromone receptor.

*15–17   In a cell line derived from normal rat thyroid, stimulation of the α_1-adrenergic receptor increases both inositol trisphosphate (IP_3) formation and release of arachidonic acid. IP_3 elevates intracellular Ca^{2+}, which mediates thyroxine efflux, whereas arachidonic acid serves as a source of prostaglandin E_2, which stimulates DNA synthesis. How is arachidonic acid release connected to the adrenergic receptor? Arachidonic acid could arise by cleavage from the diacylglycerol that accompanies IP_3 production. Alternatively, it could arise through an independent effect of the receptor on phospholipase A_2, which can directly release arachidonic acid from intact phosphoglycerides. Consider the following experimental observations:

    1.   Addition of noradrenaline to cell cultures stimulates production of both IP_3 and arachidonic acid.

    2.   If the α_1-adrenergic receptors are made unresponsive to noradrenaline by treatment with phorbol esters (which act through protein kinase C to cause phosphorylation, hence inactivation, of the receptor), addition of noradrenaline causes no increase in either IP_3 or arachidonic acid.

    3.   When cells are made permeable to GTPγS (a nonhydrolyzable analogue of GTP), production of both IP_3 and arachidonic acid is increased.

    4.   If cells are treated with neomycin (which blocks the action of phospholipase C), subsequent treatment with GTPγS stimulates arachidonic acid production but causes no increase in IP_3.

    5.   If cells are treated with pertussis toxin, subsequent treatment with GTPγS stimulates production of IP_3 but causes no increase in arachidonic acid.

  A.   Which of the two proposed mechanisms for arachidonic acid production in these cells do the observations support?

  B.   Describe a molecular pathway for activation of arachidonic acid production that is consistent with the experimental results.

15–18   Unlike myosin from skeletal muscle, smooth muscle myosin interacts with actin only when its light chains are phosphorylated. Phosphorylation is controlled by variations in the intracellular concentration of Ca^{2+}, which is mediated through calmodulin. You have purified myosin light-chain kinase from a smooth muscle, but the kinase activity is the same in the presence or absence of Ca^{2+}/calmodulin. A colleague suggests that you add protease inhibitors to ensure that the kinase remains intact through the purification. Under these conditions the kinase in the extract shows no activity unless Ca^{2+}/

Signaling via G-Protein-linked Cell-Surface Receptors

Table 15–2 Activities of Myosin Light-Chain Kinase Purified in the Presence and Absence of Protease Inhibitors (Problem 15–18)

Purification Scheme	Additions to Assay Mix	Relative Activity
Minus inhibitors	none	50
Minus inhibitors	Ca^{2+}	50
Minus inhibitors	calmodulin	50
Minus inhibitors	Ca^{2+}/calmodulin	50
Plus inhibitors	none	1
Plus inhibitors	Ca^{2+}	1
Plus inhibitors	calmodulin	1
Plus inhibitors	Ca^{2+}/calmodulin	100

calmodulin is present.

You partially purify the kinase by gel filtration and ion-exchange chromatography and then pass it over a calmodulin-affinity column in the presence of Ca^{2+}. To your delight, all the kinase activity sticks to the column and then elutes quantitatively when the column is washed with the calcium chelator EGTA. As shown in Table 15–2, the purified kinase now behaves as expected; it shows an absolute dependence on Ca^{2+}/calmodulin. Buoyed by this result, you attempt to rescue a sample of the original purified enzyme by passing it directly over the calmodulin-affinity column. To your surprise, the kinase passes straight through the column, whether or not Ca^{2+} is present.

A. What is your explanation for why the original purified enzyme was active independent of Ca^{2+}/calmodulin and was not retained on the calmodulin-affinity column?

B. Outline the sequence of molecular events that leads to contraction of smooth muscles. Begin with the entry of Ca^{2+} into the cytoplasm.

C. Do you think it is possible to use calmodulin-affinity chromatography as the first step in purification of myosin light-chain kinase?

15–19   The primary role of platelets is to control blood clotting. When they encounter the exposed basement membrane (collagen fibers) of a damaged blood vessel or a newly forming fibrin clot, they change their shape from round to spiky and stick to the damaged area. At the same time they begin to secrete serotonin and ATP, which accelerate similar changes in newly arriving platelets, leading to the rapid formation of a clot. The platelet response is regulated by protein phosphorylation. Significantly, platelets contain high levels of two protein kinases: C-kinase, which initiates serotonin release, and myosin light-chain kinase, which mediates the change in shape.

When platelets are stimulated with thrombin, the light chain of myosin and an unknown protein of 40,000 daltons are phosphorylated. When platelets are treated with a calcium ionophore, only the myosin light chain is phosphorylated; when they are treated with diacylglycerol, only the 40-kd protein is phosphorylated. Experiments using combinations of calcium ionophore and diacylglycerol show that the extent of phosphorylation of the 40-kd protein depends only on the concentration of diacylglycerol (Figure 15–10A); however, serotonin release depends on diacylglycerol and the calcium ionophore (Figure 15–10B).

A. Based on these experimental observations, describe the normal sequence of molecular events that leads to phosphorylation of the myosin light chain and the 40-kd protein. Indicate how the calcium ionophore and diacylglycerol treatments interact with the normal sequence of events.

B. Why do you think serotonin release requires both calcium ionophore and diacylglycerol?

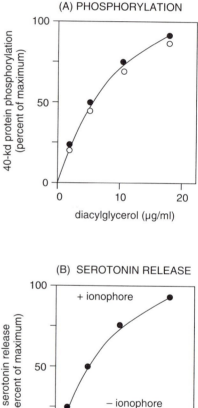

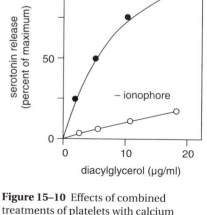

Figure 15–10 Effects of combined treatments of platelets with calcium ionophore and diacylglycerol (A) on phosphorylation of the 40-kd protein and (B) on serotonin release (Problem 15–19). Filled circles indicate experiments in which calcium ionophore was included; open circles indicate experiments in which calcium ionophore was absent.

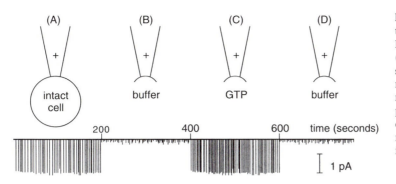

Figure 15–11 Experimental setup and typical results of patch-clamp analysis of K$^+$ channel activation by acetylcholine (Problem 15–20). The buffer is a buffered salts solution that does not contain nucleotides or Ca^{2+}. In all these experiments, acetylcholine is present inside the pipette, as indicated by the plus sign. The current through the membrane is measured in picoamps (pA). In (C) the GTP is added to the buffer.

15–20 Acetylcholine acts on muscarinic receptors in the heart to open K$^+$ channels, thereby slowing the heart rate. Treatment of heart cells with pertussis toxin blocks this physiological response, suggesting that a G protein is responsible for coupling receptor stimulation to channel activation. This process can be directly studied using the inside-out membrane patch-clamp technique. In this technique a patch of membrane is pulled from a cell with a pipette. The external surface of the membrane is in contact with the solution in the bore of the pipette, and the cytoplasmic surface faces outward and can be exposed readily to a variety of solutions (Figure 15–11). Receptors, G proteins, and K$^+$ channels remain associated with the membrane patch.

The status of the K$^+$ channel can be assessed by measuring the current through the membrane. A typical experiment is illustrated in Figure 15–11. When acetylcholine is added to a pipette (indicated by a plus sign) with a whole cell attached, K$^+$ channels open as indicated by the flow of current (Figure 15–11A). Under similar circumstances with a patch of membrane inserted into a buffered salts solution, no current flows (Figure 15–11B). When GTP is added to the buffer, however, current resumes (Figure 15–11C). Subsequent removal of GTP stops the current (Figure 15–11D). The results of several similar experiments to test the effects of different combinations of components are summarized in Table 15–3.

A. As shown in Table 15–3, line 4, addition of GppNp (a nonhydrolyzable analogue of GTP) caused the K$^+$ channel to open in the absence of acetylcholine. The flow of current, however, rose very slowly and reached its maximum only after a minute (compare with the immediate rise in Figure 15–11A and C). How do you think GppNp caused the channels to open in the absence of acetylcholine?

B. Why do you think it is that G$_\alpha$ activated the channel when the complete G protein did not?

C. Do these experiments suggest that the K$^+$ channel is likely to be activated by an intracellular messenger?

Table 15–3 Responses of K$^+$ Channel to Various Experimental Manipulations (Problem 15–20)

	Acetylcholine	Small Molecules Added to Buffer	Purified G-Protein Components Added to Buffer	K$^+$ Channel
1.	+	none	none	closed
2.	+	GTP	none	open
3.	–	GTP	none	closed
4.	–	GppNp	none	open
5.	–	none	G protein	closed
6.	–	none	G$_\alpha$	open
7.	–	none	G$_{\beta\gamma}$	closed
8.	–	none	boiled G protein	closed

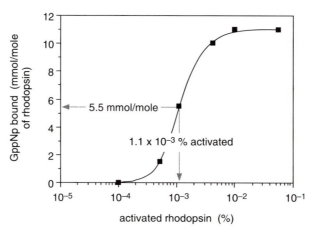

Figure 15–12 Binding of GppNp to rod cell membranes as a function of the fraction of activated rhodopsin (Problem 15–21). Background binding of GppNp to rod cell membranes in the dark has been subtracted from the values shown.

D. To the extent these experiments allow, draw a scheme for the activation of K^+ channels in heart cells in response to acetylcholine.

*15–21 Visual excitation begins with the photoisomerization of 11-*cis* retinal (the chromophore of rhodopsin) to its all-*trans* form. This activates rhodopsin, which then catalyzes an exchange of GTP for GDP in transducin (a G protein). The activated α subunit of transducin then stimulates a cyclic GMP phosphodiesterase, which rapidly hydrolyzes cyclic GMP, thereby removing a cofactor required to keep Na^+ channels open. The resulting hyperpolarization of the membrane is conveyed to the synapse.

It is estimated that one activated rhodopsin leads to hydrolysis of 5×10^5 cyclic GMP molecules per second. One stage in this enormous signal amplification is achieved by cyclic GMP phosphodiesterase, which hydrolyzes 1000 molecules of cyclic GMP per second. The additional factor of 500 could arise because one activated rhodopsin activates 500 transducins, or because one activated transducin activates 500 cyclic GMP phosphodiesterases, or through a combination of both effects. One experiment to address this question measured the amount of GppNp (a nonhydrolyzable analogue of GTP) that is bound by transducin in the presence of different amounts of activated rhodopsin. As indicated in Figure 15–12, 5.5 mmol of GppNp were bound per mole of total rhodopsin when 0.0011% of the rhodopsin was activated.

A. Assuming each transducin molecule binds one molecule of GppNp, calculate the number of transducin molecules that are activated by each activated rhodopsin molecule. Which mechanism of amplification does this measurement support?

B. Binding studies have shown that transducin-GDP has a high affinity for activated rhodopsin and that transducin-GTP has a low affinity; conversely, transducin-GTP has a high affinity and transducin-GDP has a low affinity for cyclic GMP phosphodiesterase. Are these affinities consistent with the mechanism of amplification you deduced from the above experiment? How so?

Signaling via Enzyme-linked Cell-Surface Receptors
(MBOC 759–771)

15–22 Fill in the blanks in the following statements.

A. _____ use cyclic GMP as an intracellular mediator in the same way that some G-protein-linked receptors use cyclic AMP, except the linkage between ligand binding and cyclase activity is direct rather than via a trimeric G protein.

B. Most growth factors bind to _____, which are transmembrane tyrosine-specific _____.

C. _____ recognize phosphorylated tyrosines and enable proteins that contain them to bind to intracellular signaling proteins that have been transiently phosphorylated on tyrosines.

D. The large _____ of _____ contains two other subfamilies: (1) the Rho and Rac proteins and (2) the Rab family.

E. The _____ help relay signals from receptor tyrosine kinases to the nucleus to stimulate cell proliferation or differentiation.

F. _____ increase the rate of hydrolysis of bound GTP by Ras, thereby inactivating it; these negative regulators are counteracted by _____, which promote the exchange of bound nucleotide by stimulating loss of GDP and uptake of GTP, thereby activating Ras.

G. Many serine/threonine kinases are involved in the cascades that relay signals from receptor tyrosine kinases downstream to the nucleus, but the family of _____, which contains at least five members, seems to play an especially important role.

H. _____ are thought to function in much the same way as receptor tyrosine kinases except that their kinase domain is encoded by a separate gene and is noncovalently associated with the receptor.

I. Members of the _____ of nonreceptor protein tyrosine kinases each contain SH2 and SH3 domains and are all located on the cytoplasmic side of the plasma membrane.

J. The _____ is composed of three polypeptide chains, which are thought to assemble after binding of interleukin-2 to form a functional receptor complex.

K. The tyrosine residues that are phosphorylated by protein tyrosine kinases are very rapidly dephosphorylated by _____.

L. _____ constitute a family of local mediators that regulate the proliferation and functions of most vertebrate cell types.

M. Members of the TGF-β superfamily of extracellular signaling proteins bind to receptors that are the first _____ to be identified.

N. _____ is a single-pass transmembrane receptor that plays a crucial part in the cell-cell interactions that control the fine-grained pattern of cell diversification during *Drosophila* development.

15–23 Indicate whether the following statements are true or false. If a statement is false, explain why.

__ A. Whereas a G-protein-linked receptor protein has seven transmembrane segments, each subunit of a catalytic receptor usually has only one.

__ B. The binding of atrial natriuretic peptides to a receptor quanylyl cyclase activates the cyclase to produce cyclic GMP, which in turn binds to and activates a cyclic GMP-dependent protein kinase (G-kinase), which phosphorylates specific proteins on serine and threonine residues.

__ C. It is thought that extracellular ligand binding to a receptor tyrosine kinase activates the intracellular catalytic domain by propagating a conformational change across the lipid bilayer through the single transmembrane α helix.

__ D. The small SH adaptor proteins have no intrinsic catalytic function and serve to couple tyrosine-phosphorylated proteins such as activated receptor tyrosine kinases to other proteins that do not have their own SH2 or SH3 domains.

__ E. In the signaling system that leads to the proper differentiation of the R7 photoreceptor in the *Drosophila* compound eye, and in most others that have been studied, the activation of Ras by receptor tyrosine kinases depends on the activation of a GNRP rather than on the inactivation of a GAP.

__ F. An unusual feature of the MAP-kinases is that their full activation requires phosphorylation of a threonine and a tyrosine, each of which is phosphorylated by MAP-kinase-kinase.

___ G. Receptor tyrosine kinases and tyrosine-kinase-associated receptors activate nonoverlapping sets of signaling pathways.

___ H. Receptor protein tyrosine phosphatases, when activated by ligand binding, catalyze the removal of phosphate groups from phosphotyrosines on all cellular proteins.

___ I. In principle, any mutation that results in the production of an abnormally active protein anywhere along the signaling pathways that lead from growth factor to the nucleus could promote cancer by encouraging the cell to proliferate in the absence of the appropriate extracellular signals.

___ J. Proteins in the TGF-β superfamily activate receptors that are serine/threonine protein kinases.

___ K. Typically, a cell that lacks Notch is unresponsive to lateral inhibition—the inhibitory signals from its immediate neighbors that would normally cause it to differentiate in a way different from them.

15–24   Rous sarcoma virus (RSV) carries an oncogene called *v-src* that encodes a constitutively active homologue of the normal cellular Src protein tyrosine kinase. Infection by RSV causes transformation of cells because its activated Src constitutively stimulates intracellular signaling pathways that lead to cell proliferation. You have isolated a mutant of RSV in which the Src protein is not myristoylated, as it normally would be, because of a change at its amino terminus. To determine the importance of myristoylation for transformation, you infect cells with the mutant and wild-type viruses and compare the properties of the infected and uninfected cells. As shown in Table 15–4, the cells infected with the mutant virus show almost none of the classical characteristics of transformation.

Analysis of the mutant-infected cells shows that they contain high levels of viral Src, but it is free in the cytosol rather than membrane bound. Nevertheless, the overall levels of tyrosine kinase activity in mutant-infected and wild-type-infected cells are the same; both are elevated 100-fold over the activity in uninfected cells.

A. Given that the number of molecules of the normal RSV Src protein and the nonmyristolylated version in infected cells is the same, calculate the difference in their concentration in the neighborhood of the plasma membrane. For the purposes of this calculation, assume the cell is spherical with a radius of 10 µm and that unmyristoylated Src is evenly distributed throughout, while membrane-bound Src is confined to a 4-nm-thick layer immediately beneath the membrane. (The volume of a sphere is $^4/_3\pi r^3$.)

B. Propose a molecular explanation for why the nonmyristoylated Src protein does not lead to a transformed phenotype.

*15–25   Interferon-γ (IFN-γ), which is a cytokine produced by activated T lymphocytes, binds to surface receptors on macrophages and stimulates their efficient scavenging of invading viruses and bacteria. A number of genes are

---

Table 15–4 Transformation-associated Properties of Uninfected Cells and Cells Infected with Wild-Type or Mutant RSV (Problem 15–24)

Transformation Parameters	Uninfected Cells	Wild-Type Infected Cells	Mutant Infected Cells
Growth in soft agar	–	+++	–
Surface fibronectin	+++	–	+++
Plasminogen activator	–	+++	–
Adhesion plaques	+++	–	+++
Glucose uptake	+	+++	+
Saturation density (cells/plate)	2×10^6	1×10^7	3×10^6

Figure 15–13 Sequence elements in the transcription factor that responds to IFN-γ (Problem 15–25).

SH3

P

SH2

heptad repeats

activated in response to IFN-γ binding, all of which contain a DNA sequence element with partial dyad symmetry (TTCCXGTAA) that is required for the IFN-γ response.

You have purified a 91-kd protein that binds to this sequence and have cloned the gene. The sequence of the gene indicates that the protein contains several heptad repeat sequences near its N terminus—a common element in many transcription factors that dimerize—and an SH3 domain and an SH2 domain adjacent to a site for tyrosine phosphorylation near the C terminus (Figure 15–13). By making antibodies to the protein, you show that it is normally located in the cytosol; however, after 15 minutes of exposure of cells to IFN-γ, the protein becomes phosphorylated on the tyrosine residue and moves to the nucleus.

Suspecting that tyrosine phosphorylation is the key to the regulation of this transcription factor, you assay its ability to bind the DNA sequence element in the presence of high concentrations of free phosphotyrosine or when mixed with anti-phosphotyrosine antibodies. Both treatments inhibit binding of the protein to DNA, as does treatment with a protein phosphatase. Finally, you measure the molecular weight of the cytosolic and nuclear forms of the protein: the cytosolic form is a monomer, whereas the nuclear form is a dimer.

A. Do you think that phosphorylation of the transcription factor is necessary for the factor to bind to DNA, or do you think phosphorylation is required to create an acidic activation domain to promote transcription?

B. Bearing in mind that SH2 domains bind phosphotyrosine, how do you think free phosphotyrosine interferes with the activity of the transcription factor?

C. How might tyrosine phosphorylation of the protein promote its dimerization? How do you think dimerization enhances its binding to DNA?

*15–26 The binding of platelet-derived growth factor (PDGF) stimulates dimerization of the PDGF receptor and phosphorylation of multiple tyrosine residues in the receptor's cytoplasmic domain. Most of the phosphorylation events occur in the segment of the receptor from residues 684 to 765. Deletion of this segment abolishes its ability to stimulate cell growth in response to PDGF.

Within seconds of PDGF binding to the wild-type receptor, an enzyme called phosphatidylinositol 3'-kinase (PI3-kinase), which phosphorylates the inositol ring of phosphatidylinositol on the cytoplasmic membrane face, can be immunoprecipitated by antibodies specific for the PDGF receptor. PI3-kinase contains SH2 domains in one of its subunits, which presumably bind to the phosphotyrosine residues in the activated PDGF receptor.

To assess the nature of the interaction between PI3-kinase and the activated PDGF receptor, you synthesize a set of pentapeptides with N-terminal tyrosines corresponding to tyrosines 684, 708, 719, 731, 739, 743, 746, and 755 of the PDGF receptor in both their phosphorylated and nonphosphorylated forms. You assay the association of PI3-kinase with the receptor by incubating immunoprecipitates of the PDGF receptor with PDGF to promote phosphorylation and then by mixing them with a cell extract that contains PI3-kinase in the presence of the various pentapeptides. You then wash the immunoprecipitates free of unbound PI3-kinase and assay them for their ability to transfer ^{32}P from ATP to phosphatidylinositol. Radioactivity in phosphatidylinositol was measured by autoradiography (Figure 15–14).

pentapeptides with tyrosine

684 708 719 731 739 743 746 755

pentapeptides with phosphotyrosine

684 708 719 731 739 743 746 755

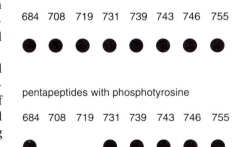

Figure 15–14 Effects of pentapeptides on phosphorylation of phosphatidylinositol (Problem 15–26). Black circles indicate the presence of radioactive phosphatidylinositol.

A. Do your results support the notion that PI3-kinase interacts with the activated PDGF receptor by binding of its SH2 domains to phosphotyrosine residues in the receptor? Why or why not?

B. The amino acid sequences of the PDGF pentapeptides and of the peptide segments in other activated receptors that are known to bind PI3-kinase are shown below.

684	YSNAL	YMMMR	(FGF receptor)
708	YMDMS	YTHMN	(insulin receptor)
719	YVPML	YEVML	(hepatocyte growth factor receptor)
731	YADIE	YMDMK	(steel factor receptor)
739	YMAPY	YVEMR	(CSF-1 receptor)
743	YDNYE		
746	YEPSA		
755	YRATL		

What are the common features of peptide segments that form binding sites for PI3-kinase?

C. Purified PI3-kinase is able to phosphorylate phosphatidylinositol in the absence of the PDGF receptor. Why do you suppose its association with the activated PDGF receptor is required for phosphorylation of phosphatidyl-inositol inside a cell?

15–27 The SH3 domain comprises approximately 60 amino acids and is found in a wide variety of proteins, including the nonreceptor tyrosine kinases—Src and Abl—and Sem5, a small adaptor protein involved in nematode signal trans-duction. Like SH2 domains, SH3 domains are thought to recognize and bind to structural motifs in other proteins.

To identify the structural motifs to which SH3 domains can bind, you construct a fusion protein between an SH3 domain and glutathione-S-transferase (GST). This enables you to produce large amounts of the fusion protein in bacteria and to purify it easily using a glutathione-affinity column. You tag the purified GST-SH3 protein with biotin (which allows you to detect the fusion protein with high sensitivity when it binds to an interacting protein immobilized on a filter) and use it to screen a cDNA library expressed in *E. coli*. As a control, you probe a duplicate filter with biotin-tagged GST.

You isolate two different clones in this screen, which you name 3BP-1 and 3BP-2. Further analysis verifies that these clones encode proteins that bind to SH3 domains. In both proteins the region responsible for binding to SH3 domains maps to a short proline-rich amino acid sequence.

A. What do you think is the point of probing a duplicate filter with biotin-tagged GST?

B. Could you use this method to find cDNAs for proteins that bind to SH2 domains? Why or why not?

C. Many protein-protein interactions are known in which a reasonably large domain of one protein specifically recognizes a short segment of another protein: for example, the recognition of a phosphorylated pentapeptide by PI3-kinase discussed in the previous problem. How do you think these kinds of interactions differ from the kinds of interactions found between the protein subunits of multisubunit enzymes?

*15–28 Since its discovery in 1981, the *ras* oncogene, which encodes a monomeric GTPase, has occupied a central place in human cancer studies. Activating mutations in the Ras protein usually prolong the lifetime of the active, GTP-bound state, and such Ras mutants are powerfully oncogenic. Ras seems to form a link between receptor tyrosine kinases and the MAP-kinase cascade, but it is unclear how the connection between the monomeric GTPase and the MAP-kinase is made.

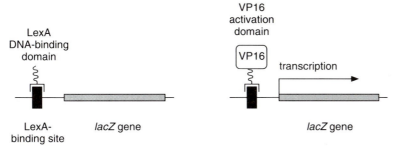

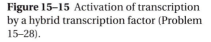

Figure 15–15 Activation of transcription by a hybrid transcription factor (Problem 15–28).

You decide to use the yeast two-hybrid system to find the proteins with which Ras interacts. The two-hybrid system is a clever scheme for detecting protein-protein interactions in such a way that a known protein such as Ras can be used to screen a random cDNA library to pick up potential interacting partners. This approach depends on the modular nature of many transcription factors, which have one domain that binds to DNA and another domain that activates transcription. Domains can be readily interchanged by recombinant DNA methods so that hybrid transcription factors can be constructed: thus, the DNA-binding domain of the *E. coli* LexA repressor can be combined with the powerful VP16 activation domain from herpes virus to activate transcription of genes downstream of a LexA DNA-binding site (Figure 15–15).

It is not even necessary that the two domains be covalently linked; if they can be brought into proximity by noncovalent protein-protein interactions, they will still activate transcription. This is the key feature of the two-hybrid system. Thus, if one member of an interacting pair of proteins is fused to the DNA-binding domain of LexA and the other is fused to the VP16 activation domain, transcription will be activated when the two hybrid proteins interact inside a yeast cell. If the gene whose transcription is turned on by these artificial means is essential for growth of the yeast, like the *HIS3* gene, or can give rise to a colored product, as can the *lacZ* gene (which encodes β-galactosidase), it is possible to design powerful screens for protein-protein interactions.

To check out the system, you construct two hybrid genes containing the LexA DNA-binding domain, one fused to Ras (LexA-Ras) and the other fused to nuclear lamin (LexA-lamin). You make a second pair of constructs with the VP16 activation domain alone (VP16) or fused to the adenylyl cyclase gene (VP16-CYR). Adenylyl cyclase is known to interact with Ras and serves as a positive control; nuclear lamins do not interact with Ras and serve as a negative control. These plasmid constructs are introduced into a strain of yeast containing modified chromosomal copies of the *HIS3* gene and the *lacZ* gene, both with LexA-binding sites positioned immediately upstream. Individual transformed colonies are tested for growth on a plate lacking histidine and for ability to form blue colonies (as compared to the normal white colonies) when grown in the presence of an appropriate substrate (XGAL) for β-galactosidase. The setup for the experiment is outlined in Table 15–5.

Table 15–5 Experiment to Test the Two-Hybrid System (Problem 15–28)

Plasmid Constructs	Growth on Plates Lacking Histidine	Color on Plates with XGAL
LexA-Ras		
LexA-lamin		
VP16		
VP16-CYR		
LexA-Ras + VP16		
LexA-Ras + VP16-CYR		
LexA-lamin + VP16		
LexA-lamin + VP16-CYR		

A. Fill in Table 15–5 with your expectations: in column 2, use a plus sign to indicate growth on plates lacking histidine and a minus sign to indicate no growth; in column 3, write "blue" or "white" to indicate the color of colonies grown in the presence of XGAL.

B. For any entries in the table that you expect to grow in the absence of histidine and form blue colonies with XGAL, sketch the structure of the active transcription factor on the *lacZ* gene.

C. If you want two proteins to be expressed in a single polypeptide chain, what must you be careful to do when you fuse the two genes together?

15–29 To use the two-hybrid system to screen for proteins that interact with Ras, you make a cDNA library in which the cDNA inserts are positioned at the C terminus of the *VP16* gene segment. You then transfect the library into yeast that already contain the LexA-Ras plasmid described in the preceding problem. You grow the transformed cells on plates in the presence of XGAL and in the absence of histidine, and you isolate blue colonies for further testing.

The LexA-Ras plasmid is eliminated from such cells by genetic selection, and the "cured" cells, which contain only a VP16-cDNA plasmid, are checked for growth in the absence of histidine and for color when grown in the presence of XGAL. Cured cells that do not grow in the absence of histidine and form white colonies in the presence of XGAL are then transformed with the LexA-lamin plasmid and tested again for growth in the absence of histidine and colony color in the presence of XGAL. Only cells that do not grow in the absence of histidine and form white colonies in the presence of XGAL are analyzed further.

Out of 1.4×10^6 original transformants only 19 clones meet your stringent criteria for detailed analysis. You sequence the cDNA inserts downstream of the *VP16* gene segment in each and find that nine clones have cDNA inserts that correspond to the N-terminal domain of known Raf serine/threonine protein kinases: six correspond to c-Raf and three correspond to A-Raf.

This is a very exciting finding because other work shows that immunoprecipitates of Raf can phosphorylate and activate MAP-kinase-kinase. Thus, Raf, if it truly binds to Ras, could provide the missing link between the monomeric GTPase and the MAP-kinase cascade.

To test whether the interaction suggested by your findings is real, you make fusion proteins between the maltose-binding protein and Raf (MBP-Raf) and between glutathione-S-transferase and Ras (GST-Ras). You then produce the two fusion proteins in bacteria and test their ability to bind together by passing the mixture over an amylose affinity column, which will bind MBP-Raf. When the bound proteins are eluted with maltose, you find that a significant fraction of the input GST-Ras protein is eluted along with the MBP-Raf protein. This result provides a biochemical confirmation of the genetic results with the two-hybrid system. You are ecstatic!

A. Since its introduction in 1989, the two-hybrid system has been modified considerably to deal with the problem of false positives, that is, to eliminate cDNA clones that do not really encode a protein that interacts with the protein of interest. Several of these modifications are built into your selection scheme.

   1. When you initially transfect the cDNA library into yeast containing the LexA-Ras plasmid, you observe, in addition to blue colonies, many white colonies when the cells are grown in the presence of XGAL and in the absence of histidine. What type of VP16-cDNA might give rise to such white colonies?

   2. Among the colonies that have been cured of the LexA-Ras plasmid—and contain only a VP16-cDNA plasmid—some grow in the absence of histidine and turn blue when grown in the presence of XGAL. What type of VP16-cDNA might give this result in the absence of the LexA-Ras plasmid?

3. When the LexA-lamin plasmid is introduced into cells that contain a VP16-cDNA plasmid (ones that alone do not grow in the absence of histidine and do not turn blue in the presence of XGAL), some now form blue colonies in the absence of histidine and in the presence of XGAL. What type of VP16-cDNA might give these results in the presence of the LexA-lamin plasmid?

B. A friend of yours is concerned about the construction of the cDNA library. As she correctly points out, in your construction scheme the cDNA can go into the plasmid in either orientation relative to the *VP16* gene segment, and only one out of three inserts in the correct orientation will be in the correct reading frame. Does this matter for the success of the two-hybrid approach? Why or why not?

C. You were completely convinced by the results of the two-hybrid screen, but your advisor pressed you to carry out the biochemical studies described in the last paragraph of the problem. Why do you think your advisor felt it necessary to demonstrate a direct interaction of Ras and Raf by these biochemical studies?

Target-Cell Adaptation
(MBOC 771–778)

15–30 Fill in the blanks in the following statements.

A. Cells reversibly adjust their sensitivity to a stimulus by a process of
_____.

B. Receptor-mediated endocytosis of ligands is often associated with a decrease in the total number of cell-surface receptors as a result of enhanced receptor degradation, a mechanism known as _____.

C. In bacteria the presence of attractants and repellents in the environment is transmitted across the membrane by four types of _____.

15–31 Indicate whether the following statements are true or false. If a statement is false, explain why.

___ A. Most receptors, when they are endocytosed, deliver their ligands to lysosomes for degradation but are themselves repeatedly retrieved and recycled.

___ B. The activation of any type of receptor in the target cell that activates adenylyl cyclase can desensitize the β_2-adrenergic receptor by A-kinase-dependent phosphorylation—an example of heterologous desensitization, where one ligand desensitizes target cells to another.

___ C. Addicts need higher doses of morphine for pain relief than do normal people—not because of receptor down-regulation, but because their opiate-sensitive cells increase their expression of the genes for A-kinase and adenylyl cyclase.

___ D. Owing to an inherent handedness in the construction of their flagella, *E. coli* and *S. typhimurium* swim in straight lines when their flagella rotate counter-clockwise, but they tumble more or less randomly when their flagella rotate clockwise.

___ E. If the concentration of an attractant remains constant, bacteria continue swimming in a straight line.

___ F. The binding of an attractant to a chemotaxis receptor increases the activity of the receptor, which stimulates the intracellular pathway and leads to tumbling. The binding of a repellent has the opposite effect: it decreases the activity of the receptor and leads to smooth swimming.

___ G. When chemotaxis receptors bind an attractant, their exposed methylation sites can be methylated, which increases the activity of the receptor to its former level and increases the frequency of tumbling.

15–32 After prolonged exposure to hormones that bind to β-adrenergic receptors, cells become refractory and cease responding. To examine this desensitization phenomenon, you use a newly developed reagent, CGP-12177, which is a hydrophilic molecule that specifically binds to β-adrenergic receptors. In contrast to the binding of dihydroalprenolol, which is hydrophobic, CGP-12177 binding exactly parallels the decrease in hormone-dependent adenylyl cyclase activity observed in extracts from cells treated with isoproterenol for increasing times (Figure 15–16). To understand the difference in receptor binding by these two molecules, you lyse untreated cells and isoproterenol-treated (desensitized) cells, fractionate them by centrifugation through sucrose-density gradients, and measure binding by dihydroalprenolol and CGP-12177 (Figure 15–17). In addition to ligand binding, you also measure 5′-nucleotidase activity, which is a marker enzyme for the plasma membrane.

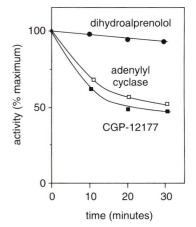

Figure 15–16 Adenylyl cyclase activity, ^{3}H-CGP-12177 binding, and ^{3}H-dihydroalprenolol binding at various times after treatment with isoproterenol (Problem 15–32). All activities are expressed as a percentage of the values at time zero.

   A. Give an explanation for the differences in binding by dihydroalprenolol and CGP-12177.

   B. What do you think might be the basis for isoproterenol-induced desensitization in these cells?

15–33 The nicotinic acetylcholine receptor is a neurotransmitter-dependent ion channel, which is composed of four types of subunit. Phosphorylation of the receptor by A-kinase attaches one phosphate to the (γ subunit and one phosphate to the δ subunit. Fully phosphorylated receptors are desensitized much more rapidly than unmodified receptors. To study this process in detail, you phosphorylate two preparations of receptor to different extents (0.8 mole phosphate/mole receptor and 1.2 mole phosphate/mole receptor) and measure desensitization over several seconds (Figure 15–18). Both preparations behave as if they contain a mixture of receptors; one form that is rapidly desensitized (the initial steep portion of the curves) and another form that is desensitized at the same rate as the untreated receptor.

   A. Assuming that the γ and δ subunits are independently phosphorylated at equal rates, calculate the percentage of receptors that carry zero, one, and two phosphates per receptor at the two extents of phosphorylation.

   B. Do these data suggest that desensitization requires one phosphate or two phosphates per receptor? If you decide that desensitization requires only one phosphate, indicate whether the phosphate has to be on one specific subunit or can be on either of the subunits.

*15–34 Four types of chemotaxis receptors have been identified in *E. coli*. These receptors mediate chemotactic responses to two different amino acids, to sugars, and to dipeptides. As part of a practical demonstration in bacterial chemotaxis, your instructor has given you a wild-type strain with all four receptors intact and four mutant strains with one or more of the receptors

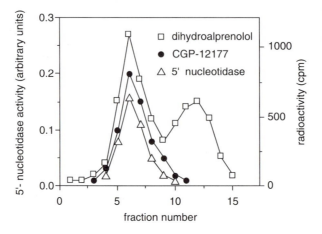

Figure 15–17 Ligand binding in sucrose-gradient fractions from untreated and isoproterenol-treated cells (Problem 15–32).

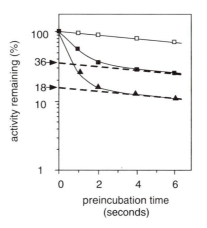

Figure 15–18 Desensitization rates of untreated acetylcholine receptor and two preparations of phosphorylated receptor (Problem 15–33). Open squares represent untreated receptors; filled squares represent receptors with 0.8 mole phosphate/mole receptor; and filled triangles represent receptors with 1.2 mole phosphate/mole receptor. Arrows indicate the fractions of the phosphorylated preparations that behaved like the untreated receptor.

missing. Your assignment is to identify which receptor mediates the response to which attractant. The experimental assay is very simple. You fill a capillary pipette with a solution of the attractant, dip it into a buffered solution containing bacteria, remove it after 5 minutes, and count the number of bacteria in the capillary. Your results are shown in Table 15–6. Identify each attractant and its appropriate receptor.

15–35   To clarify the relationship between the structure of a chemotaxis receptor and the functions of stimulus recognition, signal transduction, and adaptation, you have cloned the gene for the aspartate receptor from *Salmonella typhimurium*. Inadvertently, you also cloned a mutant form of the gene that is missing its 35 C-terminal amino acids. By introducing the wild-type and truncated forms of the gene back into a mutant of *Salmonella* that is missing the normal gene for the aspartate receptor, you can test for functional differences between the two cloned genes. To your surprise, even though the truncated receptor still contains the peptide sequences that are the targets for methylation, it is not methylated in cells. The binding of aspartate by the wild-type and truncated receptors, however, is identical to the normal cellular receptor. The cloned receptors are about 15 times more abundant than normal.

To test for signal transduction by the cloned receptors and to assess their adaptive properties, you expose bacteria containing the receptors to a sudden change in the concentration of aspartate. Wild-type bacteria in the absence of an attractant change their direction of rotation (tumble) every few seconds. Upon exposure to an attractant, however, the changes in direction of rotation are suppressed, which leads to a period of smooth swimming. If the concentration of attractant remains constant (even if high), wild-type bacteria quickly adapt and begin again to tumble every few seconds. To observe these behavioral changes experimentally, you tether bacteria by their flagella to coverslips so that you can observe their direction of rotation. You then expose

Table 15–6 Chemotaxis in Wild-Type and Mutant Strains of *E. Coli* (Problem 15–34)

Strain	Intact Receptors	Number of Cells (1000s) in Capillary				
		Serine	Aspartate	Ribose	Prolylglycine	None
1	Tap, Tar, Trg, Tsr	59	105	95	6.6	0.5
2	Tap, Tar, Trg	0.7	84	77	13	0.8
3	Trg, Tsr	34	0.7	59	0.6	0.6
4	Tap, Trg, Tsr	55	0.6	65	4.1	0.5
5	Tar, Trg, Tsr	70	59	85	0.9	0.8

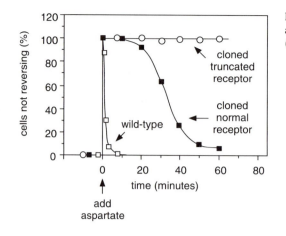

Figure 15–19 Behavior of wild-type cells and cells containing the cloned receptors (Problem 15–35).

them to aspartate and count the number of bacteria that do not reverse their direction of rotation in 1-minute intervals after addition of aspartate. The results for wild-type cells and for cells containing the cloned aspartate receptors are shown in Figure 15–19.

A. Is signal transduction by the two cloned receptors normal?
B. Are the adaptive properties of the cloned receptors normal?
C. Suggest molecular explanations for why the cloned normal receptor and the cloned truncated receptor, when introduced into bacteria, respond differently from the normal receptor in wild-type cells (Figure 15–19).

The Cytoskeleton

- **The Nature of the Cytoskeleton**
- **Intermediate Filaments**
- **Microtubules**
- **Cilia and Centrioles**
- **Actin Filaments**
- **Actin-binding Proteins**
- **Muscle**

The Nature of the Cytoskeleton
(MBOC 788–795)

16–1 Fill in the blanks in the following statements.

A. The complex network of protein filaments that extends throughout the cytoplasm is called the _____.

B. In most cells, the minus ends of microtubules are stabilized by embedding them in a structure called the _____, which generally lies next to the nucleus, at the center of the cell.

C. Directional intracellular movements in eucaryotic cells are generated by _____, which bind to either an actin filament or a microtubule and use the energy derived from repeated cycles of ATP hydrolysis to move steadily along it.

16–2 Indicate whether the following statements are true or false. If a statement is false, explain why.

__ A. Actin filaments, microtubules, and intermediate filaments each have a different arrangement in a eucaryotic cell and, in combination with a variety of accessory proteins, provide a different function.

__ B. Microtubules are highly dynamic structures that can shorten as well as lengthen; in general, they grow by addition of subunits to the plus end and shrink by removal of subunits from their minus end.

__ C. In an isolated cell fragment, microtubules will reorganize themselves to form a starlike array emanating from a new microtubule organizing center at the center of the cell fragment.

__ D. The normal locations of the ER and Golgi are thought to be determined by receptor proteins on their cytosolic surface that bind specific microtubule-dependent motors—a plus-end-directed kinesin for the ER and a minus-end-directed dynein for the Golgi apparatus.

__ E. Actin filaments lying just beneath the plasma membrane are cross-linked into a network by various actin-binding proteins that form the cell cortex, which functions to control cell-surface movements.

__ F. As a result of the contact between cytotoxic T cells and their target cells, the centrosomes in both cells become localized near the site of contact.

__ G. The cytoskeleton exerts forces and generates movements without any major chemical change, making it especially difficult to assay its function.

*16–3 In addition to conducting impulses in both directions, nerve axons carry vesicles to and from the cell body. The vesicles appear to move along micro-

*Problems with an asterisk (*) are answered in the Instructor's Manual.

tubule tracks. Do outbound vesicles move along microtubules that are oriented in one direction and incoming vesicles move along microtubules that are oriented in the opposite direction? Or are microtubules all oriented in the same direction with different protein "motors" providing the directionality of vesicle movement?

To distinguish between these possibilities, you prepare a cross-section through a nerve axon and decorate the microtubules with tubulin, which binds to the tubulin subunits of the microtubule to form hooks. The decorated microtubules are illustrated in Figure 16–1. Do all the microtubules run in the same direction or not? How can you tell?

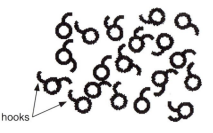

Figure 16–1 Tubulin-decorated microtubules in a cross-section through a nerve axon (Problem 16–3). The hooks represent the tubulin decoration.

16–4   Phosphorylation seems to control the intracellular movement of particles in the pigment cells (melanophores) of the African fish *Tilapianossambica*. Melanophores, which are on the surface of the fish allow it to change color. These cells contain small granules of black pigment that can be dispersed throughout the cell (making the cell darker) or aggregated in a small spot in the middle of the cell (making it appear lighter) (Figure 16–2). Changes from one state to the other occur in a matter of minutes in response to hormonal stimulation. The pigment granules run along microtubule tracks.

Melanophores can be made permeable to small molecules by washing them gently in a low concentration of detergent. Pigment aggregates when ATP is added to the treated cells, but when cyclic AMP is also included, the pigment disperses. Melanophores can be made to undergo repeated cycles of pigment dispersion and aggregation by adding and removing cyclic AMP. Cyclic AMP often affects cellular processes by stimulating the activity of A-kinase. Could a similar system be responsible for melanophore pigment mobility?

To test this hypothesis, various agents known to affect phosphorylation and desphosphorylation were tested for their effects on pigment dispersal and aggregation. The results are summarized below:

1. Addition of a protein kinase inhibitor inhibits the cyclic AMP-induced dispersion.
2. Sodium vanadate, a potent inhibitor of protein phosphatases, inhibits aggregation but does not affect dispersal.
3. Addition of ATPγS without any ATP or cyclic AMP causes a slow dispersion (see Figure 16–19 for structure of ATPγS). ATPγS can be used by many protein kinases, but it usually does not serve as a source of energy for systems that depend on the free energy of ATP hydrolysis (see Problem 16–31).
4. Pigment that has been dispersed with ATPγS reaggregates extremely slowly when new buffer containing ATP is added, in contrast to pigment that has been dispersed with ATP and cyclic AMP.
5. When ATPγ^{32}P is added in the presence of cyclic AMP, a 57-kd protein is labeled; when cyclic AMP is removed, the protein loses its label.

A. Draw a plausible molecular pathway that describes the control of pigment dispersal and aggregation.
B. Do pigment aggregation, pigment dispersal, or both require the free energy of ATP hydrolysis?

*16–5   Yeast cells choose bud sites on their surface in two distinct spatial patterns: axial for **a** and α cells and bipolar for **a**/α cells (Figure 16–3A). The selection of a new bud site establishes a cell polarity that involves the cytoskeleton and determines the site of new cell growth (Figure 16–3B). To find the genes responsible for bud-site selection, mutagenized α cells were examined visually to identify mutant cells with altered budding patterns. Five genes, *BUD1–5*, were identified this way: all are nonessential genes that have no effect on cell growth or morphology. Further genetic tests indicated that these genes are involved in a single pathway for bud-site selection. In the absence of *BUD1, BUD2,* or *BUD5* the budding pattern is random; in the absence of

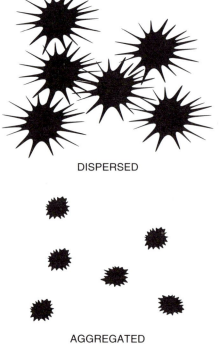

DISPERSED

AGGREGATED

Figure 16–2 Dispersed and aggregated pigment granules in melanophore cells (Problem 16–4).

BUD3 or BUD4 the budding pattern is bipolar. All five genes are required for axial budding.

$$RANDOM \xrightarrow[\substack{BUD1 \\ BUD2 \\ BUD5}]{} BIPOLAR \xrightarrow[\substack{BUD3 \\ BUD4}]{} AXIAL$$

Sequence analysis of the genes and biochemical characterization of the proteins indicate that *BUD1* encodes a monomeric GTPase, *BUD2* encodes a GTPase-activating protein (GAP) that stimulates GTP hydrolysis by the Bud1 protein, and *BUD5* encodes a guanine nucleotide releasing protein (GNRP) that catalyzes exchange of GTP for the GDP bound to the Bud1 monomeric GTPase.

A. By analogy to the use of monomeric GTPases, GAPs, and GNRPs in directing vesicular transport between membrane-bounded compartments, design a plausible scheme by which the Bud1, Bud2, and Bud5 proteins might be used to deliver critical cytosolic proteins to a bud site.

B. Does your scheme account for the initial selection of the site for bud formation?

Intermediate Filaments
(MBOC 796–803)

16–6 Fill in the blanks in the following statements.

A. The most diverse family of intermediate filament proteins comprises the _____, which form _____, with over 20 distinct forms in human epithelia and an additional 8 in hair and nails.

B. The intermediate filament protein _____ is widely distributed in cells of mesodermal origin, such as fibroblasts, endothelial cells, and white blood cells.

C. The intermediate filament protein _____ is found in both smooth and striated muscle cells.

D. Glial filaments in astrocytes and some Schwann cells are composed of _____.

E. Three _____ assemble into _____, which extend along the length of an axon and form its primary cytoskeletal components, especially in mature nerve cells.

F. The _____, which is composed of _____, is a meshwork of intermediate filaments that lines the interior surface of the inner nuclear membrane.

16–7 Indicate whether the following statements are true or false. If a statement is false, explain why.

__ A. Intermediate filaments are named for their size, which is intermediate between actin filaments and microtubules.

__ B. Intermediate filaments are soluble only in solutions containing high salt and nonionic detergents.

__ C. Like microtubules and actin filaments, intermediate filaments form polarized structures and their function depends on this polarity.

__ D. A single epithelial cell can make a variety of keratins, all of which co-polymerize into a single keratin filament system.

__ E. Keratins and vimentin-related proteins do not co-polymerize with each other: when they are expressed in the same cell, they form separate filament systems.

__ F. The most dramatic example of the importance of phosphorylation in the disassembly of intermediate filaments is provided by the nuclear lamins, which are phosphorylated and disassembled each time the cell enters mitosis.

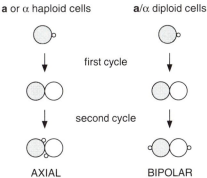

(A) BUDDING PATTERNS

a or α haploid cells   a/α diploid cells

first cycle

second cycle

AXIAL   BIPOLAR

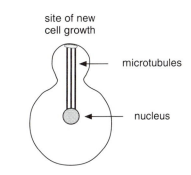

(B) CELL POLARIZATION

site of new cell growth

microtubules

nucleus

Figure 16–3 Patterns of budding (A) and cell polarization during bud formation (B) (Problem 16–5).

Figure 16–4 The amino acid sequence of nuclear lamin C (Problem 16-8).

```
                        coil 1A
METPSQRRATRSGAQASSTPLSPTRITRLQEKEDLQELNDRLAVYIDRVRSLETENA
              coil 1B
GLRLRITESEEVVSREVSGIKAAYEAELGDARKTLDSVAKERARLQLELSKVREEFK

ELKARNTKKEGDLIAAQARLKDLEALLNSKEAALSTALSEKRTLEGELHDLRGQVAK

LEAALGEAKKQLQDEMLRRVDAENRLQTMKEELDFQKNIYSEELRETKRRHETRLVE
               coil 2
IDNGKQREFESRLADALQQLRAQHEDQVEQYKKELEKTYSAKLDNARQSAERNSNLV

GAAHEELQQSRIRIDSLSAQLSQLQKQLAAKEAKLRDLEDSLARERDTSRRLLAEKE

REMAEMRARMQQQLDEYQQLLDIKLALDMQIHAYRKLLEGEEERLRLSPSPTSQRSR

GRASSHSSQTQGGGSVTKKRKLESTESRSSPSQHARTSGRVAVEEVDEEGKFVRLRN

KSNEDQSMGNWQIKRQNGDDPLLTYRFPPKFTLKAGQVVTIWAAGAGATHSPPTDLV

WKAQNTWGCGNSLRTALINSTGEEVAMRKLVRSVTVVEDDEDEDGDDLLHHHHVSGS

RR
```

___ G. Cytoplasmic intermediate filaments are thought to be essential for cell survival.

___ H. A major function of cytoplasmic intermediate filaments is to resist mechanical stress.

16–8 You have just deduced the amino acid sequence of nuclear lamin C from the nucleotide sequence of a cDNA clone. Since nuclear lamin C is a member of the intermediate filament family, it should show regions of the coiled-coil heptad repeat motif AbcDefg, where A and D are hydrophobic amino acids and b, c, e, f, and g can be almost any amino acid. Your sequence is shown in Figure 16–4 with potential coiled-coil regions in bold. Examine the segment marked "coil 1A." Does it conform to the heptad repeat? (The mnemonic "FAMILY VW" will help you recognize hydrophobic amino acids.)

*16–9 The nuclear envelope is strengthened by a fibrous meshwork of lamins (the nuclear lamina), which supports the membrane on the nuclear side. When cells enter mitosis, the nuclear envelope breaks down and the nuclear lamina disassembles. Assembly and disassembly of the nuclear lamina may be controlled by reversible phosphorylation of lamins A, B, and C, since the lamins from cells that are in mitosis carry significantly more phosphate than do the lamins from cells that are in interphase.

To investigate the role of phosphorylation, you label cells with ^{35}S-methionine and then purify lamins A, B, and C from mitotic cells and from interphase cells. You then analyze each of the purified lamins and a mixture of the lamins from mitotic and interphase cells by two-dimensional gel electrophoresis (Figure 16–5A). You also treat identical samples with alkaline phosphatase, which removes phosphates from proteins, and analyze them in the same way (Figure 16–5B).

A. Why does treatment with alkaline phosphatase reduce the number of lamin spots to three regardless of the number seen in the absence of phosphatase treatment?

B. How many phosphate groups are attached to lamins A, B, and C during interphase? How many are attached during mitosis? How can you tell?

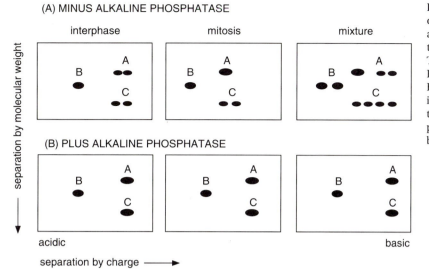

(A) MINUS ALKALINE PHOSPHATASE

interphase   mitosis   mixture

separation by molecular weight

(B) PLUS ALKALINE PHOSPHATASE

acidic   basic

separation by charge ⟶

Figure 16–5 Two-dimensional separation of nuclear lamins from cells in interphase and mitosis (Problem 16–9). (A) No treatment with alkaline phosphatase. (B) Treatment with alkaline phosphatase. Letters identify the positions of lamins A, B, and C. The purified lamins from interphase and mitotic cells were added together to create the mixture. Acidic proteins are more negatively charged; basic proteins are more positively charged.

C. Why was ^{35}S-methionine rather than ^{32}P-phosphate used to label lamins in experiments designed to measure phosphorylation differences? How would the autoradiograms have differed if ^{32}P-phosphate had been used instead?

D. Do you think that these results prove that lamin disassembly during mitosis is caused by their reversible phosphorylation?

Microtubules
(MBOC 803–815)

16–10 Fill in the blanks in the following statements.

A. The fast-growing end of a microtubule is defined as the _____ and the other as the _____.

B. When depolymerized microtubules in a cell are observed to repolymerize, the new microtubules grow out from the centrosome to form a starlike structure called an _____ and then elongate toward the cell periphery.

C. The _____ is the major microtubule organizing center in almost all animal cells.

D. Surrounding each centriole pair is a specialized region of the cytoplasm termed the _____, or _____, and it is the part of the centrosome that nucleates microtubule polymerization.

E. The fluctuations in length of microtubules, called _____, requires an input of energy to shift the chemical balance between polymerization and depolymerization—energy that comes from the hydrolysis of GTP.

F. The most versatile modifications of microtubules are conferred by the binding of proteins called _____, which serve both to stabilize microtubules against disassembly and to mediate their interaction with other cell components.

G. _____ are microtubule-dependent motor proteins involved in organelle transport and mitosis and are closely related to the motor protein in cilia and flagella.

H. _____ are a diverse family of microtubule-dependent motor proteins that are involved in organelle transport, in mitosis, in meiosis, and in transport of synaptic vesicles along axons.

16–11 Indicate whether the following statements are true or false. If a statement is false, explain why.

___ A. Each microtubule is built from 13 linear protofilaments, each composed of alternating α- and β-tubulin subunits and bundled in parallel to form a cylinder with two distinct ends.

___ B. The antimitotic drug colchicine binds to the tubulin subunits in spindle microtubules, causing them to disassemble into free subunits.

___ C. Because there is a substantial kinetic barrier to the nucleation of new microtubules in solution, tubulin polymerization in the cell occurs only at specific nucleation sites, usually the centrosome.

___ D. The structural polarity of a microtubule, which reflects the regular orientation of its tubulin subunits, makes the two ends of the polymer different in ways that have a profound effect on their rates of growth.

___ E. All microtubule organizing centers contain centrioles that help nucleate microtubule polymerization.

___ F. Individual microtubules grow toward the cell periphery at a constant rate for some period and then suddenly shrink rapidly back toward the centrosome; they may shrink partially and then recommence growing, or they may disappear completely.

___ G. The delayed hydrolysis of GTP after tubulin assembly results in the presence of a GTP cap on the end of the microtubule, and because tubulin molecules carrying GTP bind to one another with higher affinity than tubulin molecules carrying GDP, the GTP cap will encourage a growing microtubule to keep growing.

___ H. Cell polarity is thought to be determined by unknown structures or factors localized in particular regions of the cell cortex that capture the plus ends of microtubules and prevent their depolymerization.

___ I. Acetylation of α-tubulin and removal of its carboxyl-terminal tyrosine are modifications of microtubules that mark them for rapid disassembly.

___ J. MAPs have two structural domains: one binds to microtubules, and the other helps link the microtubule to other cell components.

___ K. Axons and dendrites of nerve cells are packed with microtubules, and although microtubules are long in axons and short in dendrites, they all are oriented with their plus ends pointing away from the cell body.

___ L. Each heavy chain of a microtubule-dependent motor protein contains an ATP-binding head, which is the ATPase motor and binds to the microtubule, and a tail composed of a string of rodlike domains that specify the type of cargo that the protein transports.

___ M. Most known motor proteins move in only one direction along microtubules—either toward the plus end or toward the minus end.

16–12 The function of microtubules is thought to depend on their specific spatial organization within the cell. How are specific arrangements created, and what determines the formation and disappearance of individual microtubules?

To address these questions, investigators have studied the *in vitro* assembly of tubulin into microtubules. In solutions of tubulin below 15 µM no microtubules are formed, but when the concentration is raised above 15 µM, microtubules form readily (Figure 16–6A). If centrosomes are added to the solution of tubulin, microtubules begin to form at less than 5 µM (Figure 16–6B). (Different assays were used in the two experiments—total weight of microtubules in Figure 16–6A and number of microtubules per centrosome in Figure 16–6B—but the lowering of the critical concentration for microtubule assembly in the presence of centrosomes is independent of the method of assay.)

A. Why do you think that the concentration at which microtubules begin to form (the critical concentration) is different in the two experiments?

B. Why do you think that the plot in Figure 16–6A increases linearly with increasing tubulin concentration above 15 mM, whereas the plot in Figure 16–6B reaches a plateau at about 25 mM?

C. The concentration of tubulin dimers (the subunits for assembly) in a typical

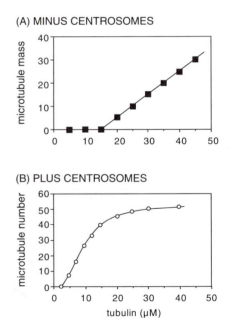

(A) MINUS CENTROSOMES

(B) PLUS CENTROSOMES

Figure 16–6 Growth of microtubules in the absence (A) and presence (B) of centrosomes as a function of tubulin concentration (Problem 16–12). Concentrations refer to tubulin dimers, which are the subunit of assembly.

cell is 1 mg/ml and the molecular weight of a tubulin dimer is 110,000. What is the molar concentration of tubulin in cells? How does the cellular concentration compare with the critical concentrations in the two experiments in Figure 16–6? What are the implications for assembly of microtubules in cells?

*16–13 In addition to centrosomes, flagellar axonemes and kinetochores also can serve as nucleation sites for microtubule assembly. Do these structures nucleate microtubule growth by binding to the plus end or to the minus end of the nascent microtubule? The following experiment was designed to determine which end of a microtubule is attached to centrosomes and kinetochores. Flagellar axonemes were included as a control since their plus and minus ends can be distinguished. Centrosomes and kinetochores (and flagellar axonemes) were incubated briefly in unlabeled tubulin to nucleate microtubule growth. A high concentration of biotin-labeled tubulin was then added and the incubation was continued for 10 minutes. At that point the preparations were fixed and the biotin-labeled segments were visualized by adding fluorescein-labeled antibodies specific for biotin. The lengths of the biotin-labeled segments were measured and plotted as shown in Figure 16–7.

A. Which end of a newly assembled microtubule is attached to the plus end of the flagellar axoneme?
B. Which end of a microtubule assembled on a flagellar axoneme grows faster?
C. Which end of an assembled microtubule is attached to a centrosome? To a kinetochore? Explain your reasoning.

16–14 The complex kinetics of microtubule assembly make it hard to predict the behavior of individual microtubules. Some microtubules in a population can grow, even as the majority shrink to nothing. One simple hypothesis to explain this behavior is that growing ends are protected from disassembly by a GTP cap and that faster growing ends have a longer GTP cap. Recent advances in video techniques now allow real-time observations of individual microtubules. Typical observations on the changes in length with time are shown for two microtubules in Figure 16–8. Measurements of their rates of growth and

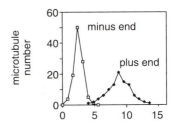

(A) AXONEMES

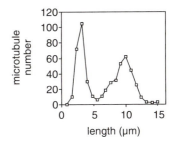

(B) CENTROSOMES

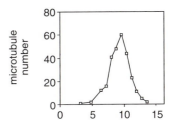

(C) KINETOCHORES

Figure 16–7 Length distributions of microtubules attached to (A) axonemes, (B) centrosomes, and (C) kinetochores (Problem 16–13).

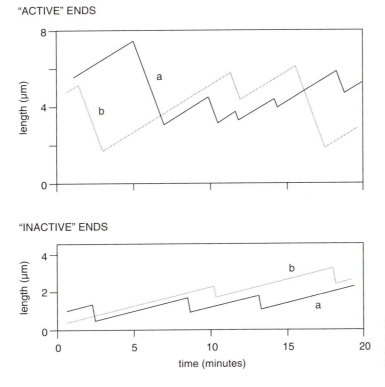

Figure 16–8 Changes in length at the ends of individual microtubules (Problem 16–14). Results from the individual microtubules are indicated with a and b.

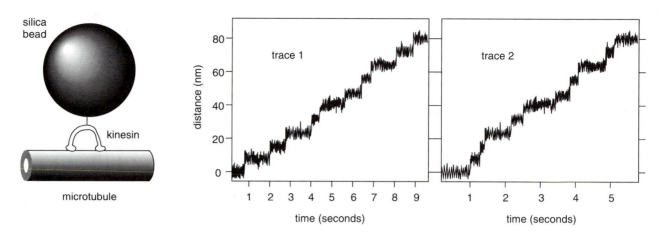

shrinkage show that one end of each microtubule (the "active" end) grows three times faster and shrinks at half the rate of the other end (the "inactive" end). The active end is thought to correspond to the plus end.

A. Are changes in length at the two ends of a microtubule dependent or independent of one another? How can you tell?
B. What does the simple GTP-cap hypothesis predict about the rate of switching between growing and shrinking states at the fast-growing end relative to the slow-growing end? Does the outcome of this experiment support the simple GTP-cap hypothesis?
C. These observations were made with pure tubulin. How do you think the observations would change if you added centrosomes to the reaction? If you added microtubule-associated proteins (MAPs) to the reaction?

Figure 16–9 Movement of kinesin along a microtubule (Problem 16–15). (A) Experiment setup with kinesin, which is linked to a silica bead, moving along a microtubule. (B) Position of kinesin (as visualized by position of silica bead) relative to center of interference pattern as a function of time of movement along the microtubule. The jagged nature of the traces results from brownian motion of the bead. The movements of two different kinesin molecules are shown.

*16–15   Advances in optics and electronics allow the movements of single motor-protein molecules to be analyzed. Using polarized laser light, it is possible to create interference patterns that exert a centrally directed force, ranging from zero at the center to a few piconewtons at the periphery (about 200 nm from the center). Individual molecules that enter the interference pattern are rapidly pushed to the center, allowing them to be captured and moved at the experimenter's discretion.

Using such "optical tweezers," single kinesin molecules can be positioned on a microtubule that is fixed to a coverslip. Although a single kinesin molecule cannot be seen optically, its movement can be tracked by attaching a larger silica bead to it and following the bead (Figure 16–9A). In the absence of ATP the kinesin molecule remains at the center of the interference pattern, but with ATP it moves toward the plus end of the microtubule. As kinesin moves along the microtubule, it encounters the force of the interference pattern, which simulates the load kinesin carries during its actual function in the cell. Moreover, the pressure against the silica bead counters the effects of brownian (thermal) motion so that the position of the bead more accurately reflects the position of the kinesin molecule on the microtubule.

Traces of the movements of two kinesin molecules along a microtubule are shown in Figure 16–9B.

A. As shown in Figure 16–9B, all movement of kinesin is in one direction (toward the plus end of the microtubule). What supplies the free energy needed to ensure a unidirectional movement along the microtubule?
B. What is the average rate of movement of each kinesin along the microtubule?
C. What is the length of each step a kinesin takes as it moves along a microtubule?
D. From other studies it is known that kinesin has two globular domains that each can bind to β-tubulin and that kinesin moves along a single protofilament

of a microtubule. In each protofilament the β-tubulin subunit repeats at 8-nm intervals. Given the step length and the interval between β-tubulin subunits, how do you suppose a kinesin molecule moves along a microtubule?

E. Is there anything in the data in Figure 16–9B that tells you how many ATP molecules are hydrolyzed per step?

Cilia and Centrioles

(MBOC 815–820)

16–16 Fill in the blanks in the following statements.

A. _____ are tiny hairlike appendages that extend from the surface of many kinds of cells and function to move fluid over the surface of the cell or to propel single cells through a fluid.

B. Sperm swim by means of _____, which are long, thin processes that propagate quasi-sinusoidal waves.

C. The core of a cilium is a complex structure called an _____, which is composed entirely of microtubules and their associated proteins.

D. The accessory protein that is responsible for the sliding of outer microtubule doublets against one another to produce ciliary bending is called _____.

16–17 Indicate whether the following statements are true or false. If a statement is false, explain why.

___ A. Eucaryotic cilia and flagella have an internal structure composed of an outer ring of nine doublet microtubules surrounding two single microtubules.

___ B. Like cytoplasmic dynein, ciliary dynein has a motor domain, which hydrolyzes ATP to move along a microtubule toward its minus end, and a tail region that carries a cargo, which in this case is an adjacent microtubule.

___ C. The bending forces that lead to ciliary motion are generated by the sliding between the central pair of microtubules and the outer ring of doublet microtubules.

___ D. During the formation of the cilium, each doublet microtubule of the axoneme grows from two of the microtubules in the triplet microtubules of the basal body so that the ninefold symmetry of the basal body is preserved in the ciliary axoneme.

___ E During centriole duplication, each member of the pair of centrioles forms a new daughter centriole at right angles to each original centriole, and then the two daughter centrioles pair together to form the new pair of centrioles.

*16–18 An electron micrograph of a cross-section through a flagellum is shown in Figure 16–10.

A. Assign the following components to the indicated positions on the figure.

A tubule	Inner sheath
B tubule	Nexin
Outer dynein arm	Radial spoke
Inner dynein arm	Single microtubule

B. Which of the above structures are composed of tubulin?

C. Which, if any, of the structures are continuous with components of the basal body?

16–19 On ciliated cells the beating of individual cilia is usually coordinated so that the cilia move in the same direction, thereby imparting unidirectional motion to the cell (or to the surrounding fluid). In principle, the coordinated, unidirectional beating of adjacent cilia could be determined by some feature

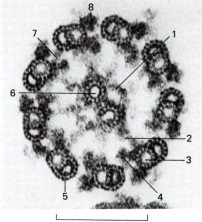

Figure 16–10 Electron micrograph of a cross-section through a flagellum of *Chlamydomonas reinhardtii* (Problem 16–18). (Courtesy of Lewis Tilney.)

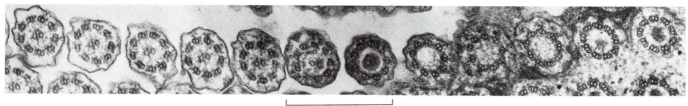

of their structure or, alternatively, by some cellular control mechanism that is independent of ciliary structure.

An electron micrograph of a cross-section through the cortex of the ciliated protozoan *Tetrahymena* is shown in Figure 16–11. The plane of section grazes the surface of the cell, showing in successive sections how the "9 + 2" arrangement of microtubules in the axoneme leads into the nine triplet microtubules of the basal body. Are there any clues in the micrograph that allow you to decide whether axoneme structure or cellular control is the basis for the *unidirectional* beating of adjacent cilia? Explain your reasoning.

Figure 16–11 Electron micrograph of a cross-section through the cortex of a ciliated protozoan, *Tetrahymena* (Problem 16–19). (Courtesy of Keith Roberts.)

*16–20 When analyzed in detail, the rhythmic beating of a cilium is revealed as a series of precisely repeated movements. In *Chlamydomonas* the flagellar beat cycle is straightforward (Figure 16–12A). The beat cycle begins with a power stroke, which is initiated by a bend at the base of the flagellum (arrow at base of flagellum 1 in Figure 16–12A). The power stroke ends when the bent segment of flagellum extends roughly through half the circumference of a circle (flagellum 5 in Figure 16–12A). The return stroke is formed by the movement of the semicircular segment of the flagellum outward toward the tip, which is accomplished by further bending at the leading edge of the semicircle and relaxation at the trailing edge (flagella 6 to 8 in Figure 16–12A). Any complete molecular mechanism for axoneme function must be able to account for these gross movements of the flagellum.

Figure 16–12 Beat cycles for wild-type (A) and mutant (B) flagella from *Chlamydomonas* (Problem 16–20). The beating flagella at *left* were photographed with stroboscopic illumination under a microscope. Individual frames from these pictures represent flagella at successive stages in the beat cycle. Arrows show the leading edge of a bent segment of flagellum as it progresses from the bottom of the flagellum to the tip during the beat cycle.

A. How much sliding of microtubule doublets against one another is required to account for the observed bending of the flagellum into a semicircle? Calculate how much farther the doublet on the inside of the semicircle protrudes

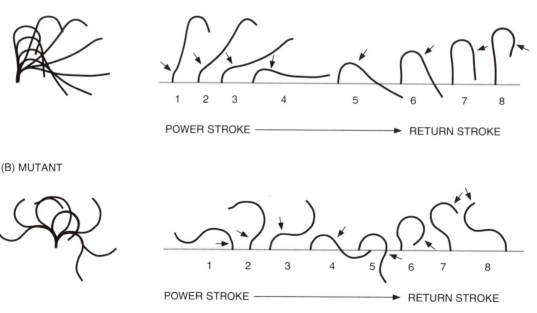

(A) WILD-TYPE

1 2 3 4 5 6 7 8

POWER STROKE ⟶ RETURN STROKE

(B) MUTANT

1 2 3 4 5 6 7 8

POWER STROKE ⟶ RETURN STROKE

beyond the doublet on the outside of the semicircle at the tip of the flagellum (Figure 16–13A).

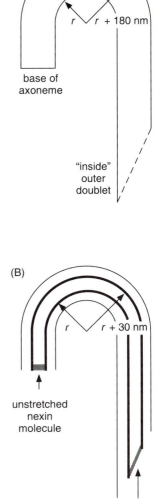

B. The elastic nexin molecules that link adjacent outer doublets must stretch to accommodate the bending of a flagellum into a semicircle. If the length of an unstretched nexin molecule at the base of a flagellum is 30 nm, what is the length of a stretched nexin molecule at the tip of a flagellum (Figure 16–13B)?

C. *Chlamydomonas* mutants that are missing radial spokes have paralyzed flagella. The paralysis can be overcome by mutations in a second gene (called *sup*$_{pf}$, for *sup*pressor of *p*aralyzed *f*lagella), which encodes a component of the outer dynein arm. Although the flagella now move, their beat pattern is aberrant (Figure 16–12B). At the gross level, how does the beat stroke of the mutant strain differ from that of wild-type strain? What does this gross difference suggest for the function mediated by the radial spokes in *Chlamydomonas*?

16-21 The sliding microtubule mechanism for ciliary bending is undoubtedly correct. The consequences of sliding are straightforward when a pair of outer doublets is considered in isolation. It is confusing however, to think about sliding in the circular array of outer doublets in the axoneme. The dynein arms are arranged so that, when activated, they push their neighboring outer doublet outward toward the tip of the cilium. In a circular array if all the dynein arms were equally active, there could be no significant relative motion. (The situation is equivalent to a circle of strongmen, each trying to lift his neighbor off the ground; if they all succeeded, the group would levitate.)

Devise a pattern of dynein activity (consistent with axoneme structure and the directional pushing of dynein) that can account for bending of the axoneme in one direction. How would this pattern change for bending in the opposite direction?

16–22 The structure of the ciliary axoneme, which is composed of more than 200 different proteins, is exceedingly complex. *Chlamydomonas reinhardtii*, which bears two flagella, is an extremely useful organism for analyzing axoneme structure: physiological and microscopic observation are straightforward, but even more important is the ease of genetic and biochemical analysis. It is a simple matter to isolate mutants with paralyzed flagella because they cannot move. These mutants can then be assigned to specific genes by genetic crosses, which are routine with *Chlamydomonas*. Finally, wild-type and mutant flagella can be detached readily (for example, by pH shock) and recovered in highly purified fractions for easy biochemical analysis.

Many of the mutants with paralyzed flagella are missing one or another of the major substructures of the axoneme, such as the radial spokes, the outer dynein arms, the inner dynein arms, or one or both central microtubules. In most cases loss of the axonemal substructure is caused by a mutation that affects a single gene, and yet biochemical analysis shows that the defective axoneme is missing multiple proteins. Consider, for example, the mutant *pf*14 (*p*aralyzed *f*lagella); electron micrographs of its flagella show a complete absence of radial spokes (Figure 16–14A), and two-dimensional electrophoretic analysis shows that it lacks 17 different proteins (Figure 16–14B).

A single-gene defect that results in multiple protein deficiencies could have two underlying explanations: (1) the defect is in a regulatory gene that controls the synthesis of the missing proteins or (2) the defect is in a gene whose product must be present in the structure before the other proteins can be added. Two approaches have been used to evaluate these possibilities.

A. The first method takes advantage of a feature of the regular mating cycle of *Chlamydomonas*. In the mating reaction, biflagellate gametes fuse efficiently to give a population of temporary dikaryons with four flagella. In a mating of *pf*14 with wild-type gametes, the paralyzed flagella recover function after fusion, indicating that the defective structures can be repaired without

Figure 16–13 Flagella bent into half-circles (Problem 16–20). (A) Representation showing the "inside" and "outside" doublets, which are 180 nm apart. (B) Representation showing adjacent doublets, which are 30 nm apart, and the nexin molecules that link them.

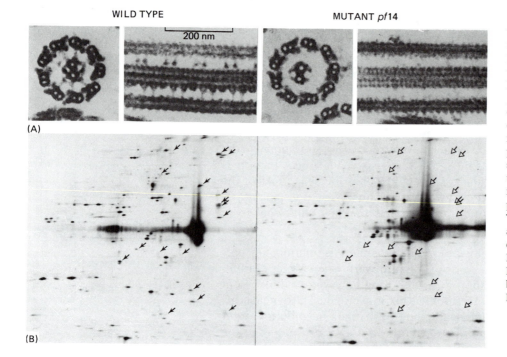

WILD TYPE   MUTANT *pf*14

200 nm

(A)

(B)

Figure 16–14 Comparison of flagella from wild-type *Chlamydomonas* with those from the mutant *pf*14 (Problem 16–22). (A) Electron micrographs showing flagella in transverse and longitudinal sections. Notice the absence of radial spokes in *pf*14. (B) Analysis of flagella by two-dimensional gel electrophoresis. The first dimension (horizontal) is separation by isoelectric focusing with the more acidic proteins on the right; the second dimension is separation by molecular weight using SDS-gel electrophoresis. Arrows indicate the positions of proteins that are present in wild-type but missing in *pf*14. The highly exposed areas correspond to α- and β-tubulin, which far outnumber the other axonemal proteins. (From G. Piperno, B. Huang, Z. Ramanis, and D. Luck, *J. Cell Biol.* 88:73–79, 1981. Reprinted by permission of the Rockefeller University Press.)

completely rebuilding them. (In normal cells there is a pool of flagellar components sufficient to rebuild an entire flagellum in the absence of protein synthesis.) To distinguish between the possible explanations for flagellar defects, investigators labeled mutant cells by growth in $^{35}SO_4$ and then fused the labeled gametes to nonradioactive wild-type gametes in the presence of an inhibitor of protein synthesis. After recovery of function, the flagella were isolated and the radioactive proteins were analyzed by two-dimensional gel electrophoresis followed by autoradiography.

Predict the expected electrophoretic pattern of *radioactive* proteins from the dikaryon if the affected gene controlled the synthesis of the missing proteins. How would it differ from the electrophoretic pattern that would be expected if the affected gene product participated in assembly?

B. The second approach was to expose *pf*14 to a mutagen to generate revertants that regained flagellar function not because the original defect has been corrected, but because a second alteration within the gene compensates for the first one. (Depending on the gene, such intragenic revertants can be very common.) The proteins from several such revertants were compared to those in the wild-type by two-dimensional gel electrophoresis.

How might this method distinguish between the two possible explanations for the defect in *pf*14?

Actin Filaments

(MBOC 821–834)

16–23 Fill in the blanks in the following statements.

A. _____ appear in electron micrographs as threads about 8 nm wide; they consist of a tight helix of uniformly oriented _____ monomers.

B. The most abundant of the special proteins that bind actin monomer and inhibit its addition to the ends of actin filaments is an unusually small protein called _____

C. _____, which is present in all cells, is thought to play a part in controlling actin polymerization in response to extracellular stimuli.

D. The leading edge of a crawling fibroblast regularly extends a thin, sheetlike process known as a _____, which contains a dense meshwork of actin filaments.

E. In addition to trimeric G proteins that are implicated in signaling processes that activate the cellular cortex, two Ras-related monomeric GTPases known as _____ and _____ have been shown to have distinct effects on the actin cytoskeleton in fibroblasts.

16–24 Indicate whether the following statements are true or false. If a statement is false, explain why.

___ A. Like a microtubule, an actin filament is a polar structure, with two structurally different ends—a relatively inert and slow-growing minus end and a faster-growing plus end.

___ B. Shortly after polymerization, the terminal phosphate of the ATP bound to the actin monomer is hydrolyzed, and the resulting phosphate and ADP are released from the polymer.

___ C. The role of ATP hydrolysis in actin polymerization is similar to the role of GTP hydrolysis in tubulin polymerization: both serve to weaken the bonds in the polymer and thereby promote depolymerization.

___ D. Cytochalasins prevent actin from polymerizing by binding to the plus end of actin filaments; phalloidins stabilize filaments against depolymerization by binding all along the side of actin filaments.

___ E. In a fibroblast cell approximately 50% of the actin is in filaments and 50% is in monomer, as expected from the intracellular concentration of actin and its critical concentration for polymerization and depolymerization.

___ F. The actin filaments in the lamellipodium of a moving cell appear to be more organized than they are in other regions of the cell cortex: many of the filaments project outward in an orderly array, with their plus ends inserted into the leading edge of the plasma membrane.

___ G. Observations of the leading edge of motile cells suggest that actin filaments are continuously lengthening by polymerization at internal sites and in that way pushing the leading edge forward.

___ H. The actin-assisted movement of the *Listeria* bacterium through the cytosol suggests that the bacterium may be using actin to propel itself forward in the same way that the plasma membrane of a eucaryotic cell uses actin to propel itself forward during the formation of a normal lamellipodium.

___ I. Eucaryotic chemotaxis involves detecting a spatial gradient of attractant concentration directly, in contrast to bacterial chemotaxis, which uses a time-dependent variation in concentration to detect gradients.

___ J. Rac and Rho not only control the polymerization of actin into filaments, but also govern the organization of these filaments into specific types of structures.

___ K. One consequence of the binding of α-factor to its receptors on an **a**-cell is to cause the cell to become polarized so that it adopts a shape known as a shmoo, with the shmoo tip directed at the highest concentration of α-factor.

16–25 Your ultimate goal is to understand human consciousness—however, your advisor wants you to understand some basic facts about actin assembly first. He tells you that ATP binds to actin monomers and is required for assembly but that ATP hydrolysis is not necessary for polymerization, since ADP can, under certain circumstances, substitute for the ATP requirement. However, ADP filaments are much less stable than ATP filaments, thus supporting your secret suspicion that the free energy of ATP hydrolysis really is used to drive actin assembly.

Your advisor suggests that you make careful measurements of the quantitative relationship between the number of ATP molecules hydrolyzed and the number of actin monomers linked into polymer. The experiments are straightforward. To measure ATP hydrolysis, you add ATPγ³²P to a solution of polymerizing actin, take samples at short intervals, and determine how much radioactive phosphate has been produced. To measure polymerization, you follow (in a spectrophotometer) the increase in light scattering that is caused by formation of the actin filaments. Your results are shown in Figure 16–15. Your light-scattering measurements indicate that 20 micromoles of actin monomers were polymerized. Since the number of polymerized actin monomers matches exactly the number of ATP molecules hydrolyzed, you conclude that one ATP is hydrolyzed as each new monomer is added to an actin filament.

When you show your advisor the data and tell him your conclusions, he smiles and very gently tells you to look more closely at the graph. He says your data prove that actin can polymerize without ATP hydrolysis.

A. What does your advisor see in the data that you have overlooked?
B. What do your data imply about the distribution of ATP and ADP in polymerizing actin filaments?

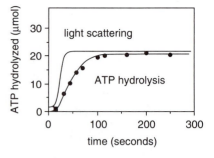

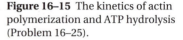

Figure 16–15 The kinetics of actin polymerization and ATP hydrolysis (Problem 16–25).

*16–26 One of the most striking examples of a purely actin-based cellular movement is the extension of the acrosomal process of a sea cucumber sperm. The sperm contains a store of unpolymerized actin in its head. When a sperm makes contact with a sea cucumber egg, the actin polymerizes rapidly to form a long spearlike extension. The tip of the acrosomal process penetrates the egg, and it is probably used to pull the sperm inside.

Are actin monomers added to the base or to the tip of the acrosomal bundle of actin filaments during extension of the process? If the supply of monomers to the site of assembly depends on diffusion, it should be possible to distinguish between these alternatives by measuring the length of the acrosomal process with increasing time. If actin monomers are added to the base of the process, which is inside the head, the rate of growth should be linear with time because the distance between the site of assembly and the pool of monomers does not change with time. On the other hand, if the subunits are added to the tip, the rate of growth should decline progressively as the acrosomal process gets longer because the monomers must diffuse all the way down the shaft of the process. In this case the rate of extension should be proportional to the square root of time. Plots of the length of the acrosomal process versus time and the square root of time are shown in Figure 16–16.

A. Are the ascending portions of the plots in Figure 16–16 more consistent with addition of actin monomers to the base or to the tip of the acrosomal process?
B. Why does the process grow so slowly at the beginning and at the end of the acrosomal reaction?

16–27 Cytochalasin B strongly inhibits certain forms of cell motility, such as cytokinesis and the ruffling of growth cones, and it dramatically decreases the viscosity of gels formed with mixtures of actin and a wide variety of actin-binding proteins. These observations suggest that cytochalasin B interferes with the assembly of actin filaments; however, it does not bind to actin monomers. How then does cytochalasin B interfere with motility?

Consider the following experiment. Short lengths of actin filaments were decorated with myosin heads and then monomeric actin was added to the decorated filaments in the presence or absence of cytochalasin B. Assembly of actin filaments was measured by assaying the viscosity of the solution (Figure 16–17) and also by examining samples in an electron microscope (Figure 16–18).

A. What does the electron microscope reveal about the way actin filaments normally grow?

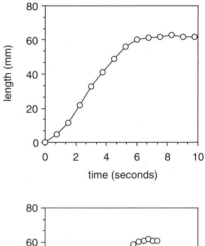

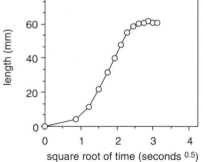

Figure 16–16 Plots of the length of the acrosome versus time and the square root of time (Problem 16–26).

B. Suggest a plausible mechanism to explain how cytochalasin B inhibits actin filament assembly. Account for both the viscosity measurements and the appearance of the filaments in the electron micrographs.

C. The normal growth characteristics of an actin filament and the actin-binding properties of cytochalasin B argue that actin monomers undergo a conformational change upon addition to an actin filament. How so?

Actin-binding Proteins

(MBOC 834–847)

16–28 Fill in the blanks in the following statements.

A. The length of actin filaments, their stability, and the number and geometry of their attachments depend on a large retinue of _____, which bind to actin filaments and modulate their properties and functions.

B. The proteins _____ and _____ were first discovered as prominent components of the membrane-associated cytoskeleton of mammalian red blood cells.

C. _____ is a bundling protein that is enriched in the parallel filament bundles at the leading edge of cells, particularly in microspikes and filopodia.

D. _____ is a bundling protein that is concentrated in stress fibers, where it is thought to be partly responsible for the relatively loose cross-linking of actin filaments in these contractile bundles.

E. _____ is a widely distributed gel-forming protein, which, although it is not present in stress fibers or the leading edge, is enriched in the cortex.

F. _____, when activated by the binding of Ca^{2+}, severs an actin filament and forms a cap on the newly exposed plus end of the filament, thus breaking up the cross-linked network of actin filaments.

G. All of the actin filament motor proteins identified to date are _____, which were originally isolated on the basis of their ability to hydrolyze ATP to ADP and P_i when stimulated by binding to actin filaments.

H. _____ is thought to be more like the original, more primitive myosin from which _____ evolved.

I. Cell division in animal cells is made possible by a beltlike bundle of actin filaments and myosin-II filaments known as the _____.

J. _____, which are prominent components of the cytoskeleton of fibroblast cells in culture, are a temporary contractile bundle of actin filaments and myosin-II.

K. When fibroblasts grow on a culture dish, most of their cell surface is separated from the substratum by a gap of more than 50 nm; but at _____, this gap is reduced to 10 to 15 nm.

L. _____ are fingerlike extensions found on the surface of many animal cells.

M. The bundling protein _____, which is found only in microvilli, cross-links actin filaments into tight parallel bundles.

N. The _____, which are rigid rod-shaped proteins, are widespread members of a class of actin-binding proteins that binds along the length of the filament, stabilizing and stiffening it.

16–29 Indicate whether the following statements are true or false. If a statement is false, explain why.

___ A. Although the actin filaments in the erythrocyte cortex are very short, those in a more typical cell cortex are much longer and thus project into the cytoplasm, where they form the basis of a three-dimensional actin filament network.

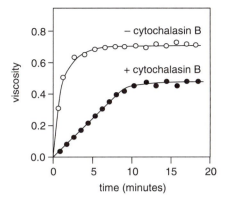

Figure 16–17 Increase in the viscosity of actin solutions in the presence and absence of cytochalasin B (Problem 16–27).

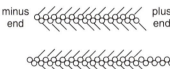

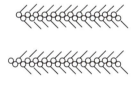

Figure 16–18 Appearance of typical actin filaments formed in the presence and absence of cytochalasin B (Problem 16–27). The decorated actin filaments present before addition of actin monomers are shown at the top. Filaments present after increasing times of incubation in the presence of actin monomers are shown below.

___ B. All actin bundling proteins cross-link actin filaments in parallel bundles with the filaments all oriented with the same polarity.

___ C. The actin-binding proteins responsible for parallel bundles, contractile bundles, and gel-like networks each have two actin-binding domains, and all the actin-binding domains are related.

___ D. One of the postulated functions of severing proteins is to help loosen or locally liquefy the cell cortex to allow membrane fusion events.

___ E. Both myosin-I and myosin-II have a rodlike tail that allows them to polymerize into bipolar filaments, which is crucial for moving groups of oppositely oriented actin filaments past each other.

___ F. In epithelia, actin filament bundles spanning the cytoplasm from one cell-cell junction to another can be linked end to end via the cell-cell junctions and can form cables that transmit and generate tension along lines of particular stress in the multicellular sheet.

___ G. The main transmembrane linker proteins of focal contacts are members of the integrin family, whose external domain binds to an extracellular matrix component while the cytoplasmic domain is linked indirectly to actin filaments in stress fibers.

___ H. At the base of the microvillus the actin filament bundle is anchored into a specialized region of cortex known as the terminal web.

___ I. The binding of tropomyosin to an actin filament increases the binding of myosin-II to the filament—an example of a cooperative interaction.

___ J. Three distinct processes can be identified in the crawling movements of animal cells: protrusion of lamellipodia or microspikes from the front of the cell, attachment of the actin cytoskeleton to the substratum, and traction, where the body of the cells moves forward.

___ K. *Dictyostelium* cells in which the gene for myosin-II has been deleted are immotile.

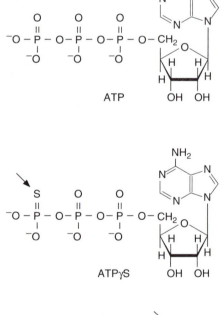

16–30 Although a large number of actin-binding proteins are known, their cellular roles are not entirely clear. The haploid slime mold *Dictyostelium discoideum* is a very useful organism for studying these factors because it is motile, has a well-defined developmental pathway, and is amenable to reverse genetics.

You have isolated mutant strains of *Dictyostelium* that fail to express either α-actinin or gelation factor, which are the major actin filament cross-linking proteins in *Dictyostelium*. Surprisingly, neither of these single mutants shows any defect in cell growth, motility, or development. You decide to see if double mutants are impaired in any way and target disruption of the gelation-factor gene in the α-actinin mutant cells, and vice versa, to give double mutants that are defective for both α-actinin and gelation factor.

The doubly mutant strains grow very well as amoebae, which can move and feed and aggregate into a normal slug in response to starvation. But then things go wrong. The slugs do not move the way they normally do and stay put wherever they first form. And instead of forming a spore head atop a stalk, the double mutants fail to form a stalk and develop slightly abnormal spores within the body of the slug.

Your studies with the single mutant defective in either α-actinin or gelation factor suggest that neither gene is essential. Why then do you think that the loss of both genes causes such dramatic defects in the cell movements associated with development?

*16–31 ATP plays two roles in the contraction of the terminal web, which controls the movement of the microvilli on the surface of intestinal cells. These two functions of ATP have been differentiated using the ATP analogues ATPγS and inosine triphosphate (ITP), which differ from ATP as indicated by the arrows in Figure 16–19. Intestinal membranes with microvilli and their associated terminal webs were isolated and used in two types of experiments. In one the

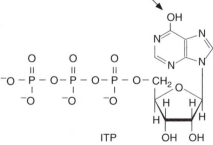

Figure 16–19 ATP and two ATP analogues, ATPγS and ITP (Problem 16–31).

Table 16–1 Use of ATP Analogues to Study Contraction of the Terminal Web in Intestinal Membranes (Problem 16–31)

Experiment	Preincubation	Incubation	Contraction
1	no analogue	ATP	yes
2	no analogue	ITP	no
3	no analogue	ATPγS	no
4	ATPγS	ATP	yes
5	ATPγS	ITP	yes
6	ATPγS	ATPγS	no

intestinal membranes were preincubated with or without ATPγS in the presence of Ca^{2+}, and then ATP, ITP, or ATPγS was added and contraction of the terminal web was assayed (Table 16–1). In the second type of experiment, intestinal membranes were incubated with labeled ATP analogues, and the resulting phosphorylated proteins were analyzed: the light chain of myosin was shown to be phosphorylated in the presence of ATPγ^{32}P and ATPγ^{35}S, but it was not phosphorylated in the presence of ITPγ^{32}P.

A. What two roles does ATP hydrolysis play in the contraction of the terminal web? For which role can ATPγS substitute? For which role can ITP substitute?

B. Both functions of ATP depend on the hydrolysis of the terminal (γ) phosphodiester bond. Why do you think it is that the ATP analogues behave differently in the two types of experiment?

Muscle

(MBOC 847–858)

16–32 Fill in the blanks in the following statements.

A. _____, which are the contractile elements of the muscle cell, are cylindrical structures 1 to 2 μm in diameter and are often as long as the muscle cell itself.

B. Each myofibril consists of a chain of tiny contractile units, or _____, each about 2.2 μm long, which give the vertebrate myofibril its striated appearance.

C. The signal initiated in muscle by a nerve impulse is relayed from the transverse tubules across a small gap of about 15 nm to the _____, an adjacent sheath of anastomosing flattened vesicles that surrounds each myofibril like a net stocking.

D. Like skeletal muscle, _____ is striated, reflecting a very similar organization of actin filaments and myosin filaments.

E. The most "primitive" muscle, in the sense of being most like nonmuscle cells, has no striations and is therefore called _____.

16–33 Indicate whether the following statements are true or false. If a statement is false, explain why.

__ A. Z discs are located at the ends of each myofibril.

__ B. Sarcomere shortening is caused by the myosin filaments sliding past the actin filaments with no change in the length of either type of filament.

__ C. The cycle of conformational changes by which a myosin molecule walks along an actin filament is made unidirectional by the large favorable change in free energy associated with the hydrolysis of ATP in each cycle.

__ D. A signal from the motor nerve triggers an action potential in the muscle cell plasma membrane, which opens voltage-sensitive Ca^{2+} release channels in the plasma membrane, allowing Ca^{2+} to flow into the cytosol from outside the cell.

__ E. When the level of Ca^{2+} is raised, troponin C causes troponin I to release its hold on actin, thereby allowing the tropomyosin molecules to shift their position slightly so that the myosin heads can bind to the actin filament.

__ F. The two extraordinarily large proteins, titin and nebulin, act as molecular rulers that specify the length of myosin filaments and actin filaments, respectively.

__ G. All three types of muscle cell—skeletal muscle, heart muscle, and smooth muscle—contract by an actin and myosin-II sliding filament mechanism.

__ H. Because contraction in smooth muscle is initiated by phosphorylation of one of the two myosin-II light chains, it is not controlled by the level of cytosolic Ca^{2+}.

16–34 Two electron micrographs of striated muscle in longitudinal section are shown in Figure 16–20. The sarcomeres in these micrographs are in two different stages of contraction.

A. Using the micrograph in Figure 16–20A identify the location of the following:

1.	Dark band	5.	Actin filaments (show plus and minus ends)
2.	Light band	6.	α-Actinin
3.	Z disc	7.	Nebulin
4.	Myosin-II filaments	8.	Titin

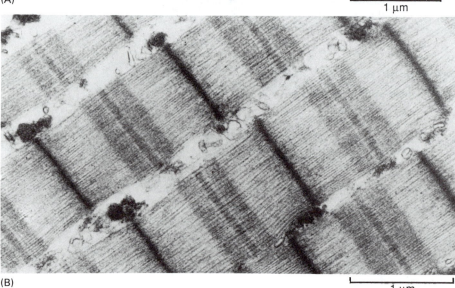

(A)

1 µm

(B)

1 µm

Figure 16–20 Two electron micrographs of striated muscle seen in longitudinal section (Problem 16–34). The micrographs have been photographed at different exposures. (Courtesy of Hugh Huxley.)

Figure 16–21 Schematic illustration of (A) a myosin-II molecule and (B) a helix of myosin-II molecules (Problem 16–35). For simplicity, the two myosin heads are shown as one protrusion. The six myosin molecules that form the first helical turn are numbered.

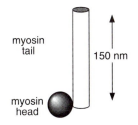

(A) MYOSIN-II MOLECULE

myosin tail

150 nm

myosin head

B. Locate the same features on the micrograph in Figure 16–20B. Be careful!

16–35 Protein assembly principles are nowhere better illustrated than in muscle and the cytoskeleton. If a protein has a binding site that is complementary to a region of its own surface, it will assemble spontaneously into an aggregate. Depending on the geometric relationship of the complementary sites, the resulting structure can be a ring (with the simplest ring being a dimer) or a helix. These simple assemblies can interact to form more complex ones, using the same principle of complementary binding. Aggregates built in this way possess geometric symmetry as a necessary consequence. Biological symmetry is relatively straightforward, but many people find it abstract and somewhat daunting. Nevertheless, the underlying principles are enormously powerful conceptual tools for analyzing biological structure.

Consider, for example, a myosin thick filament, which is a bipolar helical structure constructed from a single type of subunit (the rodlike myosin-II molecule). The hexagonal packing of actin and myosin-II filaments in muscle indicates that the myosin heads are arrayed around the helix so that they can point to the vertices of the hexagon for optimal interactions with actin filaments. A helix with six myosin molecules (Figure 16–21A) per turn would satisfy this requirement, as suggested by the schematic diagram in Figure 16–21B, which shows one half of a bipolar thick filament. The actual structure of a myosin-II filament is more complex with three chains of myosin-II molecules coiled around one another like strands of a rope. The diagram in Figure 16–21, however, is easier to think about and illustrates the important features of the true structure.

How well does this simple picture account for the appearance of a myosin-II thick filament?

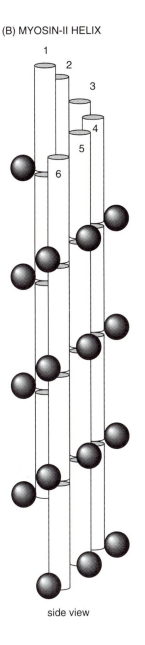

(B) MYOSIN-II HELIX

side view

A. The bipolar appearance of myosin-II thick filaments with a bare zone in the middle arises because two myosin helices are joined end to end (top end to top end as drawn in Figure 16–21B). Can two myosin helices of the type illustrated in Figure 16–21 actually fit together in this manner? Would the two ends dovetail nicely or not?

B. How long would the bare zone be in the joined helices? Assume the bare zone extends from the first myosin head on one helix to the first myosin head on the oppositely oriented helix.

C. How would you explain the tapering that can be seen at the two ends of a myosin-II thick filament? (The tapering can be seen clearly in the micrograph shown in MBOC Figure 16–88.)

D. Myosin-II thick filaments from striated muscle tend to be quite uniform in overall length with the bare zone located at the exact middle of the filament. Is there any feature of the helix in Figure 16–21 that could explain this uniformity of filament length?

*16–36 As a laboratory exercise, you and your classmates are carrying out experiments on isolated muscle fibers using a new compound, called "caged ATP" (Figure 16–22). Since caged ATP does not bind to muscle components, it can be added to a muscle fiber without stimulating activity. Then, at some later time it can be split by laser illumination to release ATP instantly throughout the muscle fiber.

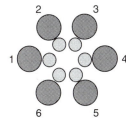

top view

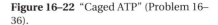
Figure 16–22 "Caged ATP" (Problem 16–36).

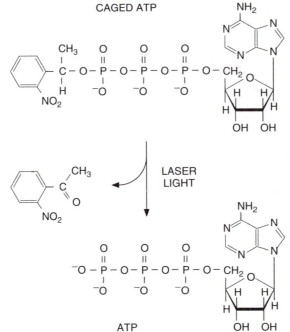

CAGED ATP

LASER LIGHT

ATP

To begin the experiment, you treat an isolated, striated muscle fiber with glycerol to make it permeable to nucleotides. You then suspend it in a buffer containing ATP in an apparatus that allows you to measure any tension generated by fiber contraction. As illustrated in Figure 16–23, you measure the tension generated after several experimental manipulations: removal of ATP by dilution, addition of caged ATP, and activation of caged ATP by laser light. You are somewhat embarrassed because your results are very different from everyone else's. In checking over your experimental protocol, you realize that you forgot to add Ca^{2+} to your buffers. The teaching assistant in charge of your section tells you that your experiment is actually a good control for the class but you will have to answer the following questions to get full credit.

A. Why did the ATP in the suspension buffer not cause the muscle fiber to contract?

B. Why did the subsequent removal of ATP generate tension? Why did tension develop so gradually? (If our muscles normally took a full minute to contract, we would all move very slowly.)

C. Why did laser illumination of a fiber containing caged ATP lead to relaxation?

*16–37 The change in sarcomere length during muscle contraction was one of the key observations that suggested a sliding filament model. The degree of tension generated at different sarcomere lengths also changes in a way that is consistent with the model. Detailed measurements of the relationship of sarcomere length to the tension generated during isometric contraction in a striated muscle are shown in Figure 16–24. In this muscle the length of the myosin filament is 1.6 µm and the lengths of the actin thin filaments that project from the Z discs are 1.0 µm.

Based on your understanding of the sliding filament model and the structure of a sarcomere, present a molecular explanation for the relationship of tension to sarcomere length in the segments marked I, II, III, and IV in the curve in Figure 16–24.

16–38 Living systems continually transform chemical free energy into motion. Muscle contraction, ciliary movement, cytoplasmic streaming, cell division, and active transport are examples of the ability of cells to transduce chemical free energy into mechanical work. In all these instances a protein motor

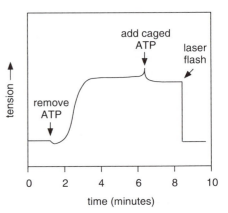

Figure 16–23 Tension in a striated muscle fiber as a result of various experimental manipulations (Problem 16–36).

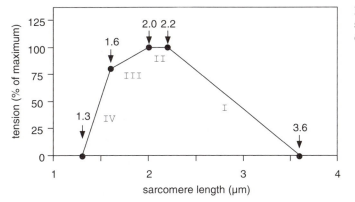

Figure 16–24 Tension as a function of sarcomere length during isometric contraction (Problem 16–37).

harnesses the free energy released in a chemical reaction to drive an attached molecule (the ligand) in a particular direction. The analysis of free-energy transduction in favorable biological systems suggests that a set of general principles governs the process in cells.

1. A cycle of reactions is used to convert chemical free energy into mechanical work.
2. At some point in the cycle a ligand binds very tightly to the protein motor.
3. At some point in the cycle the motor undergoes a major conformational change that changes the physical position of the ligand.
4. At some point in the cycle the binding constant of the ligand markedly decreases, allowing the ligand to detach from the motor.

These principles are illustrated by the two examples of cycles for free-energy transduction shown in Figure 16–25: (1) the sliding of actin and myosin filaments against each other and (2) the active transport of Ca^{2+} from inside the cell, where its concentration is low, to the cell exterior, where its concentration is high. An examination of these cycles underscores the principles of free-energy transduction.

A. What is the source of chemical free energy that powers these cycles, and what is the mechanical work that each cycle accomplishes?
B. What is the ligand that is bound tightly and then released in each of the cycles? Indicate the points in each cycle where the ligand is bound tightly.
C. Identify the conformational changes in the protein motor that constitute the "power stroke" and "return stroke" of each of the cycles.

Figure 16–25 Transduction of chemical free energy into mechanical work (Problem 16–38). (A) Sliding of actin filaments relative to myosin filaments. (B) Active transport of Ca^{2+} from the inside to the outside of the cell. In both cycles arrows are drawn in only one direction to emphasize their normal operation. The phosphorylation and dephosphorylation steps in the active transport cycle are catalyzed by enzymes that are not shown in the diagram.

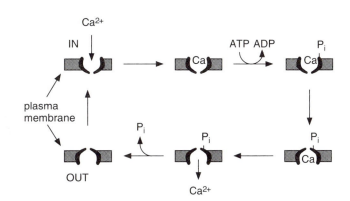

(A) SLIDING FILAMENT

(B) ACTIVE TRANSPORT

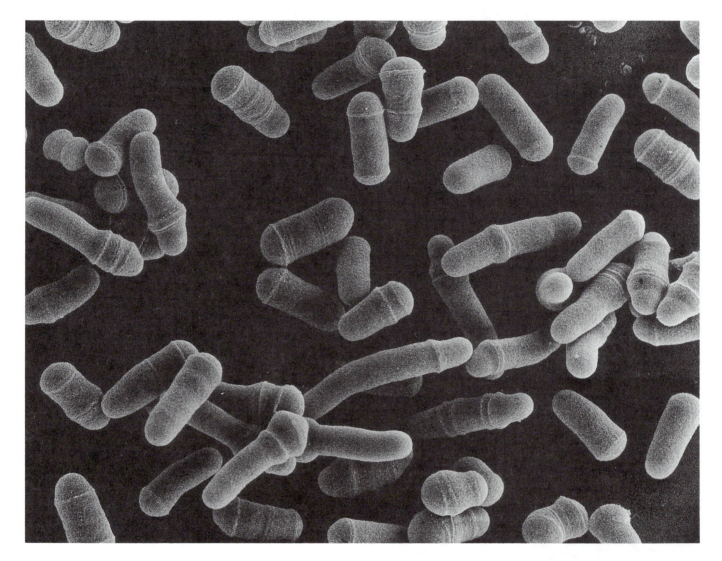

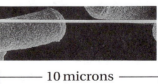

├─────── 10 microns ───────┤

Scanning electron micrograph of the fission yeast *S. pombe.* Can you tell by looking
at the picture how this classic organism for cell-cycle studies grows? Do you think it
grows at one end, or at both ends, or in the middle, or all over, or by certain
combinations of the above? And how does it maintain its cylindrical shape?
(Courtesy of Nasser Hajibagheri.)

The Cell-Division Cycle

- **The General Strategy of the Cell Cycle**
- **The Early Embryonic Cell Cycle and the Role of MPF**
- **Yeasts and the Molecular Genetics of Cell-Cycle Control**
- **Cell-Division Controls in Multicellular Animals**

The General Strategy of the Cell Cycle
(MBOC 864–870)

17–1 Fill in the blanks in the following statements.

   A. Cells reproduce by duplicating their contents and then dividing in two in a _____ that is the fundamental means by which all living things are propagated.
   B. A _____ system coordinates the cycle as a whole.
   C. The cell cycle is traditionally divided into several distinct phases, of which the most dramatic is _____, the process of cell division itself.
   D. The cell is pinched in two by a process called _____, which is traditionally viewed as the end of the mitotic phase, or _____, of the cell cycle.
   E. Replication of the nuclear DNA usually occupies only a portion of interphase, called the _____ of the cell cycle.
   F. The interval between the completion of mitosis and the beginning of DNA synthesis is called the _____, and the interval between the end of DNA synthesis and the beginning of mitosis is called the _____.
   G. One of the key proteins of the cell-cycle control system is _____, which induces downstream processes by phosphorylating selected proteins on serines and threonines.
   H. There are two main classes of cyclins: _____, which bind to Cdk during G_2 and are required for entry into mitosis, and _____, which bind to Cdk during G_1 and are required for entry into S phase.

17–2 Indicate whether the following statements are true or false. If a statement is false, explain why.

___ A. Although the lengths of all phases of the cycle are variable to some extent, by far the greatest variation occurs in the duration of G_1.

___ B. By briefly labeling asynchronously growing cells with a DNA precursor, it is possible to measure the fraction of the population in S phase, but this gives no indication of the duration of S phase in the cell cycle.

___ C. For most of the constituents of the cell, growth is a steady, continuous process, interrupted only briefly at M phase, when the nucleus and then the cell divide in two.

___ D. Each of the major essential processes in the cell cycle directly triggers the next process, as in a chain of falling dominoes.

___ E. When circumstances forbid cell division, higher eucaryotic cells generally halt at the G_2 checkpoint until they receive a signal to divide.

_____ F. Cyclins bind to Cdk and inhibit its protein kinase activity; when conditions are favorable, cyclins are rapidly degraded releasing active Cdk, which then triggers progress through a cell cycle checkpoint.

*17–3 The frequency of cells in a population that are undergoing mitosis (the mitotic index) is a convenient way to estimate the length of the cell cycle. You have decided to measure the cell cycle in the liver of the adult mouse by measuring the mitotic index. Accordingly, you have prepared liver slices and stained them to make cells in mitosis easy to recognize. After 3 days of counting, you have found only 3 mitoses in 25,000 cells. Assuming that M phase lasts 1 hour, calculate the length of the cell cycle in the liver of an adult mouse.

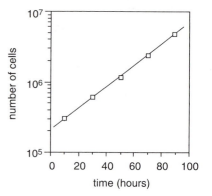

Figure 17–1 Increase in the number of mouse L cells with time (Problem 17–4).

17–4 The overall length of the cell cycle and the portions allotted to G_1, S, G_2, and M can be determined by using rather straightforward microscopic and autoradiographic analyses. Consider, for example, the following set of experiments that were used to define the cell cycle in mouse L cells.

A. The overall length of the cell cycle was measured from the growth rate of a population of exponentially growing cells. The growth rate was determined by counting the number of cells in samples of culture fluid at various times (Figure 17–1). What is the overall length of the cell cycle in mouse L cells?

B. With the exception of mitosis, which is clearly visible in the light microscope and lasts about 1 hour, defining the phases of the cell cycle requires careful experimental analysis. ^{3}H-thymidine was added to an asynchronously growing population of cells (randomly distributed throughout the cell cycle); at various times thereafter cells were stained and prepared for autoradiography. Cells that incorporated ^{3}H-thymidine exposed the photographic emulsion and were covered by silver grains. In Figure 17–2A the fraction of *mitotic cells* that are labeled is plotted as a function of time after addition of ^{3}H-thymidine. In Figure 17–2B the average number of silver grains above *mitotic cells* is plotted as a function of time after addition of ^{3}H-thymidine. From this and other information in the problem, deduce the duration of the G_1, S, and G_2 phases of the cell cycle in mouse L cells and give your reasoning.

(A) LABELED MITOSES

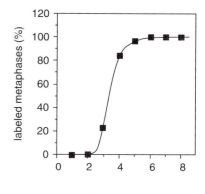

(B) NUMBER OF SILVER GRAINS

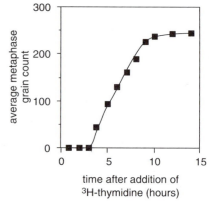

time after addition of ^{3}H-thymidine (hours)

Figure 17–2 Labeled mitotic cells as a function of time after addition of ^{3}H-thymidine (Problem 17–4). (A) Fraction of labeled mitotic cells. (B) Average number of silver grains above labeled mitotic cells.

*17–5 For many experiments it is desirable to have a population of cells that are traversing the cell cycle synchronously. One of the first, and still often used, methods for synchronizing cells is the so-called double thymidine block. If high concentrations of thymidine are added to the culture fluid, cells in S phase stop DNA synthesis, though other cells are not affected. The excess thymidine blocks the enzyme ribonucleotide reductase, which is responsible for converting ribonucleotides into deoxyribonucleotides. When this enzyme is inhibited, the supply of deoxyribonucleotides falls and DNA synthesis stops. When the excess thymidine is removed by changing the medium, the supply of deoxyribonucleotides rises and DNA synthesis resumes normally.

For a cell line with a 22-hour cell cycle divided so that M = 1 hour, G_1 = 10 hours, S = 7 hours, and G_2 = 4 hours, a typical protocol for synchronization by a double thymidine block would be as follows:

1. At 0 hours (t = 0 hours) add excess thymidine.
2. After 18 hours (t = 18 hours) remove excess thymidine.
3. After an additional 10 hours (t = 28 hours) add excess thymidine.
4. After an additional 16 hours (t = 44 hours) remove excess thymidine.

A. At what point in the cell cycle is the cell population when the second thymidine block is removed?

B. Explain how the times of addition and removal of excess thymidine synchronize the cell population.

17–6 What determines the length of S phase? One possibility is that its length

*Problems with an asterisk (*) are answered in the Instructor's Manual.

232   Chapter 17 : The Cell-Division Cycle

Table 17–1 Correlation Between Length of S Phase and DNA Content (Problem 17–6)

Organism	DNA Content of Nucleus (pg)	Length of S Phase (hours)
Lizard	3.2	15
Frog	15	26
Newt	45	41

depends on how much DNA the nucleus contains. As a test, you measure the length of S phase in dividing cells of a lizard, a frog, and a newt, each one of which has a different amount of DNA. As shown in Table 17–1, the length of S phase does increase with increasing DNA content.

Even though these organisms are similar in that in that they are all cold-blooded, they are different species. You recall that it is possible to obtain haploid embryos of frogs and repeat your measurements with haploid and diploid frog cells. Haploid frog cells have the same length S phase as diploid frog cells. Further research in the literature show that in plants, tetraploid strains of beans and oats have the same length S phase as their diploid cousins.

Propose an explanation to reconcile these apparently contradictory results. Why is it that the length of S phase increases with increasing DNA content in different species but remains constant with increasing DNA content in the same species?

The Early Embryonic Cell Cycle and the Role of MPF
(MBOC 870–879)

17–7 Fill in the blanks in the following statements.

A. _____ occur without growth and at extraordinary speed; they reveal the workings of the cell-cycle control system stripped down and simplified to the bare minimum.

B. _____ is the activity identified in *Xenopus* egg cytoplasm that, when injected into a G_2-phase oocyte, forces the oocyte into M phase to complete its maturation.

C. Whereas most proteins in sea urchin eggs accumulate continuously after fertilization, _____ shows a periodic pattern; it accumulates steadily during each interphase until the metaphase–anaphase transition, and then it is suddenly destroyed.

D. A self-limiting device, termed a _____, is somehow built into the process of DNA replication itself so as to prevent portions of the chromosomes from being replicated twice in one cell cycle.

17–8 Indicate whether the following statements are true or false. If a statement is false, explain why.

__ A. The cell division cycle in early *Xenopus* embryos has no detectable G_1 or G_2 phases.

__ B. The activities that drive the *Xenopus* oocyte to maturation and that drive the interphase mammalian cell into mitosis are the same.

__ C. The surge of MPF activity that occurs every 30 minutes in the cleaving *Xenopus* embryo is generated by a cytoplasmic oscillator that operates even in the absence of a nucleus.

__ D. Each division cycle in the cleaving *Xenopus* embryo requires synthesis of new cyclin mRNA and protein.

___ E. In early frog embryos cyclin has to be made to activate MPF and thereby induce entry into mitosis, and it has to be destroyed to inactivate MPF and thereby permit exit from mitosis.

___ F. The accumulation of cyclin to a critical threshold explains the explosive all-or-none activation of MPF.

___ G. Active MPF induces chromosome condensation, nuclear envelope breakdown, and reorganization of the cytoskeleton to form the mitotic spindle.

___ H. As in the standard cell cycle, a feedback signal from incompletely replicated DNA protects the cleaving embryo from a suicidal mitosis with its chromosomes only partially replicated.

___ I. Upon fusion with an S-phase cell the nucleus in a G_2-phase cell resumes DNA synthesis.

___ J. Removal of the block to re-replication in the early embryonic cell cycle appears to depend on the breakdown of the nuclear envelope at mitosis.

17–9 Frog oocytes mature into eggs when incubated with progesterone. This maturation is characterized by disappearance of the nucleus (termed germinal vesicle breakdown) and formation of a meiotic spindle. The requirement for progesterone can be bypassed by microinjecting 50 nl of egg cytoplasm directly into a fresh oocyte (1000 nl), which then matures normally (Figure 17–3). The control experiment of microinjecting cytoplasm from untreated oocytes into other oocytes causes no maturation, as expected. MPF activity in the egg cytoplasm is responsible for maturation.

Progesterone-induced maturation requires protein synthesis, as indicated by its sensitivity to cycloheximide; however, MPF-induced maturation does not. By placing progesterone-stimulated oocytes into cycloheximide at different times after stimulation, it can be shown that maturation becomes cycloheximide independent (no longer inhibited by cycloheximide) a few hours before the oocytes become eggs. In addition, the time at which the oocytes become cycloheximide independent corresponds to the appearance of MPF activity.

Is synthesis of MPF itself the cycloheximide-sensitive event? To test this possibility, you transfer MPF serially from egg to oocyte to test whether its activity diminishes with dilution. You first microinject 50 nl of cytoplasm from an activated egg into an immature oocyte as shown in Figure 17–3; when the oocyte matures into an egg, you transfer 50 nl of its cytoplasm into another immature oocyte; and so on. Surprisingly, you find that you can continue this process for at least 10 transfers, even if the recipient oocytes are bathed in cycloheximide! Moreover, the apparent MPF activity in the last egg is equal to that in the first egg.

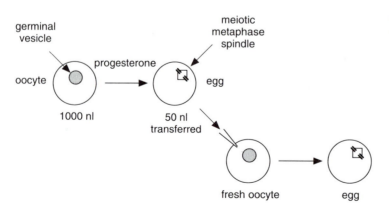

Figure 17–3 Progesterone- and MPF-induced maturation of oocytes (Problem 17–9).

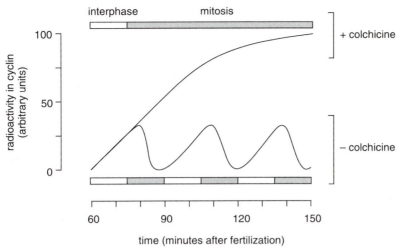

Figure 17–4 Effects of colchicine treatment on cyclin destruction and cell-cycle progression (Problem 17–10).

A. What dilution factor is achieved by 10 serial transfers of 50 nl into 1000 nl? Do you consider it likely that a molecule might have an undiminished biological effect over this concentration range?

B. How can MPF activity, which is due to a protein, be absent from immature oocytes yet appear in activated eggs, even when protein synthesis has been blocked by cycloheximide?

C. Propose a means by which MPF might maintain its activity through repeated serial transfers.

D. Propose a role for the cycloheximide-sensitive factor that is required for the appearance of MPF activity in a progesterone-stimulated oocyte.

17–10   You are studying the synthesis and destruction of cyclins in dividing clam embryos. You suspect that colchicine, a drug that binds to tubulin and arrests cells in mitosis, might work by inhibiting the normal destruction of cyclin at the metaphase–anaphase transition. To test this idea, you add ^{35}S-methionine to a suspension of fertilized clam eggs, divide the suspension in two and add colchicine to one. You take duplicate samples at 5-minute intervals, one for analysis of cyclin by gel electrophoresis and the other for analysis of mitotic chromosomes by fixing and staining the cells.

As shown in Figure 17–4, untreated cells alternated between interphase and mitosis every 30 minutes, whereas colchicine-treated cells entered mitosis normally but remained there for hours. Moreover, colchicine treatment abolished the normal disappearance of cyclin that precedes the metaphase–anaphase transition.

To get a clearer picture of how colchicine inhibits cyclin destruction, you repeat the experiment, but add the protein synthesis inhibitor emetine just before cells enter mitosis. As shown in Figure 17–5, emetine-treated cells in

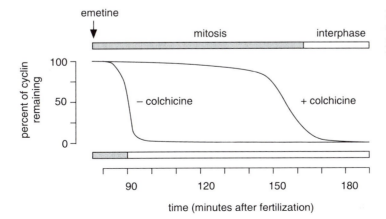

Figure 17–5 Effects of emetine treatment on cyclin destruction and cell-cycle progression in normal cells and colchicine-treated cells (Problem 17–10).

the absence of colchicine entered and exited mitosis normally, and divided into two daughter cells, which then remained indefinitely in interphase. In the presence of colchicine, the emetine-treated cells stayed in mitosis for about 2 hours and then decondensed their chromosomes and re-formed nuclei without dividing. In the presence and absence of colchicine, the exit from mitosis coincided with the disappearance of cyclin.

A. What effect does colchicine have on cyclin synthesis and destruction? Can these effects explain how colchicine causes metaphase arrest?
B. How does inhibition of protein synthesis eventually reverse the metaphase arrest produced by colchicine?

*17–11   At the transition from metaphase to anaphase MPF is inactivated and chromosomes begin to separate into sister chromatids. MPF is inactivated by cyclin-B protease, which destroys the cyclin-B component of MPF and eliminates its kinase activity.

You want to know how the separation of sister chromatids is related to MPF inactivation. To answer this question, you make cell-free extracts from unfertilized frog eggs. When nuclei are added to the extract, they spontaneously form spindles with condensed chromosomes aligned on the metaphase plate. Anaphase and the separation of sister chromatids can be triggered by addition of Ca^{2+}, which activates cyclin-B protease and turns off MPF.

To determine the reason for sister-chromatid separation, you make use of two mutant forms of cyclin B (Figure 17–6). Cyclin BΔ90 is missing the destruction box, a sequence of amino acids required for inactivation by cyclin-B protease, but it retains its ability to bind to Cdc2 kinase and make functional MPF. Cyclin B13-110 retains the destruction box but cannot bind to Cdc2 kinase. When either protein is added in excess to the extract, MPF kinase activity remains high after addition of Ca^{2+}. The two proteins differ, however, in their effects on chromatid separation. In the presence of cyclin BΔ90, sister chromatids separate normally; in the presence of cyclin B13-110, sister chromatids remain linked.

A. Why does MPF remain active in the presence of Ca^{2+} when cyclin BΔ90 is added to the extract?
B. Why does MPF remain active when cyclin B13-110 is added to the extract?
C. How is the separation of sister chromatids related to MPF inactivation? Do sister chromatids separate because a linker protein must be in its phosphorylated state in order to hold the chromatids together? Or do chromatids separate because cyclin-B protease degrades the linker protein?

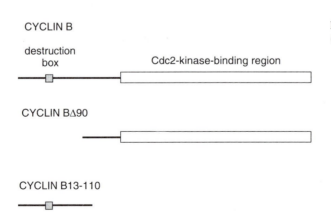

CYCLIN B

destruction box

Cdc2-kinase-binding region

CYCLIN BΔ90

CYCLIN B13-110

Figure 17–6 Cyclin B and two mutants (Problem 17–11).

Yeasts and the Molecular Genetics of Cell-Cycle Control

(MBOC 880–891)

17–12 Fill in the blanks in the following statements.

A. The organism *Saccharomyces cerevisiae* is a _____.

B. The organism *Schizosaccharomyces pombe* is a _____.

C. Each _____ mutant is typically deficient in a gene product required to get the cell past the specific point in the cell cycle at which the mutant cells arrest.

D. _____ mutants, which divide at a smaller size than normal, are expected to be deficient in a product that normally inhibits passage through a size checkpoint.

E. The fission yeast homologue of the protein kinase subunit of MPF is encoded by a gene called _____.

F. In budding yeast the G_1 checkpoint, called _____, is the major size checkpoint.

G. In G_1 the Cdc2 protein associates with G_1 cyclin to form a complex referred to as _____.

H. A safety device that operates in mammalian cells to restrain them from entering S phase with damaged DNA depends on a protein called _____, which accumulates in the cell in response to DNA damage and halts the cell-cycle control system in G_1.

17–13 Indicate whether the following statements are true or false. If a statement is false, explain why.

___ A. To maintain a constant average cell size, the length of the cell cycle must exactly match the time it takes a cell to double in size.

___ B. The G_2 checkpoint is the most important size checkpoint and also the one at which most environmental controls act in budding yeasts and mammalian cells.

___ C. Yeasts, like all eucaryotic cells, disassemble their nuclear envelope at mitosis and reassemble it after chromosome segregation.

___ D. Since a mutant that cannot complete a division cycle cannot be propagated, *cdc* mutants can be selected and maintained only if their phenotype is conditional, that is, if the gene product functions normally under one set of conditions but fails to function under another set of conditions.

___ E. In both yeasts and vertebrates entry into mitosis is driven by essentially the same kinase, which has to be complexed with cyclin to become active.

___ F. Active MPF is generated by the binding of cyclin to Cdc2.

___ G. It seems to be a general rule that cell size is roughly proportional to ploidy, suggesting that the control mechanism depends on some sort of titration— direct or indirect—of the quantity of a cytoplasmic component against the quantity of DNA.

___ H. Start kinase and MPF phosphorylate different sets of target proteins because the protein kinase component of each complex is encoded by different, though related, genes.

___ I. The cell-cycle control system in budding yeast appears to be regulated by mechanisms that control the rate of accumulation of G_1 cyclins.

___ J. Strong circumstantial evidence suggests that Start kinase activates transcription of some of the proteins required for chromosome replication.

___ K. The cell cycle controls in mammalian cells and early frog embryos are similar in that inhibitors of DNA replication delay entry into mitosis.

___ L. Mutants of budding yeast that lack the *rad9* gene product still possess the machinery for DNA repair, but they fail to delay in G_2 when they have been irradiated.

___ M. Feedback controls in the cell cycle are based on positive signals that move the control system forward rather than on negative signals that arrest the cell-cycle control system.

*17–14 A common first step in characterizing cell-division-cycle (*cdc*) mutants is to define the phase of the cell cycle at which the mutational block stops the cell's progress. Temperature-sensitive *cdc* mutants are particularly useful because they grow normally at one temperature (the permissive temperature) but express a mutant phenotype when grown at a higher temperature (the restrictive temperature). One method for characterizing temperature-sensitive *cdc* mutants uses the drug hydroxyurea, which blocks DNA synthesis by inhibiting ribonucleotide reductase (which provides deoxyribonucleotide precursors). Hydroxyurea blockade of DNA synthesis can be reversed simply by changing the incubation medium. Consider the following results with the hypothetical mutants *cdc*101 and *cdc*102.

You incubate a culture of a yeast *cdc*101 mutant at its restrictive temperature (37°C) for 2 hours (the approximate length of the cell cycle in yeasts) so that its mutant phenotype is expressed. Then you transfer it to medium containing hydroxyurea at the permissive temperature (20°C). None of the cells divide.

You now reverse the order of treatment. You incubate *cdc*101 at 20°C for 2 hours in medium containing hydroxyurea and then transfer it to medium without hydroxyurea at 37°C. The cells undergo one round of division.

You repeat these two experiments with the *cdc*102 mutant. The cells do not divide in either case.

A. In what phase of the cell cycle is *cdc*101 blocked at the restrictive temperature? Explain the results of the reciprocal temperature-shift experiments.
B. In what phase of the cell cycle is *cdc*102 blocked at the restrictive temperature? Explain the results of the reciprocal temperature-shift experiments.

17–15 You have isolated a temperature-sensitive mutant of a budding yeast. It grows well at 25°C, but at 35°C all the cells develop a large bud and then stop growing. The characteristic morphology of the cells at the time they stop growing is known as the landmark morphology.

It is very difficult to synchronize the growth of this yeast, but you would like to know as exactly as possible at what point in the cell cycle the temperature-sensitive gene product must function in order for the cell to complete the cycle. The critical point in the cycle at which a gene product functions is its execution point, in the terminology of the field. A clever friend, who has a good microscope with a heated stage and a time-lapse video recorder, suggests that you take pictures of a field of cells as they experience the temperature increase

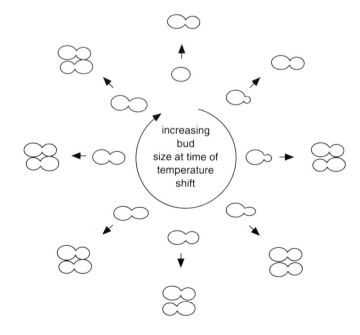

Figure 17–7 Time-lapse photography of a temperature-sensitive mutant of yeast (Problem 17–15). Cells on the inner circle are arranged in order of their bud sizes, which corresponds to their position in the cell cycle. After 6 hours at 35°C, they have given rise to the cells shown on the outer circle. No further growth or division occurs.

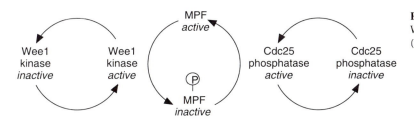

Figure 17–8 Control of MPF activity by Wee1 kinase and Cdc25 phosphatase (Problem 17–16).

and follow the behavior of individual cells as they stop growing. Since the cells do not move much, it is relatively simple to study individual cell behavior. To make sense of the what you see, you arrange a circle of photos of cells at the start of the experiment in order of the size of their daughter buds. You then find the corresponding photos of those same cells 6 hours later, when growth has completely stopped. The results with your mutant are shown in Figure 17–7.

A. Indicate on the diagram in Figure 17–7 where the execution point for your mutant lies.

B. Does the execution point correspond to the time at which the cell cycle is arrested in your mutant? How can you tell?

17–16   One critically important aspect of cell cycle control is ensuring that transitions such as entry into mitosis occur rapidly and completely. The regulation of MPF by Wee1 tyrosine kinase and Cdc25 tyrosine phosphatase gives a good appreciation of how this control can be achieved. The balance of the activities of Wee1 and Cdc25 determines the state of phosphorylation of tyrosine 15 in the Cdc2 component of MPF. When tyrosine 15 is phosphorylated, MPF is inactive; when tyrosine 15 is not phosphorylated, MPF is active (Figure 17–8). Just as the activity of MPF itself is controlled by phosphorylation, so too are the activities of Wee1 kinase and Cdc25 phosphatase.

The regulation of these various activities can be studied in extracts of frog oocytes. In such extracts Wee1 kinase is active and Cdc25 phosphatase is inactive. As a result MPF is inactive because its Cdc2 component is phosphorylated on tyrosine 15. MPF in these extracts can be rapidly activated by addition of okadaic acid, which is a potent inhibitor of serine/threonine protein phosphatases. Using antibodies specific for each component, it is possible to examine their phosphorylation state by changes in mobility upon gel electrophoresis (Figure 17–9). (Phosphorylated proteins generally run slower than their nonphosphorylated counterparts.)

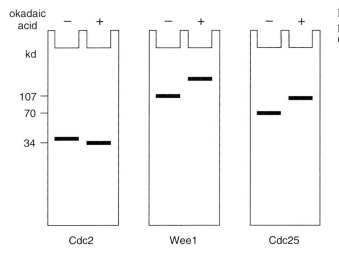

Figure 17–9 Effects of okadaic acid on the phosphorylation states of Cdc2, Wee1, and Cdc25 (Problem 17–16).

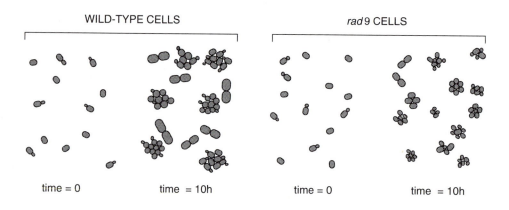

WILD-TYPE CELLS   *rad*9 CELLS

time = 0   time = 10h   time = 0   time = 10h

Figure 17–10 Time-lapse pictures of wild-type and *rad*9 cells before and after x-irradiation (Problem 17–17).

A. Based on the results with okadaic acid decide whether the active forms of Wee1 kinase and Cdc25 phosphatase are phosphorylated or nonphosphorylated. In Figure 17–8 indicate the phosphorylated forms of Wee1 and Cdc25 and label the arrows to show which are protein kinases and which are protein phosphatases.

B. Are the protein kinases and phosphatases that control Wee1 and Cdc25 specific for serine/threonine residues or tyrosine residues? How do you know?

C. How does addition of okadaic acid cause an increase in phosphorylation of Wee1 and Cdc25, but a decrease in phosphorylation of Cdc2?

D. If you assume that Cdc25 and Wee1 are targets for phosphorylation by Cdc2 kinase in active MPF, can you explain how the appearance of a small amount of active MPF would lead to its rapid and complete activation?

17–17   You have found a new way to study radiation-sensitive yeast mutants. By growing cells on a thin layer of agar on a microscope slide you can use time-lapse photography to follow the fate of individual cells. When you irradiate wild-type cells with x-rays, which cause chromosome breaks, and follow their growth for the next 10 hours, you find that most of the cells arrest temporarily at the large-bud (dumbbell) stage, but about half the cells eventually recover and have formed small viable colonies after 10 hours (Figure 17–10). The fraction that are still arrested at the dumbbell stage after 10 hours is equal to the fraction of nonviable cells (Table 17–2).

You repeat these experiments with seven different radiation-sensitive (*rad*) mutants. For six of the mutants you observe a similar equality of 10-hour arrested cells and nonviable cells, as shown in Table 17–2 for the *rad* 52 mutant, although relative to wild-type cells a higher fraction of *rad* cells are still arrested in the dumbbell stage.

One of the seven *rad* mutants, however, has a strikingly different phenotype. Many fewer *rad*9 cells arrest even temporarily at the dumbbell stage, and after 10 hours only 20 percent are still arrested there (Table 17–2). Although many of the cells divide once or twice, they mostly form nonviable microcolonies (Figure 17–10, Table 17–2).

Table 17–2 Fractions of Arrested and Nonviable Cells After X-ray Treatment (Problem 17–17)

Strain	Arrested at 10 hours (%)	Nonviable (%)
Wild-type	50	50
*rad*52	90	95
*rad*9	20	70

Table 17–3 Mutant Strains That Affect the Mitotic Entry Checkpoint (Problem 17–18)

Mutant Strains	Mitotic Delay in Response to	
	Damaged DNA	Unreplicated DNA
rad24	No	Yes
cdc2-3w	Yes	No
hus1	No	No
hus2	No	No
rad1	No	No
cdc2-F15	Yes	No

A. For the wild-type cells decide which cells in the population appear most likely to remain arrested at the dumbbell stage after 10 hours. Given that the cells used in these experiments are haploid and that x-ray-induced breaks are repaired by homologous recombination, decide which stage of the cell cycle the sensitive cells are in.

B. By staining with DNA-binding reagents and tubulin-specific antibodies, you show that cells in the dumbbell stage have a single nucleus stuck in the neck and no visible spindle. Given what you know about cell cycle checkpoints, in what stage of the cell cycle do you think the cells are arrested?

C. Why do half of the wild-type cells temporarily arrest at the dumbbell stage but then go on to form viable colonies after 10 hours?

D. Why do you think that so many more rad52 mutant cells (relative to wild-type cells) are arrested at the dumbbell stage after 10 hours?

E. Why do you think that so few rad9 mutant cells arrest even temporarily at the dumbbell stage? Why are so many of the rad9 cells nonviable?

F. Would you expect the fraction of nonviable rad9 cells to increase, decrease, or stay the same if the cells were artificially delayed in G_2 for a couple of hours using a microtubule inhibitor that reversibly prevents spindle formation? Why?

*17–18 Fission yeast respond to damaged DNA and unreplicated DNA by delaying entry into mitosis. You want to know how the signals from such DNA interact with the mitotic entry checkpoint. You screen a large number of mutant yeast strains and find six that do not delay mitosis in response to DNA damage, unreplicated DNA, or both, as shown in Table 17–3. Which of the signaling pathways shown in Figure 17–11 is supported by your data? On the pathway you choose, indicate where each of the mutant genes acts.

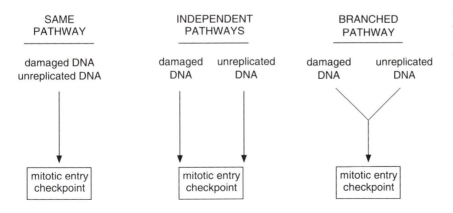

Figure 17–11 Possible pathways by which signals from damaged DNA and unreplicated DNA interact with the mitotic entry checkpoint (Problem 17–18).

Cell-Division Controls in Multicellular Animals

(MBOC 891–906)

17–19 Fill in the blanks in the following statements.

   A. The essential components of serum that allow mammalian cells in culture to continue dividing are certain highly specific proteins called _____, most of which are required only in very low concentration.

   B. The principal factor in serum that enables fibroblasts in tissue culture to divide is _____.

   C. The mutant form of a proliferation gene that results in the excessive cell proliferation characteristic of cancer is called an _____, and the normal proliferation gene is called a _____.

   D. The antiproliferation genes found in normal cells are often referred to as _____.

   E. The genes that growth factors induce fall into two classes: _____ genes are induced within 15 minutes of growth-factor treatment; _____ genes, by contrast, are not induced until at least 1 hour after growth-factor treatment.

   F. The _____ gene was identified originally through studies of a cancer that occurs in the eyes of children who lack a functional copy of the gene.

   G. Fibroblasts taken from a normal human fetus will go through only about 50 population doublings before proliferation slows down and then halts, a phenomenon known as _____.

17–20 Indicate whether the following statements are true or false. If a statement is false, explain why.

   __ A. Positive signals from growth factors act by overriding intracellular negative controls that otherwise restrain growth and block progress of the cell-cycle control system.

   __ B. Like yeasts, higher animals have multiple cyclins that interact with a single cyclin-dependent kinase to control passage past various cell cycle checkpoints.

   __ C. Cells that escape senescence and divide indefinitely provide an unlimited source of a standardized, genetically homogeneous type for use in cell-cycle studies.

   __ D. In intact animals proliferation of most cell types depends on a specific growth factor.

   __ E. Most growth factors originate from cells in the neighborhood of the affected cell and act as local mediators.

   __ F. In mammalian cells, as in yeasts, there is a relatively rigid rule that couples cell size and cell division.

   __ G. The time taken for a cell to progress from the beginning of S phase through mitosis is usually brief (typically 12 to 24 hours in mammals) and remarkably constant, irrespective of the interval from one division to the next.

   __ H. Serum deprivation causes proliferating cells to stop wherever they are in the cell cycle and enter G_0.

   __ I. Quiescent G_0 cells are severely depleted in Cdk protein and in all of the G_1 cyclins, whereas Cdk proteins and some of the G_1 cyclins are present at a nearly constant level during all phases of the cycle in cycling cells.

   __ J. Cells growing in a monolayer stop dividing when they contact one another, in a way that is independent of the concentration of growth factors in the medium.

   __ K. The cycle of attachment and detachment as cells pass through M phase presumably allows cells to rearrange their contacts with other cells and with the extracellular matrix so as to accommodate the daughter cells and then bind them securely into a tissue.

___ L. Both copies of a tumor-suppressor gene must be lost or inactivated to bring about the loss of growth control, whereas only one copy of a proto-oncogene need be activated to bring about a similar effect.

___ M. A protein growth factor initiates its effect by binding to a transmembrane receptor at the surface of the target cell.

___ N. Some of the essential components of the cell-cycle control system itself, including Cdk proteins and several cyclins, are found among the products of the delayed-response genes, which are not transcribed until well after the addition of growth factor.

___ O. In its inactive, dephosphorylated state, Rb binds a set of regulatory proteins that favor cell proliferation, holding them sequestered and out of action; the phosphorylation of Rb makes it release these proteins, allowing them to act.

___ P. Although cell senescence occurs at a predictable time for a given cell population, it is not strictly programmed at the level of the individual cell.

*17–21   Vertebrate cells pause in the G_1 phase of the cell cycle until growth conditions are appropriate for their entry into S phase with subsequent cell division. Some of the growth-factor requirements for the passage of fibroblasts through G_1 have been defined using mouse 3T3 cells, which are a fibroblastlike cell line. In the absence of serum these cells do not enter S phase. If serum is added to a culture of such arrested cells, they progress through G_1 and begin to enter S phase 12 hours later. The serum requirement can be met by supplying three growth factors: PDGF, EGF, and Somatomedin C. When these growth factors are mixed with appropriate nutrients and added to quiescent cells, the cells begin to enter S phase 12 hours later. If any one of the growth factors is left out, the cells do not enter S phase.

Do all three growth factors have to be present at the same time? Is their stimulation of cells independent of one another? Or do they stimulate cells in an ordered sequence? To address these questions, you pretreat cells with the growth factors in a defined order and then add complete medium (containing serum and nutrients) in the presence of ^{3}H-thymidine. At various times thereafter you fix cells and subject them to autoradiography. You define the time of appearance of the first labeled nuclei as the time of entry into S phase. The results of these experiments are given in Table 17–4.

Do the cells require these growth factors simultaneously, independently, or in an ordered sequence? Explain your answer.

17–22   EGF stimulates the proliferation of many types of epithelial cells by binding to EGF receptors on their surface. The role of the EGF receptor in propagating the proliferation signal is being clarified by study of the receptor itself. The

---

Table 17–4 Effect of Growth-Factor Pretreatment of 3T3 Cells on the Timing of Entry into S Phase (Problem 17–21)

| Experiment | Order of Addition | | | | Entry into S Phase |
	1	2	3	4	
1	EGF	PDGF	SomC	medium	12 hours
2	EGF	SomC	PDGF	medium	12 hours
3	PDGF	EGF	SomC	medium	1 hour
4	PDGF	SomC	EGF	medium	6 hours
5	SomC	EGF	PDGF	medium	12 hours
6	SomC	PDGF	EGF	medium	6 hours

Cells were treated for 6 hours with the indicated growth factors in the order listed. They were thoroughly washed to remove one growth factor before the next one was added. After the regimen of growth-factor pretreatment, complete medium with ^{3}H-thymidine was added and the time before labeled nuclei appeared was determined.

Cell-Division Controls in Multicellular Animals

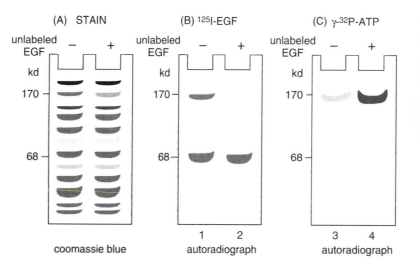

Figure 17–12 Analysis of the EGF receptor (Problem 17–22). (A) SDS gel of a membrane preparation from A-431 cells. (B) SDS gel of a membrane preparation from A-431 cells incubated with ^{125}I-EGF and a protein cross-linking agent in the presence and absence of excess unlabeled EGF. (C) SDS gel of antibody-precipitated EGF receptor incubated with γ-^{32}P-ATP in the presence and absence of EGF.

mouse fibroblast A-431 cell line, which fortuitously carries enormously increased numbers of EGF receptors, makes their characterization much easier. Consider the following set of experiments.

1. Plasma membrane preparations from A-431 cells contain many proteins as shown on the SDS gel in Figure 17–12A. However, when ^{125}I-EGF is added to such a preparation in the presence of a protein cross-linking agent, two proteins become labeled (Figure 17–12B, lane 1). When excess unlabeled EGF is included in the incubation mixture, the labeled band at 170 kd disappears (Figure 17–12B, lane 2).
2. If the membrane preparation is incubated with γ-^{32}P-ATP, several proteins, including the 170 kd protein, become phosphorylated. This reaction is significantly stimulated by including EGF in the incubation mixture.
3. When antibodies specific for the 170 kd protein are used to precipitate the protein and the incubation with γ-^{32}P-ATP is repeated with the precipitate, the 170 kd protein is phosphorylated in an EGF-stimulated reaction (Figure 17–12C, lanes 3 and 4).
4. If the antibody-precipitated protein is first run on the SDS gel and then renatured in the gel, subsequent incubation with γ-^{32}P-ATP in the presence and absence of EGF yields the same pattern shown in Figure 17–12C.

A. Which of these experiments demonstrates most clearly that the 170-kd protein is the EGF receptor?
B. Is the EGF receptor a substrate for an EGF-stimulated protein kinase? How do you know?
C. Which experiments show that the EGF receptor is a protein kinase?
D. Is it clear whether the EGF receptor is a substrate for its own protein kinase activity?

17–23   You have been led by a bizarre accident to a productive line of experimentation. You lost one of your contact lenses while transferring a line of tissue culture cells; a few days later you found the lens on the bottom of a petri dish. Interestingly, the cells attached to the lens were rounded up and very sparsely distributed, whereas those on the rest of the dish were flat and had grown to near confluency with most cells touching their neighbors. Ah ha! you thought, perhaps this observation can be used to investigate the relationship between cell shape and growth control. As someone once said, "Chance favors the prepared mind."

The manufacturer graciously sends you a supply of the plastic—poly(HEMA)—from which your soft contact lenses were made. When an alcoholic mixture of poly(HEMA) is pipetted into a plastic culture dish, a thin,

Table 17–5 Incorporation of ³H-Thymidine by Cells Grown on Normal Dishes and on Poly(HEMA)-treated Dishes (Problem 17–23)

Type of Dish	Cell Density (cells/dish)	Confluency	Cell Height (μm)	³H Incorporation (cpm/dish)
Normal	60,000	subconfluent	6	15,200
Normal	200,000	confluent	15	11,000
Normal	500,000	confluent	22	3500
Poly(HEMA)	30,000	sparse	6	7500
Poly(HEMA)	30,000	sparse	15	1500
Poly(HEMA)	30,000	sparse	22	210

hard, sterile film of optically clear polymer remains bound to the plastic surface after the alcohol evaporates. Serial dilutions of the alcohol-polymer solution, introduced into each dish at a constant volume, result in decreasing thicknesses of the polymer film. The thinner the film, the more strongly cells adhere to the dish. Moreover, there is a gradual change in cell shape from round to flat with decreasing thickness of the film. Using the height of the cells as an indicator of cell shape (no mean technical feat), you demonstrate that there is a smooth relationship between cell shape and growth potential: the flatter the cells, the better they incorporate ³H-thymidine.

Now for the big question: Is density-dependent inhibition of cell growth mediated by changes in cell shape? You grow cells to different densities (different degrees of confluency) on normal plastic dishes and measure the height of the cells and their ability to incorporate ³H-thymidine. As shown in Table 17–5, the more confluent the cells, the greater their height and the lower their incorporation of ³H-thymidine. Is the decrease in growth due to the increase in cell crowding or to the change in cell shape? To answer this question, you distribute cells at a low density on plates with poly(HEMA) films, such that the height of the cells in the sparse cultures matches the height of cells in the various confluent cultures. Your measurements of ³H-thymidine incorporation in these sparse cultures are shown in Table 17–5.

Based on the results in Table 17–5, would you conclude that density-dependent inhibition of cell growth correlates completely, partially, or not at all with changes in cell shape? Explain your reasoning.

*17–24 Retinoblastoma is an extremely rare cancer of the nerve cells in the eye. The disease mainly affects children up to the age of five years because it can only occur while the nerve cells are still dividing. In some cases tumors occur in only one eye, but in other cases tumors develop in both eyes. The bilateral cases all show a familial history of the disease; most of the cases affecting only one eye arise in families with no previous history of the disease.

An informative difference between unilateral and bilateral cases becomes apparent when the fraction of still undiagnosed cases is plotted against the age at which diagnosis is made (Figure 17–13). The regular decrease with time shown by the bilateral cases suggests that a single chance event is sufficient to trigger onset of bilateral retinoblastoma. By contrast, the presence of a "shoulder" on the unilateral curve suggests that multiple events in one neuron are required to trigger unilateral retinoblastoma. (A shoulder arises because the events accumulate over time. For example, if two events are required, most affected cells at early times will have suffered only a single event and will not generate a tumor. With time the probability increases that a second event will occur in an already affected cell and, therefore, cause a tumor.)

A possible explanation for these observations is that tumors develop when both copies of the critical gene (the retinoblastoma, *Rb*, gene) are lost or

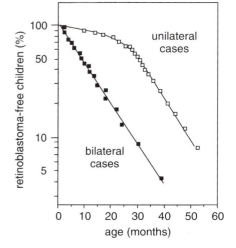

Figure 17–13 Time of onset of unilateral and bilateral cases of retinoblastoma (Problem 17–24). A population of children who ultimately developed retinoblastoma are represented in this graph. The fraction of the population that is still retinoblastoma free is plotted against the time after birth.

mutated. In the inherited (bilateral) form of the disease, a child receives a defective *Rb* gene from one parent: tumors develop in an eye when the other copy of the gene in any nerve cell in the eye is lost through somatic mutation. In fact, the loss of a copy of the gene is frequent enough that tumors usually occur in both eyes. If a person starts with two good copies of the *Rb* gene, tumors arise in an eye only if both copies are lost *in the same cell.* Since such double loss is very rare, it is usually confined to one eye.

To test this hypothesis, you use a cDNA clone of the *Rb* gene to probe the structure of the gene in cells from normal individuals and from patients with unilateral or bilateral retinoblastoma. As illustrated in Figure 17–14, normal individuals have four restriction fragments that hybridize to the cDNA probe (which means each of these restriction fragments contains at least one exon). Fibroblasts (nontumor cells) from the two patients also show the same four fragments, although three of the fragments from the child with bilateral retinoblastoma are present in only half the normal amount. Tumor cells from the two patients are missing some of the restriction fragments.

A. Explain why fibroblasts and tumor cells from the same patient show different band patterns.
B. What are the structures of the *Rb* genes in the fibroblasts from the two patients? In the tumor cells from the two patients?
C. Are these results consistent with the hypothesis that retinoblastoma is due to the loss of the *Rb* gene?
D. Suggest a plausible explanation for how the loss of the *Rb* gene product might cause retinoblastoma.

17–25 The formation of tumors is a multistep process that may involve the successive activation of several oncogenes. This notion is supported by the discovery that certain pairs of oncogenes, of which *ras* and *myc* are among the best-studied examples, transform cultured cells more efficiently than either alone. Similar experiments with pairs of oncogenes have now been carried out in transgenic mice. In one experiment the *ras* oncogene was placed under the control of the MMTV (mouse mammary tumor virus) promoter and incorporated into the germ lines of several transgenic mice. In a second experiment

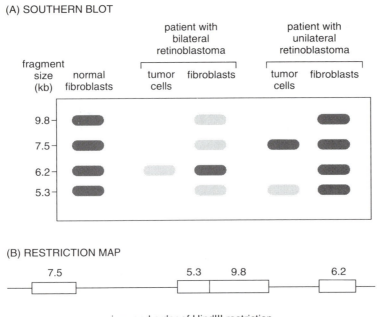

(A) SOUTHERN BLOT

Figure 17–14 Patterns of blot hybridization of restriction fragments from the retinoblastoma gene (Problem 17–24). (A) Southern blot for normal individuals and for patients with unilateral and bilateral retinoblastoma. Lighter shading of some bands indicates half the normal number of copies. (B) The order of the restriction fragments. Fragments that contain exons (shown as rectangles) hybridize to the cDNA clone that was used as a probe in these experiments.

the *myc* oncogene under control of the same MMTV promoter was incorporated into the germ lines of several transgenic animals. In a third experiment mice containing the individual oncogenes were mated to produce mice with both oncogenes.

All three kinds of mice developed tumors at a higher frequency than normal animals. Female mice were most rapidly affected because the MMTV promoter, which is responsive to steroid hormones, turns on the transferred oncogenes in response to the hormonal changes at puberty. In Figure 17–15 the rate of appearance of tumors is plotted as the percent of tumor-free females as a function of time after puberty.

A. Assume that the lines drawn through the data points are an accurate representation of the data. How many events in addition to expression of the oncogenes are required to generate a tumor in each of the three kinds of mice? (You may wish to reread Problem 17–25.)

B. Is activation of the cellular *ras* proto-oncogene the event required to trigger tumor formation in mice that are already expressing the MMTV-regulated *myc* oncogene (or vice versa)?

C. Why do you think the rate of tumor production is so high in the mice containing both oncogenes?

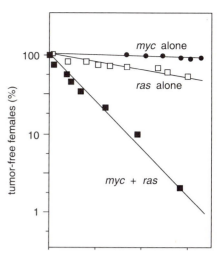

Figure 17–15 Fraction of tumor-free female mice as a function of time after puberty (Problem 17–25).

*17–26 It has been suggested that normal cells are limited to about 50 cell divisions to restrict the maximum size of tumors, thus affording some protection against cancer. Assuming that 10^8 cells weigh 1 gram, calculate the weight of a tumor that originated from 50 doublings of a single cancerous cell.

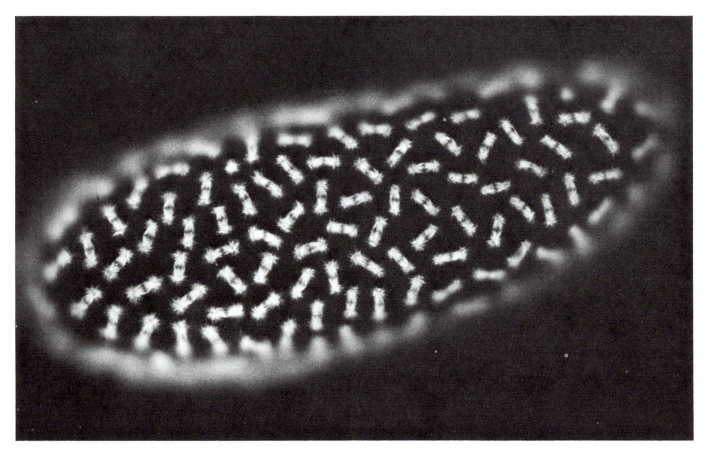

An early *Drosophila* embryo stained with an anti-tubulin antibody reveals hundreds of mitotic spindles, thanks to the synchrony of division at this stage of development. (Courtesy of Bruce Alberts.)

The Mechanics of Cell Division

- An Overview of M Phase
- Mitosis
- Cytokinesis

An Overview of M Phase
(MBOC 911–919)

18–1 Fill in the blanks in the following statements.

A. The progressive compaction of the dispersed interphase chromatin into threadlike chromosomes that occurs at the onset of M phase is called _____.

B. There are two distinct mechanical processes in M phase: (1) _____, which is the segregation of the chromosomes and the formation of two nuclei in place of one, and (2) _____, which is the splitting of the cell as a whole into two.

C. The bipolar _____, composed of microtubules and their associated proteins, aligns the replicated chromosomes in a plane that bisects the cell.

D. A _____ of actin filaments and myosin-II forms just beneath the plasma membrane in a plane perpendicular to the spindle and pulls the membrane inward so as to divide the cell into two.

E. The principal microtubule organizing center in most animal cells is the _____, a cloud of poorly defined material associated with a pair of _____.

18–2 Indicate whether the following statements are true or false. If a statement is false, explain why.

__ A. M phase is turned on by protein phosphorylation and terminated by protein dephosphorylation.

__ B. M phase is very similar in all cells from bacteria to humans.

__ C. In yeasts, two sister chromosomes make the error of going to the same pole (leaving the other daughter cell without a copy of the chromosome) about once in every 100,000 cell divisions.

__ D. Before a eucaryotic cell divides, it must duplicate its centrosome to provide one for each of its two daughter cells.

__ E. The six stages of M phase—prophase, prometaphase, metaphase, anaphase, telophase, and cytokinesis—occur in strict sequential order.

__ F. Daughter nuclear envelopes re-form before cytokinesis is complete.

__ G. The Golgi apparatus and the ER break up into a set of smaller fragments and vesicles during mitosis, presumably because in this highly vesiculated form they can be more evenly distributed when cells divide.

*18–3 When cells divide after mitosis, their surface area increases—a natural consequence of dividing a constant volume into two compartments. The increase in

*Problems with an asterisk (*) are answered in the Instructor's Manual.

surface requires an increase in the amount of plasma membrane. One can estimate the increase in plasma membrane by making certain assumptions about the geometry of cell division. Assuming that the parent cell and the two progeny cells are spherical, one can apply the familiar equations for the volume and surface area of a sphere.

$$\text{Volume} = \frac{4}{3}\pi r^3$$
$$\text{Area} = 4\pi r^2$$

A. Assuming that the progeny cells are equal in size, calculate the increase in plasma membrane that accompanies cell division. (Although this problem can be solved algebraically, you may find it easier to substitute real numbers. For example, let the volume of the parent cell equal 1.) Do you think that the magnitude of this increase is likely to cause any problem for the cell? Explain your answer.

B. During early development many fertilized eggs undergo several rounds of cell division without any overall increase in total volume. For example *Xenopus* eggs undergo 12 rounds of division before growth commences and the total cell volume increases. Assuming once again that all cells are spherical and equal in size, calculate the increase in plasma membrane that accompanies development of the early embryo: in going from one large cell (the egg) to 4096 small cells (12 divisions).

18–4 One of the least well understood aspects of the cell cycle is the reproduction of the spindle poles. As illustrated in Figure 18–1, the centrosome normally splits at the beginning of mitosis to form the two spindle poles, which orchestrate chromosome segregation. During the next interphase, the centriole pair within the centrosome is duplicated so that the centrosome can split at the next mitosis. The cycles of centrosome duplication and splitting normally keep step with cell division so that all cells have the capacity to produce bipolar spindles. However, it is possible to throw the two cycles out of phase as indicated by experiments first performed in the late 1950s.

If a fertilized sea urchin egg at the metaphase stage of the first mitotic division is exposed to mercaptoethanol (MSH), the mitotic spindle disassembles (Figure 18–2A). (It is not known how mercaptoethanol causes this,

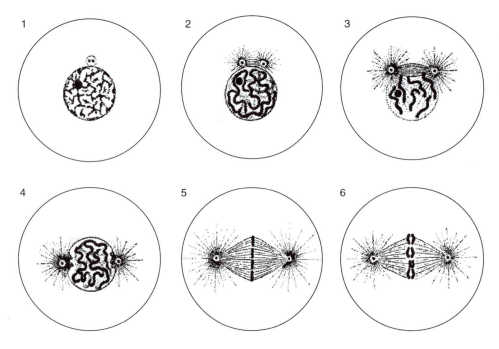

Figure 18–1 Normal process of centrosome splitting during mitosis to form a bipolar spindle (Problem 18–4). (Adapted from E.B. Wilson, The Cell in Development and Inheritance, 1st ed., 1896. Figures 19 and 20. New York and London: Johnson Reprint Corporation, 1966.)

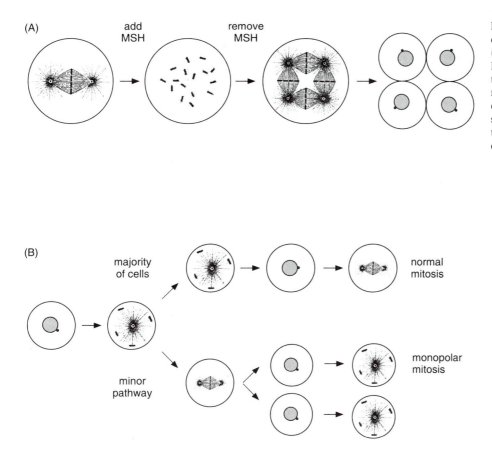

(A)

add MSH → remove MSH →

Figure 18–2 Abnormal spindles and cell divisions induced by mercaptoethanol (MSH) treatment (Problem 18–4). (A) Formation of tetrapolar spindles upon treatment of fertilized sea urchin eggs with mercaptoethanol followed by four-way division. (B) Major and minor pathways for spindle formation and cell division among the daughter cells from a four-way division.

(B)

majority of cells → → normal mitosis

minor pathway → → monopolar mitosis

but the effect is reversible.) While the eggs are kept in mercaptoethanol, the nucleus does not re-form, no DNA synthesis occurs, and the chromosomes stay condensed. When mercaptoethanol is washed out, the spindle re-forms and cell division takes place. However, although some eggs re-form a bipolar spindle and divide normally, the majority of eggs form a tetrapolar spindle and divide into four daughter cells (Figure 18–2A). No matter how long the eggs are arrested in mercaptoethanol, they never divide into more than four cells.

The daughter cells from a four-way division re-form the nucleus and traverse the next cell cycle; however, at mitosis they form a monopolar spindle. In a majority of cases these cells stay in mitosis a little longer than usual and then decondense their chromosomes, disassemble the spindle, and re-form a nucleus (Figure 18–2B). At the next mitosis these cells form a normal bipolar spindle and divide normally (Figure 18–2B). More rarely, the cells with monopolar spindles stay in mitosis much longer, the monopole splits to form a bipolar spindle, and the cell divides normally (Figure 18–2B). However, the daughter cells from such a division once again form a monopolar spindle at the next mitosis (Figure 18–2B).

Describe patterns of centriole duplication and splitting that can account for the observations shown in Figure 18–2.

18–5 You are studying the genetics of color variegation in corn. One reproducible way to produce speckled kernels involves crosses with a strain that carries an x-ray-induced rearrangement of chromosome 9 (Figure 18–3). This chromosome carries a color marker (*C*, colored; recessive form *c*, colorless), which allows you to follow its inheritance in individual kernels of an ear of corn. When strains carrying the rearranged chromosome 9 bearing the dominant *C* allele are crossed with wild-type corn bearing the recessive *c* allele, a small number of kernels in the progeny ears of corn have a speckled appearance.

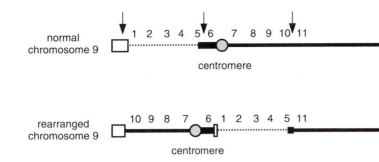

This color variegation arises by an interesting mechanism. In meiosis, recombination within the rearranged segment generates a chromosome with two centromeres as shown in Figure 18–4. In a fraction of these meioses the recombined chromosome with two centromeres gets strung out between the two poles at the first meiotic anaphase, forming a bridge between the two meiotic poles. Sometime in anaphase to telophase the strained chromosome breaks at a random position between the duplicated centromeres. The broken ends of the chromosome tend to fuse together after the next S phase, which generates a new dicentric chromosome whose structure depends on where the previous break occurred (Figure 18–4). This chromosome in turn will be broken by the forces acting during the subsequent mitosis, and the fusion-bridge-breakage cycle will repeat itself in the next cell cycle, unless a repair mechanism adds a telomere to the broken end.

Figure 18–4 also shows the chromosomal location of another genetic marker on chromosome 9 that can cause a "waxy" alteration to the starch deposited in the kernels, which can be detected by staining with iodine. Waxy (*wx*) is recessive to the normal, nonwaxy (*Wx*) allele. By following the inheritance of the *C* and *Wx* markers in the kernels, you can gain an understanding of the behavior of broken chromosomes. You inspect the speckled kernels and stain them with iodine to see if there are any rules about the topology of the patches. You observe three types of patches within the otherwise colored, nonwaxy (*C-Wx*) kernels: colorless, nonwaxy (*c-Wx*) patches; colorless, nonwaxy (*c-Wx*) patches containing one or more colorless, waxy (*c-wx*) spots; and intensely colored, nonwaxy (*?-Wx*) patches (Figure 18–5).

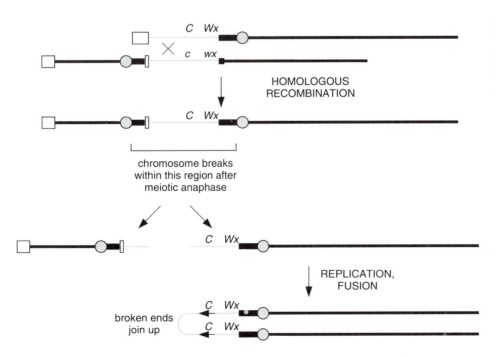

Figure 18–4 Recombination between a normal and a rearranged chromosome 9 to give a dicentric chromosome (Problem 18–5). Breakage of the original dicentric chromosome followed by replication and fusion of the ends gives rise to a second dicentric chromosome.

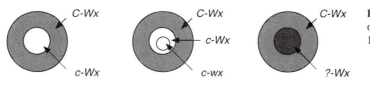

A. Patches arise because cells with a different genetic constitution divide to give identical neighbors that remain together in a cluster. Starting with the dicentric chromosome shown at the bottom of Figure 18–4, show how fusion-bridge-breakage cycles might account for the three types of patches shown in Figure 18–5. What is the genetic constitution of the intensely colored patches? (In these crosses, the dominant alleles—C and Wx—are carried on the rearranged chromosome, and the recessive alleles—c and wx—are carried on the normal chromosome.)

B. Would you ever expect to see colored spots within colorless patches? Why or why not?

C. Would you ever expect to see colorless spots within the intensely colored patches? Why or why not?

18–6 By the turn of the century it was clear that chromosomes were the carriers of hereditary information. But a fundamental question remained: Does each chromosome carry the total hereditary information or does each chromosome carry a different portion of the hereditary information? According to the first view, the multiple chromosomes in cells were required to raise the total quantity of hereditary material above a threshold value needed for proper development. According to the second view, multiple chromosomes were required so that all portions of the hereditary information would be represented. This question was answered definitively in a classic series of experiments carried out by Theodor Boveri from 1901 to 1905.

As an experimental system, Boveri followed the development of sea urchin eggs that had been fertilized by two sperm, which occurs relatively frequently in artificial fertilization when large quantities of sperm are added. In contrast to the normal bipolar mitotic spindle and division into two cells, eggs fertilized by two sperm form a tetrapolar mitotic spindle and then divide into four cells. The three sets of chromosomes—one from the egg and two from the sperm—are distributed randomly among the four spindles as shown for four chromosomes in Figure 18–6A. If the dispermic eggs are gently shaken immediately after fertilization, one of the spindle poles often fails to form, resulting in a tripolar mitotic spindle followed by division into three cells (Figure 18–6B).

The species of sea urchin that Boveri studied, *Echinus microtuberculatus,* has a diploid chromosome number of 18, but will develop normally to a pluteus larva—a free swimming stage in sea urchin development—with a haploid number of chromosomes (9). For a tripolar or tetrapolar egg to develop to a normal pluteus, each of the cells resulting from the initial three-way or four-way division must have either at least 9 total chromosomes or at least 9 different chromosomes—depending on which view of chromosomes is correct. Boveri followed the development of 695 tripolar eggs and found that 58 developed into a normal pluteus. Among 1170 tetrapolar eggs, he observed none that formed a normal pluteus.

A. Consider first the hypothetical case in which the egg and two sperm each contribute a single chromosome. If the three chromosomes are distributed randomly among the mitotic spindles, what fraction of tripolar eggs are expected to give rise to three cells that each have at least one chromosome? What fraction of tetrapolar eggs are expected to give rise to four cells that each have at least one chromosome? (Examination of Figure 18–6 will help in your thinking about this question.)

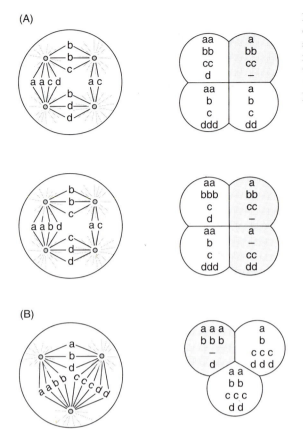

Figure 18–6 Examples of the random distribution of three sets of four chromosomes among the four mitotic spindles in a tetrapolar egg (A) or among the three mitotic spindles in a tripolar egg (B) (Problem 18–6). Shaded cells lack at least one type of chromosome.

B. If the total number of chromosomes is the critical factor, the number of tripolar and tetrapolar eggs in which each cell gets the minimum number of chromosomes will increase above that calculated in part A with increasing numbers of chromosomes. If the distribution of chromosomes is the critical factor, the number of tripolar and tetrapolar eggs in which each cell gets at least one of each different chromosome will decrease according to the fraction calculated in part A raised to a power equal to the number of different chromosomes (9 in this case). How well do Boveri's observations match with these two expectations?

Mitosis
(MBOC 919–934)

18–7 Fill in the blanks in the following statements.

A. Three sets of spindle microtubules can be distinguished: the _____ microtubules, which overlap at the midline of the spindle; the _____ microtubules, which attach to the specialized structure that forms at the centromere of each duplicated chromosome; and the _____ microtubules, which radiate in all directions from the centrosomes.

B. At the start of M phase, each chromosome consists of two sister chromatids joined together along their length, with a constriction in a unique region called the _____.

C. Replicated chromosomes bind to the mitotic spindle via structures called

_____.

D. Prophase ends and _____ begins when the rapid phosphorylation of the nuclear lamina triggers the breakdown of the nuclear envelope and the microtubules gain access to the chromosomes.

E. The alignment of chromosomes at an equal distance from the two spindle poles defines the beginning of _____.

F. In mitosis, _____ begins abruptly with the synchronous splitting of each chromosome into its sister chromatids.

G. The poleward movement of chromosomes and concomitant shortening of kinetochore microtubules is known as _____; the separation of the poles themselves ·accompanied by elongation of the polar microtubules is known as _____.

H. During the final stage of mitosis, which is called _____, a nuclear envelope reassembles around each group of chromosomes to form the two daughter nuclei.

18–8 Indicate whether the following statements are true or false. If a statement is false, explain why.

___ A. In contrast to the highly dynamic interphase array of microtubules, the microtubules that are nucleated from the centrosomes are much more stable.

___ B. The proteins that cross-link oppositely oriented sets of polar microtubules are thought to be plus-end-directed microtubule motors that stabilize the spindle and push the spindle poles apart.

___ C. Only at anaphase, when the sister chromatids separate, do the kinetochore microtubules exert a tension on the chromosomes.

___ D. Unlike yeasts, which have very small centromeres, mammalian centromeric DNA encodes the proteins of the kinetochore, which accounts for their much greater length.

___ E. After the nuclear envelope breaks down, microtubules gain access to the chromosomes and, every so often, a randomly probing microtubule passes close to a kinetochore and captures the chromosome.

___ F. Kinetochore microtubules, much like polar microtubules, are in a state of dynamic flux.

___ G. Stable attachment of microtubules to kinetochores appears to require tension so that only chromosomes attached to both poles will be stably attached to the spindle.

___ H. Equal and opposite pulls toward the two poles of the spindle are sufficient to position chromosomes on the metaphase plate.

___ I. Because motor proteins continually associate and dissociate from the microtubule as they go through their catalytic cycle, they are ideally designed to hold onto a microtubule whose subunits are moving.

___ J. Once the kinetochore is split apart at anaphase, the sister chromatids can be pulled apart by the mechanical force exerted by the kinetochore microtubules.

___ K. It is thought that any kinetochore that is not attached to microtubules generates a diffusible signal that somehow stabilizes MPF, guaranteeing that all of the chromosomes are properly aligned before anaphase begins.

___ L. During anaphase kinetochore microtubules become shorter and polar microtubules elongate.

___ M. Depolymerization of kinetochore microtubules at anaphase occurs almost entirely at the spindle poles.

___ N. Spindle poles are thought to be separated by two forces: overlapping polar microtubules are pushed apart by plus-end-directed motors and astral microtubules are pulled toward the cell cortex by minus-end-directed motors.

___ O. Cell-free extracts of *Xenopus* eggs can assemble nuclei around DNA from any source, provided that it contains a eucaryotic centromeric sequence.

18–9 DNA sequences involved in centromere function have been isolated from many yeast chromosomes. These centromeric sequences confer two chromosomelike properties on autonomously replicating circular plasmids in yeasts: (1) they lower the number of copies per cell to one or two, and (2) they promote correct segregation at mitosis, one plasmid to each daughter cell.

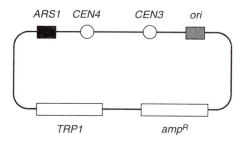

Figure 18–7 Structure of a dicentric plasmid (Problem 18–9). *CEN3* and *CEN4* refer to centromeric sequences from yeast chromosomes 3 and 4, respectively.

What would happen if two such sequences were present on the same DNA molecule? In higher eucaryotes, rare chromosomes containing two centromeres at different locations are highly unstable: they are literally torn apart at anaphase when the chromosomes separate. Yeast chromosomes are too small, however, to analyze microscopically. So other means (in this case cloning and restriction mapping) must be used to answer the question.

You construct a plasmid with two centromeric sequences as shown in Figure 18–7. Growth of this plasmid in bacteria requires the bacterial origin of replication (*ori*) and a selectable marker (*ampR*); its growth in yeast requires the yeast origin of replication (*ARS1*) and a selectable marker (*TRP1*). You prepare a stock of this plasmid by growing it in *E. coli*. This plasmid transforms yeast with about the same efficiency as a plasmid that contains a single centromeric sequence. However, the individual colonies selected after transformation with the dicentric plasmid vary considerably in size, unlike the uniform colonies arising from transformation with a monocentric plasmid. You find that the larger colonies all contain plasmids with a single centromeric sequence, whereas the smaller colonies contain plasmids that have lost both centromeric sequences. In no case did you recover a plasmid that still contained the original two centromeric sequences. By contrast, colonies transformed with the monocentric plasmid invariably contained intact plasmids.

A. Considering their extreme instability in yeasts, why are dicentric plasmids stable in bacteria?
B. Why do you think that the dicentric plasmid is unstable in yeasts?
C. Suggest a mechanism for deletion of one of the centromeric sequences from a dicentric plasmid grown in yeast. Can this mechanism account for loss of both centromeric sequences from some of the plasmids?

*18–10   Circular yeast plasmids that contain an origin of replication but no centromere are distributed among individual cells in a peculiar way. In cultures grown under conditions that require a plasmid-encoded product, only 5% to 25% of the cells harbor the plasmids. However, in these plasmid-bearing cells the plasmid copy number ranges from 20 to 50 copies per cell. To investigate the apparent paradox of a high average copy number but only a small fraction of plasmid-bearing cells, you perform a pedigree analysis to determine the pattern of plasmid segregation during mitosis. You use a yeast strain that requires histidine for growth and a plasmid that carries the histidine gene missing from the host cell. The strain carrying the plasmid grows well under selective conditions, that is, when histidine is absent from the medium. By micromanipulation you separate mother and daughter cells for five divisions under selective conditions and then score for those cells that can form a colony. In Figure 18–8 cells that formed colonies are indicated with heavy lines and cells that failed to form colonies are shown with lighter lines.

A. From the pedigree analysis, it is apparent that cells that lack the plasmid can grow for several divisions in selective medium. How can this be?
B. Does this plasmid segregate equally to mother and daughter cells?

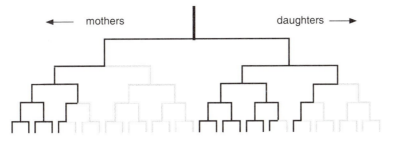

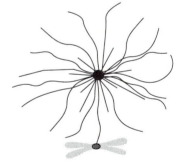

Figure 18–8 A pedigree analysis showing the inheritance of a plasmid that contains an origin of replication and a selectable histidine marker (Problem 18–10). The heavy lines show cells containing the plasmid, and the lighter lines show cells lacking the plasmid. At each division mother cells are shown to the left and daughter cells are shown to the right.

C. Assuming that plasmids in yeast cells replicate only once per cell cycle as the chromosomes do, how can there be 20 to 50 molecules of the plasmid per plasmid-bearing cell?

D. When grown under selective conditions, cells containing plasmids with one centromere (1–2 plasmids per cell) form large colonies, whereas cells containing plasmids with no centromere (20–50 plasmids per cell) form small colonies (see Problem 18–9). Does the pedigree analysis help to explain the difference?

*18–11 Kinetochore microtubules, which link centrosomes to kinetochores, guide chromosomes to the poles during mitosis. How are microtubules arrayed to carry out this process? What manner of connection allows relative movement of spindles and chromosomes? After many clever experiments, the picture is still not clear.

Nevertheless, some aspects of the connections between centrosomes and chromosomes have been clarified by experiment. Consider the following observation. Centrosomes were used to initiate microtubule growth, and then chromosomes were added. The chromosomes bound to the free ends of the microtubules, as illustrated in Figure 18–9. The complexes were then diluted to very low tubulin concentration and examined again (Figure 18–9). As is evident, only the kinetochore microtubules were stable to dilution.

A. Why do you think kinetochore microtubules are stable?

B. Explain the disappearance of the astral microtubules after dilution. Do they detach from the centrosome, depolymerize from an end, or disintegrate along their length at random?

C. How would a time course after dilution help to distinguish among these possible mechanisms of disappearance?

18–12 It was discovered some time ago that kinetochores can be functionally inactivated by a nearby strong promoter that directs transcription toward the centromere. Your advisor recently discovered that kinetochores provide a strong block to transcription. Evidently, centromere function and transcription are mutually interfering processes. Based on this mutual interference, your advisor has devised a clever scheme that exploits the transcriptional block to measure the strength of DNA-protein interactions in the kinetochore.

To carry out this scheme, you construct a test system consisting of the yeast actin gene fused to the *E. coli* β-galactosidase gene, with the hybrid gene under control of the strong galactose-inducible *GAL10* promoter, as shown in Figure 18–10. Into the intron in the actin gene you insert a 165-base pair fragment containing either a functional yeast centromere (*CEN6*wt) or a nonfunctional version that carries a single nucleotide change in conserved element III of the centromere (*CEN6*CDEIII). When you introduce these plasmids into cells growing on galactose and measure β-galactosidase activity, you are encouraged to find that the wild-type centromere blocks transcription and that the mutant centromere partially relieves the transcription block (Figure 18–10).

You next test a collection of chromosome-transmission-fidelity (*ctf*) mutants, which were isolated on the basis of showing elevated rates of chromosome loss at 37°C. You expect that if any of these *ctf* mutants are defective in

before dilution

after dilution

Figure 18–9 Arrangements of centrosomes, chromosomes, and microtubules before and after dilution to low tubulin concentration (Problem 18–11).

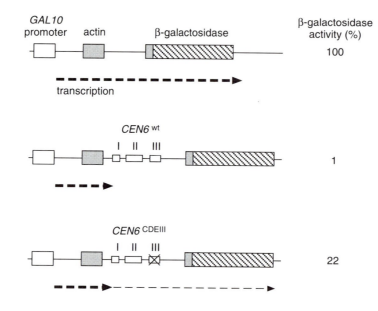

Figure 18–10 Constructs for testing centromere function (Problem 18–12). β-galactosidase activity observed when these constructs are introduced into cells is shown on the *right*.

the assembly or function of kinetochores, they should show enhanced expression of β-galactosidase in the same way as does the CDEIII mutation that inactivates the centromere. As shown in Table 18–1, some but not all of these *ctf* mutants give considerably elevated levels of β-galactosidase.

A. The test plasmid used in these experiments already contained its own centromere. Why does the introduction of a second centromere in the intron of the hybrid gene not lead to breakage of the plasmid and extreme plasmid instability as described in Problem 18–9?

B. Why do you think a functional kinetochore imposes a block to transcription?

C. Some of the *ctf* mutants (*ctf9*, for example) do not show enhanced β-galactosidase activity, yet they do show elevated rates of chromosome loss. What other mechanisms, apart from failure to assemble a functional kinetochore, might lead to the *ctf* phenotype?

18–13   The results obtained from the transcription assay described in the previous problem are very suggestive that *ctf13* encodes a kinetochore protein, but it is by no means conclusive. As a second and independent check on kinetochore function, you set up a functional assay for kinetochores using dispensible dicentric minichromosomes. One of the two centromeres on the minichromosome is perfectly normal, while the other is conditionally active: it is inactive when transcription is directed toward it by the galactose-inducible *GAL10* promoter, and it is fully active when transcription from the promoter is turned off by addition of glucose (Figure 18–11). As controls, you make two

Table 18–1 Quantitation of Transcription Read Through in *ctf* Mutants (Problem 18–12)

Host Strain	Transcription Block	β-Galactosidase Activity (nmol/min/mg protein)
Wild-type	$CEN6^{wt}$	22
Wild-type	$CEN6^{CEDIII}$	135
ctf7	$CEN6^{wt}$	86
ctf8	$CEN6^{wt}$	50
ctf9	$CEN6^{wt}$	19
ctf13	$CEN6^{wt}$	110
ctf17	$CEN6^{wt}$	160

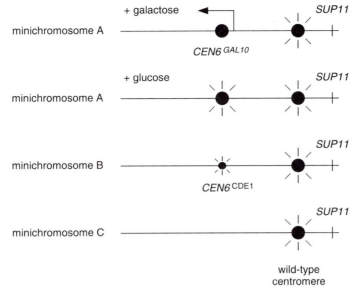

Figure 18–11 Minichromosomes used to test kinetochore function (Problem 18–13).

other minichromosomes: one with a weak second centromere due to a small deletion in the centromere sequence (minichromosome B) and the other with a single centromere (minichromosome C) (Figure 18–11).

If both centromeres are functional, they will cause the minichromosomes to break. The loss of an intact minichromosome can be followed visually due to a suppressor tRNA gene (*SUP11*) located next to the normal centromere (Figure 18–11). When the minichromosome breaks, the daughter cell that does not inherit the suppressor tRNA gene will turn red, owing to a suppressible mutation (that is no longer suppressed) in the adenine biosynthetic pathway in the host strain. This gives rise to white colonies containing red sectors.

To test the effects of the *ctf13* mutation on the stability of dicentric minichromosomes, you introduce the three minichromosomes into yeast carrying the *ctf13* mutation and into yeast with the normal wild-type gene. In all cases you grow the cells in the presence of galactose and then test for sectoring on agar plates in the presence of glucose (Figure 18–12).

A. Explain why you placed the second centromere in minichromosome A under control of the *GAL10* promoter, and why you grew the yeasts in galactose

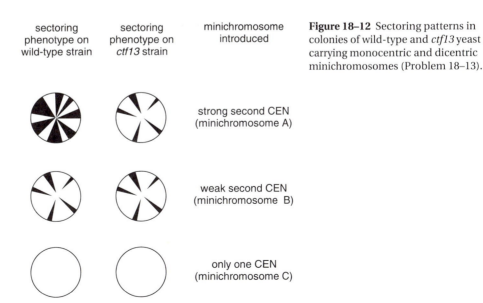

Figure 18–12 Sectoring patterns in colonies of wild-type and *ctf13* yeast carrying monocentric and dicentric minichromosomes (Problem 18–13).

Table 18–2 DNA Content, Number of Chromosomes, and Microtubules per Chromosome in a Variety of Organisms (Problem 8–14)

Type of Organism	Species	DNA Content (bp)	Number of Chromosomes	Microtubules/ Chromosome
Yeast	*S. cerevisiae*	1.4×10^7	16	1
Yeast	*S. pombe*	1.4×10^7	3	3
Protozoan	*Chlamydomonas*	1.1×10^8	19	1
Fly	*Drosophila*	1.7×10^8	4	10
Human	*Homo sapiens*	3.9×10^9	23	25
Plant	*Haemanthus*	1.1×10^{11}	18	120

before switching to glucose for the sectoring assay.

B. Why do you think minichromosome B, which has a weak second centromere, gives less sectoring in the wild-type strain than does minichromosome A, which has two fully functional centromeres?

C. Do your results support the notion that the *ctf13* gene encodes a kinetochore protein? In your answer account for the difference in sectoring observed when minichromosome A is introduced into the *ctf13* strain versus the wild-type strain.

D. The *ctf13* mutant strain was isolated on the basis of an elevated rate of chromosome loss. Yet in your experiments it apparently lowers the rate of loss of minichromosome A. How can it be that a mutation that enhances the loss of normal chromosomes actually stabilizes dicentric chromosomes?

*18–14 How much DNA does a single microtubule carry in mitosis? From the information in Table 18–2, calculate the average length of chromosomes in each organism in base pairs and in millimeters (1 bp = 0.34 nm) and then how much DNA (in base pairs) each microtubule carries on average in mitosis. Do microtubules carry about the same amount of DNA or does it vary widely in different organisms?

Cytokinesis
(MBOC 934–943)

18–15 Fill in the blanks in the following statements.

A. During _____ the cytoplasm divides by a process called cleavage.

B. The first visible sign of cleavage in animal cells is a puckering and _____ of the plasma membrane during anaphase.

C. The cleavage furrow is defined by the _____, which consists of circumferentially oriented filaments of actin and myosin-II bound to the cytoplasmic face of the plasma membrane.

D. The two daughter cells of a mitotic division remain tethered by a structure called the _____, which is composed of the remains of the polar microtubules embedded in a dense matrix.

E. In plants the new cross-wall formed between two daughter cells after division is known as the _____.

F. The residual polar microtubules in plant cells that have almost completed division form an open cylindrical structure called the _____, which guides vesicles containing cell-wall precursors to deposit their contents at the newly forming cell wall.

G. The circumferential band of microtubules that appears and forms a ring around the entire plant cell just below the plasma membrane is called the _____.

18–16 Indicate whether the following statements are true or false. If a statement is false, explain why.

__ A. The plane of cell division at cytokinesis is determined by the position of the metaphase plate during mitosis, which somehow aligns the actin filaments that constitute the contractile ring.

__ B. Whether cells divide symmetrically or asymmetrically, the mitotic spindle positions itself centrally in the cytoplasm.

__ C. The musclelike sliding of actin and myosin filaments in the contractile ring generates the force required for cleavage.

__ D. The mechanism of cytokinesis in plant cells is fundamentally different from that in animal cells: in plants a new cell wall is constructed inside the mother cell to divide it into the two daughter cells.

__ E. After the microtubules in the preprophase band depolymerize, the radial actin strands remain and provide a memory of the predetermined division plane.

__ F. In many unicellular organisms mitosis occurs without breakdown of the nuclear envelope.

18–17 Cytokinesis—the actual process of cell division—has attracted theorists for well over 100 years. Indeed, it has been said that all possible explanations of cytokinesis have been proposed; the problem is to decide which one is correct. Consider the following three hypotheses about cytokinesis:

1. *Chromosome signaling:* When chromosomes split at anaphase, they emit a signal to the nearby cell surface to initiate furrowing.
2. *Polar relaxation:* Asters relax the tension in the nearest region of the cell surface (the polar region), allowing the region of the membrane farthest from the poles (the equatorial plane) to contract and initiate furrowing.
3. *Aster stimulation:* The asters stimulate contraction in the region of the cell surface where oppositely oriented spindle fibers overlap (that is, the equatorial region), thereby initiating furrowing.

These hypotheses have been tested in a number of ways. One particularly informative experiment involved pressing a glass ball onto the center of a dividing sand dollar egg so as to deform it into a torus (donut shape). As illustrated in Figure 18–13, at the first division the egg divided into a single sausage-shaped cell; at the second division it divided into four cells.

A. What does the chromosome-signaling hypothesis predict for the result of this experiment? Do the predictions match the experimental observations?

B. What does the polar-relaxation hypothesis predict for the experimental outcome? Do the predictions match the experimental observations?

C. What does the aster-stimulation hypothesis predict? Do the predictions match the experimental observations?

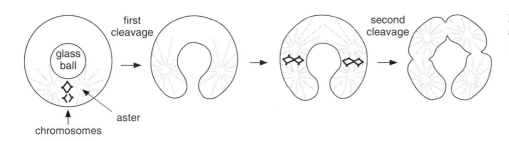

Figure 18–13 First and second divisions in a torus-shaped sand dollar egg (Problem 18–17).

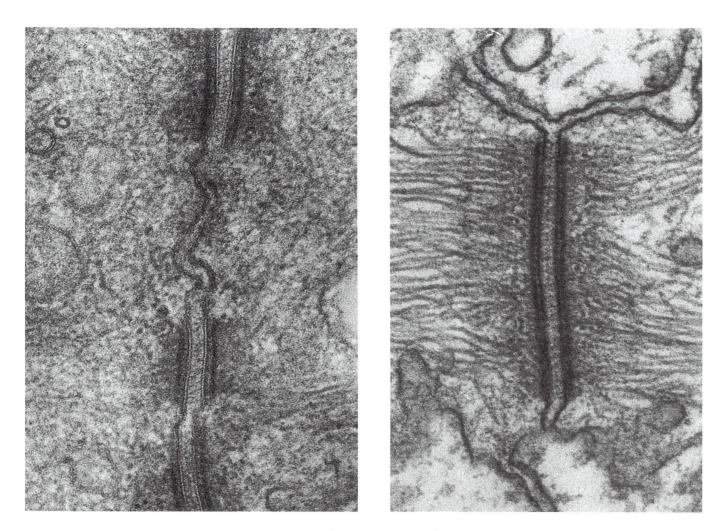

Two examples of cell junctions: (*left*) from rat intestine; (*right*) between two embryonic newt cells. What are these structures called? See Figure 19–12 on page 957 of MBOC, 3rd Edition, for a detailed explanation. (*Left*, from N.B. Gilula, in Cell Communication [R.P. Cox, ed.], pp. 1–29. New York: Wiley, 1974. Reprinted by permission of John Wiley & Sons, Inc.; *right*, from D.E. Kelly. *J. Cell Biol.* 28:51–59, by copyright permission of the Rockefeller University Press.)

Cell Junctions, Cell Adhesion, and the Extracellular Matrix

- Cell Junctions
- Cell-Cell Adhesion
- The Extracellular Matrix of Animals
- Extracellular Matrix Receptors on Animal Cells: The Integrins
- The Plant Cell Wall

Cell Junctions

(MBOC 950–963)

19–1 Fill in the blanks in the following statements.

A. In _____ extracellular matrix is plentiful and cells are sparsely distributed within it.

B. A sheet of cells is called an _____.

C. Specialized junctions between cells can be classified into three general groups: _____ junctions make molecule-tight seals between cells; _____ junctions mechanically connect cells and their cytoskeletons to each other or to the extracellular matrix; and _____ junctions mediate the passage of small molecules from cell to cell.

D. The calcium-dependent junctions that seal epithelia so molecules cannot leak from one side of the sheet to the other are known as _____ junctions.

E. Epithelial cells in the gut have one set of transport proteins on their _____ surface, which faces the lumen of the gut, and a second set of transport proteins on their _____ surface, which faces away from the gut.

F. _____ junctions connect the actin filaments of neighboring cells; they are composed of transmembrane linker proteins, which hold the cells together, and intracellular attachment proteins, which connect the actin filaments to the linker proteins.

G. Epithelial cells are connected by a beltlike structure called the _____, which is thought to mediate the folding of cell sheets into tubes during morphogenesis in animals.

H. _____ junctions enable cells to get a hold on the extracellular matrix by connecting their actin filaments to the matrix.

I. Fibroblasts in culture adhere to the substratrum at specialized regions of the plasma membrane called _____, where bundles of actin filaments terminate.

J. _____ junctions, which are widespread in invertebrate tissues, share a number of features with adhesion belts.

K. _____ are buttonlike points of intercellular contact that serve as anchoring sites for intermediate filaments and help hold adjacent cells together.

L. The basal surface of an epithelial cell is joined to the basal lamina at _____, which link intermediate filaments to the extracellular matrix.

M. The most common type of communicating junction between cells is the _____; it allows substances with molecular weights under 1000 to pass freely between cells.

N. With a few specialized exceptions, every living cell in a higher plant is connected to its living neighbors by _____, which form fine cytoplasmic channels through the intervening cell walls.

19–2 Indicate whether the following statements are true or false. If a statement is false, explain why.

___ A. Tight junctions get their name from their property of holding cells together so tightly that they cannot be separated by mechanical forces.

___ B. Directional pumping of nutrients across epithelia would be impossible if the proteins on the apical and basolateral surfaces were the same.

___ C. Epithelial cell sheets differ markedly in the permeability of their tight junctions; for example, bladder epithelium is much more leaky to ions than is the intestinal epithelium.

___ D. Gap junctions connect the cytoskeletal elements of one cell to a neighboring cell or to the extracellular matrix.

___ E. A desmosome bears the same relationship to a hemidesmosome that an adhesion belt does to a focal contact.

___ F. In cell culture a cell expressing one type of connexin rarely forms a functional gap junction with a cell expressing a different connexin.

___ G. The permeability of gap junctions is regulated by the extracellular Ca^{2+} and pH.

___ H. In spite of the radical difference of structure between plasmodesmata and gap junctions, they seem to function in remarkably similar ways, allowing exchange of low molecular weight molecules between cells.

19–3 Examine the three protein monomers in Figure 19–1. From the arrangement of complementary binding domains, which are indicated by similarly shaped protrusions and invaginations, decide which monomer might assemble into a strand of a tight junction, which monomer might assemble into a desmosome, and which monomer could not assemble into either.

*19–4 Two structures for tight junctions have been proposed. In one ingenious model, which is based on observations from freeze-fracture electron microscopy, each sealing strand in a tight junction results from a membrane fusion that forms cylinders of lipid at the points of fusion (Figure 19–2A). The other model proposes that each strand of a tight junction is formed by a chain of transmembrane proteins whose extracellular domains bind to one another to seal the epithelial sheet (Figure 19–2B).

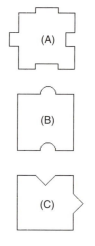

Figure 19–1 Three monomeric membrane proteins (Problem 19–3).

Figure 19–2 Schematic representations of the lipid and protein models for tight-junction structure (Problem 19–4). (A) Three views of the cylinder of lipids that is proposed to form a tight junction according to the lipid model. (B) Schematic representation of the protein model for tight-junction structure.

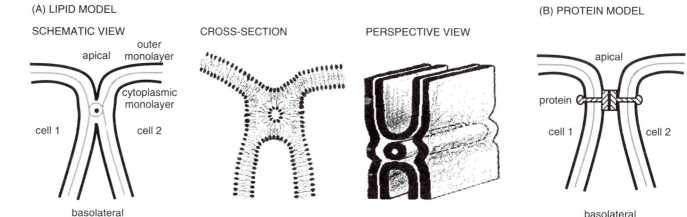

(A) LIPID MODEL

SCHEMATIC VIEW   CROSS-SECTION   PERSPECTIVE VIEW

(B) PROTEIN MODEL

*Problems with an asterisk (*) are answered in the Instructor's Manual.

As you compare these lipid and protein models for tight junctions, you realize that they might be distinguishable on the basis of lipid diffusion between the apical and basolateral surfaces. Both models predict that lipids in the cytoplasmic monolayer will be able to diffuse freely between the apical and the basolateral surfaces. However, the models suggest different fates for lipids in the external monolayer. In the lipid model, lipids in the outer monolayer will be confined to either the apical surface or the basolateral surface, since the cylinder of lipids interrupts the external monolayer, preventing diffusion through it. By contrast, in the protein model the apical and basolateral surfaces appear to be connected by a continuous external monolayer, suggesting that lipids in the external monolayer should be able to diffuse freely between the two surfaces.

You have exactly the experimental tools to resolve this issue! You have been working with a line of dog kidney cells that forms an exceptionally tight epithelium with well-defined apical and basolateral surfaces. In addition, after infection with influenza virus, the cells express a fusogenic protein only on their apical surface. This feature allows you to fuse liposomes specifically to the apical surface of infected cells very efficiently by brief exposure to low pH, which activates the fusogenic protein. Thus, you can add fluorescently labeled lipids to the apical surface and detect their migration to the basolateral surface using fluorescence microscopy.

For the experiment you prepare two sets of labeled liposomes: one with a fluorescent lipid only in the outer monolayer, the other with the fluorescent lipid equally distributed between the inner and outer monolayers. You fuse these two sets of liposomes to epithelia in which about half the cells were infected with virus. By adjusting the focal plane of the microscope, you examine the apical and basolateral surfaces for fluorescence. As a control, you remove Ca^{2+} from the medium—a treatment that disrupts tight junctions—and reexamine the basolateral surface. The results are shown in Figure 19–3.

You are delighted! These results show clearly the lipids in the external monolayer are confined to the apical surface, whereas lipids in the cytoplasmic monolayer diffuse freely between the apical and basolateral surfaces.

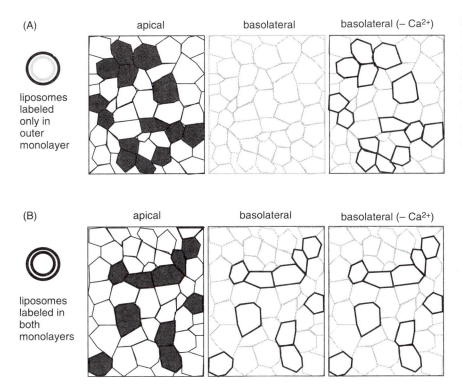

Figure 19–3 Experimental test of the lipid and protein models for tight-junction structure (Problem 19–4). (A) Liposomes labeled in outer monolayer. (B) Liposomes labeled in both monolayers. Only about half the cells in the epithelium in each experiment were infected with virus. Only infected cells are competent to fuse with the labeled liposomes under the conditions of the experimental protocol.

Triumphantly, you show these results to your advisor as proof that the lipid model for tight-junction structure is correct. He examines your results carefully, shakes his head knowingly, gives you that penetrating look of his, and tells you that, although the experiments are exquisitely well done, you have drawn exactly the wrong conclusion. These results prove that the lipid model is incorrect.

What has your advisor seen in the data that you have overlooked? How do your results disprove the lipid model? If the protein model is correct, why do you think it is that the fluorescent lipids are confined to the apical surface?

19–5   Most of the current carried by small ions across an epithelium must pass through the gaps between cells because the cellular membranes are excellent electrical insulators. Thus, the electrical resistance of an epithelium depends on the properties of the tight junctions that seal them (Figure 19–4).

You and your advisor some time ago observed that there is a rough correlation between the electrical resistance of an epithelium and the number of sealing strands in the tight junction. You can imagine two ways that resistance might depend on the number of sealing strands. If each sealing strand provided a given resistance, then the overall resistance of a tight junction would be linearly related to the number of sealing strands (like electrical resistors in series). On the other hand, suppose each sealing strand could exist in two states: a closed, high-resistance state and an open, low-resistance state. The resistance of the tight junction would then be related to the probability that all strands in a given pathway through the junction would be open at the same time. In that case the overall resistance would be logarithmically related to the number of sealing strands.

To put the data on a quantitative basis so that you can distinguish between these two possibilities, you measure the resistance of four different epithelia from a rabbit: the very leaky proximal tubule of the kidney, the less leaky gall bladder, the tight distal tubule of the kidney, and the very tight bladder epithelium. In addition, you prepare freeze-fracture electron micrographs, from which you determine the average number of sealing strands in the tight junctions that surround each cell in these epithelia. The results are shown in Table 19–1. Which of the two proposed interpretations of the correlation between electrical resistance of the epithelium and the number of sealing strands in a tight junction is supported by your measurements?

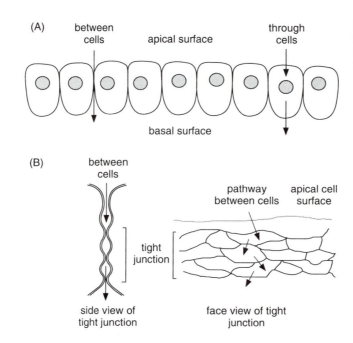

(A) Side view of an epithelium.

between cells   apical surface   through cells

basal surface

(B) between cells

tight junction

pathway between cells   apical cell surface

side view of tight junction   face view of tight junction

Figure 19–4 (A) Side view of an epithelium. (B) Two views of the tight junctions that link cells in an epithelium (Problem 19–5).

Table 19–1 Correlation Between Electrical Resistance and the Number of Strands in the Tight Junctions from Various Epithelia (Problem 19–5)

Epithelium	Mean Number of Strands in Tight Junction	Specific Resistance of Tight Junction
Rabbit proximal tubule	1.2	1.2×10^4
Rabbit gall bladder	3.3	5.6×10^4
Rabbit distal tubule	5.3	6.2×10^5
Rabbit urinary bladder	8.0	5.6×10^6

*19–6   When properly maintained in culture, mouse heart cells beat regularly with a characteristic frequency that can be increased by addition of cyclic AMP or by inhibitors of cyclic AMP hydrolysis. Addition of noradrenaline increases the beat frequency by raising the intracellular cyclic AMP level.

Rat ovary cells in culture respond to follicle stimulating hormone (FSH) by increasing the synthesis of a protease activator called tissue-plasminogen activator (TPA). This response is also mediated by an increase in intracellular cyclic AMP levels.

FSH has no effect on the beating of heart cells, and noradrenaline has no effect on synthesis of TPA by ovary cells. However, when the two cell types are mixed together and co-cultured, addition of FSH or noradrenaline increases both the beat frequency of the heart cells and TPA synthesis in the ovary cells. Adding an enzyme to the medium that hydrolyzes cyclic AMP to 5′ AMP has no effect on the collaboration.

Suggest an explanation for these results. How might you test your hypothesis?

19–7   Fertilized mouse eggs divide very slowly at first. They reach two cells after about 24 hours and eight cells by 48 hours. At the eight-cell stage they undergo a process known as compaction, as illustrated in Figure 19–5. Although the mechanism is not clear, the cells appear to adhere to one another more strongly; consequently, they change from being a clump of loosely associated cells to a tightly sealed ball. You wish to know what kinds of intercellular junctions are present before and after this change in adhesion.

To study this question, you use very fine glass micropipettes, which allow you to measure electrical events and at the same time to microinject either the enzyme horseradish peroxidase (HRP), 40,000 daltons, or the fluorescent dye fluorescein, 330 daltons. Fluorescein glows bright yellow under UV illumination, while HRP can be detected by fixing the cells and incubating them with appropriate substrates.

You inject embryos at various stages of development with the two marker substances. At both the two-cell and eight-cell stages, different results are obtained, depending on whether the injections are made immediately after

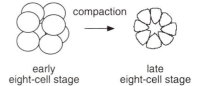

Figure 19–5 Compaction of the eight-cell mouse embryo (Problem 19–7).

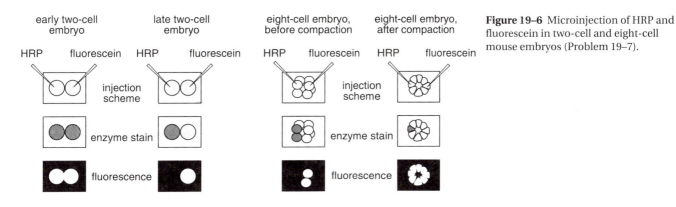

Figure 19–6 Microinjection of HRP and fluorescein in two-cell and eight-cell mouse embryos (Problem 19–7).

cell division or later (Figure 19–6); some of this difference can be attributed to the cytoplasmic bridges that linger for a while before cytokinesis is truly completed.

A. Why do both HRP and fluorescein enter neighboring cells early, but not late, at the two-cell stage?

B. Why does fluorescein enter all cells in the compacted eight-cell embryo, whereas HRP is confined to the injected cell?

C. In which of the four stages of development diagrammed in Figure 19–6 would you detect electrical coupling if you injected current from the HRP injection electrode and recorded voltage changes in the fluorescein electrode?

19–8 You are studying cell junctions in insects, and the first electron micrographs just came back from the photolab (Figure 19–7). Unfortunately, the labels were lost. Can you deduce the type of junction that is illustrated in each micrograph?

(A)

(B)

Figure 19–7 Electron micrographs of a variety of cell junctions (Problem 19–8). (Courtesy of Dr. Nancy Lane.)

(C)

(D)

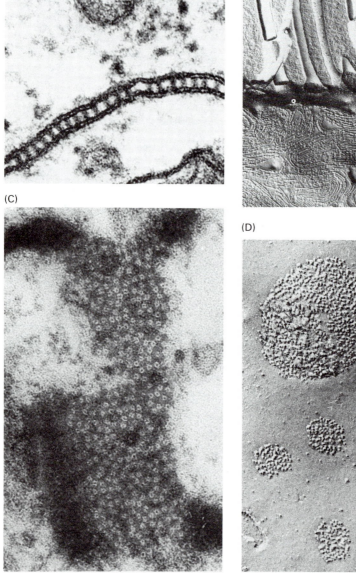

Cell-Cell Adhesion

(MBOC 963–971)

19–9 Fill in the blanks in the following statements.

A. Two distinct classes of _____ govern selective cell adhesion in most multicellular animals.

B. The _____ are responsible for Ca^{2+}-dependent cell-cell adhesion in vertebrate tissues.

C. Antibodies against _____, which is also called uvomorulin, prevent the compaction of cleavage-stage mouse embryos.

D. If cells adhere to one another using one kind of molecule, the cell adhesion is said to occur by _____ binding; if cells adhere to one another through interactions between different kinds of molecules, the cell adhesion is said to occur by _____ binding.

E. A family of cell-surface carbohydrate-binding proteins called _____ function in a variety of transient cell-cell adhesion interactions in the blood-stream.

F. _____, which is a member of the immunoglobulin superfamily, is expressed by a variety of cell types, including most nerve cells, and is the most prevalent of the Ca^{2+}-independent cell-cell adhesion molecules in vertebrates.

G. _____, which is an Ig-like protein that mediates Ca^{2+}-independent cell-cell adhesion in *Drosophila,* has five Ig-like domains and operates by homophilic binding.

19–10 Indicate whether the following statements are true or false. If a statement is false, explain why.

___ A. "Were the various types of cells to lose their stickiness for one another and for the supporting extracellular matrix, our bodies, would at once disintegrate and flow off into the ground in a mixed stream of cells." Warren Lewis, 1922. (Quoted by J.P. Trinkaus, Cells into Organs, 2nd ed., Englewood Cliffs, NJ: Prentice-Hall, 1984.)

___ B. Tissues form from specialized cells either by growth and division of founder cells whose progeny stay together or by the guided migration of cells that only stop moving when they recognize specific target sites.

___ C. The sorting out of cell types that occurs when dissociated cells from two different organs are mixed mimics similar sorting-out processes that occur during the development of most tissues.

___ D. One of the difficulties in studying cell-adhesion mechanisms is that almost any antibody directed against cell-surface components will block cell adhesion.

___ E. Cadherins usually link cells together using a heterophilic mechanism.

___ F. The cell-surface receptors responsible for binding cells to each other and to the extracellular matrix have much lower affinity for their ligands than do most hormone receptors.

___ G. Migrating embryonic cells do not form junctional contacts.

___ H. Cell-adhesion molecules are dispersed over the surface of a cell until the cell makes contact with another cell, whereupon they concentrate at particular sites.

*19–11 The cellular slime mold *Dictyostelium discoideum* is a eucaryote that lives on the forest floor as independent motile cells called amoebae, which feed on bacteria and yeast. When their food supply is exhausted, the amoebae stop dividing and gather together to form tiny, multicellular, wormlike structures, which crawl about as glistening slugs and leave trails of slime behind them. In forming the slug, individual *Dictyostelium* amoebae aggregate by chemotaxis using cyclic AMP as a chemical signal to attract other amoebae.

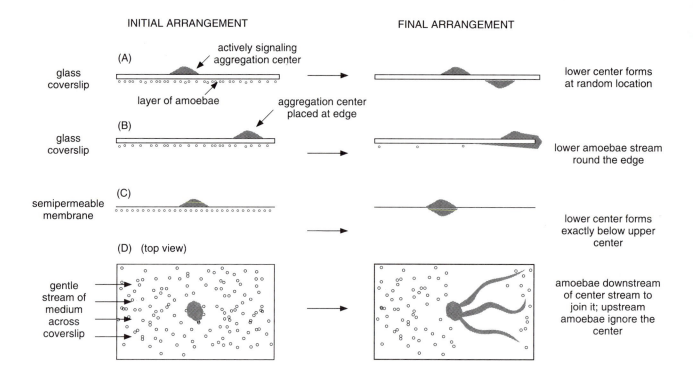

(A)

glass
coverslip

actively signaling
aggregation center

layer of amoebae

lower center forms
at random location

(B)

glass
coverslip

aggregation center
placed at edge

lower amoebae stream
round the edge

(C)

semipermeable
membrane

lower center forms
exactly below upper
center

(D)   (top view)

gentle
stream of
medium
across
coverslip

amoebae downstream
of center stream to
join it; upstream
amoebae ignore the
center

Not all slime molds use cyclic AMP to signal aggregation. You have just discovered such a species living in a compost heap in your garden. It is indifferent to cyclic AMP, yet still shows strong aggregation when starved. Since the usual chemical signal is ineffective, you wonder whether something other than chemical signaling might be used.

To investigate the nature of the signal, you repeat some of the classic work that was done with *Dictyostelium*. You find that the amoebae will aggregate if placed on a glass coverslip underwater, provided that simple salts are present. The center of the aggregation pattern can be removed with a pipette and placed in a field of fresh amoebae, which immediately start streaming toward it. Thus, the center is emitting some sort of attractive signal.

You now prepare four experiments using an existing center of aggregation as the source of the signal and previously unexposed amoebae as the target cells. The arrangements of aggregation centers and test amoebae at the beginning and end of the experiments are shown in Figure 19–8.

Do these results show that your species of slime mold aggregates by chemical signaling? How so?

Figure 19–8 Four experiments to study the nature of the attractive signal generated by aggregation centers (Problem 19–11).

19–12   The attachment of bacteriophage T4 to *E. coli* K is an instructive model for cell-cell adhesion. It illustrates the value of multiple weak interactions, which allow relative motion until fixed connections are made. During infection, T4 first attaches to the surface of *E. coli* by the tips of its six tail fibers. It then wanders around the surface until it finds an appropriate place for attachment of its baseplate. When the baseplate is securely fastened, the tail sheath contracts, injecting the phage DNA into the bacterium (Figure 19–9). The initial tail-fiber cell-surface interaction is critical for infection: phages that lack tail fibers are totally noninfectious.

Analysis of T4 attachment is greatly simplified by the ease with which resistant bacteria and defective viruses can be obtained. Bacterial mutants resistant to T4 infection fall into two classes: one lacks a major outer membrane protein called ompC (**o**uter **m**embrane **p**rotein C); the other contains alterations in the long polysaccharide chain normally associated with bacterial lipopolysaccharide (LPS, a lipid with a long polysaccharide chain attached to its head group). The infectivity of T4 on wild-type and mutant cells is

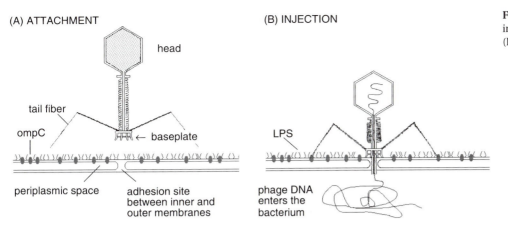

(A) ATTACHMENT

head

tail fiber

ompC

← baseplate

periplasmic space

adhesion site
between inner and
outer membranes

(B) INJECTION

LPS

phage DNA
enters the
bacterium

Figure 19–9 T4 attachment (A) and injection (B) of its DNA into a bacterium (Problem 19–12).

indicated in Table 19–2. These results suggest that each T4 tail fiber has two binding sites: one for LPS and one for ompC. Electron micrographs showing the interaction between isolated tail fibers and LPS suggest that individual associations are not very strong, since only about 50% of the fibers are seen bound to LPS.

A. Assume that at any instant each of the six tail fibers has a 0.5 probability of being bound to LPS and the same probability of being bound to ompC. With this assumption, the fraction of the phage population on the bacterial surface that will have none of its six tail fibers attached in a given instant is $(0.5)^{12}$ (which is the probability of a given binding site being unbound, 0.5, raised to number of binding sites, two on each of six tail fibers). In light of these considerations, what fraction of the phage population will be attached at any one instant by at least one tail fiber? (The attached fraction is equal to one minus the unattached fraction.) Suppose that the bacteria were missing ompC. What fraction of the phage population would now be attached by at least one tail fiber at any one instant?

B. Surprisingly, the above comparison of wild-type and *ompC*- bacteria suggests only a very small difference in the attached fraction of the phage population at any one instant. However, as shown in Table 19–2, phage infectivities on these two strains differ by a factor of 1000. Can you suggest an explanation that might resolve this apparent paradox?

*19–13 Nerve cells attach to muscle cells during development. Interactions between different types of cells can occur in ways that are analogous to those for interactions between identical cells (Figure 19–10). To study the mechanism of attachment, you use a line of nerve cell precursors that differentiate into nerve cells under appropriate conditions. You assay their attachment to muscle cells by mixing labeled nerve cells with collagenase-digested muscle cells, which are still surrounded by an intact basal lamina. Each muscle cell binds about 50 nerve cells. The adhesion is fairly specific; for example, embryonic fibroblasts and liver cells do not show any adhesion to these muscles cells.

To define the interacting cell-surface components, you treat the nerve and muscle cells with neuraminidase, which removes *N*-acetylneuraminic acid (sialic acid) from sugar polymers, and pronase, which cleaves proteins. The results are shown in Table 19–3. In addition, you find that sialic acid alone among all sugars tested inhibits nerve cell binding when added to the assay. Mucin, which contains a large proportion of sialic acid, is even more effective than the monomeric sugar. Nonsialic-acid-rich polymers such as heparin, hyaluronic acid, and chondroitin sulfate have no effect. When the muscle cells are preincubated with mucin and then washed, they no longer bind nerve cells. When nerve cells are pretreated with mucin and then tested for muscle binding, however, they still adhere perfectly.

Table 19–2 Infectivity of Phage T4 on Various Bacterial Mutants (Problem 19–12)

Bacterial Strain	Phage T4 Infectivity Relative to Nonmutant Bacteria
ompC+LPS+	1
ompC-LPS+	10^{-3}
ompC+LPS-	10^{-3}
ompC-LPS-	10^{-7}

DIRECT BINDING   LINKER MOLECULE

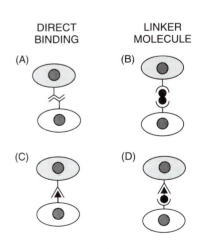

Figure 19–10 Four different types of interaction involving nonidentical cells (Problem 19–13).

Table 19–3 Effects of Enzyme Treatments on Adhesion of Nerve Cells to Muscle Cells (Problem 19–13)

	Treatment		Result
Experiment	Nerve Cells	Muscle Cells	Nerve Cells/Muscle Cell
1	none	none	58
2	neuraminidase	none	4
3	none	neuraminidase	56
4	neuraminidase	neuraminidase	2
5	pronase	none	55
6	none	pronase	5

Describe the interaction between nerve cells and muscle cells as completely as these data allow.

19–14   Another example of interactions between nonidentical cells occurs in yeasts. The yeast *Hansenula wingei* has two haploid mating types, called type 5 and type 21. Cells of one mating type do not aggregate with themselves; however, when the two mating types are mixed together, they do aggregate. Although aggregation assists mating, it is not essential. To determine the nature of the binding, you treat the two mating types with trypsin, which cleaves proteins, and mercaptoethanol, which reduces disulfide bonds. The results are summarized in Table 19–4.

In other experiments, you have purified a factor from type 5 cells by treating them with an enzyme that cleaves polysaccharides. This factor when added to type 21 cells promotes their aggregation. In the presence of mercapto-ethanol the factor dissociates into high and low molecular weight components, which are individually inactive in promoting aggregation of type 21 cells. When these components are recombined under oxidizing conditions, which allow reformation of disulfide bonds, they are once again active in promoting aggregation of type 21 cells.

Describe the interaction between mating types 5 and 21.

Table 19–4 Effects of Treatments on Aggregation of Type 5 and Type 21 Yeast Cells (Problem 19–14)

	Treatment		
Experiment	Type 5	Type 21	Aggregation
1	none	none	+
2	trypsin	none	+
3	none	trypsin	−
4	mercaptoethanol	none	−
5	none	mercaptoethanol	+

The Extracellular Matrix of Animals
(MBOC 971–995)

19–15   Fill in the blanks in the following statements.

A. The network of interacting protein and polysaccharide molecules in contact with the outside surface of most cells in multicellular organisms is called the _____ .

B. Skin and bone are composed mainly of _____ , a term that is often used to describe the extracellular matrix plus the cells found in it.

C. In most connective tissues the matrix macromolecules are secreted largely by cells called _____.

D. _____ are long, unbranched polysaccharide chains composed of repeating disaccharide units that always contain an amino sugar.

E. _____ is an abundant polysaccharide of the extracellular matrix of developing animals; its large size, lack of sulfation, and simple repeating disaccharide structure distinguish it from other glycosaminoglycans.

F. _____ are synthesized much like glycoproteins; however, the polysaccharide chains are attached to serine residues, and the molecular weight of the carbohydrate can exceed that of the core protein by a factor of 10 to 20.

G. The _____ are the most abundant proteins in mammals; their distinguishing feature is a triple-stranded helix rich in glycine and proline.

H. After being secreted into the extracellular space, type I collagen molecules assemble into ordered polymers called _____.

I. The _____ retain their propeptides and do not aggregate with one another; instead they bind in a periodic manner to the surface of collagen fibrils.

J. The main component of elastic fibers is _____, which is a highly hydrophobic, nonglycosylated protein that contains little hydroxyproline and no hydroxylysine.

K. The best characterized of the extracellular adhesive proteins is _____, which helps cells adhere to their substratum by binding both to cell-surface receptors and to various other components of the extracellular matrix.

L. A specific tripeptide sequence, the _____, is found in one of the _____ repeats and is a central feature of the site responsible for cell binding by fibronectin.

M. _____ collagen molecules interact via their uncleaved terminal domains to assemble extracellularly into a flexible, sheetlike, multilayered network.

N. The continuous thin layer of specialized extracellular matrix that underlies all epithelial cell sheets and tubes and surrounds individual muscle cells and fat cells is called the _____.

O. _____, which is a glycoprotein composed of three polypeptides arranged in the shape of a cross, binds to type IV collagen, heparan sulfate, and cell surfaces.

19–16 Indicate whether the following statements are true or false. If a statement is false, explain why.

___ A. The extracellular matrix is a relatively inert scaffolding that stabilizes the structure of tissues.

___ B. The main chemical difference between glycoproteins and proteoglycans lies in the structure of their carbohydrate side chains: glycoproteins contain short, highly branched oligosaccharides, whereas proteoglycans contain much longer, unbranched polysaccharide side chains.

___ C. Although the heterogeneity of glycosaminoglycans makes it difficult to classify proteoglycans in terms of their sugars, recombinant DNA techniques have revealed that the core proteins fall into a few families.

___ D. Proteoglycans in the basal lamina of the kidney glomerulus play a critical role in regulating the passage of macromolecules from the blood into the urine.

___ E. Breakdown and resynthesis of collagen must be important in maintaining the extracellular matrix; otherwise, vitamin C deficiency in adults would not cause scurvy, which is characterized by a progressive weakening of connective tissue due to inadequate hydroxylation of collagen.

___ F. During maturation of the fibrillar collagens, three peptide bonds in each monomeric polypeptide must be broken.

ADDITIONS TO CULTURE MEDIUM

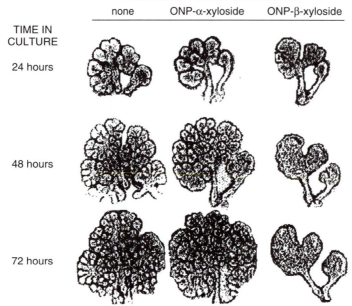

Figure 19–11 Development of mouse salivary glands in culture (Problem 19–17).

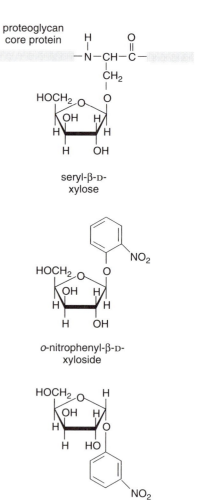

seryl-β-D-xylose

o-nitrophenyl-β-D-xyloside

o-nitrophenyl-α-D-xyloside

Figure 19–12 Structures of the protein-xylose linkage, ONP-β-D-xyloside, and ONP-α-D-xyloside (Problem 19–17).

__ G. The elasticity of elastin derives from its high content of α helices, which act as molecular springs.

__ H. All forms of fibronectin are produced from one large gene by alternative splicing.

__ I. Although most types of collagen assemble into fibrils, type IV collagen assembles into a sheetlike network that forms the core of all basal lamina.

__ J. When tissues such as muscles, nerves, and epithelia are damaged, the basal lamina is destroyed and must be resynthesized by cells as they regenerate.

__ K. A localized degradation of matrix components is required when cells migrate through a basal lamina.

19–17   Embryonic mouse salivary glands are composed of an epithelium that under-goes a series of repetitive branching and folding events to form the ducts and lobes that make up the mature gland. Salivary glands in culture undergo much the same development as they would in the animal, typically forming 7 to 10 times more lobes after 3 days than were present at the end of day 1 (Figure 19–11). The process of lobe development is known as branching morphogenesis.

The basal lamina is critical for branching morphogenesis. For example, inhibitors of collagen synthesis such as the proline analogue azetidine-2-carboxylic acid inhibit branching. One of the other major constituents of the basal lamina is proteoglycan. What would happen if its synthesis was blocked? Glycosaminoglycans are built one sugar at a time on the core protein, starting with the unusual sugar β-D-xylose, which is linked to a serine residue in the polypeptide backbone (Figure 19–12). The xyloside analogue o-nitrophenyl-β-D-xyloside (ONP-β-D-xyloside, Figure 19–12) acts as a competitor of gly-cosaminoglycan addition to proteins in the Golgi. When this analogue is added to growing salivary glands, the glands grow but do not form lobes or branches (Figure 19–11). Treatment with ONP-α-D-xyloside has no effect on gland development (Figure 19–11). This shows that the effect is specific for the naturally occurring isomer and that the inhibition is not due to nonspecific toxicity.

To determine how ONP-β-D-xyloside affects proteoglycan synthesis, the incorporation of $^{35}SO_4$ into newly made glycosaminoglycans was measured under a variety of conditions, as shown in Table 19–5. All the $^{35}SO_4$ was incorporated into glycosaminoglycans, as judged by its sensitivity to hyal-uronidase digestion.

Table 19–5 Effects of Xylosides on Glycosaminoglycan Synthesis (Incorporation of $^{35}SO_4$) in Mouse Salivary Glands in Culture (Problem 19–17)

Cell Treatment	Counts in Tissue	Counts in Medium	Total Counts
Control	1000	300	1300
ONP-α-D-xyloside	1000	300	1300
ONP-β-D-xyloside	200	3000	3200
Cycloheximide	ND*	ND	200
Cycloheximide + ONP-α-D-xyloside	ND	ND	200
Cycloheximide + ONP-β-D-xyloside	ND	ND	3000

*ND = not determined.

A. Why does cycloheximide inhibit $^{35}SO_4$ incorporation in the absence of ONP-β-D-xyloside but not in its presence?

B. Why are such large amounts of $^{35}SO_4$-labeled material found in the medium when ONP-β-D-xyloside is present?

C. Why does ONP-β-D-xyloside, but not ONP-α-D-xyloside, cause such profound changes in $^{35}SO_4$ incorporation?

D. How might ONP-β-D-xyloside suppress branching morphogenesis of mouse salivary glands in tissue culture?

*19–18 Defects in collagen genes are responsible for several inherited diseases. For example, osteogenesis imperfecta, a disease characterized by brittle bones, can result from a defect in the gene encoding the type 1 α1 collagen chain. Similarly, Ehlers-Danlos syndrome, which can lead to sudden death due to ruptured internal organs or blood vessels, can result from a defect in the type III α1 collagen gene.

In both diseases the medical problems arise because the defective gene in some way compromises the function of collagen fibrils. For example, homozygous deletions of the type I α1 gene eliminates α1(I) collagen entirely, thereby preventing formation of any type I collagen fibrils. Such homozygous mutations are usually lethal in early development. The more common situation is for an individual to be heterozygous for the mutant gene, having one normal gene and one defective gene. Here the consequences are less severe.

A. Calculate the fraction of type I collagen molecules, $[α1(I)]_2α2(I)$, that will be normal in an individual who is heterozygous for a deletion of the entire α1(I) gene. Repeat the calculation for an individual who is heterozygous for a point mutation in the α1(I) gene.

B. Calculate the fraction of type III collagen molecules, $[α1(III)]_3$, that will be normal in an individual who is heterozygous for a deletion of the entire α1(III) gene. Repeat the calculation for an individual who is heterozygous for a point mutation in the α1(III) gene.

C. Which kind of collagen gene defect—deletion or point mutation—is more likely to be dominant (that is, cause the heterozygote to display a mutant phenotype)?

19–19 The formation of a mature collagen molecule is a complex process. Procollagen is formed first from three collagen chains, which have extensions at the N- and C-termini. Once the chains have been properly wound together, the terminal propeptides are removed. A principal function of the terminal propeptides may be to align the chains correctly to facilitate their assembly.

Two forms of type III collagen molecules have been used to study collagen assembly: (1) the mature type III collagen molecule and (2) a precursor with the N-terminal propeptides still attached, type III pN-collagen (Figure 19–13).

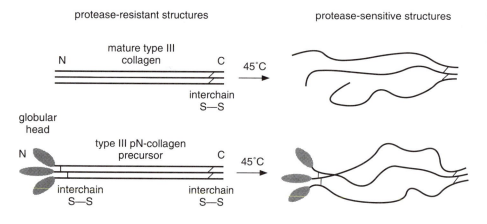

protease-resistant structures   protease-sensitive structures

Figure 19–13 Native and denatured forms of type III collagen and type III pN-collagen (Problem 19–19).

In both types of collagen molecule the individual chains are held together by disulfide bonds. These bonds hold the chains in register, even after they have been completely unwound, so that they reassemble collagen molecules correctly (Figure 19–13).

Reassembly of these two collagen molecules was studied after denaturation at 45°C under conditions that leave the disulfide bonds intact. The temperature was then reduced to 25°C to allow reassembly, and samples were removed at intervals up to 60 minutes. Each sample was immediately digested with trypsin, which rapidly cleaves denatured chains but does not attack the collagen helix. When the samples were analyzed by SDS-gel electrophoresis in the presence of mercaptoethanol, which breaks disulfide bonds, both collagens yielded identical patterns (Figure 19–14). Significantly, if the disulfide bonds were left intact during SDS-gel electrophoresis, all bands were replaced by bands with three times the molecular weight.

A. Why did all the resistant peptides increase threefold in molecular weight when the disulfide bonds were left intact?

B. Which set of disulfide bonds in type III pN-collagen (N-terminal or C-terminal) is responsible for the increase in molecular weight of its resistant peptides?

C. Does reassembly of these collagen molecules begin at a specific site or at random sites? If they reassemble from a specific site, deduce its location.

D. Does reassembly follow a zipperlike or an all-or-none mechanism? How do these results distinguish between these mechanisms?

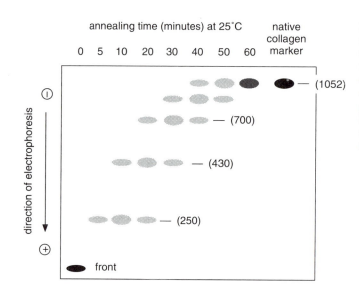

Figure 19–14 Results of reassembly analysis of type III collagen or type III pN-collagen (Problem 19–19). After various times of reannealing at 25°C, samples were digested with trypsin and subjected to electrophoresis. Shaded areas indicate the positions of the trypsin-resistant peptides. Numbers in parentheses indicate the approximate number of amino acids in the trypsin-resistant peptides.

Table 19–6 Fibronectin-related Peptides Tested for Their Ability to Promote Cell Sticking (Problem 19–20)

Peptide	Sequence	Concentration Required for 50% Cell Attachment (nmol/ml)
Fibronectin		0.10
Peptide 1	YAVTGRGDSPASSKPISINYRTEIDKPSQM(C)*	0.25
Peptide 2	VTGRGDSPASSKPI(C)	1.6
Peptide 3	SINYRTEIDKPSQM(C)	>100
Peptide 4	VTGRGDSPA(C)	2.5
Peptide 5	SPASSKPIS(C)	>100
Peptide 6	VTGRGD(C)	10
Peptide 7	GRGDS(C)	3.0
Peptide 8	RGDSPA(C)	6.0
Peptide 9	RVDSPA(C)	>100

*The (C) at the C-terminus indicates the cysteine linkage to the carrier protein.

*19–20 Fibronectin is a large glycoprotein component of the extracellular matrix. It contains several binding domains along its length, one of which binds to fibronectin receptors on cell surfaces.

Fibronectin can stick cells to surfaces to which they would otherwise not bind, forming the basis of a simple assay for the part of the molecule recognized by the fibronectin receptor. Fibronectin is coated on the surface of plastic dishes. A suspension of cells is then added and left to incubate for 30 minutes. Finally, the dishes are washed, and the number of cells that stick are counted. Without fibronectin, no cells stick; with it, 80–90% of the cells stick. Other proteins, such as serum albumin, will stick to the plastic, but they do not promote cell sticking.

Very small fragments of fibronectin will work in this assay, although they have to be chemically attached to larger molecules such as albumin in order for them to adhere to the plastic dishes. By making fragments of fibronectin, it was possible to identify the cell-binding domain as a 108-amino-acid segment about three-quarters of the way from the N-terminus.

Synthetic peptides corresponding to different portions of the 108-amino-acid segment were then tested in the cell-binding assay to localize the active region precisely. Two experiments were conducted. In the first, peptides were attached covalently to protein-coated plastic via a cysteine residue and tested for their ability to promote cell sticking. The results are shown in Table 19–6.

In the second experiment, plastic dishes were coated with native fibronectin, and cells were incubated in the dishes for 30 minutes in the presence of the synthetic peptides indicated in Table 19–7. The dishes were washed and the number of stuck cells was counted.

A. The two experiments use different assays to detect the cell-binding segment of fibronectin. Does the sticking of cells to the dishes mean the same thing in both assays? Explain the difference between the assays.

B. From the results in Tables 19–6 and 19–7, deduce what amino acid sequence in fibronectin is recognized by cells.

C. How might you make use of these results to design a method for isolating the fibronectin receptor?

19–21 One way to obtain mutations in mice is to infect embryos with retroviruses, which integrate randomly into the genome. If they integrate into a cell that becomes part of the germ line, they can be passed on to future generations. Using this method, you have isolated a particularly interesting strain in which

Table 19–7 Fibronectin-related Peptides Tested for Their Ability to Block Cell Sticking (Problem 19–20)

Peptide	Percent of Input Cells Sticking
GRGDSPC	2.0
GRGDAPC	1.9
GKGDSPC	48
GRADSPC	49
GRGESPC	44
None	47

The Extracellular Matrix of Animals

(A) STRUCTURE OF GENES

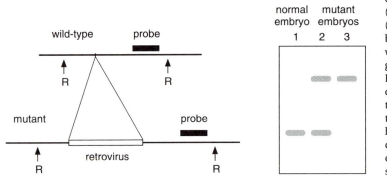

(B) GEL ANALYSIS

normal   mutant
embryo   embryos
1   2   3

Figure 19–15 Analysis of DNA from embryos from a brother-sister mating (Problem 19–21). The schematic diagram (A) illustrates the structural relationship between the normal gene and the gene with the retroviral insert. The autoradiogram (B) was obtained by digesting the DNA with a restriction enzyme, R, which does not cut the retrovirus, and running the fragments on a gel. The DNA was then transferred to a nitrocellulose filter, hybridized to a radioactive probe that corresponds to cellular sequences just outside the site of integration, and subjected to autoradiography.

the retrovirus appears to have integrated in a gene that is crucial for early embryonic development (Figure 19–15A). The crucial nature of the gene was revealed when you mated a brother and sister, each heterozygous for the retroviral insert, and analyzed DNA from 12-day embryos. Using a probe that lies just outside the site of integration, you find three distinct patterns of hybridization (Figure 19–15B): pattern 1 is homozygous for the normal gene; pattern 2 is heterozygous; and pattern 3 is homozygous for the mutant gene.

Although a quarter of the 12-day embryos show pattern 3, you have never observed this pattern in a live birth. You find that around day 14 these embryos die, apparently from massive hemorrhaging from large blood vessels. In all other respects the major tissues and cell types appear normal. This result suggests that the mutation, when homozygous, causes a stage-specific developmental defect.

Upon further study, you find that the mRNA corresponding to the affected gene is made only by fibroblasts, myoblasts, and chondrocytes. In addition, this mRNA is readily detectable at day 12 and then increases markedly as development proceeds. The levels of expression, the cell types involved, and the burst blood vessels lead you to suspect that the affected gene may encode collagen or some other component of the extracellular matrix. Accordingly, you borrow two collagen-gene clones from other researchers and use them to analyze the mRNA present in 12-day-old normal and homozygous mutant embryos, as shown in Figure 19–16.

A. What gene did the retrovirus integration interrupt? Given the apparent sensitivity of blood vessels in the mutant embryos, does the identity of the affected gene surprise you?

B. In matings between heterozygous brothers and sisters, what fraction of 12-day embryos do you expect to display patterns 1, 2, and 3 in Figure 19–15? What fraction of newborn mice will display these patterns?

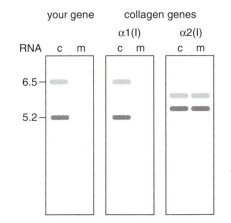

Figure 19–16 Analysis of mRNA from control (c) and mutant (m) embryos (Problem 19–21). RNA was isolated from normal mouse embryos and from homozygous mutant embryos, separated by electrophoresis on a gel, and probed with radiolabeled DNA from your gene and from the collagen clones. Numbers indicate the sizes of the mRNAs in kb. Two bands of hybridizing RNA are seen in several lanes; it is common for single genes in eucaryotic cells to produce multiple species of mRNA.

C. If you are familiar with recombinant DNA techniques, try this exercise. One distinct advantage of using retroviral insertions is that the gene it disrupts can be readily cloned using the retrovirus DNA sequences as a "tag." Outline briefly how a retrovirus might be used in this way to clone the normal, unaffected gene.

Extracellular Matrix Receptors on Animal Cells: The Integrins
(MBOC 995–1000)

19–22 Fill in the blanks in the following statements.

A. The principal receptors on animal cells for binding most extracellular matrix proteins, including collagen, fibronectin, and laminin, are the _____, a large family of homologous transmembrane linker proteins.

B. Transformed cells that make less fibronectin than normal adhere poorly to the substratum and fail to flatten out or develop the organized intracellular actin filament bundles known as _____.

19–23 Indicate whether the following statements are true or false. If a statement is false, explain why.

___ A. Integrins differ from cell-surface receptors for hormones and for other soluble signaling molecules in that they bind their ligand with relatively low affinity.

___ B. Since the same integrin molecule in different cell types always has the same ligand-binding activity, it appears that the binding specificity is contained entirely within the α- and β-integrin chains.

___ C. Integrins function as transmembrane linkers mediating the interactions between the cytoskeleton and the extracellular matrix that are required for cells to grip the matrix.

___ D. Although extracellular fibronectin filaments are connected to intracellular actin filaments in cultured fibroblasts, the two sets of filaments generally are oriented randomly relative to one another.

___ E. The matrix-binding integrins of many cells in tissues are constantly in an adhesive-competent state, but the integrins on blood cells often have to be activated before they can mediate cell adhesion.

___ F. The clustering of integrins at the sites of contact with the matrix or with another cell can activate intracellular signaling pathways to optimize cellular responses to growth factors.

19–24 K562 cells are a line of mouse erythroleukemic cells that are capable of differentiating into red blood cells in response to certain stimuli. These cells express a single type of integrin, $\alpha_5\beta_1$, on their surface, which endows the cells with the ability to bind the RGD motifs of the extracellular matrix protein fibronectin.

Your laboratory has raised a large number of high-affinity monoclonal antibodies against the separated α_5 and β_1 chains. To test their effect on the ability of K562 cells to bind fibronectin, you incubate the cells with radioactively labeled fibronectin in the presence of various antibodies and measure the amount of radioactivity that binds to the cells. Most of the antibodies against either chain reduce the amount of fibronectin that binds to the cells. However, one antibody specific for the β_1 chain gives a 20-fold increase in binding (Table 19–8).

Additional experiments using mixtures of anti-integrin antibodies, anti-fibronectin antibodies, and peptides with and without the RGD motif confirmed that the observed binding was due to interactions of fibronectin with the integrin (Table 19–8). Detailed binding studies revealed that the integrin in the presence of the anti-β_1 antibody had an increased affinity for fibronectin

Table 19–8 Effects of Antibodies on Cell Binding of Fibronectin (Problem 19–24)

Additions	Fibronectin Bound (cpm)
None	2,000
Anti-α_5 antibody	0
Anti-α_4 antibody	2,000
Anti-β_1 antibody	40,000
Anti-β_1 antibody + anti-α_5 antibody	0
Anti-β_1 antibody + anti-α_4 antibody	40,000
Anti-β_1 antibody + anti-fibronectin antibody	500
Anti-β_1 antibody + GRGDSP peptide	3,000
Anti-β_1 antibody + GRGESP peptide	40,000

(Figure 19–17). Any intracellular influences on the stimulated binding in the presence of the anti-β_1 antibody were ruled out by repeating the experiments (with the same results) in the presence of various metabolic poisons such as azide.

A. How do the experiments in Table 19–8 confirm that the observed anti-β_1-stimulated fibronectin binding is due to its interaction with the $\alpha_5\beta_1$ integrin?
B. Which experiments rule out the possibility that the apparent increase in binding affinity in Figure 19–17 is really due to increased numbers of integrin molecules on the cell surface: either newly synthesized or newly transferred from some internal compartment?
C. How do you suppose this particular anti-β_1 antibody increases the affinity of the $\alpha_5\beta_1$ integrin for fibronectin?

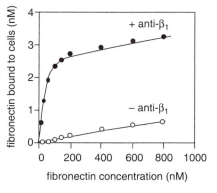

Figure 19–17 Detailed binding studies for fibronectin binding to K562 cells (Problem 19–24).

*19–25 The major surface protein of blood platelets is an integrin, known as $\alpha_{IIb}\beta_3$, which binds to fibrinogen when platelets are stimulated with clotting factors such as thrombin, allowing platelets to aggregate to form blood clots. The fascinating feature of this process is that platelets do not bind fibrinogen or aggregate until after stimulation, although $\alpha_{IIb}\beta_3$ is always present on their surface. What regulates the activity of this abundant and all-important integrin?

You have cloned both subunits of $\alpha_{IIb}\beta_3$ and are able to express them in Chinese hamster ovary (CHO) cells. When you incubate the $\alpha_{IIb}\beta_3$ transfected CHO cells with fibrinogen in the presence or absence of thrombin, they fail to aggregate. However, when you incubate the cells with an activating antibody (MAb 62, analogous to the anti-β_1 antibody described in the previous problem), the cells aggregate within minutes of adding fibrinogen. Untransfected control cells show no tendency to aggregate when treated in this manner.

You next test the effect of deleting a portion of the short cytoplasmic domains of α_{IIb} and β_3. You then express various combinations of truncated and wild-type α_{IIb} and β_3 in CHO cells, and you repeat the aggregation assays with fibrinogen. Although all combinations of the α_{IIb} and β_3 chains allow cells to aggregate in the presence of fibrinogen and MAb 62, the truncated α_{IIb} chain allows aggregation in the absence of MAb 62 (Table 19–9).

A. Why do you suppose that truncating the cytoplasmic domain of the α_{IIb} subunit increases the affinity of the integrin for fibrinogen and allows the cells to aggregate?
B. The $\alpha_{IIb}\beta_3$ integrin is accessible on the surface of the CHO cells, as revealed by the various aggregation studies. Why then does thrombin not stimulate the cells to aggregate?
C. There are two genes for α_{IIb} in human cells, which are diploid. If one of the two genes suffered a mutation of the kind described in this problem, do you think the individual would show any blood-clotting problems?

Table 19–9 Fibrinogen-dependent Aggregation of CHO Cells Expressing Various Wild-Type and Mutant α_{IIb} and β_3 Integrin Subunits (Problem 19–25)

α_{IIb} Chain	β_3 Chain	Aggregation	
		No Antibody	With MAb 62
Normal	normal	–	+++
Truncated	normal	+++	+++
Normal	truncated	–	+++
Truncated	truncated	+++	+++

The Plant Cell Wall

(MBOC 1000–1006)

19–26 Fill in the blanks in the following statements.

A. To accommodate subsequent cell growth, newly formed cells are covered by _____, which are thin and only semirigid; once growth stops, a much more rigid _____ is usually laid down around each cell.

B. The osmotic imbalance between the fluid in the cell wall and the cell's interior causes the cell to develop a large internal hydrostatic pressure, or _____, which pushes outward on the cell wall.

C. Adjacent cellulose molecules adhere to one another in overlapping parallel arrays, forming highly ordered crystalline aggregates called _____.

D. _____ are a heterogeneous group of branched polysaccharides that bind tightly to components of the cell wall and thereby cross-link them into a complex network.

19–27 Indicate whether the following statements are true or false. If a statement is false, explain why.

__ A. Each cell wall consists of a thin, semirigid primary cell wall adjacent to the cell membrane and a thicker, more rigid secondary cell wall outside the primary wall.

__ B. Turgor pressure is the main driving force for cell expansion during growth, and it provides much of the mechanical rigidity of living plant tissues.

__ C. Unlike the extracellular matrix of animals cells, which contains a large amount of protein, plant cell walls are composed entirely of polysaccharides.

__ D. If the entire cortical array of microtubules is disassembled by drug treatment, new cellulose microfibrils are laid down in random orientations.

__ E. Cortical microtubules can rapidly reorient in response to external stimuli, including such growth factors as ethylene and gibberellic acid.

*19–28 In plant cells the cortical array of microtubules determines the orientation of cellulose microfibrils, which in turn fixes the direction of cell expansion. The plant growth factors ethylene and gibberellic acid have opposite effects on the orientation of microtubule arrays in epidermal cells of young pea shoots. Gibberellic acid promotes a net orientation of the cortical microtubule array that is perpendicular to the long axis of the cell, whereas ethylene treatment causes the microtubule arrays to orient parallel to the long axis of the cell (Figure 19–18).

Which treatment do you think would produce short, fat shoots and which treatment would produce long, thin shoots?

19–29 The ripening of fruit is a complicated process of development, differentiation, and death (except for the seeds, of course). The process is triggered by minute amounts of ethylene gas. (This was discovered by accident many years ago; the paraffin stoves used to heat greenhouses in the olden days gave off enough

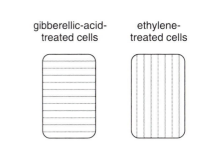

gibberellic-acid-treated cells   ethylene-treated cells

Figure 19–18 Effects of gibberellic acid and ethylene on the orientation of cortical arrays of microtubules (Problem 19–28).

ethylene to initiate the process.) The ethylene is normally produced by the fruits themselves in a biochemical pathway, the rate-limiting step of which is controlled by ACC synthase, which converts S-adenosylmethionine to a cyclopropane compound that is the immediate precursor of ethylene. Ethylene initiates a program of sequential gene expression that includes production of several new enzymes, including polygalacturonase, which probably contributes to softening the cell wall.

Your company, Agribucks, is trying to make mutant tomatoes that cannot synthesize their own ethylene. Such fruit could be allowed to stay longer on the vine, developing their flavor while remaining green and firm. They could be shipped in this robust unripe state and exposed to ethylene upon arrival at market. This should allow them to be sold at the peak of perfection, and the procedure involves no artificial additives of any kind.

You decide to use an antisense approach, which works especially well in plants. You place an ACC synthase cDNA into a plant expression vector so that the gene will be transcribed backward, introduce it into tomato cells, and regenerate whole tomato plants. Sure enough, ethylene production is inhibited by 99.5% in these transgenic tomato plants, and their fruit fails to ripen. But when placed in air containing a small amount of ethylene, they turn into beautiful, tasty ripe red fruit in about 2 weeks.

A. How do you imagine that transcribing the ACC synthase gene backward blocks the production of ethylene?
B. Will you be a millionaire before you are 30?

Answers

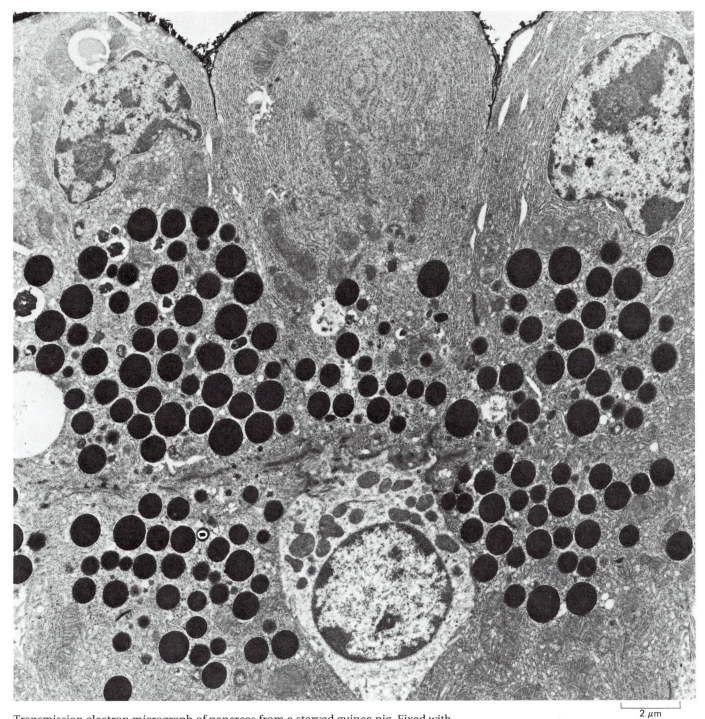

Transmission electron micrograph of pancreas from a starved guinea pig. Fixed with glutaraldehyde and osmium tetroxide, stained with uranyl acetate and lead citrate. (Courtesy of Dr. Brij Gupta.)

2 μm

Basic Genetic Mechanisms

RNA and Protein Synthesis

6–1

A. RNA polymerase, DNA transcription
B. promoter, stop (termination) signal
C. consensus sequences
D. anticodon, codon
E. aminoacyl-tRNA synthetases, aminoacyl-tRNA
F. ribosome, peptidyl-tRNA-binding site, aminoacyl-tRNA-binding site
G. peptidyl transferase, rRNA
H. stop
I. reading frames
J. initiation factors (IFs)
K. initiator tRNA, start, methionine

6–2

A. False. Binding to the promoter orients RNA polymerase and the choice of template strand because the RNA chain, which is synthesized in the 5′-to-3′ direction, must be antiparallel to the template strand.
B. True
C. True
D. False. Modified nucleotides are produced by covalent modification of the standard nucleotides after RNA synthesis.
E. False. A single-base change would alter tRNA^{Tyr} to recognize two serine codons, UCU/C (due to wobble pairing). In a cell-free system competition between the modified tRNA^{Tyr} and the normal tRNA^{Ser} would probably produce a protein with a mixture of serine and tyrosine at positions specified by UCU and UCC. However, at the positions in the protein specified by the other four serine codons, there would be serine alone.
F. True
G. False. Wobble base-pairing occurs between the third position in the codon and the first position in the anticodon.
H. True
I. False. Protein synthesis consumes much more total energy than transcription because hundreds to thousands of protein molecules are made from each mRNA.
J. False. AUG also encodes methionine in the interior of the mRNA; the selection of one AUG as the initiation site depends on other features of the mRNA nucleotide sequence.
K. True
L. True

6–3

A. 5′-GUAGCCUACCCAUAGG-3′
B. If translation begins at the 5′ end of the RNA, the synthesized protein would be valine-alanine-tyrosine-proline (VAYP).

Only after a peptide bond has been formed between alanine and tyrosine will tRNA^{Ala} leave the ribosome. Thus, the next tRNA that will bind to the ribosome after tRNA^{Ala} has left is tRNA^{Pro}.

When the amino group of alanine forms a peptide bond, the ester bond between valine and tRNA^{Val} is broken, tRNA^{Val} is expelled from the ribosome, and tRNA^{Ala} moves from the A-site to the P-site.

C. This short mRNA encodes three different peptides, since there are three different reading frames. In the second reading frame, the first codon is the stop codon UAG; however, the subsequent codons can be translated.

        5'-GUAGCCUACCCAUAGG-3'

Frame 1	V	A	Y	P	*
Frame 2	*	P	T	H	R
Frame 3		S	L	P	I

The other possible mRNA from this DNA would read

        5'-CCUAUGGGUAGGCUAC-3'

Frame 1	P	M	G	R	L
Frame 2	L	W	V	G	Y
Frame 3		Y	G	*	

Thus, the sequence of the peptides would be completely different. Be very careful to keep the polarity of the strands correct; it is very easy to fall into the trap of thinking that the complementary sequence of the first mRNA is 5'-CAUCGGAUGGGUAUCC-3', which is incorrect because the strands of DNA run in opposite directions.

D. Since there is no AUG or even GUG (which is sometimes used as a start signal for translation in *E. coli*), this sequence cannot come from the beginning of the coding region of a gene. It could be from the end of a gene, if the first or second reading frames were used, or from the middle of a gene, if the third reading frame were used. More information is required to distinguish between these possibilities.

*6–4

6–5

A. The sequence data for the *Tetrahymena* protein is unusual because it indicates that UAG and UAA, which are stop codons in other organisms, specify glutamine (Q) in *Tetrahymena*.

B. The very minor protein from the pure TMV mRNA is produced by reading through the normal stop codon. Although the mechanism of such a rare event is difficult to know, it is thought to represent the frequency with which the reticulocyte translation system mistakenly inserts an amino acid at the site of the stop codon instead of terminating properly. It is a little surprising that a second termination codon is not encountered for 506 codons.

C. Given that *Tetrahymena* uses UAG and UAA as codons for glutamine, the increase in the proportion of the readthrough TMV protein is most likely due to the presence of tRNA^{Gln} species with anticodons complementary to the normal TMV stop codon, which is UAG. The addition of *Tetrahymena* RNA causes a small shift in the proportions because it contains some charged tRNA^{Gln}. The cytoplasm causes a larger shift because it also contains the appropriate aminoacyl tRNA synthetase. (The additional shift with the cytoplasm suggests that the tRNA synthetases in the reticulocyte lysate cannot recharge the special *Tetrahymena* tRNA.) These results suggest that at least two components from *Tetrahymena*—a special tRNA and its cognate aminoacyl tRNA synthetase— must be added to a reticulocyte lysate to allow *Tetrahymena* mRNA to be translated efficiently. These components compete effectively with the reticulo- cyte release factors, allowing the *Tetrahymena* mRNAs to be read.

*Problems with an asterisk (*) are answered in the Instructor's Manual.

D. Although slight variations in the genetic code were discovered several years ago in mitochondrial genomes, they were not so surprising as the *Tetrahymena* changes. After all, mitochondrial genomes are small and encode relatively few proteins, so it is less difficult to imagine how changes might occur. However, the *Tetrahymena* genome encodes thousands of proteins. It is much more surprising that it managed to survive the presumptive transition from the standard code to its present-day code.

References: Horowitz, S.; Gorovsky, M.A. An unusual genetic code in nuclear genes of *Tetrahymena*. *Proc. Natl. Acad. Sci. USA* 82:2452–2455, 1985.

Andreasen, P.H.; Dreisig, H.; Kristiansen, K. Unusual ciliate-specific codons in *Tetrahymena* mRNAs are translated correctly in a rabbit reticulocyte lysate supplemented with a subcellular fraction from *Tetrahymena*. *Biochem. J.* 244:331–335, 1987.

6–6

A. The data in Figure 6–2 indicate that the N-terminus of the protein is synthesized first. The linear gradient of radioactivity from the start of the chain—the N-terminus—to the finish is exactly what you would expect according to the model shown in Figure 6–36A and B. Here, all the ribosomes carry the lysine at position 8, but the ribosome at the 5′ end of the mRNA has not yet reached the lysine at position 16, so there is less radioactivity in it, and so on down the line. Almost none of the ribosomes [actually (147–144)/147, or just over 2%] will carry the lysine at position 144.

B. The slopes of the lines for the α and β chains are very similar, indicating that roughly equal *numbers* of each chain are being synthesized. However, there is not enough information to decide whether the numbers of α- and β-globin

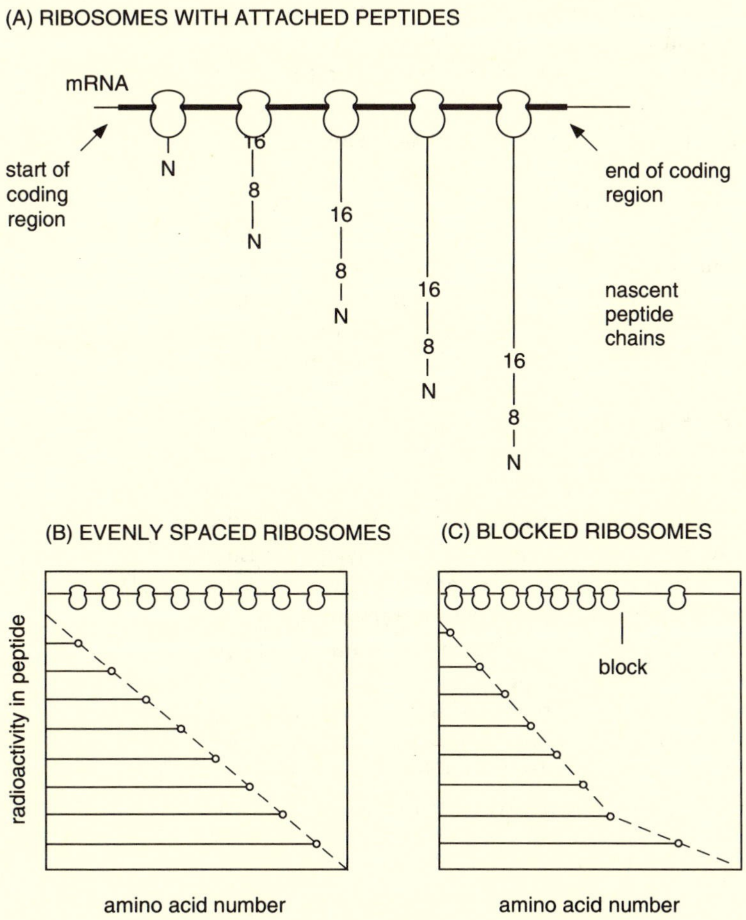

Figure 6–36 Relationship of ribosome position to peptide length and labeling pattern (Answer 6–6). (A) Lengths of peptides associated with ribosomes at various positions along β-globin mRNA. Numbers refer to positions of the first two lysine residues. (B) Pattern of peptide labeling for evenly spaced ribosomes. (C) Pattern of peptide labeling for ribosomes whose movement is inhibited at a point midway down the mRNA. Peptides associated with each ribosome are shown on the graph as lines. Small circles correspond to the C-terminus of the polypeptide and are aligned immediately below the ribosome on which they are synthesized. Dashed lines through the small cirles show the expected patterns of peptide labeling.

mRNA molecules are equal. You would need to know how many ribosomes there were on each mRNA—the average polyribosome size for α- and β-globin mRNAs—to deduce the relative abundance of the two mRNAs from these kind of data. Actually, there is about twice as much α-globin mRNA as β-globin mRNA, but the α-globin mRNA is less efficiently translated; that is, fewer ribosomes initiate synthesis on α-globin mRNA per unit time than on β-globin mRNA. These factors cancel out to give a fairly balanced production of the two chains.

C. The graph hits zero right at the end of the coding region, which indicates that chains are released from ribosomes as soon as they encounter the stop codon—or at least they do so without a measurable pause on this time scale.

D. If there were a significant roadblock to ribosome movement, the data would resemble that in Figure 6–3A. A roadblock would result in more densely packed ribosomes in front of the block and less densely packed ribosomes beyond the block. The consequences of inhibited ribosome movement are illustrated schematically in Figure 6–36C.

*6–7

6–8

A. The DNA sequence GGG TAT CTT *TGA* CTA CGA CGC should not encode the protein sequence of RF2, since UGA is a termination codon. It appears that this sequence must break the usual rules of the triplet code, with a leucyl tRNA decoding the italicized quadruplet.

GGG TAT CTTT GAC TAC GAC GCC

In essence, the ribosome must shift its frame of reading in the middle of the gene!

Frameshift mutations were originally isolated by Benzer in his work on the r_{II} genes of bacteriophage T4 and exploited by Crick in the discovery of the triplet code. Later, mutant tRNA molecules that could read four bases at a time were isolated by clever genetic selection and shown to suppress certain frameshift mutations. It comes as a great surprise, however, to find occasional *natural* examples of frameshift suppression. The first example was found in bacteriophage T7 gene 10. Since then, several retroviruses and retroposons have been found to use frameshift suppression of termination codons as a way of making minor gene products. The mechanism of suppression in these cases is not clear.

B. The occurrence of an inframe suppressible UGA codon (which is recognized uniquely by RF2) in the sequence of RF2 immediately suggests a novel form of gene control. Although the mechanism of frameshifting is undefined, there is very likely a competition between frameshifting and termination at the UGA codon. When the level of RF2 in the cell is high, termination should occur more frequently at the UGA codon than when the level of RF2 is low. Thus, very little new RF2 would be synthesized when its levels were already adequate, but if the levels fell, the chances of ribosomal frameshifting would increase and more RF2 would be made. Thus, this situation seems to be a very cleverly appropriate autoregulation. Notice also that the frameshift occurs near the beginning of the gene, so not too much energy is wasted making a useless polypeptide.

Although this is a very attractive hypothesis to explain the regulation of RF2 synthesis, there is no direct evidence for or against it.

References: Craigen, W.J.; Cook, R.G.; Tate, W.P.; Caskey, C.T. Bacterial peptide chain release factors: conserved primary structure and possible frameshift regulation of release factor 2. *Proc. Natl. Acad. Sci. USA* 82:3616–3620, 1985.

Jacks, T.; Varmus, H. Expression of the Rous sarcoma virus *pol* gene by ribosomal frameshifting. *Science* 230:1237–1242, 1985.

6–9

A. Since the bacteria were labeled for one generation, which represents a doubling in mass, 4 µg of the 8 µg of flagellin isolated from the gel were synthesized in the presence of ^{35}S cysteine. The amount of radioactivity in the sample indicates that about 1 out of every 1670 flagellin (flgn) molecules contains a cysteine.

$$\frac{Cys}{flgn} = \frac{300\ cpm\ Cys}{4\ µg\ flgn} \times \frac{pmol\ Cys}{5 \times 10^3\ cpm} \times \frac{4 \times 10^4\ µg\ flgn}{µmol\ flgn} \times \frac{µmol}{10^6\ pmol}$$

$$= \frac{6 \times 10^{-2}\ pmol\ Cys}{100\ pmol\ flgn}$$

$$\frac{Cys}{flgn} = 6 \times 10^{-4}$$

which is equal to 1 cysteine per 1670 flagellin molecules ($1/6 \times 10^{-4}$).

B. The normal codons for cysteine are UGU and UGC. Thus, the error in anti-codon-codon interaction is a mistake at the first position of the codon (third position of the anticodon). The experiment described, and other experiments too, suggest that ribosomes tend to mistake U for C and C for U in the first two positions of the codon, and C and U for A in the first position.

C. Assuming that all six arginine codons are equally likely, there should be six sensitive (CGC and CGU) arginine codons ($2/6 \times 18$) in a flagellin molecule. Therefore, the actual error frequency per codon-at-risk is

$$error\ frequency = \frac{1\ cysteine}{1670\ flagellin\ molecules} \times \frac{flagellin\ molecule}{6\ sensitive\ codons}$$

$$error\ frequency = 10^{-4}$$

D. If the probability of making a mistake at each codon is 10^{-4}, the probability of not making a mistake at each codon is $(1 - 10^{-4})$. The probability of not making a mistake at n codons is then $(1 - 10^{-4})^n$. Thus, the percentage of correctly synthesized molecules 100 amino acids in length is $(1 - 10^{-4})^{100}$, or 99%. For a protein 1000 amino acids long, 90% are correct. For a molecule 10,000 amino acids long, only 37% are correct. Given these sorts of estimates, it is perhaps not surprising that proteins more than 3000 amino acids long are rare.

Reference: Edelman, P.; Gallant, J. Mistranslation in *E. coli. Cell* 10:131–137, 1977.

*6–10 **Reference:** Safer, B.; Kemper, W.; Jagus, R. Identification of a 48S preinitiation complex in reticulocyte lysate. *J. Biol. Chem.* 253:3384–3386, 1978.

DNA Repair

6–11

A. DNA repair, mutation
B. fibrinopeptides
C. depurination, deamination
D. DNA repair nucleases, DNA polymerase, DNA ligase
E. base excision repair, DNA glycosylases
F. nucleotide excision repair
G. SOS response

6–12

A. True

B. True

C. False. The nucleotide sequences of histone H4 genes from different species vary considerably; however, the nucleotide changes are confined for the most part to third positions in codons so that they encode the same amino acid sequence.

D. True

E. False. Repair depends on the two copies of genetic information contained in the two strands of the DNA double helix.

F. True

G. True

H. False. The principal function of the SOS response is to relax the fidelity of DNA synthesis so that replication can proceed past a block. Since a cell will die if replication remains blocked, the benefit is survival; the cost is mutation.

I. True. Deamination of A, G, or C (T does not carry an amino group) leads to formation of a base that is unnatural in DNA. For example, C is deaminated to U—which may be why DNA uses T instead of U. A deamination problem does arise in eucaryotic cells in which many CG dinucleotides are methylated at the 5 position of the C ring. Deamination of 5-methyl C gives a T.

6–13

A. The extreme UV sensitivity of *uvrArecA* double mutants relative to pairs of *uvr* mutants suggests that there are two separate pathways for dealing with UV damage. The *uvr* gene products are involved in one pathway, whereas the *recA* gene product is involved in a different pathway. As a rule of thumb, if a combination of defective genes produces no more mutant a phenotype than the individual defective genes, the gene products are likely to act in the same biochemical pathway.

  The *uvr* gene products form the uvrABC endonuclease, which specifically removes a 12-nucleotide-long oligonucleotide that encompasses a pyrimidine dimer. The RecA protein is involved in two pathways for handling UV damage: the SOS response and recombinational repair. The *uvr* and *recA* pathways are not entirely independent since expression of the *uvr* genes is substantially increased as a part of the SOS response.

B. A lethal hit in the *uvrArecA* strain corresponds to about one pyrimidine dimer. The number of pyrimidine dimers per lethal hit can be calculated as follows: Since *E. coli* is 50% GC, all four bases are equally represented in the genome. If they are arranged randomly (which they are not, but this assumption is a reasonable approximation), then of the 16 possible dinucleotide pairs in DNA, one-quarter are pyrimidine pairs; hence, the *E. coli* genome (4×10^6 base pairs) contains 10^6 possible UV targets. Given that a dose of 400 J/m^2 converts 1% of the pyrimidine (pyr) pairs into pyrimidine dimers, the number of pyrimidine dimers per lethal hit in *E. coli* is

$$\frac{\text{pyr dimers}}{\text{lethal hit}} = \frac{10^6 \text{ pyr pairs}}{E.\ coli} \times \frac{0.04 \text{ J/m}^2}{\text{lethal hit}} \times \frac{1 \text{ pyr dimer}}{100 \text{ pyr pairs}} \times \frac{1}{400 \text{ J}/\text{m}^2}$$

$$\frac{\text{pyr dimers}}{\text{lethal hit}} = 1$$

*6–14

*6–15

6–16

A. The even distribution of frameshift mutations indicates that UV damage is distributed throughout the gene, since a frameshift mutation anywhere in the gene would be detected by the gene-fusion assay (which is independent of repressor function). Since UV damage is evenly distributed, the nonrandom distribution of missense mutations in the *lacI* gene presumably reflects the functional importance of the ends of the LacI protein. Most mutations in the ends yield a nonfunctional protein; this enables one to detect them as mutants. However, the middle of the gene, being less critical for function, can accommodate some alterations and still produce a functional protein. These "silent" mutations would not be detected in an assay that depends on loss of function.

B. The common deletion of one nucleotide in response to UV damage is thought to occur as a mistake during error-prone DNA synthesis opposite a pyrimidine dimer. Presumably, in response to the abnormal spacing of bases caused by the pyrimidine dimer, error-prone synthesis inserts a single nucleotide rather than two. The frameshift hot spots presumably are hot spots because they contain runs of T's and therefore multiple possibilities for dimer formation.

Reference: Miller, J.H. Mutagenic specificity of ultraviolet light. *J. Mol. Biol.*182:45–65, 1985.

6–17

The variable in these experiments is light. For a given UV dose, the brighter the light, the less the observed killing. Thus, visible light can reverse the effects of UV irradiation. Direct reversal of UV damage is common in microorganisms and is called enzymatic photoreactivation. Although the mechanistic details are unclear, the energy of visible light in some way is harnessed to split apart pyrimidine dimers, thereby reversing the damage.

The account here is not so much different than the original discovery of photoreactivation by Albert Kelner in the 1940s. While investigating the effects of postirradiation temperature on UV survival, Kelner was plagued by another variable. In his own words:

> Careful consideration was made of variable factors which might have accounted for such tremendous variation. We were using a glass-fronted water bath placed on a table near a window, in which were suspended transparent bottles containing the irradiated spores. The fact that some of the bottles were more directly exposed to light than others suggested that light might be a factor. . . . Experiments showed that exposure of UV-irradiated suspensions to light resulted in an increase in survival rate or a recovery of 100,000- to 400,000-fold. Controls kept in the dark . . . showed no recovery at all.

Reference: Friedberg, E.C. DNA repair. New York: W.H. Freeman, 1984.

*6–18 Reference: Teo, I.; Sedgwick, B.; Kilpatrick, M.W.; McCarthy, T.V.; Lindahl, T. The intracellular signal for induction of resistance to alkylating agents in *E. coli. Cell* 45:315–324, 1986.

6–19

A. As shown in Figure 6–13, untreated bacteria and bacteria adapted to low levels of MNNG differ only in the amount of O^6-methylguanine. The absence of O^6-methylguanine in adapted bacteria correlates with the low level of mutation, suggesting that O^6-methylguanine is the mutagenic lesion. O^6-methylguanine is thought to be mutagenic because it can mispair with T during replication.

B. The kinetics of removal of the methyl group from O^6-methylguanine are peculiar because the amount removed does not increase with time as one might expect for a typical enzyme. In addition, the amount of O^6-methylguanine that is demethylated is directly proportional to the amount of purified protein added to the reaction. One possible explanation for such behavior is that the enzyme is very unstable; however, the identical end points at 5°C and 37°C argue against this explanation.

C. A calculation of the number of mutagenic bases that are demethylated per enzyme molecule indicates that each enzyme molecule removes only one methyl group. This calculation shows that the protein is used stoichiometrically instead of catalytically, which explains the peculiar kinetics.

For example, 2.5 ng of protein removes 0.13 pmol (0.5×0.26 pmol) of methyl groups from DNA. Thus, the number of enzyme molecules is

$$\text{enzyme molecules} = 2.5 \text{ ng} \times \frac{\text{nmol}}{19,000 \text{ ng}} \times \frac{6.0 \times 10^{14} \text{ molecules}}{\text{nmol}}$$

$$= 7.9 \times 10^{10} \text{ molecules}$$

and the number of methyl groups is

$$\text{methyl groups} = 0.13 \text{ pmol} \times \frac{6.0 \times 10^{11} \text{ methyl groups}}{\text{pmol}}$$

$$= 7.8 \times 10^{10} \text{ methyl groups}$$

It turns out that methyl groups are transferred to one particular cysteine residue in the protein. Once methylated, the protein is dead and ultimately is degraded. Because the methyltransferase inactivates itself during the reaction, it is not an enzyme in the usual sense. (An enzyme is a catalyst, which by definition is not consumed during the reaction.) Note that repair of O^6-methylguanine is similar to enzymatic photoreactivation (Problem 6–17) in that both proceed by direct reversal of DNA damage.

Reference: Lindahl, T.; Demple, B.; Robins, P. Suicide inactivation of the *E. coli.* O^6-methylguanine-DNA methyltransferase. *EMBO J.* 1:1359–1363, 1982.

DNA Replication

6–20

A. DNA polymerase
B. DNA replication fork
C. DNA ligase
D. leading strand, lagging strand
E. proofreading exonuclease
F. RNA primers, DNA primase
G. DNA helicase
H. single-stranded DNA-binding (SSB) proteins
I. primosome
J. mismatch proofreading (mismatch repair)
K. replication origins
L. DNA topoisomerases

6–21

A. True. (If the replication fork moves forward at 500 nucleotide pairs per second, the DNA ahead of it must rotate at $500/10.5 = 48$ revolutions per second, or 2880 revolutions per minute.)
B. True
C. False. The sequence of nucleotides in the progeny strand is very different from that in the parental strand, even though the two strands are related by complementary base-pairing.
D. True
E. False. All DNA synthesis occurs in the 5′-to-3′ direction. DNA on the lagging strand is made in pieces that are joined together later so that the lagging strand lengthens in the 3′-to-5′ direction.
F. True
G. False. In the absence of the 3′-to-5′ proofreading exonuclease activity, DNA synthesis will be much more error prone.

H. False. Single-strand binding proteins keep DNA strands apart by binding to the phosphate backbone; they leave the bases exposed so that they can serve as a template for DNA synthesis.

I. False. The methylation-dependent repair system relies on methyl groups in the parent strand and their absence in the progeny strand in order to distinguish the two strands.

J. True

K. True

L. True

6–22

A. The indicated phosphate (P) is at the 5′ end of the fragment to which it is attached.

B. The gap will be filled in by continuous DNA repair synthesis, starting at the indicated OH on the bottom strand and proceeding in the 5′-to-3′ direction (leftward) until it reaches the phosphate on the adjacent fragment.

C. In the absence of DNA ligase, the two pieces on the bottom strand will remain unlinked even after the gap is filled in.

*6–23 Reference: Inman, R.B.; Schnos, M. Structure of branch points in replicating DNA: presence of single-stranded connections in lambda DNA branch points. *J. Mol. Biol.* 56:319–325, 1971.

6–24

A. The different labels used for the T and C nucleotides make it easy to measure their respective losses from the polymer. The radioactive disintegrations from the energetic ^{32}P and from the rather weak ^{3}H can be distinguished using a liquid scintillation counter. (The measurement could have been done using a single isotope and chromatographic separation of the released nucleotides, but the procedure is more time-consuming.)

B. Because the nuclease activity of DNA polymerase I is an exonuclease (that is, it removes nucleotides from the ends of strands), the T nucleotides cannot be released until all the C nucleotides have been removed, hence the lag.

C. When dTTP is added to the reaction, polymerization will begin just as soon as a proper AT nucleotide pair is found. Polymerization cannot occur from a mismatched AC pair. Since the rate of polymerization exceeds the rate of exonuclease by two or three orders of magnitude, the labeled T residues will be buried quickly and will be unavailable to the exonuclease.

D. The results will not be affected by the presence of dCTP. Since the template is poly (dA), C nucleotides cannot be incorporated. If they are incorporated by mistake, the mismatched C will not serve as a primer for polymerization.
   Reference: Brutlag, D.; Kornberg, A. Enzymatic synthesis of deoxyribonucleic acid. 36. A proofreading function for the 3′ to 5′ exonuclease activity in deoxyribonucleic acid polymerases. *J. Biol. Chem.* 247:241–248, 1972.

*6–25 Reference: Cha, T.A.; Alberts, B.M. Studies of the DNA helicase-RNA primase unit from bacteriophage T4. A trinucleotide sequence on the DNA template starts RNA primer synthesis. *J. Biol. Chem.* 261:7001–7010, 1986.

6–26

A. ATP hydrolysis is required for unwinding because energy is needed to melt DNA. Strand separation is energetically unfavorable because stacking interactions between the planar base pairs are largely lost upon strand separation. In addition, the H bonds that link the bases present a kinetic barrier to strand separation.

B. Since DnaB melts off only the 3′ half-fragment of substrate 3, it must bind to the long single strand and move along it in the 5′-to-3′ direction until it reaches the double-stranded region formed by the 3′ half-fragment, at which point it unwinds the fragment. The 5'-to-3' movement of DnaB suggests that it unwinds the parental duplex at the replication fork by moving along the lagging strand.

If DnaB moves in the 5'-to-3' direction, why does it not melt the 5' half-fragment off of substrate 3 by binding to the short 5' tail? Pat yourself on the back if you wondered about this. In real experiments a small amount of the 5' half-fragment is melted off. Melting due to the short 5' tail is presumed to be inefficient because of the difference in target size: DnaB is much more likely to bind to the long single strand rather than the short one.

C. If SSB is added first, it inhibits DnaB-mediated unwinding because it coats the single-stranded DNA, preventing DnaB from binding. By contrast, if SSB is added after DnaB has bound, it stimulates unwinding by preventing the unwound DNA from reannealing.

Reference: LeBowitz, J.H.; McMacken, R. The *Escherichia coli* DnaB replication protein is a DNA helicase. *J. Biol. Chem.* 261:4738–4748, 1986.

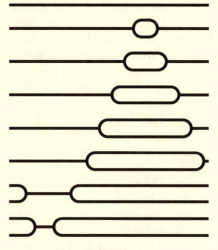

Figure 6–37 Bidirectional replication from a unique origin (Answer 6–27).

6–27

As always, you come through with flying colors. Although you were initially bewildered by the variety of structures, you quickly realized that H forms were just like the bubbles except that cleavage occurred within the bubble instead of outside it. Next you realized that by reordering the molecules according to the increasing size of the bubble (and flipping some structures end-for-end), you could present a convincing visual case for bidirectional replication away from a unique origin of replication (Figure 6–37). Finally, you remind your lab mates that this experiment does not define the location of the origin, since it could be clockwise or counterclockwise from the restriction site at which the circle was cleaved. You are planning to repeat the experiment using a different restriction enzyme. Your advisor is pleased.

*6–28

Genetic Recombination

6–29

A. general recombination
B. heteroduplex joint (staggered joint)
C. DNA renaturation (hybridization), helix nucleation
D. RecA protein
E. branch migration
F. cross-strand exchange, Holliday junction
G. gene conversion
H. site-specific

6–30

A. True
B. True
C. False. This statement would be true for the RecBCD protein. The RecA protein is required for the pairing of homologous duplexes.
D. True
E. True
F. False. Crossing and noncrossing pairs of strands can be interconverted by rotational movements that do not require strand breakage.
G. False. Gene conversion has nothing to do with sex change; it refers to the unequal recovery of parental alleles in the progeny.
H. True
I. True

6–31

The recombination substrates and products are shown in Figure 6–38. The first rule for deducing the recombination products is to align the homologous segments, that is, to draw the arrows one above the other so that they are pointing in the same direction. Alignment requires a twisting of substrates 3, 4, and 6. Alignment is necessary in order to form a Holliday junction, as would become apparent if real sequences were used instead of arrows.

Substrates 3 and 4 illustrate a useful rule. Recombination between direct repeats in a chromosome (as in substrate 3) deletes one copy of the repeat and the intervening DNA. Recombination between inverted repeats in a chromosome (as in substrate 4) simply inverts the DNA between the repeats.

*6–32 **Reference:** Ponticelli, A.S.; Schultz, D.W.; Taylor, A.F.; Smith, G.R. Chi-dependent DNA strand cleavage by recBCD enzyme. *Cell* 41:145–151, 1985.

6–33

A. If binding saturates at a 1:12 weight ratio of nucleotides to SSB protein, then the ratio of nucleotides to SSB molecules is 8.8:1.

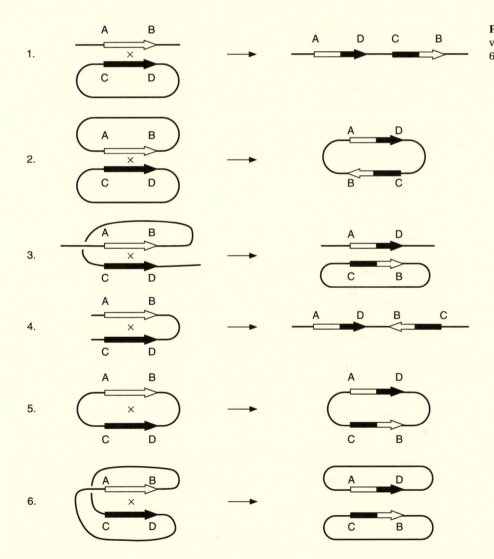

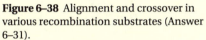

Figure 6–38 Alignment and crossover in various recombination substrates (Answer 6–31).

$$\frac{\text{nucleotides}}{\text{SSB molecule}} = \frac{1 \text{ d nucleotide}}{12 \text{ d SSB}} \times \frac{35{,}000 \text{ d SSB}}{\text{SSB molecule}} \times \frac{1 \text{ nucleotide}}{330 \text{ d nucleotide}}$$

$$= 8.8 \text{ nucleotides / SSB molecule}$$

B. If there are 10 nucleotides per 3.4 nm of single-stranded DNA, then 8.8 nucleotides would stretch about 3 nm. (If the single-stranded DNA were fully extended, it would stretch about twice as far.) The 12-nm length of an SSB molecule suggests that at saturation SSB proteins are very likely to be in contact with one another and probably overlap considerably.

C. The absence of significant binding at a low concentration of SSB protein, but nearly quantitative binding at a tenfold higher concentration, suggests that binding of SSB protein to DNA is cooperative. In essence, cooperative binding means that once one monomer has bound, additional monomers can bind more readily. If the monomers actually overlap with one another when they are bound, as suggested by the calculation in part B, cooperativity is easy to understand, for it suggests that each monomer has two binding sites—one for DNA and one for other monomers. Under these conditions binding of the first monomer to DNA will be weaker than the binding of subsequent monomers. That is because the first monomer can bind to the DNA only through its DNA-binding site, whereas the second monomer can bind adjacent to an already bound monomer and, thereby, make use of both of its binding sites. Mathematically, this type of interaction leads to a steep dependence of binding on concentration.

Reference: Alberts, B.M.; Frey, L. T4 bacteriophage gene 32: a structural protein in the replication and recombination of DNA. *Nature* 227:1313–1318, 1970.

*6–34   Reference: Cox, M.; Lehman, I.R. The polarity of the recA protein-mediated branch migration. *Proc. Natl. Acad. Sci. USA* 78:6023–6027, 1981.

6–35

Formation of a Holliday junction between the two parental duplexes along with its resolution and subsequent replication are shown in Figure 6–39. Note that the lower duplex was rotated to put the 3′ strand on top so that the crossover strands (in this case the 3′ strands) could be represented more simply. The crossover strands must be corresponding strands (both 3′ or both 5′); they cannot be complementary strands (one 3′ and one 5′). Do you understand why?

Resolution of the Holliday junction by breaking the 3′ strands is easy to see; however, resolution by breaking the 5′ strands is more difficult. One way to visualize the resolution is to isomerize the structure so that the 5′ strands become the crossover strands, as shown in MBOC Figure 6–65. Another way to see the relationship between the Holliday junction and the products is as follows: In the Holliday structure with the broken 5′ strands, cover the lower duplex segment to the left of the crossover and the upper duplex segment to the right of the crossover. What remains visible is the upper of the two product duplexes; it can be converted to the upper product by rotating the bottom duplex segment to put the 5′ strand on top.

Replication of the initial products of the recombination event resolves the heteroduplex segments and gives a truer picture of the recombinants as they would be detected in an actual experiment. Notice that if all four strand breaks are on the same strand (the 3′ strand in this case), the final result is insertion of a short segment of information from one duplex into the other (left-hand pathway). If two breaks are in 5′ strands and two breaks are in 3′ strands, the recombination products will be duplexes that have crossed over (right-hand pathway).

*6–36   Reference: Potter, H.; Dressler, D. On the mechanism of genetic recombinaton: electron microscopic observation of recombination intermediates. *Proc. Natl. Acad. Sci. USA* 73:3000–3004, 1976.

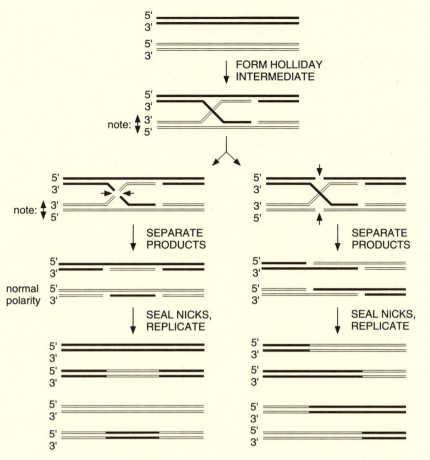

Figure 6–39 Formation and resolution of a Holliday intermediate followed by replication (Answer 6–35).

Viruses, Plasmids, and Transposable Genetic Elements

6–37

A. bacteriophages
B. lyse
C. capsid
D. enveloped
E. RNA-dependent RNA polymerase, replicase
F. envelope proteins
G. lysogenic, provirus
H. neoplastic transformation
I. reverse transcriptase, retroviruses
J. transposable elements, transposases
K. retrotransposon
L. plasmids
M. viroids

6–38

A. False. When T4 DNA enters a cell, it is completely dependent on host RNA polymerase to express its "early" genes. Later it modifies the host RNA polymerase so that it recognizes other viral promoters, leading to the coordinated expression of the T4 genome.

B. False. Although the host replication machinery provides the majority of the enzymatic apparatus, even small viruses make special proteins needed for initiation of DNA synthesis. If they did not, they would be subject to the strict rules of chromosome replication, which permit only one round of replication per cell generation, and multiple rounds of virus replication would be impossible.

C. False. It is true that the RNA genomes of negative-strand viruses do not serve directly as mRNA, but their complements do; thus, they contain genes.

D. True

E. True

F. True

G. False. Many viruses that reproduce by budding do not cause cancer; it is also possible to have cells transformed by oncogenic viruses that produce no virus particles at all.

H. False. The growth of some RNA viruses is inhibited by actinomycin D. Retroviruses, for example, are inhibited by actinomycin D because they are first converted to DNA by reverse transcriptase and then are transcribed in the usual manner, which is sensitive to actinomycin D. If the statement had been "If the growth of the virus is not inhibited by actinomycin D, it must be an RNA virus," it would have been true.

I. False. Transposable elements integrate nearly randomly and genes often are destroyed or altered by the integration event.

J. True

K. False. Overlapping genes are thought to have evolved in response to selective pressure for optimal use of small genomes that were limited by capsid size.

L. True

6–39

A. The phage suspension was diluted 1000-fold (0.1 ml/100 ml) upon mixing with the bacterial culture; therefore the initial phage titer was 10^7 phage/ml (10^{10} phage/ml $\times$ 1/1000).

B. The phage titer after 5 minutes was about tenfold lower that it was initially. This initial decrease in titer occurs because phages adsorb to the bacteria and inject their DNA. Since infectivity depends on an intact phage, the titer decreases. The low titer at 5 minutes represents phages that have not yet adsorbed to bacteria.

   Historically, this initial loss of titer (which was termed the eclipse phase of infection) was puzzling because it implied that the parental phages must be destroyed before new phages could be built. It is difficult now to appreciate the puzzle, but remember at that time the principal model for biological growth was cell division, where destruction of the parent would absolutely prevent formation of progeny.

C. The phage titer rises more slowly in the control samples because many phages (90% of the total between 20 and 40 minutes) are trapped inside the chloroform-killed bacteria. Incubation with lysozyme breaks open the bacteria, releasing the trapped phages. The titer of the control samples ultimately reaches that of the lysozyme-treated samples because the bacteria break open (lyse) naturally at the end of the infection. Lysis is controlled by phages so that a maximum number of phages can be made before the host is destroyed.

D. The initial mixture contains equal numbers of bacteria and phages (10^7/ml), so there are enough phages to infect every bacterium. Since there are 10^7 bacteria/ml initially and 10^9 phage/ml at the end, there are 100 phages per bacterium.

   The actual calculation of phages per *infected* bacterium is not quite so straightforward in this example. It is described briefly here so that you will be aware of an important fact: a one-to-one mixture of bacteria and phages does not mean that every bacterium will be infected by one phage. Most bacteria will be infected by one phage, but some will be infected by two or more phages, and some will not be infected at all. The fraction of bacteria in any particular class can be calculated from the Poisson distribution. From the Poisson distribution, the fraction of bacteria infected with zero phage is

$$P_0 = e^{-x}$$

where P_0 is the probability of not being infected and x is the ratio of phages to bacteria. At x = 1, P_0 = 0.37. Thus, only 63% of the bacteria will be infected. As a result, the number of phages per *infected* bacterium is 160.

The principle illustrated by this example arises in a number of different guises throughout biology. See, for example, Problem 6–13, in which delivery of one lethal UV "hit" per bacterium allowed 37% of the population to survive.

*6–40

6–41

A. The spot test described here is a classic version of a complementation test, which allows one to decide whether two mutants are defective in the same gene or in different genes. If a pair of mutants is defective in the same gene, they cannot help each other during infection (since they are both missing the same gene product) and, therefore, grow no better in a mixed infection than they do individually. By contrast, if a pair of mutants is defective in different genes, they can help one another: between them they have at least one functional copy of every gene and therefore can grow as a mixture of mutants.

The results of the spot test with the r_{II} mutants indicate that they fall into two complementation groups. Mutants 2, 5, and 8 are defective in one gene, and mutants 1, 3, 4, 6, and 7 are defective in a second gene. Classical experiments showed in much the same way that there are two r_{II} genes—r_{IIA} and r_{IIB}.

B. As suggested by the description in part A, if you repeated the spot test using *E. coli* K infected with mutant 3, you would see the same pattern as the spot test using *E. coli* K infected with mutant 1, which is defective in the same gene.

C. The small fraction of wild-type T4 in the mixture of mutant 1 and mutant 5 arises by genetic recombination. Viral genomes growing in the same cell will occasionally recombine. If a recombination event occurs between the mutations, a wild-type genome can be produced, as illustrated in Figure 6–40. Since the frequency of recombination is roughly proportional to the separation of the mutations, careful measurements of the fraction of wild-type in genetic crosses between mutants can be (and were) used to determine the order and spacing of mutations along the chromosome.

Reference: Crick, F.H.; Brenner, S. The absolute sign of certain phase-shift mutants in bacteriophage T4. *J. Mol. Biol.* 26:361–363, 1967.

*6–42

6–43

A. All colonies must have arisen by transposition of Tn10 into the bacterial genome, because survival depends on the presence of the tetracycline-resistance gene carried by Tn10. The presence of mixed colonies with blue and white sectors is the key observation. Since the frequency of sectored colonies is high but transposition is rare, sectored colonies must arise commonly in a

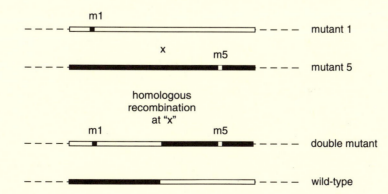

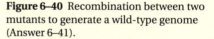

Figure 6–40 Recombination between two mutants to generate a wild-type genome (Answer 6–41).

single transposition event. A replicative mechanism can transfer only one strand of the parent heteroduplex and, thus, can generate only white or blue colonies, depending on which strand is transferred. A nonreplicative mechanism, however, transfers both strands of the heteroduplex, which upon replication and segregation into daughter bacteria will produce a sectored colony. (Once the bacteria are spread onto a petri dish, all the descendants of the original infected cell are confined to the immediate vicinity and, thus, grow together to form the colony. If two different daughters are produced at the first division, their descendants will grow together to produce a single colony with sectors containing the two different kinds of bacteria.) The pure blue and pure white colonies arise from transposition events that involve the homoduplexes. The proportions of blue, white, and sectored colonies are as expected from the equal mixture of heteroduplexes (which give rise to the sectored colonies) and homoduplexes (which give rise to the pure colonies).

B. Each heteroduplex contains a mismatched region of DNA corresponding to the position of the mutation in the *lacZ* gene. If these heteroduplexes were introduced into bacteria that could repair such mismatches, then the frequency of sectored colonies would decrease. In essence, each repair event would convert a heteroduplex into a homoduplex. If the mismatch repair is unbiased, the frequencies of blue colonies and white colonies would each increase equally.

C. If you used an integrating form of the phage genome, then the surviving colonies would have resulted from site-specific recombination, which is much more frequent than transposition. Site-specific recombination integrates both strands of the phage genome (see MBOC Figure 6–68). Thus, the proportion of blue, white, and sectored colonies would be the same as for nonreplicative transposition. Indeed, in the actual experiments that were done to define the mechanism of Tn10 transposition, an integrating (but nonreplicating) form of bacteriophage lambda was used as a positive control against which to compare the results of transposition. The two experiments gave identical results.

Reference: Bender, J.; Kleckner, N. Genetic evidence that Tn10 transposes by a nonreplicative mechanism. *Cell* 45:801–815, 1986.

*6–44   **Reference:** Sanger, F., et al. The nucleotide sequence of bacteriophage φX174. *J. Mol. Biol.* 125:225–246, 1978.

Recombinant DNA Technology

The Fragmentation, Separation, and Sequencing of DNA Molecules

7–1

    A. restriction nucleases

    B. restriction map

    C. DNA footprinting

7–2

    A. False. The recognition sequences for the nuclease, where they occur in the genome of the bacterium itself, are protected from cleavage by methylation at an A or a C residue.

    B. True

    C. False. When DNA molecules travel end-first through the gel in a snakelike configuration, their rate of movement is independent of length. In pulsed-field gel electrophoresis the direction of the field is changed periodically, which forces the molecules to reorient before continuing to move snakelike through the gel. This reorientation takes much more time for larger molecules, so that progressively larger molecules move more and more slowly.

    D. False. Polynucleotide kinase transfers a single phosphate from ATP to the 5′ end of each DNA chain.

    E. True

    F. True

7–3

    A. The 5′ and 3′ ends of the cut molecules are indicated in Figure 7–26 (next page). It is standard practice to represent DNA sequences so that the 5′ end of the top strand is on the left.

    B. As indicated in Figure 7–26, the BamHI ends can be filled in by DNA polymerase, but the PstI ends cannot. These different fates follow from the requirements of DNA polymerase: a primer with a 3′-OH to which dNTPs can be added and a template strand to specify correct addition. These requirements are met by the BamHI ends but not by the PstI ends, which have recessed 5′ ends that cannot serve as primers and, thus, cannot be filled in.

    NOTE: A standard technique in recombinant DNA technology is to use T4 DNA polymerase to blunt both types of ends. It will blunt BamHI ends by filling them in, as indicated in Figure 7–26. However, it will also blunt PstI ends by virtue of

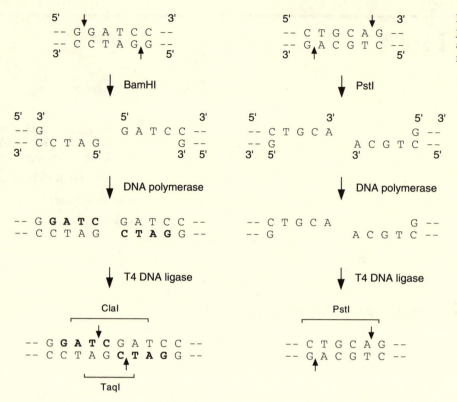

Figure 7–26 Cleavage, modification, and joining of DNA containing a BamHI site or a PstI site (Answer 7–3). Brackets indicate recognition sites for restriction enzymes.

an associated 3′-to-5′ exonuclease activity, which removes the 3′ extension, leaving a blunt end.

C. As indicated in Figure 7–26, the blunted BamHI ends and the unmodified PstI ends can both be joined by T4 DNA ligase.

D. Joining of the treated ends regenerates the PstI site but not the BamHI site. Joining of the filled-in BamHI ends generates two new restriction sites, as indicated in Figure 7–26. Cleavage, filling in the ends, and rejoining often generates new restriction sites that sometimes are useful for further manipulation of the DNA.

*7–4

7–5

A. The complexity of the original ligation pattern arises because any two BamHI ends can join together. Thus, the 0.4-kb fragments can join together to generate a set of fragments with sizes 0.8 kb, 1.2 kb, 1.6 kb, 2.0 kb, and so forth. Similarly, the 0.9-kb fragments will produce a set of fragments with sizes 1.8 kb, 2.7 kb, 3.6 kb, and so forth. Finally, combinations of the two fragments generate a third set of fragments with sizes 1.3 kb, 1.7 kb, 2.1 kb, 2.2 kb, and so forth. (The actual pattern often is more complicated still, since these fragments can circularize by joining their ends.)

B. Since any two BamHI ends can join, even one size of fragment can have several different structures, as shown in Figure 7–27 for the 1.3-kb fragment. Digestion of the population of 1.3-kb fragments with EcoRI generates a variety of fragments, which range from 0.1 kb (from the left ends of structures 3 and 4, Figure 7–27) to 1.0 kb (the internal fragment in structure 4, Figure 7–27).

*7–6 **Reference:** Anderson, J.E. Restriction endonucleases and modification methylases. *Curr. Opin. Struct. Biol.* 3:24–30, 1993.

*Problems with an asterisk (*) are answered in the Instructor's Manual.

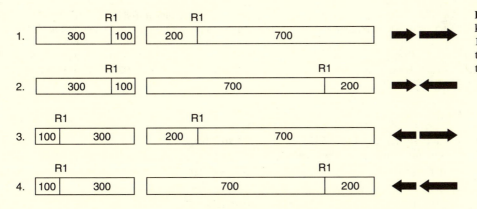

Figure 7–27 Various arrangements of 0.4-kb and 0.9-kb BamHI fragments to form a 1.3-kb fragment (Answer 7–5). Arrows on the right show the four arrangements of the large and small fragments.

7–7

As indicated in Figure 7–28, the restriction enzyme Sau96I will cleave the cDNA sequence at the segment corresponding to adjacent codons for glycine and proline, regardless of the particular codon used for glycine. AluI might cut this DNA if the right codons were used for both proline and alanine.

7–8

The restriction map for the BamHI fragment is shown in Figure 7–29. A map that is flipped end for end is equally valid. One approach to deriving this map is as follows. Draw a line of appropriate length to represent the 3.0-kb BamHI fragment. Because HpaII cuts the fragment only once, the HpaII site can be located unequivocally; mark its position 1.6 kb from one end (either end) of the fragment. EcoRI cuts the fragment twice. If you position an EcoRI site 1.7 kb from either end of the fragment, you will find that its distance from the HpaII site cannot be reconciled with the sizes of bands from the double digest. Therefore, the 1.7-kb band must come from the middle of the fragment and the two EcoRI sites must be 0.4 kb and 0.9 kb from the two ends. Only the arrangement shown in Figure 7–29 is consistent with the fragment sizes from the double digest.

7–9

The lengths of the restriction fragments do not sum to 21 kb and the electrophoretic patterns change with denaturation and reannealing because the ribosomal minichromosome is an inverted dimer, or a palindrome (Figure 7–30). That is to say, the sequences on the left half of the minichromosome are repeated in the opposite orientation on the right half. Thus, a restriction

```
K    I    G    P    A    C    F
AAA  ATT  GGT  CCT  GCT  TGT  TTT
G    C    C    C    C    C    C
          A    A    A    A
          G         G    G
```

Sau96I cleavage site | potential AluI cleavage site

Figure 7–28 Representation of the nucleotide sequences that could encode the peptide KIGPACF (Answer 7–7). Only the DNA strand that corresponds to the mRNA sequence is shown.

Figure 7–29 Restriction map of the 3.0-kb BamHI fragment (Answer 7–8). Numbers indicate fragment sizes in kilobases.

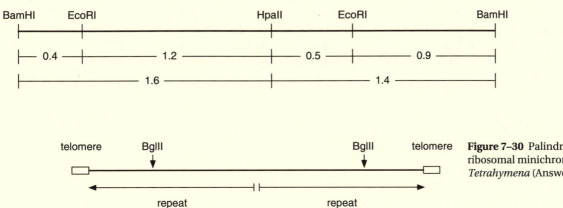

Figure 7–30 Palindromic structure of the ribosomal minichromosome from *Tetrahymena* (Answer 7–9).

```
         Y   K   L   D   N   Q   F   E   L   V   F   V   V   G   F   Q   K   I   L   T
GTATAAACTGGACAACCAGTTCGAGCTGGTGTTCGTGGTCGGTTTTCAGAAGATCCTAAC
  3                                                              2

     L   T   Y   V   D   K   L   I   D   D   V   H   R   L   F   R   D   K   Y
GCTGACGTACGTAGACAAGTTGATAGATGATGTGCATCGGCTGTTTCGAGACAAGTAC
  2         2         2   2   3
```

Figure 7–31 Nucleotide and amino acid sequence of a cloned gene (Answer 7–11). Stop codons are underlined. The amino acids encoded in reading frame 1 are centered over their codons.

enzyme cuts each arm (that is, each repeated segment) at sites that are the same distance from the ends, thereby generating one central fragment and two identical copies of each terminal fragment. For BglII digestion, the central fragment is 13.4 kb and the two terminal fragments are each 3.8 kb. The fragments on the gel did not add up to 21 kb because the terminal fragment should have been counted twice: 2 times 3.8 kb plus 13.4 kb does sum to 21 kb.

Denaturing and reannealing the minichromosome or its digestion products alters the electrophoretic pattern because the left and right halves of the single strands derived from the central fragment are complementary and, therefore, can reanneal internally. Because the two complementary halves are part of the same molecule, they reanneal internally very much faster than they do with other molecules. Self-reannealing reduces the apparent size of the fragment by a factor of 2. Thus, the uncut minichromosome, when denatured and reannealed, is reduced in size from 21 kb to 10.5 kb. Similarly, the 13.4-kb central BglII fragment is reduced to 6.7 kb. Of course, the single strands from the terminal fragments are not self-complementary and so can reanneal only with other complementary single strands to re-form their usual double-stranded structure.

Most unexpectedly, the ribosomal genes of *Tetrahymena* exist in a micronuclear chromosome as one-half of the inverted repeat that makes up the macronuclear minichromosome. Since the micronucleus generates a new macronucleus after conjugation, the ribosomal genes must be cleaved out of the chromosome in a way that specifically generates an inverted dimer. You might try to imagine how this could be accomplished before looking up the speculation by scientists involved in some of these studies (Yao, M-C.; Zhu, S-G.; Yao, C-H. *Mol. Cell. Biol.* 5:1260, 1985).

Reference: Karrer, K.M.; Gall, J.G. The macronuclear ribosomal DNA of *Tetrahymena pyriformis* is a palindrome. *J. Mol. Biol.* 104:421–453, 1976.

*7–10

7–11

The sequence of the cloned DNA and the amino acids encoded by the open reading frames are shown in Figure 7–31. Stop codons are underlined and labeled 2 or 3 to indicate their reading frame. As you can appreciate from this exercise, the sequencing gel must be read very carefully; omission of a single nucleotide would have disastrous consequences for determining the open reading frame. To minimize this problem, it is best to determine the sequence of both strands of DNA. Since the two strands are complementary, any mistakes will be readily apparent.

Nucleic Acid Hybridization

7–12

A. DNA renaturation (DNA hybridization)
B. Northern blotting
C. Southern blotting

D. restriction fragment length polymorphisms (RFLPs)
E. *in situ* hybridization

7–13

A. True
B. False. If the DNA:RNA hybrid is treated with appropriate nucleases, the protected portions of the DNA probe give information about transcription start and stop sites as well as the positions of introns.
C. True
D. True
E. False. Genetic markers that have only a 50-50 chance of being inherited together are genetically unlinked. Markers that are on different chromosomes are unlinked, but markers that are far apart on the same chromosome may also be unlinked because of the high probability that they will be separated during the frequent crossing-over that occurs during meiosis in the development of eggs and sperm.
F. True
G. False. Although members of a gene family are close relatives, they do not have identical sequences. For this reason hybridization must be carried out at reduced stringency.
H. True

7–14

A. Gene A is expressed only from active X chromosomes, as indicated by equal expression in males and females and by the pattern of expression in the hybrid cells. Gene B is expressed from both active and inactive X chromosomes, as indicated by the results with the hybrid cell lines and by the levels of expression in the other cells, which correlates with the total number of X chromosomes. Gene C is expressed only from inactive X chromosomes, as indicated by its expression in females but not males and by the pattern of expression in the hybrid cell lines.

The most common pattern of expression is like that of gene A; the vast majority of genes on the inactive X chromosome are turned off. A few genes like gene B are expressed from both active and inactive X chromosomes. The pattern indicated for gene C is very surprising. Only one gene, called XIST (for X_i-specific transcripts), is now known with this expression pattern. Remarkably, the gene is located near the X-inactivation center, and it may play a direct role in the inactivation process.

B. The rule for X inactivation is that only one X chromosome remains active—all the rest are inactivated. This is apparent in the Northern analysis in Figure 7–10. Gene A, which is expressed from the active X, is expressed at uniform levels regardless of the number of X chromosomes, indicating that only one X remains active. By contrast, gene C, which is expressed only from the inactive X, is expressed in all cells that have more than one X chromosome at levels that depend on the number of X chromosomes.

These rules for X inactivation were worked out from cytological observations long before the advent of molecular biology techniques. The inactive X chromosome is highly condensed and easily visualized as a distinct entity in the nucleus—the so-called Barr body. It was noted early on that female cells have one Barr body and male cells have none. Abnormal individuals with extra X chromosomes have a number of Barr bodies equal to the number of X chromosomes minus one.

Reference: Brown, C.J.; Ballabio, A.; Rupert, J.L.; Lafreniere, R.G.; Grompe, M.; Tonlorenzi, R.; Willard, H.F. A gene from the region of the human X inactivation centre is expressed exclusively from the inactive X chromosome. *Nature* 349:38–44, 1993.

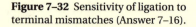

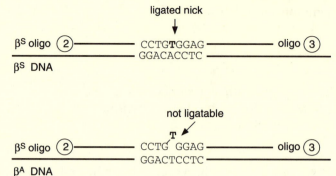

*7–15

7–16

A. Both oligonucleotides will hybridize to β^A and β^S DNA. The β^A oligo, for example, is a perfect (20 of 20) match for β^A DNA and a 19-of-20 match for β^S DNA. The situation is equivalent for the β^S oligo. The difference in hybridization between a 19-of-20 match and a 20-of-20 match is difficult to detect, except under very carefully chosen hybridization conditions, especially when the mismatch is at one end of the oligonucleotide. Under normal hybridization conditions both oligonucleotides hybridize equally well to β^A and β^S DNA.

B. Although hybridization itself is not affected by the mismatch, the reaction catalyzed by DNA ligase is exquisitely sensitive to it. As shown in Figure 7–32 for the β^S oligo, ligation can occur only if the bases on both sides of the nick are properly paired with the complementary bases in the target DNA. Ligation is essential if the radioactive oligo is to be linked to the biotin-labeled oligo and bound to the solid support where it can expose the x-ray film.

Reference: Landegren, U.; Kaiser, R.; Sanders, J.; Hood, L. A ligase-mediated gene detection technique. *Science* 241:1077–1080, 1988.

*7–17 Reference: Gelehrter, T.D.; Collins, F.S. Principles of Medical Genetics, p. 80. Baltimore: Williams & Wilkins, 1990.

7–18

The restriction pattern which indicates that the fetus is likely to have inherited the genetic disease will have bands at 1, 2, and 9 kb only.

Most of the difficulty in sorting out these restriction patterns arises because each individual is diploid and, therefore, carries two copies of the genetic locus. It is easiest to begin to define the patterns of single chromosomes by identifying individuals who are homozygous for one pattern. Homozygous individuals can be identified in two ways: they have patterns composed of equal intensity bands and the sum of the lengths of the bands is 12 kb (the size of the region being analyzed—Figure 7–15). By those criteria the maternal grandmother and the paternal grandfather are homozygous. By necessity they must have contributed a chromosome with their characteristic patterns to their progeny. Therefore, the mother must have one chromosome with a pattern identical to her mother's, and the father must have one chromosome with a pattern identical to his father's. Knowledge of one chromosome makes it simpler to deduce the pattern of cutting on the other chromosome.

Within the family tree you have analyzed, there are three different patterns of cutting: Pattern I is + at sites A, B, and C, which leads to bands at 1, 2, 4, and 5 kb; Pattern II is + at sites A and C, but – at site B, which leads to bands at 1, 5, and 6 kb; and Pattern III is + at sites A and B, but – at site C, which leads to bands at 1, 2, and 9 kb. Since Pattern I is homozygous in the maternal grandmother and Pattern II is homozygous in the paternal grandfather, neither of these patterns is likely to be associated with the genetic disease. The father is heterozygous for Patterns I and III, whereas the mother is heterozygous for

Patterns II and III. The unaffected child is heterozygous for Patterns I and II. Only the homozygous Pattern III is not represented among living members of the family. Also, Pattern III is present in the DNA of the paternal grandmother, whose brother was affected by the disease. Collectively, these observations indicate that Pattern III (bands at 1, 2, and 9 kb) is most likely the one associated with the genetic disease.

*7–19

DNA Cloning

7–20

    A. DNA cloning
    B. plasmid vectors
    C. genomic DNA clone, genomic DNA library
    D. cDNA clone, cDNA library
    E. subtractive hybridization
    F. chromosome walking
    G. polymerase chain reaction (PCR)

7–21

    A. True
    B. False. A cDNA library will contain only those genes that were expressed in the cell type used to make the library. In addition, each expressed gene will be represented at different levels in the library, depending on its level of expression.
    C. False. cDNA clones contain the uninterrupted coding sequence of a gene, not the complete sequence. During production of mRNA molecules, the intron sequences are removed.
    D. True
    E. True
    F. True
    G. False. Chromosome walks are generally too long for complete DNA sequencing to be practicable. The gene of interest is usually identified by transgenic experiments or by a variety of other criteria, especially in the case of human genes.
    H. True
    I. False. While the approach is straightforward, it presently requires 10 to 100 person-years to isolate a human gene this way.
    J. True

7–22

    A. The KpnI-BamHI junction and its oligonucleotide splint are shown in Figure 7–33A. The oligonucleotide splint (5′-GATCGTAC) is identical to the one in Figure 7–16. Thus one splint works for both junctions.

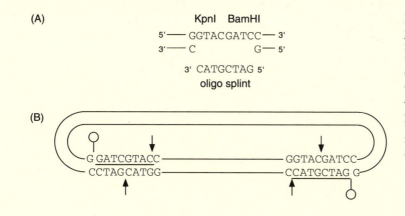

Figure 7–33 Scheme to splint BamHI-KpnI junctions (Answer 7–22). (A) Oligo splint required for KpnI-BamHI junction. It has the same sequence as the first oligonucleotide. (B) Splint-mediated ligation of KpnI-cut fragment into a BamHI-cut vector. Arrows indicate nicks that can be ligated, and open circles indicate nicks that cannot be ligated without further enzymatic repair. The two oligonucleotide sequences are underlined.

B. Treatment of the mixture with DNA ligase produces the recombinant molecule illustrated in Figure 7–33. As indicated by the arrows in the figure, two of the three nicks at each BamHI-KpnI junction will be ligated. This is sufficient to link the KpnI-cut fragment to the BamHI-cut vector.

C. Your friend's scheme works fine, in theory and in practice. The remaining nicks are repaired when the DNA is transformed into cells. Note that it is possible to add a phosphate to the 5′ end of the oligonucleotide using the enzyme polynucleotide kinase and ATP. Such a treated oligo would allow all nicks to be sealed by DNA ligase.

7–23

A. The ratio of vector molecules to cDNA molecules is about 1 to 2 in the ligation mixture. The vector is about 40 times longer than the cDNA (43 kb vs. 1 kb), but there is only 20 times as much vector as cDNA by weight (2 µg vs. 0.1 µg). Thus there are half as many vector molecules as cDNA molecules in the ligation mixture.

B. The number of vector molecules in 2 µg is 4.2×10^{10}.

$$\text{molecules} = \frac{1\,\text{molecule}}{43\,\text{kb}} \times \frac{\text{kb}}{1000\,\text{bp}} \times \frac{\text{bp}}{660\,\text{d}} \times \frac{6 \times 10^{17}\,\text{d}}{\mu\text{g}} \times 2\,\mu\text{g}$$
$$= 4.2 \times 10^{10}$$

Since this number of vector molecules gave 4×10^7 recombinants, the efficiency of generating recombinants is about 0.1%.

$$\text{efficiency} = \frac{4 \times 10^7 \text{ recombinants}}{4.2 \times 10^{10} \text{ vector molecules}} = 0.1\%$$

C. Incubation of the XhoI-cut vector DNA with dTTP and DNA polymerase adds a single T to each end.

```
5′-CT              TCGAG-3′
3′-GAGCT              TC-5′
```

The adaptor oligonucleotides have single-strand tails that are complementary to the modified vector ends.

```
5′-CT      CGAGATTTACC-3′
3′-GAGCT      CTAAATGG-5′
```

This allows the modified vector ends to be ligated to the adaptor-modified cDNAs.

After modification the vector ends are not complementary to each other and thus cannot be ligated together. The modified cDNAs are also not self-complementary and cannot be ligated together.

D. After modification of the vector and cDNAs, the vector molecules can ligate only to the cDNAs and vice versa. This greatly improves the efficiency of formation of recombinant molecules because it eliminates unproductive vector-to-vector ligations.

E. The ligation of the modified vector ends to the adaptors regenerates XhoI sites (CTCGAG) at both junctions of the cDNA with the vector. Thus XhoI will cut the cDNA out of the recombinant plasmid. Because XhoI has a six-base-pair recognition site, and thus cuts DNA relatively rarely, most cDNAs will be cut out intact.

Reference: Elledge, S.J.; Mulligan, J.T.; Ramer, S.W.; Spottswood, M.; Davis, R.W. λYES: a multifunctional cDNA expression vector for the isolation of genes by complementation of yeast and *Escherichia coli* mutations. *Proc. Natl. Acad. Sci. USA* 88:1731–1735, 1991.

*7–24

The appropriate pair of PCR primers are 5'-GACCTGTGGAAGC and 5'-CAATCCCGTATG. The first primer will hybridize to the bottom strand and prime synthesis in the rightward direction. The second primer will hybridize to the top strand and prime synthesis in the leftward direction.

The middle two primers in each list would not hybridize to either strand. The remaining pair of primers would hybridize, but would prime synthesis in the wrong direction—that is, outward, away from the central segment of DNA. Each of these wrong choices has been made at one time or another in most laboratories that use PCR. In most cases the confusion arises because the conventions for writing nucleotide sequences have been ignored. By convention, nucleotide sequences are written 5' to 3' so the 5' end is on the left. For double-stranded DNA the 5' end of the top strand is on the left.

*7–26 **Reference:** Chamberlain, J.S.; Gibbs, R.A.; Ranier, J.E.; Caskey, C.T. Multiplex PCR for the diagnosis of Duchenne muscular dystrophy. In PCR Protocols (M.A. Innis, D.H. Gelfand, J.J. Sninsky, T.J. White, eds.), pp. 272–281. San Diego, CA: Academic Press, 1990.

7–27

A. The consensus binding sequence for your transcription factor is 5'-ATGCCCATATATGG. As shown in Table 7–4, where the sequences are aligned according to the consensus binding site, the consensus is very good for these 14 nucleotides, but drops off dramatically in the flanking nucleotides.

B. A portion of the consensus sequence (5'-CCATATATGG) is palindromic. The double-stranded sequence has a twofold axis of symmetry between the middle two A-T base pairs. Thus this portion of the consensus is composed of two equivalent "half-sites," 5'-CCATA-3'. This feature of a binding site usually means that it is bound by a dimeric protein, in which each monomer recognizes one half-site. The four base pairs that are not part of the symmetric portion of the consensus sequence indicate that there is an asymmetry in the way the two monomers interact with the DNA.

C. In the 0.2-ng starting sample there are 4.8×10^9 oligonucleotides that are 76 nucleotides long.

Table 7–4 Sequences of Selected and Amplified DNAs Aligned According to the Consensus Sequence (Answer 7–27)

```
GAATTCGCCTCGAGCACATCATTGCCCATATATGGCACGACAGGATCC
    GAATTCGCCTCTTCTAATGCCCATATATGGACTTGCTCGACAGGATCC
  GGATCCTGTCGGTCCTTTATGCCCATATATGGTCATTGAGGCGAATTC
      GAATTCGCCTCATGCCCATATATGGCAATAGGTGTTTCGACAGGATCC
      GAATTCGCCTCTATGCCCATATAAGGCGCCACTACCCCGACAGGATCC
GAATTCGCCTCGTTCCCAGTATGCCCATATATGGACACGACAGGATCC
    GGATCCTGTCGACACCATGCCCATATTTGGTATGCTCGAGGCGAATTC
GAATTCGCCTCATTTATGAACATGCCCTTATAAGGACCGACAGGATCC
GAATTCGCCTCTAATACTGCAATGCCCAAATAAGGAGCGACAGGATCC
      GAATTCGCCTCATGCCCAAATATGGTCATCACCTACACGACAGGATCC
```

	A	T	G	C	C	C	A	T	A	T	A	T	G	G		
A	1	3	9	0	0	0	0	9	2	10	0	9	3	0	0	4	3
T	4	3	1	10	0	0	0	0	1	8	0	10	1	7	0	0	3	0
C	4	4	0	0	0	10	10	10	0	0	0	0	0	0	0	0	3	5
G	1	0	0	0	10	0	0	0	0	0	0	0	0	0	10	10	0	2

consensus: A T G C C C A T A T A T G G

$$\text{oligonucleotides} = 0.2 \text{ ng} \times \frac{\text{g}}{10^9 \text{ ng}} \times \frac{6 \times 10^{23} \text{d}}{\text{g}} \times \frac{\text{nucl.}}{330 \text{ d}} \times \frac{\text{oligo}}{76 \text{ nucl.}}$$
$$= 4.8 \times 10^9$$

D. Yes. The total number of different 14-mers is $4^{14} = 2.7 \times 10^8$, and there were 4.8×10^9 molecules in the starting sample. Moreover, each molecule (in its double-stranded form) contains 26 14-base-pair-long sequences in its central 26 random nucleotides. There are 13 14-bp-long sequences in each orientation of the duplex. Thus, in the starting sample each possible 14-bp sequence is likely to be represented several hundred times.

Reference: Pollock, R.; Treisman, R. A sensitive method for the determination of protein-DNA binding specificities. *Nucleic Acids Res.* 18:6197–6204, 1990.

*7–28   **Reference:** Ochman, H.; Medhora, M.M.; Garza, D.; Hartl, D.L. Amplification of flanking sequences by inverse PCR. In PCR Protocols (M.A. Innis, D.H. Gelfand, J.J. Sninsky, T.J. White, eds.), pp. 219–227. San Diego, CA: Academic Press, 1990.

DNA Engineering

7–29

A. genetics
B. reverse genetics
C. fusion proteins
D. epitope tagging
E. dominant-negative
F. antisense RNA
G. transgenic, transgenes

7–30

A. True
B. False. The best path to a pure RNA species is to clone a DNA segment that encodes the RNA next to the promoter for an efficient viral polymerase and transcribe the gene *in vitro*.
C. False. The usual method is to overexpress the protein in cells and purify it in conventional ways.
D. True
E. True
F. True
G. False. Some signals carried by proteins require that the protein fold correctly to bring distant portions of the polypeptide chain into close proximity. The "signal patch" that directs lysosomal proteins from the Golgi to lysosomes is one example.
H. True
I. False. Mutant proteins with dominant-negative effects generally function as part of larger protein complexes, which can be inactivated by inclusion of just one nonfunctional component.
J. True
K. True
L. True

7–31

A. If the cells alone give colonies, something is wrong. There are several possibilities. It could be that your fellow student mistakenly prepared competent cells from cells that already carried a cloning vector and were resistant to the antibiotic. Alternatively, the competent cells may have been contaminated with antibiotic-resistant cells. Finally, it could be that you forgot to put the antibiotic into the plates, or didn't use enough of it.

Thus the point of Control 1 is to make sure the competent cells and culture plates are working properly.

B. If the cells transfected with uncut vector give no colonies, something is wrong. Once again there are several possibilities. The cells may not have been prepared properly to make them competent to take up DNA; there could be too much antibiotic in the plates; or the plates may have been made up incorrectly so that they do not support bacterial growth.

Thus the point of Control 2, like Control 1, is to make sure that the cells and plates are working as they are supposed to.

C. Control 3 checks that the ends of the cut vector are as expected and that the DNA ligase is active. Occasionally a preparation of restriction enzyme (or the buffers used) are contaminated with exonucleases that modify the ends, making them unligatable. Alternatively, if the buffer is improperly constituted, it may prevent DNA ligase from working.

Control 4 checks that the alkaline phosphatase is active.

D. Treating the vector with alkaline phosphatase removes the 5′ phosphates, which prevents the vector from religating to itself. Religation of the vector can give a high background of colonies that carry vector but no insert. Since the cDNA is not treated with phosphatase, it retains its 5′ phosphates; thus the cDNA can still be ligated to the vector. Thus the cloning manual recommends phosphatase treatment of the vector to lower the number of colonies that contain only the vector, thereby increasing the proportion that carry recombinant clones.

*7–32

7–33

The pairs of PCR primers for adding a stretch of six histidines to either the N-terminus or C-terminus of your protein are illustrated in Figure 7–34 (next page). In both cases the left primer is complementary to the bottom strand of DNA (not shown) and primes synthesis in the rightward direction. The right primer is complementary to the top strand of DNA and primes synthesis in the leftward direction.

If you are set up to do a database search you might try to find out what protein is being modified in this problem.

*7–34

*7–35 **Reference:** Higuchi, R. Recombinant PCR. In PCR Protocols (M.A. Innis, D.H. Gelfand, J.J. Sninsky, T.J. White, eds.), pp. 177–183. San Diego, CA: Academic Press, 1990.

7–36

A. The truncated-tail phenotype induced by the NAF RNA is a striking confirmation of your hypothesis that Raf-1 is involved in posterior development in vertebrates. The reduction in frequency of the truncated-tail phenotype when normal *raf-1* RNA is coinjected with NAF RNA is an important control that strengthens the conclusion. Without this control, you would have to worry whether NAF protein had an effect, unrelated to Raf-1, that caused the truncated-tail phenotype. Reversal of the effect by bona fide Raf-1 eliminates that concern. The injection of water serves as control to check the deleterious effects of the injection process itself. Although this control shows that injection causes some abnormalities, their frequency is well below that seen when the various RNAs are injected.

B. In response to the intracellular signal triggered by FGF, some positive activator presumably binds to the regulatory domain of Raf-1 and stimulates the protein kinase activity. In the presence of NAF, which is in excess in these experiments, all of the positive activator is tied up in unproductive interactions with NAF. By soaking up the positive activator, NAF interferes with the normal activation of the Raf-1 protein in the embryos.

C. When *raf-1* RNA is injected in equal amounts with NAF RNA, the positive activator is distributed between Raf-1 and NAF, which are present in more nearly equal amounts. Thus, under these conditions some Raf-1 is activated, propagating the signal that leads to proper posterior development.

Reference: MacNicol, A.; Muslin, A.J.; Williams, L.T. Raf-1 kinase is essential for early *Xenopus* development and mediates the induction of mesoderm by FGF. *Cell* 73:571–583, 1993.

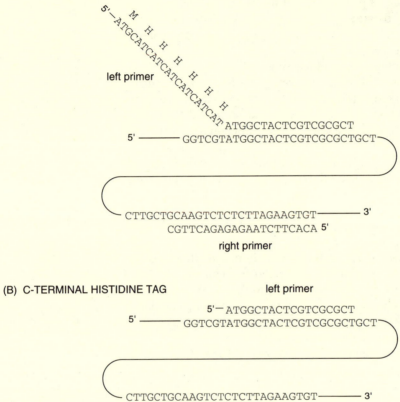

(A) N-TERMINAL HISTIDINE TAG

(B) C-TERMINAL HISTIDINE TAG

Figure 7–34 PCR primers to add hexa-histidine tags to your protein (Answer 7–33). (A) Primers to add histidines to N-terminus. (B) Primers to add histidines to C-terminus. In these primers only one codon for histidine is used, but the other histidine codon or a mixture could have been used. For the C-terminal histidine tag the complement of the histidine codon is used.

The Cell Nucleus

Chromosomal DNA and Its Packaging

8–1

A. chromosome, genome
B. DNA replication origin, centromere, telomeres
C. artificial chromosomes
D. gene
E. exons, introns
F. histones, nonhistone chromosomal, chromatin
G. nucleosomal, H1
H. nucleosome
I. histone octamer
J. nuclease-hypersensitive sites

8–2

A. True
B. False. Chromosomes are not replicated precisely and do lose nucleotides from the ends each time that they are replicated. The telomere solves this end-replication problem by confining the loss of nucleotides to the telomere itself. A special enzyme periodically extends the simple repeating sequence that constitutes the telomere, thereby *compensating* for the inevitable loss of nucleotides associated with replication of the linear chromosome.
C. True
D. False. Although it is true that introns are usually larger than exons, exons always outnumber introns by one in every gene.
E. True
F. True
G. True
H. True
I. False. Nuclease-hypersensitive sites are not found in the short stretches of linker DNA between nucleosomes but rather in long stretches of nucleosome-free DNA, which often define regulatory regions of genes.
J. False. In living cells nucleosomes are packed upon one another to generate regular arrays in which the DNA is more highly condensed, usually in the form of a 30-nm fiber.

*8–3   Reference: Szostak, J.W.; Blackburn, E.H. Cloning yeast telomeres on linear plasmid vectors. *Cell* 19:245–255, 1982.

8–4

A. The number of repeats in the terminal fragment must be quite variable, as indicated by the broad band that results from cleavage by AluI near the end of the minichromosome (observation 6). If the number of repeats were the same for every molecule, the band containing the terminal fragment would be sharp instead of broad. The difference between the leading and trailing edges of the band, which is about 160 nucleotide pairs, corresponds to about 27 repeats, which represents the variability in the number of repeats per telomere.

*Problems with an asterisk (*) are answered in the Instructor's Manual.

It is more difficult to estimate the average number of repeats per telomere from these data because the location of the AluI cleavage site relative to the beginning of the repeats is not known. However, if AluI cut exactly at the start of the repeats, then the number of repeats would vary between 60 (360/6) and 87 (520/6). It is thought that there are about 70 repeats per telomere.

B. The CCCCAA repeats are at the 5′ ends of the individual strands of the minichromosome. Since DNA synthesis that begins in the tandemly repeated DNA moves progressively toward the center of the minichromosome (observation 7), the repeats must be at the 5′ ends. If the repeats were at the 3′ ends, DNA synthesis would proceed toward the ends of the minichromosome.

C. The single-strand interruptions in the CCCCAA repeats are, for the most part, single nucleotide gaps. Since denaturation releases single-stranded fragments (observation 4), there must be breaks in the phosphodiester backbone. These discontinuities presumably have a 3′-OH since they serve as sites for DNA synthesis (observations 1, 3, and 7); however, they cannot be simple breaks with a 5′-PO_4^{2-} and a 3′-OH, since treatment with DNA ligase does not reduce incorporation by DNA polymerase (observation 2). The nature of the single-strand interruptions is revealed by replacing free 5′ phosphates with labeled phosphates and then breaking all bonds to purine nucleotides, which yields predominantly the labeled fragment CCC (observation 5). This result indicates that most of the discontinuities are single-nucleotide gaps, where the first C (5′ C) of the repeat is missing. (If the 3′ C were missing [CCC__AA], then the label at the 5′ phosphate would have been added to an A and the CCC fragment would not have been labeled.) The gap is unlikely to be longer than one nucleotide because DNA polymerase can incorporate labeled C in the absence of other nucleotides (observation 1). (If the gap were two nucleotides long, for example, then the first nucleotide incorporated by DNA polymerase would have to be an A.)

D. Since denaturation of the telomere releases single-stranded fragments corresponding to 2, 3, and 4 CCCCAA repeats (observation 4), the single-strand interruptions must be present on average about once every two to four repeats.

E. These data suggest that the structure of the telomere on the ribosomal minichromosome is as shown in Figure 8–36. This model encompasses all the information in the experimental observations.

   Reference: Blackburn, E.H.; Gall, J.G. A tandemly repeated sequence at the termini of the extrachromosomal ribosomal RNA genes in *Tetrahymena*. *J. Mol. Biol.* 120:33–53, 1978.

*8–5   **References:** Konkel, D.A.; Maizel, J.V.; Leder, P. The evolution and sequence comparison of two recently diverged mouse chromosomal β-globin genes. *Cell* 18:865–873, 1979.

Staden, R. An interactive graphics program for comparing and aligning nucleic acid and amino acid sequences. *Nucleic Acids Res.* 10:2951–2961, 1982.

*8–6   **Reference:** Prunell, A.; Kornberg, R.D.; Lutter, L.; Klug, A.; Levitt, M.; Crick, F.H. Periodicity of deoxyribonuclease I digestion of chromatin. *Science* 204:855–858, 1979.

*8–7

8–8

A. Micrococcal nuclease generates fragments whose lengths vary depending on the spacing of the internucleosomal cleavages that define the ends of the

5′ C-C-C-C-A-A $_{OH}$ C-C-C-A-A⌐C-C-C-C-A-A⌐C-C-C-C-A-A $_{OH}$ C-C-C-A-A 3′
3′ G-G-G-G-T-T-G-G-G-G-T-T⌊G-G-G-G-T-T⌋G-G-G-G-T-T-G-G-G-G-T-T 5′
                                    0–2

Figure 8–36 A portion of the terminal repeated sequence of the ribosomal minichromosome, showing positions of the specific one-nucleotide gaps (Answer 8–4).

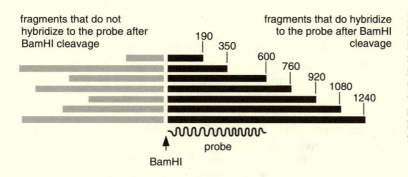

Figure 8–37 Diagram relating indirect end labeling to the fragment lengths observed in Figure 8–7 (Answer 8–8). Sites of micrococcal-nuclease cleavage are indicated by vertical lines. The small gap in the micrococcal-nuclease fragments shows the position of BamHI cleavage. Numbers refer to the lengths of the fragments that hybridize to the probe; they correspond to the lengths shown in Figure 8–7.

fragments (Figure 8–37). Since micrococcal nuclease does not cleave at precise sites within the linker DNA, there is some variability in the lengths of fragments produced by cleavage even between the same two pairs of nucleosomes. Furthermore, similar-size fragments can be produced by cleavage between several different pairs of internucleosomal sites. These sorts of variability obscure the fine-structure details of the ordering of adjacent nucleosomes.

Digestion with BamHI sharpens the pattern of bands because it precisely defines one end of each DNA fragment. As shown in Figure 8–37, only the fragments to the right of the BamHI-cleavage site hybridize to the radioactive probe. The resulting pattern is easy to interpret because the length of each fragment gives the distance from the nuclease-cleavage site to the BamHI site directly. In the absence of BamHI cleavage the bands are defined by micrococcal-nuclease cleavage at both ends. Such a pattern does not allow one to deduce the exact sites of nuclease cleavage relative to the probe, since nuclease cleavage at several pairs of sites can yield the same-size fragment. The method for mapping nuclease cut sites that is illustrated in this problem is called indirect end labeling because a defined end (the BamHI-cleavage site) is labeled indirectly through hybridization to a radiolabeled probe.

B. The sizes of the bands indicate the distances between the nuclease-cut sites and the BamHI-cleavage site (Figure 8–37). Since micrococcal nuclease cleaves between nucleosomes, the cut sites define the positions of the nucleosomes relative to the BamHI site (Figure 8–38). With the exception of the region around the centromere the cut sites are spaced at 160-nucleotide intervals, suggesting that the nucleosomes occupy about 160 nucleotides of DNA. However, the cut sites on either side of the centromere are 250 nucleotides apart, suggesting that some special (nonnucleosomal) structure covers the centromere. It is thought that centromere-specific proteins bind to and protect the centromere from nuclease digestion. The cleavage sites on either side of the centromere indicate that there is unprotected DNA (analogous to the linker DNA between nucleosomes) between the centromere and the adjacent nucleosomes on either side.

C. The naked DNA control is important because all DNA sequences are not equally susceptible to micrococcal-nuclease cleavage. It is essential to know the susceptibility of the specific DNA sequence under investigation. Otherwise, one can be fooled into thinking that a specific band results from the binding of a protein adjacent to the cleavage site, when it actually derives from the cleavage specificity of the nuclease. Indeed, the centromere itself is a preferred site of cleavage (although that was left out of the naked DNA digestion shown

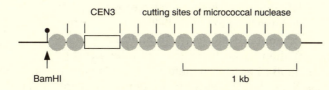

Figure 8–38 Positions of micrococcal-nuclease-cleavage sites and arrangement of nucleosomes around the centromere (Answer 8–8). Nucleosomes are shown as shaded circles.

in Figure 8–7); the absence of cutting at this sensitive site in chromatin is all the more evidence that it is specifically protected.

D. Your results answer this question very elegantly. The band patterns from the three plasmids are the key result. If the nucleosomes were ordered simply because they were lined up next to the special structure at the centromere, then it should not make any difference what DNA sequence was present beyond the centromere. Your results with plasmids 2 and 3, however, show clearly that the ordered arrangement disappears at the point where the bacterial sequences (plasmid 2) or the noncentromeric yeast sequences (plasmid 3) begin. This result argues strongly that the regular ordering of nucleosomes around the centromere is due to some feature of the sequence of the neighboring DNA itself.

Reference: Bloom, K.S.; Carbon, J. Yeast centromere DNA is in a unique and highly ordered structure in chromosomes and small circular minichromosomes. *Cell* 29:305–317, 1982.

The Global Structure of Chromosomes

8–9

A. lampbrush chromosomes
B. polytene chromosomes
C. chromosome puffs
D. active
E. heterochromatin, euchromatin
F. mitotic chromosomes
G. karyotype
H. G bands, R bands

8–10

A. False. Most of the DNA in lampbrush chromosomes is in the chromomeres, not in the loops.
B. False. Side-by-side adherence of chromatin strands gives rise to polytene chromosomes. Ploidy refers to the number of copies of the whole genome.
C. True
D. False. Classical genetic studies suggested that each band probably corresponded to a gene; however, more recent molecular data indicate that there are three times as many distinct mRNAs as bands. Thus, the "one-band, one-gene" hypothesis seems unlikely.
E. True
F. True
G. False. Although transcriptionally inactive chromatin is condensed into a relatively nuclease-resistant form, only about 10% to 20% of it is packed into the highly condensed conformation known as heterochromatin.
H. True
I. False. The bands are unrelated. Bands in insect polytene chromosomes correspond to regions of condensed chromatin; bands in mitotic chromosomes are formed by the selective staining of specific dyes.

8–11 The uniform pattern of labeling observed with the smaller chromatin loops is the expected pattern. At first, it might seem that, if transcription proceeds from one end of a loop to the other, the labeling pattern should also progress from one end to the other as it does for the large loop in Figure 8–8. However, since transcription occurs throughout a chromatin loop (by multiple RNA polymerases) as illustrated in Figure 8–9, then ^{3}H-uridine should also be incorporated throughout the length of the loop as each polymerase molecule adds a uridine nucleotide to the growing RNA chain (Figure 8–39). With increasing

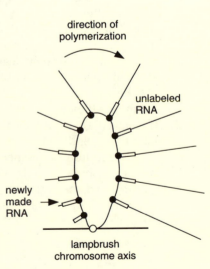

direction of polymerization

unlabeled RNA

newly made RNA

lampbrush chromosome axis

Figure 8–39 Uniform incorporation of radioactive uridine into a chromatin loop (Answer 8–11). Newly synthesized RNA is indicated by open rectangles adjacent to RNA polymerase molecules, which are represented by filled circles.

time more label will be incorporated; thus the intensity of labeling over the entire loop is expected to increase.

What then is the explanation for the labeling pattern observed in the large chromatin loops? The answer is not yet known. Because the loops do not label until after a day, it is thought that these loops may not be transcribed at all. Instead, they may be storage sites for RNA that is synthesized elsewhere—with an entry site into the loop at one end.

References: Callan, H.G. The nature of lampbrush chromosomes. *Int. Rev. Cytol.* 15:1–34, 1963.

Gall, J.G. Personal communication.

*8–12 **Reference:** Spierer, A.; Spierer, P. Similar levels of polyteny in bands and interbands of *Drosophila* giant chromosomes. *Nature* 307:176–178, 1984.

8–13

A. DNase I preferentially digests active chromatin. Red cells express globin and treatment of red cell nuclei with DNase I reduces the ability of the DNA to protect globin cDNA. Thus, the chromatin from which globin RNA is transcribed is preferentially degraded by DNase I. By contrast, fibroblasts do not express globin, and treatment of fibroblast nuclei with DNase I does not reduce the ability of the DNA to protect globin cDNA. Thus, in fibroblasts the globin genes are no more sensitive than the bulk of the chromatin.

B. DNase I digestion of red cell nuclei preferentially degrades regions of active chromatin. A comparison of the extents of protection of ^{3}H-labeled red cell DNA by total DNA (95%) and by DNase-I-digested DNA (78%) suggests that about 17% (95%–78%) of red cell DNA is sensitive to DNase I—and by that criterion, in active chromatin.

C. Trypsin treatment of nucleosome monomers affects a specific population of monomers—namely, those monomers that were present in active chromatin. This conclusion comes from a comparison of trypsin-treated monomers with DNase-I-treated red cell DNA. Digestion of trypsin-treated nucleosome monomers with micrococcal nuclease yields DNA that protects globin cDNA and red cell DNA (25% and 83%, respectively) to the same extents as DNase-I-treated red cell DNA (25% and 78%, respectively) as shown in Table 8–1.

(If a random population of nucleosomes were affected, all the DNA sequences present in the untreated nucleosomes would still be present in the trypsin-treated nucleosomes. If that were the case, both the untreated and treated monomers would behave identically in their capacity to protect globin cDNA and total red cell DNA.)

D. Since the DNA in individual nucleosome monomers shows the same sensitivity to DNase I as chromatin in nuclei, the property of active chromatin that distinguishes it from bulk chromatin must be present in individual nucleosomes. This viewpoint is supported further by the observation that trypsin treatment of nucleosome monomers renders those from active chromatin sensitive to micrococcal nuclease. Individual monomers from regions of active chromatin must be physically distinct from other nucleosome monomers.

Reference: Weintraub, H.; Groudine, M. Chromosomal subunits in active genes have an altered conformation. *Science* 193:848–856, 1976.

Chromosome Replication

8–14

A. S phase
B. T-antigen, replication bubble
C. replication units
D. telomeres, telomerase

A. True

B. True

C. False. Replication forks in eucaryotic cells travel at about one-tenth the rate they do in bacteria. This slower rate may reflect the difficulty of replicating DNA packaged into chromatin.

D. True

E. False. The active X chromosome is replicated throughout the S phase, whereas the inactive X chromosome, which is completely condensed into heterochromatin, is replicated late in S phase.

F. True

G. True

H. False. In such fusion experiments G_1-phase nuclei, not G_2-phase nuclei, are stimulated to synthesize DNA. Replication origins in G_2-phase nuclei have a block to DNA re-replication, which was acquired as they passed through S phase.

I. False. Unlike most proteins, which are made continuously throughout interphase, histones are synthesized mainly in the S phase, during which the level of histone mRNA increases about 50-fold as a result of both increased transcription and decreased mRNA degradation.

J. True

8–16

A. Hybridization at the 4.5-kb position is due to plasmid molecules that were not replicating at the time DNA was isolated. The intensity of this spot indicates that the majority of plasmid molecules were not replicating. The low frequency of replicating plasmid molecules, even during S phase, has been one of the contributing factors in the difficulty of proving that an ARS is an origin.

B. The results in Figure 8–13 indicate that ARS1 behaves as an origin of replication. The gel pattern with BglII-digested DNA looks like the pattern due to replicating molecules with two branches (Figure 8–12C). The gel pattern with PvuI-digested DNA looks like the pattern due to replication intermediates with asymmetrically located replication bubbles (Figure 8–12D). (The very short tail on the spot at 9 kb in Figure 8–13B indicates that the replication bubbles are only slightly asymmetrically situated in the replicating molecules.) These gel patterns are exactly what would be expected if replication began at ARS1. As shown in Figure 8–40, cleavage with BglII, which cuts at ARS1, would generate molecules with two branches; cleavage with PvuI, which cuts almost half way around the circle from ARS1, generates molecules with slightly asymmetrically located replication bubbles.

C. The discontinuity in the arc of hybridization of PvuI-cut plasmids (Figure 8–13B) results from the difference in migration of bubble forms and branched forms. Molecules that have just begun replicating will be converted to bubble forms by PvuI cleavage, whereas molecules in which replication has proceeded beyond the PvuI site will be converted to branched forms. Thus, a replicating molecule that is cleaved either has a bubble or it is branched—there is no intermediate. Since the two forms migrate differently, there is a gap in the electrophoretic pattern.

   Reference: Brewer, B.J.; Fangman, W.L. The localization of replication origins on ARS plasmids in *S. cerevisiae. Cell* 51:463–471, 1987.

8–17

A. In addition to its site-specific DNA binding, T-antigen possesses an ATP-dependent helicase activity. A helicase activity is necessary for unwinding DNA (see MBOC Chapter 6). The presence of T-antigen at the forks is also consistent with its activity as a helicase.

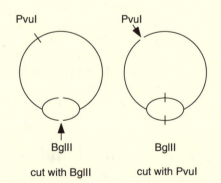

cut with BglII   cut with PvuI

Figure 8–40 Conversion of replicating plasmid molecules into linear forms with two branches by BglII or linear forms with slightly asymmetrically located replication bubbles by PvuI (Answer 8–16).

The ability of T-antigen to bind specifically to SV40 origins of replication and unwind them is a natural first step in DNA replication. These activities expose single-stranded regions so that primases and DNA polymerases can gain access to the DNA. However, T-antigen probably also guides the entry of one or more of these components: antibodies to T-antigen immunoprecipitate a DNA polymerase-primase complex, and T-antigen stimulates replication at SV40 origins only in certain cell types. Both these observations suggest that initiation depends on specific interactions between T-antigen and one or more components of the actual replication machinery.

B. To demonstrate that the unwound regions are at the origin, you can digest the DNA with a restriction enzyme that cuts at a defined location relative to the origin. (Since the two ends of a linear molecule cannot be distinguished in the electron microscope, at least two different restriction digestions are required to position unwound regions unambiguously.) If unwinding does not occur at the origin, the unwound region will not include the origin (Figure 8–41A). If unwinding occurs at the origin, the unwound region will include the origin. If unwinding occurs in one direction, one end of the unwound region will coincide with the origin (Figure 8–41B). If unwinding occurs in both directions (at the same rate), the center of the unwound region will coincide with the origin (Figure 8–41C). The actual experimental results indicate that unwinding occurs in both directions at approximately the same rate.

C. Topoisomerase I is required for unwinding closed circular DNA molecules in order to relieve overwinding strain in the duplex part of the molecule. In a covalently closed circle the removal of duplex winding in one segment causes the rest of the duplex to become overwound, which is energetically unfavorable. The transient nicking-closing activity of topoisomerase I allows this winding tension to be relieved.

D. If T-antigen initiated replication multiple times from an SV40 origin integrated in a chromosome, a complex multistranded structure would be produced. Three rounds of initiation at a chromosomal origin are illustrated in Figure 8–42.

Apparently, a chromosomal SV40 origin in the presence of T-antigen replicates in a manner similar to that shown in Figure 8–42. The process is termed "onion-skin" replication because the resulting structure has a layered appearance reminiscent of an onion.

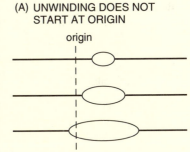

(A) UNWINDING DOES NOT START AT ORIGIN

origin

(B) UNIDIRECTIONAL UNWINDING

origin

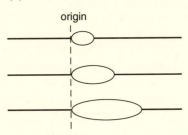

(B) BIDIRECTIONAL UNWINDING

origin

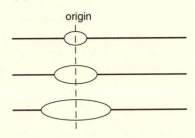

Figure 8–41 The use of restriction digestion to position the site of unwinding by T-antigen relative to the SV40 origin of replication (Answer 8–17). The ends of the linear molecule were generated by digestion with a restriction enzyme.

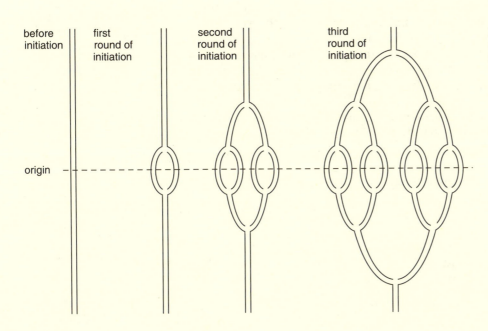

before initiation | first round of initiation | second round of initiation | third round of initiation

origin

Figure 8–42 Multiple initiation events at a chromosomal SV40 origin of replication (Answer 8–17).

Reference: Dodson, M.; Dean, F.B.; Bullock, P.; Echols, H.; Hurwitz, J. Unwinding of duplex DNA from the SV40 origin of replication by T-antigen. *Science* 238:964–967, 1987.

*8–18   **Reference:** Huberman, J.A.; Riggs, A.D. On the mechanism of DNA replication in mammalian chromosomes. *J. Mol. Biol.* 32:327–341, 1968.

*8–19   **Reference:** Orr-Weaver, T.L.; Spradling, A.D. *Drosophila* chorion gene amplification requires an upstream region regulating s18 transcription. *Mol. Cell. Biol.* 6:4624–4633, 1986.

8–20

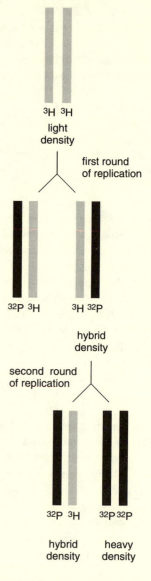

A. The three density peaks represent, from light to heavy, unreplicated DNA, once replicated DNA, and twice replicated DNA (Figure 8–43). The injected DNA is labeled with ^{3}H but is otherwise normal DNA, which is "light." Each newly synthesized strand incorporates ^{32}P label and BrdU, which increases its density. Thus, after one round of replication, the DNA will be intermediate in density, containing one "light" ^{3}H-labeled strand and one "heavy" ^{32}P-labeled strand. After the second round of replication, the hybrid DNA will give rise to one hybrid density duplex and to one duplex that contains two ^{32}P-labeled "heavy" strands. The fully "heavy" duplex will appear at the densest position in the gradient. Since the fully "light" DNA is unreplicated, it will contain no ^{32}P label; since the fully "heavy" DNA contains two new strands, it will contain no ^{3}H label.

The formation of discrete peaks in this experiment makes an important point: most of the observed labeling is due to replication and not to repair synthesis. If the incorporation of label were due to repair synthesis, which is patchy, the label would be smeared through the gradient rather than concentrated in discrete peaks.

B. The injected DNA mimics the expected behavior of chromosomal DNA in one very important way: it undergoes only a single round of replication in one cell cycle. This behavior is apparent from the lack of fully "heavy" DNA after one cell cycle. In another way, however, the injected DNA behaves very differently from chromosomal DNA: a large fraction of the injected DNA does not replicate even after two cell cycles. The lack of replication is apparent from the persistence of a fully "light" peak of DNA. It is not clear why some of the injected DNA does not replicate. Perhaps some of the eggs have been rendered incompetent for replication by the experimental protocol.

C. Since cycloheximide is an inhibitor of protein synthesis, it should have no direct effect on the synthesis of DNA. And, indeed, in the presence of cycloheximide one round of replication is completed normally, though no more occur thereafter. Cycloheximide apparently blocks further progress through the cell cycle because a key cell-cycle event depends on protein synthesis. The important point is that the DNA will not replicate again unless the cells progress through the cell cycle.

Reference: Harland, R.M.; Laskey, R.A. Regulated replication of DNA microinjected into eggs of *Xenopus laevis. Cell* 21:761–771, 1980.

Figure 8–43 Schematic representation of strand labeling (Answer 8–20).

RNA Synthesis and RNA Processing

8–21

A. DNA transcription
B. RNA polymerase, promoter
C. RNA polymerase II, RNA polymerase I, RNA polymerase III
D. heterogeneous nuclear RNA (hnRNA)
E. 5′ cap

F. poly-A tail

G. primary RNA transcript

H. RNA splicing

I. hnRNP (heterogeneous nuclear ribonucleoprotein)

J. snRNPs (small nuclear ribonucleoproteins)

K. 5′ splice site, 3′ splice site

L. spliceosome

M. thalassemia syndromes

N. nucleolus

O. nucleolar organizer

8–22

A. False. Initiation and elongation factors are not permanent subunits of the enzyme. They associate transiently with the enzyme during either the initiation or elongation phase of RNA synthesis.

B. False. RNA polymerases I, II, and III are structurally similar to one another and do have some common subunits, although other subunits are unique.

C. True

D. False. The 3′ end of most polymerase II transcripts is defined not by the termination point of transcription but by cleavage of the RNA chain 10 to 30 nucleotides downstream of the sequence AAUAAA.

E. True

F. True

G. False. Both hnRNP particles and snRNPs are composed of multiple polypeptide chains complexed with RNA. However, hnRNP particles contain mRNA, which is very unstable.

H. False. Although intron sequences are mostly dispensable, they must be removed precisely because an error of even one nucleotide would shift the reading frame in the resulting mRNA molecule and make nonsense of its message.

I. True

J. False. 28S rRNA, 18S rRNA, and 5.8S rRNA are transcribed as part of one large 45S precursor RNA, which then is cleaved to give one copy each of the three rRNA products. The derivation of these three rRNAs from the same primary transcript ensures that they will be made in equal quantities.

K. False. Ribosomal RNAs are packaged with ribosomal proteins in the nucleolus.

L. True

M. True

N. False. Interphase chromosomes tend to occupy discrete domains in interphase nuclei and are not thought to be extensively intertwined with other chromosomes.

O. True

8–23

A. Since RNA polymerase is blocked by pyrimidine dimers, the sensitivity of transcription of a gene will depend on the distance between the promoter and the probe. It is a simple function of the size of the target for UV damage. If the polymerase must go twice as far to make a transcript, the chances of encountering a block to transcription are twice as great.

B. Transcription through the VSG gene is seven times more sensitive to UV irradiation than transcription through the ribosomal transcription unit at the site of rRNA probe 4, which is about 7 kb from its promoter. Thus, the beginning of the VSG gene is located about 50 kb (7×7) away from its promoter. This calculation assumes that the DNA between the VSG promoter and the VSG gene has about the same sensitivity to UV light as the DNA in the ribosomal RNA transcription unit.

C. If the nearby gene is 20% less sensitive to UV irradiation than the VSG gene, it is inactivated at 80% the rate of the VSG gene. Therefore, its promoter is 40 kb away (0.80×50 kb). Given that the nearby gene is 10 kb in front of the VSG gene, its promoter must map very near the promoter for the VSG gene. Thus, it is likely that the two genes are transcribed from the same promoter.

Reference: Johnson, P.J.; Kooter, J.M.; Borst, P. Inactivation of transcription by UV irradiation of *T. brucei* provides evidence for a multicistronic transcription unit including a VSG gene. *Cell* 51:273–281, 1987.

*8–24 Reference: Berget, S.M.; Berk, A.J.; Harrison, T.; Sharp, P.A. Spliced segments at the 5′ termini of adenovirus-2 late mRNA: a role for heterogeneous nuclear RNA in mammalian cells. *Cold Spring Harbor Symp. Quant. Biol.* 42:523–529, 1977.

8–25

A. The 5′ ends of the RNA molecules were labeled. Only labeled fragments show up in the autoradiograph (Figure 8–21). Thus, if the shortest fragments (those that were run at the bottom of the gel) are from the 5′ end, the 5′ end must have been labeled.

B. The bands corresponding to the A's in the AAUAAA signal sequence are missing from the ladder of bands in polyadenylated and cleaved RNA (Figure 8–21, lanes 3 and 4) because modification of any one of those A's interferes with cleavage and polyadenylation. Thus, RNA molecules that carry a single modification in the signal sequence are not recognized by the components of the extract and, as a result, do not show up in the population of molecules that carry poly-A tails (lane 3) or in the population of molecules that are cleaved (lane 4).

C. The band at the arrow in Figure 8–21 is absent in the polyadenylated RNA but present in the cleaved RNA because modification of this A does not prevent cleavage, but it does prevent polyadenylation. Thus, RNA molecules with this A modified are present in the cleaved molecules (lane 4) but not present in the polyadenylated molecules (lane 3).

D. The analysis of the missing bands in parts B and C above indicates that the AAUAAA signal sequence is important for the cleavage of precursor RNAs and that the AAUAAA sequence and the single A are required for polyadenylation.

E. If the other end—the 3′ end—of the RNA molecules were labeled, it would have been possible to determine whether any of the A's or G's on the 3′ side of the cleavage site were important for polyadenylation. These experiments have been done; they show that no single modification 3′ of the polyadenylation site prevents polyadenylation.

Reference: Conway, L.; Wickens, M. Analysis of mRNA 3′-end formation by modification interference: the only modifications which prevent processing lie in AAUAAA and the poly (A) site. *EMBO J.* 6:4177–4184, 1987.

8–26

A. Your idea was to try to cleave the RNA component of the snRNP that you suspected was interacting with the conserved sequence at the 3′ end of the histone precursor. If the snRNP was interacting by hybridizing to the precursor RNA, then an oligonucleotide that matches the sequence in the precursor RNA

```
       3'  GTGTCGATGAAACCA  5' human
            ||||||////
   5'  m₃G-NNGUGUUACAGCUCUUUUAGAAUUUGUCUAGU  3' human U7 snRNA
              ||||||||||
       3'  TTGTCGAGAAAGGC   5' mouse
            ||||||||||
       3'  TGGTCGAGAAAGAAA  5' consensus
```

Figure 8–44 Pairing between the three oligonucleotides and the human U7 snRNA (Answer 8–26).

Figure 8–45 Expected products for minigene 1 (A) and minigene 2 (B) for 5'-to-3' scanning and 3'-to-5' scanning (Answer 8–29). Open boxes indicate complete exons; shaded boxes represent partial exons.

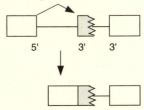

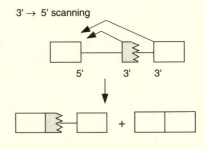

should be able to hybridize to the snRNA. Formation of a DNA-RNA hybrid would render the snRNA sensitive to cleavage by added RNase H. Cleavage of the snRNA in this critical region of interaction should render the extract incapable of processing the precursor. This result was the one you observed for the mouse and consensus oligonucleotides.

B. The inability of the human oligonucleotide to block processing was not anticipated, since you were using a human extract. However, an examination of the hybrids that can be formed between the various oligonucleotides and the U7 snRNA reveals that the mouse and consensus oligonucleotides can hybridize perfectly for a 10-nucleotide and a 9-nucleotide stretch, respectively (Figure 8–44). By contrast, hybridization to the human oligonucleotide is split by an unmatched nucleotide into two segments of 6 and 4 nucleotides (Figure 8–44). The stability of pairing of two separate segments is not as great as for a continuous pairing segment. Hence, the human oligonucleotide does not pair with sufficient stability to render the U7 snRNA sensitive to RNase H cleavage.

Reference: Mowry, K.L.; Steitz, J.A. Identification of human U7 snRNP as one of several factors involved in the 3'-end maturation of histone premessenger RNAs. *Science* 238:1682–1687, 1987.

*8–27

*8–28 **Reference:** Krause, M.; Hirsh, D. A trans-spliced leader sequence on actin mRNA in *C. elegans. Cell* 49:753–761, 1987.

8–29

A. If the splicing machinery binds to one splice site and scans across the intron to find its complementary splice site, it must use the first appropriate splice site it encounters. (Not using the first appropriate splice site is equivalent to skipping an exon.) The expected products from intron scanning in your two minigenes are shown in Figure 8–45. If the splicing machinery binds to a 5' splice site and scans toward a 3' splice site, minigene 1 should generate one product (Figure 8–45A) and minigene 2 should generate two products (Figure 8–45B). By contrast, if the splicing machinery binds to a 3' splice site and scans toward a 5' splice site, minigene 1 should generate two products (Figure 8–45A) and minigene 2 should generate one product (Figure 8–45B).

B. The results of your experiment do not match the expectations for either direction of intron scanning. Therefore, selection of splice sites by a mechanism involving unidirectional scanning through introns in either a 5'-to-3' or in a 3'-to-5' direction seems unlikely. The mechanism by which cells avoid exon skipping has not yet been defined.

Reference: Kuhne, T.; Wieringa, B.; Reiser, J.; Weissmann, C. Evidence against a scanning model for RNA splicing. *EMBO J.* 2:727–733, 1983.

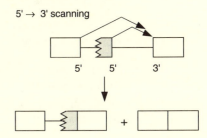

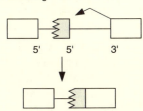

The Organization and Evolution of the Nuclear Genome

8–30

A. satellite DNAs
B. transposable elements
C. transposition bursts
D. L1 transposable element, *Alu* sequence

A. True
B. False. The sequences of tandemly repeated genes and their spacer DNAs are both homogenized by unequal crossing over and gene conversion.
C. True
D. True
E. False. Split genes are thought to be the ancient condition. Bacteria are thought to have lost their introns—after most of their proteins had evolved—in response to strong selective pressure to reproduce at the maximum rate permitted by the level of nutrients in the environment.
F. True
G. False. Although most transposable elements rarely do move, so many elements are present that their movement has a major effect on species variability. For example, more than half the spontaneous mutations in *Drosophila* are due to insertion of a transposable element in or near the mutant gene.
H. True
I. True
J. False. *Alu* sequences are transcribed by RNA polymerase III. Since polymerase III promoters are internal to the transcript, an *Alu* sequence carries the information necessary for its own transcription wherever it goes.

*8–32 **Reference:** van Arsdell, S.W.; Weiner, A.M. Human genes for U2 small nuclear RNA are tandemly repeated. *Mol. Cell. Biol.* 4:492–499, 1984.

*8–33 **References:** Nathans, J.; Thomas, D.; Hogness, D.S. Molecular genetics of human color vision: the genes encoding blue, green, and red pigments. *Science* 232:193–202, 1986.
Nathans, J.; Piantanida, T.P.; Eddy, R.L.; Shows, T.B.; Hogness, D.S. Molecular genetics of inherited variation in human color vision. *Science* 232:203–210, 1986.
Vollrath, D.; Nathans, J.; Davis, R.W. Tandem array of human visual pigment genes at Xq28. *Science* 240:1669–1671, 1988.

8–34

A. Recombination events that could have given rise to the abnormal gene structures associated with dichromats and anomalous trichromats are illustrated in Figure 8–46. These recombination events are not unique; most could have arisen by equivalent events between different chromosome arrays. Nevertheless, they serve to illustrate the principles of unequal crossing over.
   (Because the genes are sex-linked, recombination between different chromosome arrays would require that the event occur in the female. Recombination between identical arrays could also occur between sister chromatids in the male. Either event would have to occur in the germ line in order to be transmitted to the next generation.)
   The recombination event necessary to generate the green-blind dichromat (G⁻R⁺) deserves special comment. This event may have resulted from two unequal crossing-over events, or it may have been the product of a gene conversion event. Either event (and the two are indistinguishable in terms of the final product) results in the replacement of a segment of one DNA molecule by a homologous segment from another DNA molecule.
B. Control elements for transcriptional regulation of gene expression are usually located around the 5′ ends of genes. Thus, if expression of these genes is controlled at the level of initiation of transcription—as is likely to be the case—you might reasonably expect that the hybrid genes (like the normal genes) would be expressed according to the sequences at their 5′ ends. Consequently, the hybrid genes in the red-blind dichromat (G⁺R⁻) and the red-anomalous trichromat (G⁺R′) are probably expressed like red genes because their 5′ ends are derived from red genes. Similarly, the hybrid genes in the green-blind dichromat (G⁻R⁺) and in the green-anomalous trichromat (G′R⁺) are probably

expressed like green genes because their 5′ ends are derived from green genes.

C. The single gene present in the red-blind dichromat (G⁺R⁻) must have the spectral sensitivity of a green gene, even though it contains only a portion of the green gene. (Results from many other color-deficient males support the idea that the 3′ half of the pigment gene encodes the domain of the visual pigment that is responsible for the absorption properties of the pigment.) If the single hybrid gene encodes a "green" pigment, then this male would be expected to be red-blind. Interestingly, it may be that the "green" pigment is expressed in cells that would normally become red cones since the hybrid gene has the 5′ end characteristic of a red pigment gene. Cells that would normally become green cones may have no pigment at all since there is no gene with a 5′ end characteristic of a green pigment gene.

The two genes present in the green-blind dichromat (G⁻R⁺) probably behave like red genes since the 3′ half of the hybrid gene is derived from a red pigment gene. Since the 5′ end of the hybrid gene is from a green gene, the hybrid "red" pigment is likely to be expressed in cells that would normally become green cones. Thus, the normal red cones presumably express red pigment, and the normal green cones may express the hybrid "red" pigment.

The red-anomalous trichromat (G⁺R′) has two normal green genes that are presumably expressed normally in green cones. The hybrid gene, which is presumably expressed in red cones, must encode a pigment with altered spectral sensitivity in order to account for the anomalous color vision in the red-anomalous trichromat.

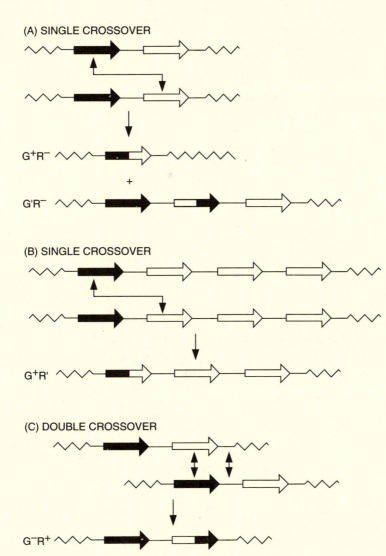

(A) SINGLE CROSSOVER

(B) SINGLE CROSSOVER

(C) DOUBLE CROSSOVER

Figure 8–46 Unequal recombination events to generate gene structures of dichromats and anomalous trichromats (Answer 8–34).

The green-anomalous trichromat (G'R⁺) has one normal red gene and one normal green gene, each of which is presumably expressed in the appropriate cones. The hybrid gene has a 3' end from a red gene and thus presumably has either red or an altered spectral sensitivity. Since its 5' end is derived from a green gene, it is presumably expressed in green cones. The mixture of normal green pigment and red or abnormal pigment in the same cone presumably alters the normal spectral sensitivity of the cone, resulting in an anomalous response to green.

Reference: Nathans, J.; Piantanida, T.P.; Eddy, R.L.; Shows, T.B.; Hogness, D.S. Molecular genetics of inherited variation in human color vision. *Science* 232:203–210, 1986.

*8–35   **Reference:** Boeke, J.D.; Garfinkel, D.J.; Styles, C.A.; Fink, G.R. Ty elements transpose through an RNA intermediate. *Cell* 40:491–500, 1985.

8–36

A. The left and right boundaries of the inserted *Alu* sequences and the mutational changes in the flanking chromosomal sequences (arbitrarily shown on the 5' side) are indicated in Figure 8–47. The boundaries can be located unambiguously using two complementary approaches. In the first approach, by comparing the sequences vertically, one can make a tentative assignment of the boundaries based on the point at which the sequences diverge. The second approach makes use of a common feature of the *Alu* insertion process, namely, the duplication of chromosomal sequences at the target of insertion. A comparison of the sequences to the left and right of each *Alu* sequence shows that they are tandemly duplicated—one end of each duplication is precisely at the tentative boundary assigned on the basis of the vertical comparison.

The mutations in the flanking sequences can be located easily by comparing the repeated sequences that were generated when each *Alu* sequence inserted into the chromosome.

B. As indicated in Figure 8–47, there are five nucleotide changes in the 120 nucleotides of duplicated DNA that flank the *Alu* sequences. Using the estimate of 3×10^{-3} substitutions per site per million years, the *Alu* sequences inserted into the human albumin-family genes about 14 million years ago.

$$\text{years after insertion} = \frac{10^6 \text{ years} \times 1 \text{ site}}{3 \times 10^{-3} \text{ mutations}} \times \frac{5 \text{ mutations}}{120 \text{ sites}}$$

$$= 14 \times 10^6 \text{ years}$$

C. These particular flanking sequences are critical for the calculation because they were generated by the *Alu* sequences when they inserted into the human albumin-family genes. Thus, the sequences that constitute the target-site duplications mark the time of *Alu* insertion.

Additional intron sequences would not help in the calculation because mutations outside the target-site duplication are unrelated to the time of *Alu*-sequence insertion. The mutations in the *Alu* sequences themselves are also not useful for estimating the time of insertion. Some of the observed mutations

◄——— *Alu* repeat (300 nucleotides) ———►

```
       TTAAATA | GGCCGGG ------- AAAAAAAAAAAA | TTAAATA
      TGTTGTGGG | GATCAGG ------- AAAAAAAAAAAA | TCTGTGGG
      TCTTCTTA | GGCTGGG ------- GAAAAAAAAAAA | TCTTCTTA
ATAATAGTATCTGTC | GGCTGGG ------- AGAAAAAAAAAA | TAAATAGTATCTGTC
     GGATGTTGTGG | GGCCGGG ------- AAAAAAAAAAAA | GGATGTTGTGG
      AGAACTAAAAG | GGCTAGG ------- AAAAAGAGAAGA | AGAACCGAAAG
```

Figure 8–47 Boundaries of *Alu* inserts and mutational alterations in the flanking target-site duplication (Answer 8–36). Boundaries are indicated by vertical lines; mutational changes are indicated by underlining.

in the *Alu* sequences very likely were generated while they sat in the genome at another location—prior to the time at which a copy inserted into the human albumin-gene family.

D. The calculation in part B indicates that these *Alu* sequences invaded the human albumin-gene family about 14 million years ago, that is, well after the mammalian radiation and the separation of the lineages leading to rats and humans.

Reference: Ruffner, D.E.; Sprung, C.N.; Minghetti, P.P.; Gibbs, P.E.M.; Dugaiczyk, A. Invasion of the human albumin-α-fetoprotein gene family by *Alu, Kpn,* and two novel repetitive DNA elements. *Mol. Biol. Evol.* 4:1–9, 1987.

Control of Gene Expression

An Overview of Gene Control

9–1

    A. transcriptional
    B. RNA processing
    C. RNA transport
    D. translational
    E. mRNA degradation
    F. protein activity

9–2

    A. False. Carrots can be grown from single carrot cells, and tadpoles can be gotten by injecting differentiated frog nuclei into frog eggs, but carrots cannot be gotten from frog eggs no matter what.
    B. True
    C. True

DNA-binding Motifs in Gene Regulatory Proteins

9–3

    A. gene regulatory
    B. helix-turn-helix
    C. homeodomain
    D. zinc finger
    E. leucine zipper
    F. combinatorial
    G. helix-loop-helix (HLH)
    H. DNA affinity chromatography

9–4

    A. True
    B. True
    C. False. Specific nucleotide sequences have local irregularities such as tilted nucleotide pairs or a helical twist angle larger or smaller than 36°.
    D. True
    E. False. Although the individual contacts are weak, there are so many that DNA-protein interactions are among the tightest and most specific known in biology.
    F. True
    G. False. Whereas the helix-turn-helix motif of bacterial gene regulatory proteins is often embedded in different structural contexts, the helix-turn-helix motif of homeodomains is always surrounded by the same structure (the homeodomain), suggesting that the motif is always presented to the DNA in the same basic manner.
    H. True
    I. False. One class of gene regulatory proteins uses two strands of beta sheet to read the information on the surface of the major groove.

J. False. Although there are many leucine zipper proteins, they cannot all form heterodimers with one another. If they did, the amount of cross-talk between the gene regulatory circuits of a cell would be so great as to cause chaos.

K. True

L. False. There is no simple amino acid–base pair recognition code. Gene regulatory proteins utilize different combinations of amino acid side chains to create surfaces that are complementary to the DNA sequences they recognize.

M. True

N. False. Although DNA affinity chromatography can achieve purifications as great as 10,000-fold, the amounts of pure protein are usually small. They are adequate, however, to obtain an amino-terminal amino acid sequence, which can then be used to clone the gene. Once the gene is cloned, the protein can be produced in virtually unlimited amounts.

9–5

A. At the point of minimum relative migration, the CAP sites are separated by 85 nucleotides (Figure 9–1C). At 10.6 nucleotides per turn, this number of nucleotides corresponds to 8 helical turns ($85/10.6 = 8$). At the point of maximum relative migration, the CAP sites are separated by 79 nucleotides, which corresponds to 7.5 helical turns.

B. The two CAP sites must be bent exactly the same way since they are identical. Therefore, they will both have the same groove of the helix facing the inside of the bend at the center of bending. In order for the DNA to bend into the *cis* configuration, the two centers of bending must be on the same side of the helix. The major grooves (or minor grooves) are on the same side of the helix at integral numbers of helical turns. Thus, it is expected that the point of minimum relative migration (the *cis* configuration) will occur after an integral number of helical turns. Similarly, the point of maximum relative migration (the *trans* configuration) will occur when the centers of bending are on opposite sides of the helix, that is, at half integral numbers of helical turns.

C. At the point of minimum migration of the construct with one CAP site and one $(A_5N_5)_4$ site, the centers of bending are separated by 101 nucleotides (Figure 9–6D). At 10.6 nucleotides per helical turn, the centers of bending are separated by 9.5 helical turns ($101/10.6 = 9.5$).

D. Since the point of minimum relative migration (the *cis* configuration) occurs at a half-integral number of turns, the two centers of bending cannot have the same groove of the helix facing the inside of the bend. As discussed in part B, if the same groove of the helix faced the inside of the bend, the centers of bending in the *cis* configuration would be separated by an integral number of turns. Therefore, the two centers of bending must have opposite grooves facing the inside of the bend. Because the $(A_5N_5)_4$ site is known to bend with the major groove facing the inside of the bend (as was stated in the problem), the CAP-binding site must be bent so that the minor groove faces the inside of the bend at the center of bending.

References: Zinkel, S.S.; Crothers, D.M. DNA bend direction by phase sensitive detection. *Nature* 328:178–181, 1987.

Gartenberg, M.R.; Crothers, D.M. DNA sequence determinants of CAP-induced bending and protein binding affinity. *Nature* 333:824–829, 1988.

*9–6 **References:** Treisman, R. Identification and purification of a polypeptide that binds to the c-fos serum response element. *EMBO J.* 6:2711–2717, 1987.

Norman, C.; Runswick, M.; Pollock, R.; Treisman, R. Isolation and properties of cDNA clones encoding SRF, a transcription factor that binds to the *c-fos* serum response element. *Cell* 55:989–1003, 1988.

How Genetic Switches Work

9–7

    A. operator, tryptophan repressor
    B. negative control
    C. positive control
    D. general transcription factors
    E. enhancer
    F. gene control, promoter, regulatory
    G. gene activator
    H. gene repressor

9–8

    A. False. When the tryptophan repressor binds two molecules of tryptophan, its helix-turn-helix motif is distorted so that it *can* bind its operator sequence and suppress transcription. In this way the presence of adequate tryptophan in the environment shuts off synthesis of the enzymes required for tryptophan synthesis.
    B. True
    C. True
    D. True
    E. False. Although all three RNA polymerases require a set of general transcription factors, only RNA polymerase II requires a TATA sequence in the promoter.
    F. True
    G. False. Although general transcription factors are abundant, gene regulatory proteins are usually present in very small amounts in a cell.
    H. False. Acidic activators accelerate assembly of the initiation complex by providing a negatively charged surface to which other proteins can bind.
    I. False. It is not yet clear how eucaryotic gene repressor proteins work, but unlike bacterial repressors they do not compete with RNA polymerase for access to the DNA.
    J. True
    K. True
    L. True
    M. True
    N. False. Inactive gene regulatory proteins in mammalian cells are activated in many different ways, including several ways, such as phosphorylation and stimulation of nuclear entry, that are not used in bacterial cells.
    O. True
    P. True

*9–9   **Reference:** Schleif, R.F. Genetics and Molecular Biology, Chap. 13. Reading, MA: Addison Wesley, 1986.

9–10

    A. The rapid bacterial growth at the beginning of the experiment results from the metabolism of glucose, and the slower growth at the end results from metabolism of lactose. The bacteria stopped growing in the middle of the experiment because they ran out of glucose but did not yet possess the enzymes necessary for lactose metabolism. Before they could utilize the lactose in the medium, they had to induce the *lac* operon. The delay in growth represents the time required for the induction.
    B. Induction of the *lac* operon requires that two conditions be met: lactose must be present and glucose must be absent. During the first part of the experiment, glucose and lactose are both present; therefore, the conditions for induction are not met. Only when glucose is exhausted are the requirements for induction satisfied.

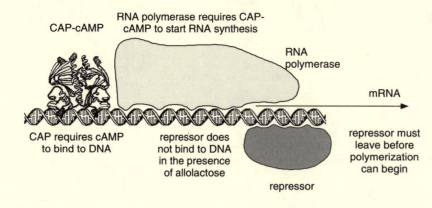

CAP-cAMP

RNA polymerase requires CAP-
cAMP to start RNA synthesis

RNA
polymerase

mRNA

CAP requires cAMP
to bind to DNA

repressor does
not bind to DNA
in the presence
of allolactose

repressor must
leave before
polymerization
can begin

repressor

Figure 9–26 Induction of the *lac* operon (Answer 9–10). The proteins and DNA are drawn approximately to scale. The three-dimensional structure of CAP is known, but the structures of RNA polymerase and repressor have not yet been determined.

The requirements for induction are mediated by CAP and the lactose repressor (Figure 9–26). For the operon to be on, CAP must be bound and the lactose repressor must not be bound. The presence of lactose in the medium increases the intracellular concentration of allolactose, which binds to the lactose repressor, thereby lowering its affinity for its binding site and causing its release from the DNA. Removal of the lactose repressor satisfies one condition for induction. The second condition is tied to the concentration of glucose. When the concentration of glucose falls, the intracellular level of cAMP rises. cAMP binds to CAP and alters its conformation so that it can bind to its binding site. When CAP is in place (and the lactose repressor is absent), RNA polymerase can bind to the promoter and initiate transcription.

Reference: Monod, J. The phenomenon of enzymatic adaptation. Growth Symposium XI:223–289, 1947. [Reprinted in Selected Papers in Molecular Biology by Jacques Monod (A. Lwoff, A. Ullmann, eds.), pp. 68–134, New York: Academic Press, 1947.]

*9–11 **Reference:** Ninfa, A.J.; Reitzer, L.J.; Magasanik, B. Initiation of transcription at the bacterial *glnAp2* promoter by purified *E. coli* components is facilitated by enhancers. *Cell* 50:1039–1046, 1987.

9–12

Neither the propagation of an altered DNA structure nor the oligomerization of a protein from the repression site would be expected to be sensitive to small changes in the spacing between the repression site and the start site of transcription. Nor is there any obvious reason in those mechanisms why some insertions and deletions would prevent repression, while other interspersed insertions and deletions would maintain repression.

The third mechanism—formation of a loop in the DNA—is consistent with the observations. Repression of the *galK* gene occurs when integral numbers of helical turns (multiples of 10.5 nucleotides, which is the number of nucleotide pairs per helical turn) are added to or deleted from the DNA. By contrast, when half-integral numbers of turns are involved, repression is prevented. This is exactly the behavior expected if DNA must bend into a tight loop to allow araC at site 2 to interact with another protein near the transcription start site. As shown in Figure 9–27, nonintegral turns would place the araC protein on the wrong face of the DNA, which would require that the DNA twist half a turn to allow proper positioning of the proteins. Although twisting a DNA helix by half a turn may not seem difficult, it actually requires about 4 kcal/mole for a DNA 200 nucleotides in length. The binding energy available from typical protein-DNA interactions is only about 10–15 kcal/mole. Since a substantial fraction of the binding energy would be required to twist the DNA, it is not unreasonable to expect that twisting the DNA could alter a delicately balanced interaction required for repression.

Other experiments suggest that araC at site 2 may interact with araC at site 1 to form the DNA loop. In the absence of arabinose, the *araBAD* genes are fully repressed, presumably by interference of the DNA loop with the binding of RNA polymerase. If araC-binding site 2 is deleted (or if the araC protein is absent because of mutation), the DNA loop cannot form and the binding of RNA polymerase is not prevented, thereby leading to a 10-fold elevation in transcription of the *araBAD* cluster of genes. When arabinose is present (and glucose is absent), the araC protein undergoes a conformational change that prevents formation of the DNA loop and also facilitates the binding of RNA polymerase or aids its conversion to an open complex, thereby increasing transcription 1000-fold.

Reference: Dunn, T.M.; Hahn, S.; Ogden, S.; Schleif, R.F. An operator at –280 base pairs that is required for repression of *araBAD* operon promoter: addition of DNA helical turns between the operator and promoter cyclically hinders repression. *Proc. Natl. Acad. Sci. USA* 81:5017–5020, 1984.

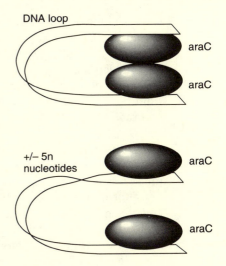

Figure 9–27 Diagram showing consequences of introducing a half-integral number of turns into a DNA loop formed by interaction between two proteins (Answer 9–12).

9–13

A. The 400-nucleotide transcript is absent from lane 4 (Figure 9–7) because GTP was included in the reaction mixture. GTP allows transcription to proceed beyond the C-minus sequence (the synthetic sequence lacking C nucleotides), thereby generating transcripts longer than 400 nucleotides. In the absence of GTP (lane 2) transcription cannot proceed beyond the C-minus sequence. In the presence of GTP and RNase T1 (lanes 6 and 8) the longer transcripts are cleaved at the first G to yield the 400-nucleotide transcript.

B. One of the difficulties in assaying promoter function *in vitro* is the high background of nonspecific initiation of transcription. It is this background that is so evident in lane 3. Its source is not altogether clear, but transcription may start at sequences in the rest of the plasmid that bear a weak resemblance to true RNA polymerase II promoters.

C. A transcript of about 400 nucleotides is present in lane 5 because cleavage with RNase T1 liberates it from any randomly initiated transcript that has traversed the C-minus sequence. It is actually a few nucleotides longer than the specifically initiated transcript since its 5′ end is defined by the first G that precedes the C-minus sequence.

   The 400-nucleotide transcript is absent from lane 7 because 3′-O-methyl GTP will terminate most transcripts that are initiated in front of the C-minus sequence. The combination of 3′-O-methyl GTP and RNase T1 eliminates virtually all the background synthesis from the control plasmid.

D. You should have no difficulty assaying specific transcription in crude extracts (that is, in the presence of GTP). As shown in Figure 9–7 (lanes 7 and 8), specific transcription can be assayed in the presence of G nucleotides if 3′-O-methyl GTP and RNase T1 are included (to inhibit background transcription and to cleave any random transcripts into small pieces).

Reference: Sawadogo, M.; Roeder, R.G. Factors involved in specific transcription by human RNA polymerase II: analysis by a rapid and quantitative *in vitro* assay. *Proc. Natl. Acad. Sci. USA* 82:4394–4398, 1985.

*9–14 Reference: Sawadogo, M.; Roeder, R.G. Factors involved in specific transcription by human RNA polymerase II: analysis by a rapid and quantitative *in vitro* assay. *Proc. Natl. Acad. Sci. USA* 82:4394–4398, 1985.

9–15

A. The ability of the transcription factor to stimulate transcription from the –61 deletion but not from the –50 deletion in the nuclear extract indicates that a critical portion of the binding site for the factor resides in the interval between –61 and –50 nucleotides upstream from the start site of transcription. Deletion studies alone do not define the end of the binding site closest to the start site since removal of a part of the binding site would prevent binding of the factor. DNA footprinting experiments, however, show that the factor actually protects a 12-nucleotide-long sequence from –63 to –52.

B. At first it seems contradictory that a *stimulatory* factor causes the transcript from the –50 deletion template to disappear from the gel. In fact, there is no contradiction: the stimulatory factor causes a ten- to twentyfold increase in transcription from all the templates that retain the binding sequence. Thus the amount of transcript from the –50 deletion template falls *relative* to the amount of transcript produced from the stimulated templates (and for that reason, is so much less intense on the gel that it seems to be absent).

C. The enormous difference in the stability of binding of the factor in the presence and absence of TFIID suggests that the factor binds not only to the DNA but also to TFIID. Thus, in the presence of TFIID the stimulatory factor is anchored in place by two interactions (with the DNA and with TFIID), whereas in the absence of TFIID it is bound by a single interaction (with the DNA).

   Reference: Sawadogo, M.; Roeder, R.G. Interaction of a gene-specific transcription factor with the adenovirus major late promoter upstream of the TATA box region. *Cell* 43:165–175, 1985.

9–16

A. Since equal amounts of transcription from each template were observed when the preincubation was carried out with the individual templates or a mixture of them (Figure 9–9C, lanes 1 to 3), Srb2 protein does not show a preference for either template.

B. Your results indicate that Srb2 acts stoichiometrically. If Srb2 acted catalytically, it should have been able to modify the second template after the two were mixed, which would have produced transcripts from both templates regardless of which one was originally included in the preincubation with the extract to which Srb2 was added. When excess Srb2 was added at the beginning of the preincubation with one template, transcription was observed from both templates after mixing. This is consistent with a stoichiometric requirement for Srb2 and rules out the possibility that Srb2 was inactivated during the preincubation and for that reason unable to act (catalytically) on the second template after mixing.

C. The production of transcripts only from the template that was preincubated with extract and Srb2 indicates that the Srb2 protein is part of the preinitiation complex. If Srb2 were able to act after transcription had begun, transcripts would have been produced from both templates regardless of which one was included in the preincubation.

D. During preincubation of the template with the extract and Srb2, a number of proteins including Srb2 bind to the promoter to form a preinitiation complex. Evidently, the preinitiation complex, once formed, is stable and does not readily exchange proteins with other templates that are added later.

E. Although the *SRB2* gene was originally identified as a suppressor of the cold-sensitive phenotype of yeast carrying an RNA polymerase II gene with a short CTD, neither the genetic results nor these transcription assays provide evidence that Srb2 binds to the CTD. (Nor do they argue against direct interaction; they simply do not speak to the issue.) The question of what binds to what has been investigated by affinity chromatography. When the CTD is attached to a column, it will selectively bind a preinitiation complex that includes TFIID and Srb2. Pure TFIID also binds to a CTD affinity column, but pure Srb2 does not. An affinity column made with TFIID binds Srb2, and an Srb2 affinity column binds TFIID. Although these results do not eliminate the possibility of a direct interaction between Srb2 and the CTD (the interaction might be very weak and require the presence of other proteins to be stable enough to be detected in this way), they suggest that Srb2 may influence the activity of the CTD indirectly through other proteins such as TFIID.

References: Koleske, A.J.; Buratowski, S.; Nonet, M.; Young, R.A. A novel transcription factor reveals a functional link between the RNA polymerase II CTD and TFIID. *Cell* 69:883–894, 1992.

Thompson, C.M.; Koleske, A.J.; Chao, D.M.; Young, R.A. A multisubunit complex associated with the RNA polymerase II CTD and TATA-binding protein in yeast. *Cell* 73:1361–1375, 1993.

*9–17 **Reference:** Godowski, P.J.; Rusconi, S.; Miesfeld, R.; Yamamoto, K.R. Glucocorticoid receptor mutants that are constitutive activators of transcriptional enhancement. *Nature* 325:365–385, 1987.

9–18

A. Construct 1 corresponds to mutant embryo B, construct 2 corresponds to mutant embryo D, and construct 3 corresponds to mutant embryo C.

B. With the exception of one aspect of embryo D, as discussed below, the results with the various constructs validate the simple rule for *eve* expression in stripe 2. When the Krüppel-binding sites are removed (construct 1), thereby eliminating the effects of the Krüppel repressor, the stripe-2 expression of β-galactosidase expands slightly in the posterior direction (Figure 9–13, mutant embryo B). This is as expected according to the rule because the activators hunchback and bicoid are both present slightly beyond the posterior end of stripe 2.

When two of the bicoid-binding sites are removed (construct 3), thereby making the construct less sensitive to the effects of the bicoid activator, the stripe-2 expression of β-galactosidase is lessened but appears at its normal position (Figure 9–13, mutant embryo C).

When the giant-binding sites are removed (construct 2), thereby eliminating the effects of the giant repressor, the stripe-2 expression of β-galactosidase expands substantially in the anterior direction (Figure 9–13, mutant embryo D). The simple rule for *eve* stripe-2 expression predicts that β-galactosidase would be expressed in the entire anterior end of the mutant embryo, since the activators bicoid and hunchback are both present but no effective repressors are present. This results suggest that something is missing from the formulation of the simple rule.

C. The reason why β-galactosidase is not expressed in the anterior end of the embryo is not yet known. There are several plausible mechanisms that fall into two general categories. It is possible that another protein whose expression is confined to the anterior end modifies one or both activators to render them nonfunctional for binding to the *eve* stripe-2 control element. The modification could take the form of a phosphorylation, for example, or occur by direct binding to the activators. The second general mechanism would be the expression of another, as yet undefined repressor that binds to the *eve* stripe-2 control element and prevents expression.

D. The overlap of repressor-binding sites and activator-binding sites in the *eve* stripe-2 control element is striking. It suggests that the repressors function primarily by preventing activator binding. The extensive competition for binding between activators and repressors presumably sharpens the boundary between *eve* expression and nonexpression, giving rise to very well defined stripes in the embryo.

Reference: Stanojevic, D.; Small, S.; Levine, M. Regulation of a segmentation stripe by overlapping activators and repressors in the *Drosophila* embryo. *Science* 254:1385–1387, 1991.

Chromatin Structure and the Control of Gene Expression

9-19

A. position effect

9-20

A. True

B. True

C. False. In the first step the chromatin of the entire globin locus is decondensed, which is presumed to allow some gene regulatory proteins access to the DNA. In the second step the remaining gene regulatory proteins assemble on the DNA and direct the expression of individual genes.

D. False. The unwinding that accompanies initiation of transcription and the movements of RNA polymerase along the DNA both alter DNA topology. Through such topological effects an event that occurs at a single DNA site can produce forces that are felt throughout an entire chromatin domain.

9-21

A. Only TFIID must be present during the preincubation to insure that the promoter remains accessible after chromatin assembly. This conclusion is based on the absence of a transcript in lane 6 of Figure 9-14, and its presence in lanes 4, 5, 7, and 8. If any other component were required, its absence would have prevented transcription under one of the other conditions. TFIID binds to the TATA element of RNA polymerase II promoters. Its binding allows nucleosome assembly in a way that permits transcription.

B. The results in Figure 9-14, lanes 9, 10, 11, and 13, indicate that TFIIA, TFIIB, and RNA polymerase II in the presence of TFIID can form a complex with the promoter that is stable to chromatin assembly and purification.

C. As indicated by the result in Figure 9-14, lane 12, only transcription factor TFIIE needs to be added during the transcription assay when all the transcription components were present during the preincubation.

Reference: Workman, J.L.; Roeder, R.G. Binding of transcription factor TFIID to the major late promoter during *in vitro* nucleosome assembly potentiates subsequent initiation by RNA polymerase II. *Cell* 51:613–622, 1987.

The Molecular Genetic Mechanisms That Create Specialized Cell Types

9-22

A. phase variation

B. mating

C. mating-type (MAT)

D. lambda repressor (cI), cro

E. DNA methylation

F. genomic imprinting

G. CG islands

9-23

A. False. Although there are a few examples of this mode of gene regulation in procaryotes, there are no known examples of regulation by reversible genetic rearrangements in mammalian cells.

B. False. Even though the original resident gene at the MAT locus is discarded during mating-type switching, the process is reversible because information for both the α-type and **a**-type genes is maintained elsewhere as silent genes.

C. True

D. True

E. True

F. True

G. False. Diffusible gene regulatory proteins cannot explain the heritability of the inactive X chromosome, since identical DNA sequences on the active and inactive chromosomes are regulated differently.

H. True

I. True

J. False. The preservation of CG islands in the promoters of housekeeping genes is thought to occur because sequence-specific DNA-binding proteins protect them from methylation in germ cells.

*9–24   **Reference:** Zieg, J.; Silverman, M.; Hilmen, M.; Simon, M. Recombinational switch for gene expression. *Science* 196:170–172, 1977.

9–25

A. In the presence of **a**1, α2 exists in two forms with different binding specificities. Your second set of experiments (lanes 5 and 6) supports this interpretation and rules out the possibility that α2 is present in a single form that can bind to both **a**-specific and haploid-specific regulatory sequences. Addition of excess unlabeled **a**-specific DNA eliminates binding to the radioactive **a**-specific fragments, but not to the radioactive haploid-specific fragments. Similarly, addition of excess unlabeled haploid-specific DNA eliminates binding to the radioactive haploid-specific fragments. If a single form of α2 were able to bind to both regulatory sequences, either unlabeled site in excess would have eliminated binding to both kinds of radioactive fragments.

   Your third experiment (lanes 7 and 8) shows that the ratio of binding activities for **a**-specific and haploid-specific sequences varies depending on the amount of **a**1 protein. If the α2 repressor in a diploid cell were shifted entirely into a form that could bind both sites, the ratio should be independent of the amount of **a**1 protein.

B. Your experiments are most easily explained if **a**1 acts stoichiometrically to alter the binding specificity of the α2 repressor, presumably by binding to it. In the fragment binding experiments shown in lanes 7 and 8 of Figure 9–16, when **a**1 is low, the binding to haploid-specific fragments is low; when **a**1 is high, the binding to haploid-specific fragments is high. The simplest explanation for the effects of the defective α2 repressor is that it binds to **a**1 protein, thereby reducing the availability of the **a**1 protein for binding to the normal α2 repressor.

   Note that neither of these experiments rules out the possibility that **a**1 protein acts catalytically on the α2 repressor. However, a catalytic mechanism for **a**1 protein would require special assumptions to account for the experimental observations. For these reasons it is considered most likely that the form of the repressor that binds to haploid-specific sequences will be found to contain both the **a**1 protein and the α2 repressor.

   Reference: Goutte, C.; Johnson, A.D. **a**1 protein alters the DNA-binding specificity of α2 repressor. *Cell* 52:875–882, 1988.

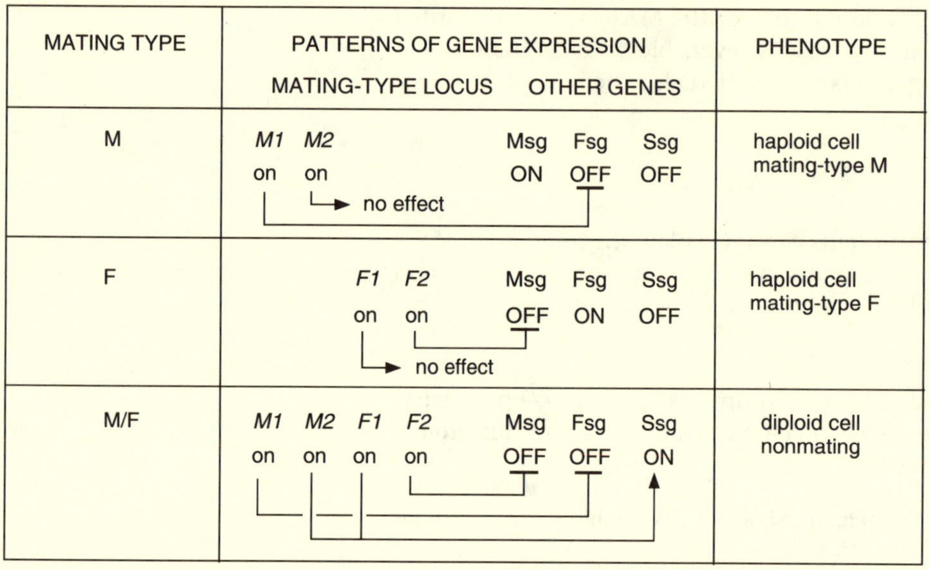

MATING TYPE	PATTERNS OF GENE EXPRESSION		PHENOTYPE
	MATING-TYPE LOCUS	OTHER GENES	
M	*M1* *M2* on on └──▶ no effect	Msg Fsg Ssg ON OFF OFF	haploid cell mating-type M
F	*F1* *F2* on on └──▶ no effect	Msg Fsg Ssg OFF ON OFF	haploid cell mating-type F
M/F	*M1* *M2* *F1* *F2* on on on on	Msg Fsg Ssg OFF OFF ON	diploid cell nonmating

Figure 9–28 Regulation of gene expression by mating-type genes in a new strain of yeast (Answer 9–26).

9–26

A regulatory scheme that accounts for the behavior of the mutants is shown in Figure 9–28. F-Specific genes are ON unless they are repressed by the product of the *M1* gene. M-specific genes are ON unless they are repressed by the product of the *F2* gene. Sporulation-specific genes are OFF unless they are activated by the combined products of the *M2* and *F1* genes, which are present only in diploid cells.

At first, such a collection of mutants and phenotypes may seem impossible to decipher. However, most models of gene expression were first developed from just such genetic data. (Only later were the predicted molecular interactions tested by biochemical experiments and elaborated in full detail.) In the genetic data in Table 9–2, there are three principal clues that allow the regulatory scheme to be deduced. In mating-type M cells, an *M1⁻* mutation allows expression of both M-specific and F-specific genes, suggesting that the product of the *M1* gene may be a repressor of the F-specific genes. In mating type F cells, an *F2⁻* mutation allows expression of both F-specific and M-specific genes, suggesting that the product of the *F2* gene may be a repressor of M-specific genes. Finally, although *M2⁻* and *F1⁻* mutations seem to have no effect in haploid cells, either mutation in diploid cells prevents expression of sporulation-specific genes, suggesting that they may function in combination as an activator of sporulation-specific genes. These three clues suggest three regulatory interactions, which provide a consistent interpretation of the entire data set. Further genetic and biochemical tests would be necessary before the regulatory scheme in Figure 9–28 could be considered proven.

9–27

A. The methylation status of the 5S RNA gene does not affect its transcriptional activity, as indicated by the equal intensities of the maxigene RNA bands and the 5S RNA bands in the mixtures of templates that were not treated with restriction enzymes (the lanes marked "none" in Figure 9–17B).

B. The pattern of transcription after cleavage is exactly as you expected. Transcription from the fully methylated 5S RNA maxigene is specifically abolished after cleavage with DpnI, which cleaves only fully methylated restriction sites. Cleavage with MboI specifically abolishes transcription from unmethylated 5S RNA genes. And cleavage with Sau3A, which is insensitive to the methylation status of the restriction site, abolishes transcription from all 5S RNA genes.

C. The pattern of transcription after replication and cleavage indicates that transcription complexes do not remain associated with the 5S RNA maxigene during replication. If the replicated molecules retained an active transcription complex, transcriptional activity would have been evident after DpnI digestion, which does not cleave the replicated molecules. The activity that remains after MboI cleavage derives from fully methylated DNA molecules, which were not replicated. If transcriptional complexes were not erased by replication, there would have been activity after DpnI cleavage.

D. If only 50% of the molecules had been assembled into active transcription complexes, the pattern would have been unchanged (except for a reduction in the intensity of the bands). You could no longer conclude, however, that replication eliminated the transcription complexes. If only 50% of the molecules were assembled into complexes and only 50% were replicated, then a skeptic (that is, a good scientist) would raise the possibility that the assembly into transcription complexes inhibits replication so that only the unassembled molecules were replicated. Thus, the experiment would fail to test what it was designed to test. The point of the objection is clearer if you imagine that 10% of the molecules were assembled into transcription complexes and 10% were replicated. For your conclusion to be valid, it is essential to prove that molecules with transcriptional complexes were replicated. Although there might be ways to test specifically for replication of transcriptional complexes, the easiest way is to show that the fraction of molecules that are replicated is significantly greater than the fraction of molecules that are not assembled into transcription complexes.

Reference: Wolffe, A.P.; Brown, D.D. DNA replication *in vitro* erases a *Xenopus* 5S RNA gene transcription complex. *Cell* 47:217–227, 1986.

*9–28   **Reference:** Murray, E.J.; Grosveld, F. Site-specific demethylation in the promoter of human γ-globin does not alleviate methylation mediated suppression. *EMBO J.* 6:2329–2335, 1987.

9–29

A. Probes 2 and 3 are useful because they are specific for mRNAs that are not present in untreated 10T½ cells. You are searching for an mRNA that is present in both kinds of myoblast but is not present in 10T½ cells. The reason for hybridizing the radioactive myoblast cDNA probes with RNA from 10T½ cells is to remove from the probe all the sequences that correspond to mRNAs that are common between the 10T½ cells and the myoblasts. This subtractive-hybridization procedure makes the probe more specific; it eliminates from analysis a large class of cDNA clones that you do not think will contain the gene of interest.

B. The A class of clones includes cDNAs corresponding to RNAs that are common to 10T½ cells and to induced and normal myoblasts. In this class are all the normal housekeeping genes present in all cells. This is the class of clones that subtractive hybridization was meant to eliminate. Since probes 2 and 3 hybridized to only 1% of the cDNA clones identified by probe 1, these house-keeping RNAs must represent the vast majority of the mRNA species in cells.

The B class of clones includes cDNAs corresponding to RNAs that are induced by 5-aza C treatment but are not present in normal myoblasts or in the 10T½ cells. Since 5-aza C causes widespread demethylation, it is not surprising that it activates some genes that are not normally expressed in myoblasts.

The C class of clones includes cDNAs corresponding to RNAs that are present in normal myoblasts but are not present in 5-aza-C-induced myoblasts. If the induced myoblasts were identical to the normal myoblasts, this class of clones should not exist. It would be natural to suspect some deficiency in the induced myoblasts; however, the problem seems to lie with the normal myoblasts. The normal myoblasts contain a small fraction of fully differentiated myotubes, whereas the induced myoblasts do not. Thus the RNA isolated from the normal myoblast cell population includes RNA from more differentiated cell types. When analyzed, the C class of clones are found to encode muscle-specific gene products, such as troponin I, and myosin heavy and light chains.

The D class of clones includes cDNAs that correspond to myoblast-specific RNAs. It is among these clones that you expect to find the regulatory gene that controls myoblast differentiation.

The actual gene that seems to control myoblast differentiation (*myoD1*) was found among this class of clones. It was selected out of this class after further tests that were based on the assumptions that (1) the regulatory gene should not be expressed at all in 10T½ cells, (2) its expression should reach a maximum in myoblasts, (3) its expression should decline in myotubes, and (4) it should not be expressed in variants of 10T½ cells that do not differentiate after treatment with 5-aza C. The final test was that the cloned cDNA, when introduced into 10T½ cells in an expressed form, should induce the cells to differentiate into myoblasts—which it does!

Reference: Davis, R.L.; Weintraub, H.; Lassar, A.B. Expression of a single transfected cDNA converts fibroblasts to myoblasts. *Cell* 51:987–1000, 1987.

Posttranscriptional Controls

9–30

    A. posttranscriptional
    B. transcription attenuation
    C. alternative RNA splicing
    D. gene
    E. *trans* RNA splicing
    F. RNA editing
    G. eIF-2
    H. negative translation
    I. translational recoding

9–31

    A. True
    B. False. Although this is true for most examples of regulated alternative splicing, constitutive alternative splicing produces several versions of the encoded protein in all cells in which the gene is expressed.
    C. True
    D. False. RNA splicing can create mRNAs that cause a segment of the original carboxyl terminus to be removed and to be replaced by a new segment of protein.
    E. True
    F. True
    G. True
    H. True
    I. True
    J. False. Hydrolysis of GTP to GDP by eIF-2 signifies that an initiating AUG in the mRNA has been located. Hydrolysis causes a conformational change in eIF-2, releasing it from the small ribosomal subunit and allowing the large ribosomal subunit to join the small subunit to form a complete ribosome.
    K. False. The stability of transferrin receptor mRNA and the translatability of ferritin mRNA are regulated by the same iron-sensitive RNA-binding protein. The consequences of binding, however, are different for the two mRNAs. Binding to the iron-response element at the 5′ end of ferritin mRNA blocks translation and decreases the level of ferritin. On the other hand, binding to the iron-response element at the 3′ end of transferrin receptor mRNA stabilizes the mRNA, allowing more transferrin receptor to be made.

L. True

M. True

N. False. Although UGA is a commonly used termination codon, the UGA codons that specify selenocysteine are in a special context that allows them to be selectively recognized.

O. True

9–32

A. Because calcitonin mRNA is produced when the cells are transfected with the wild-type gene, and CGRP mRNA is produced when they are transfected with the exon-4 splice-site mutant, the lymphocytes must contain all the processing factors necessary to generate both mRNAs.

B. If selection of a polyadenylation site was the critical choice in the expression of calcitonin mRNA in the lymphocyte cell line, then the mutant that was missing the exon-4 polyadenylation site would be expected to produce CGRP mRNA. If the splicing of exon 3 to exon 5 (to produce CGRP mRNA) is precluded by use of the polyadenylation site in exon 4, then removal of the site should permit CGRP mRNA production.

   By contrast, the mutant lacking the exon-4 splice site might still be expected to be preferentially polyadenylated at exon 4, which would prevent production of CGRP mRNA. (As explained in part D, although CGRP mRNA is not made in the exon-4 splice-site mutant, the aberrant RNA that is generated does not match this simple expectation.)

C. If selection of the exon-4 splice site was the critical choice in the expression of calcitonin mRNA in the lymphocyte cell line, then the mutant that was missing the exon 4 splice site would be expected to produce CGRP mRNA. If the splicing of exon 3 to exon 4 is favored in lymphocytes, then removal of the exon-4 splice site should permit the exon-5 splice site to be used, thus generating CGRP mRNA.

   By contrast the mutant lacking the exon-4 polyadenylation site might still be expected to splice exon 3 to exon 4 preferentially, which might be expected to lead to an aberrant RNA containing the fourth intron along with exons 5 and 6. (As explained in part D, although an aberrant RNA is made, it is not the one expected by this simple reasoning.)

D. The predictions of the splice-site-selection model best match the results from the two mutants. As explained in parts B and C, the splice-site-selection model correctly predicts that CGRP mRNA will be made by the mutant lacking the exon-4 splice site. The polyadenylation-site-selection model, by contrast, predicts incorrectly that CGRP mRNA will be made by the mutant that is missing the exon-4 polyadenylation site. Thus, these results favor splice-site selection as the critical choice that explains the ability of the lymphocyte cell line to produce calcitonin mRNA instead of CGRP mRNA.

   However, neither of these simple models predicts the structure of the aberrant RNA that is produced when the exon-4 polyadenylation mutant is transfected into the lymphocyte cell line. The aberrant RNA retains both the third and the fourth introns even though the 5′ splice site in exon 3 and the 3′ splice sites in exons 4 and 5 are both present. Neither simple model for differential processing predicts this result. For this reason, differential processing of the calcitonin/CGRP transcript is thought to be a somewhat more complex version of the splice-site-selection model, whose details are not yet understood.

   Reference: Leff, S.E.; Evans, R.M.; Rosenfeld, M.G. Splice commitment dictates neuron-specific alternative RNA processing in calcitonin/CGRP gene expression. *Cell* 48:517–524, 1987.

9–33

A. The percentage of total protein synthesis that is due to synthesis of ferritin for each sample is equal to ferritin synthesis divided by the total protein synthesis.

Table 9–6 Synthesis of Ferritin in the Rat After Various Treatments (Answer 9–33)

Injection	Actinomycin D	Fraction	Total Synthesis	Ferritin Synthesis	Percent Ferritin	Adjusted Ferritin	Distribution of Ferritin
Saline	absent	polysomes	750,000	700	0.093	0.079	49%
		supernatant	255,000	1400	0.549	0.082	51%
						0.161	
Iron	absent	polysomes	500,000	900	0.180	0.153	89%
		supernatant	400,000	500	0.125	0.019	11%
						0.172	
Saline	present	polysomes	800,000	800	0.100	0.085	53%
		supernatant	600,000	3000	0.500	0.075	47%
						0.160	
Iron	present	polysomes	780,000	1380	0.177	0.150	89%
		supernatant	550,000	700	0.127	0.019	11%
						0.169	

For the first sample, this value is $700/750,000 = 0.093$. All other values are listed in Table 9–6.

To obtain the distribution of ferritin mRNA in the polysomal and supernatant fractions, the ferritin synthesis in each sample must be adjusted to take into account the distribution of bulk mRNA in the polysomal and supernatant fractions. To do this, it is necessary to multiply the percentage of ferritin synthesis in the polysomal fraction by 0.85 and to multiply the percentage of ferritin synthesis in the supernatant fraction by 0.15. These values are listed in Table 9–6 under the column labeled "adjusted ferritin." The percentage of total ferritin mRNA in the polysomal fraction from a particular treatment is equal to the "adjusted ferritin" divided by the total "adjusted ferritin." For the first sample this value is $0.079/0.161 = 49\%$. The rest of the values are listed in the last column in Table 9–6.

B. Two features of the data show clearly that iron does not control ferritin synthesis by regulating the rate of transcription. First, injection of iron does not increase the amount of ferritin mRNA. As shown under the "adjusted ferritin" column, the total amount of ferritin mRNA present after saline injection or iron injection is nearly the same. Second, the RNA synthesis inhibitor actinomycin D has no effect on the amount of total ferritin mRNA. If the increased synthesis of ferritin induced by iron were due to an increase in the amount of ferritin mRNA, the increase in ferritin mRNA should have been detectable and it should have been blocked by actinomycin D.

C. The major effect of iron is to alter the distribution of ferritin mRNA between the polysomal and supernatant fractions. In the absence of iron about 50% of the ferritin mRNA is not bound to polysomes. When iron is present, about 90% of the ferritin mRNA is in the polysome fraction. Thus, the fraction of ferritin mRNA on polysomes increases by nearly a factor of two in the presence of iron. This shift from free mRNA to polysomal mRNA accounts nicely for the twofold increase in ferritin synthesis in the presence of iron, since only mRNAs that are in polysomes are translated into protein.

The molecular mechanism by which ferritin mRNA is kept from binding to ribosomes and entering the polysomal fraction is not yet known. Ferritin genes from rat, human, chicken, and frog have a highly conserved 28 nucleotide sequence (the iron-response element) in their 5' untranslated regions that is essential for regulation by iron. It appears that an iron-sensitive regulatory protein binds to these regions of ferritin mRNAs and prevents ribosome binding. In the presence of iron the ferritin mRNAs are released and bound by ribosomes with a consequent increase in ferritin synthesis. The protein no longer binds to ferritin mRNA, thereby permitting its translation.

References: Zahringer, J.; Baliga, B.S.; Munro, H.N. Novel mechanism for translational control in regulation of ferritin synthesis by iron. *Proc. Natl. Acad Sci. USA* 73:857–861, 1976.

Liebold, E.A.; Munro, H.N. Cytoplasmic protein binds *in vitro* to a highly conserved sequence in the 5′ untranslated region of ferritin heavy- and light-subunit mRNAs. *Proc. Natl. Acad. Sci. USA* 85:2171–2175, 1988.

9–34

A. The concentration of ribosomes in a reticulocyte lysate is 2.5×10^{-7} M, and the concentration of HCR that completely blocks protein synthesis is 5.6×10^{-9} M.

$$[\text{ribosomes}] = \frac{1\,\text{mg}}{\text{ml}} \times \frac{1000\,\text{ml}}{\text{L}} \times \frac{\text{mole}}{4 \times 10^6\,\text{g}} \times \frac{\text{g}}{10^3\,\text{mg}}$$

$$[\text{ribosomes}] = 2.5 \times 10^{-7}\,\text{M}$$

$$[\text{HCR}] = \frac{1\,\mu\text{g}}{\text{ml}} \times \frac{1000\,\text{ml}}{\text{L}} \times \frac{\text{mole}}{180,000\,\text{g}} \times \frac{\text{g}}{10^6\,\mu\text{g}}$$

$$[\text{HCR}] = 5.6 \times 10^{-9}\,\text{M}$$

Thus, when protein synthesis is completely blocked, the ratio of HCR molecules to ribosomes is 1 to 45 ($2.5 \times 10^{-7}/5.6 \times 10^{-9} = 45$). Since there are an average of 4 ribosomes per globin mRNA, the ratio of HCR molecules to globin mRNA molecules is about 1 to 10.

B. The ratios of HCR molecules to ribosomes and to globin mRNA indicate that HCR is unlikely to inhibit protein synthesis by a stoichiometric interaction with either of these components. These ratios, however, do not rule out the possibility that HCR stoichiometrically inactivates some other factor that is essential for protein synthesis, and that is present at a concentration tenfold below that of globin mRNA.

It is now known that HCR is a protein kinase that inactivates protein synthesis by a catalytic mechanism. HCR phosphorylates the initiation factor eIF-2.

Reference: Farrell, P.J.; Balkow, K.; Hunt, T.; Jackson, R.J.; Trachsel, H. Phosphorylation of initiation factor eIF-2 and the control of reticulocyte protein synthesis. *Cell* 11:187–200, 1977.

*9–35 **References:** Gay, D.A.; Yen, T.J.; Lau, J.T.Y.; Cleveland, D.W. Sequences that confer β-tubulin autoregulation through modulated mRNA stability reside within exon 1 of a β-tubulin mRNA. *Cell* 50:671–679, 1987.

Yen, T.J.; Machlin, P.S.; Cleveland, D.W. Autoregulated instability of β-tubulin mRNAs by recognition of the nascent amino terminus of β-tubulin. *Nature* 334:580–585, 1988.

*9–36 **Reference:** Powell, L.M., et al. A novel form of tissue-specific RNA processing produces apolipoprotein-B48 in intestine. *Cell* 50:831–840, 1987.

9–37

A. The mRNA from the *fos*-globin-*fos* hybrid gene, which includes the 3′ end of the *c-fos* gene, has the same stability characteristics as the mRNA from the normal *c-fos* gene (Figure 9–25A and C). By contrast, the *fos*-globin hybrid gene, which includes the same 5′ *fos* sequences as the *fos*-globin-*fos* gene (but is missing the 3′ *fos* sequences) is very stable (Figure 9–25B). Thus, the 3′ end of the human *c-fos* gene confers instability on the *c-fos* mRNA. The instability element almost certainly must be included in the mRNA to make it unstable. Therefore, the element is presumably located in the 3′ exon of the *c-fos* gene, rather than in the 3′ flanking sequences.

B. Although not obvious at first, the behavior of the mRNA from the *fos*-globin hybrid gene can be accounted for in terms of mRNA stability. If the *fos*-globin mRNA is very stable—like normal globin mRNA—then a low rate of transcription in the absence of serum is sufficient to allow the mRNA to accumulate to high levels in the 24-hour period before serum was added and the first measurement was made. If the *fos*-globin mRNA is already present at high levels, then the transient burst of transcription that follows serum addition will not appreciably increase the total amount of the *fos*-globin mRNA. Thus, the enhanced stability of the mRNA from the *fos*-globin hybrid gene can account for both the high initial levels and the lack of induction observed with the *fos*-globin hybrid gene.

These results emphasize the need for instability if a system must respond rapidly to change. This requirement has many familiar analogues in everyday life. For example, if images on a TV screen persisted for more than a fraction of a second, moving objects would be trailed by their ghosts. In a similar way, echoing acoustics blur the perception of both speech and music. Biological signaling pathways have built-in mechanisms to return the system to the starting state: old signals are continually "erased" so that they do not blur the perception of new signals.

Reference: Treisman, R. Transient accumulation of *c-fos* RNA following serum stimulation requires a conserved 5′ element and *c-fos* 3′ sequences. *Cell* 42:889–902, 1985.

Wilson, T.; Treisman, R. Removal of poly(A) and consequent degradation of c-*fos* mRNA facilitated by 3′ AU-rich sequences. *Nature* 336:396–399, 1988.

Membrane Structure

- The Lipid Bilayer
- Membrane Proteins

The Lipid Bilayer

10–1

A. lipid bilayer
B. phospholipids
C. amphipathic
D. polar, hydrophobic hydrocarbon (fatty acid)
E. micelles, bilayers
F. liposomes, black. (The black membranes are so called because destructive interference between the reflected light from the two surfaces makes them appear black.)
G. phospholipid translocators
H. phase transition
I. cholesterol
J. glycolipids
K. gangliosides, G_{M1}

10–2

A. True
B. False. Lipid bilayers are thermodynamically stable structures. Energy is required to make the various lipids in the first place, but no energy is needed to maintain the bilayer arrangement in the membrane.
C. True
D. False. The length of the fatty acid side chains and the number of *cis*-double bonds they contain are also important determinants of membrane fluidity.
E. False. Although the choline head group is indeed positively charged, it is linked to the rest of the lipid via a phosphodiester bond, and the phosphate carries a negative charge at physiological pH. Hence, phosphatidylcholine does not carry a *net* charge.
F. True
G. True. (Glycolipids are synthesized in the lumen of the Golgi apparatus and cannot flip-flop across the bilayer. The enzymes that add carbohydrates to lipids are not found free in the cytoplasm.)

*10–3

10–4

A. Only phosphatidylserine and phosphatidylethanolamine have primary amino groups, which can react with SITS. Since these phospholipids are labeled only when the red cells are made permeable (ghosts), they presumably reside in the inner monolayer. This conclusion is supported by the results from experiments with sea snake venom, which degrades phosphatidylserine and phosphatidylethanolamine only in ghosts. These results, taken together, indicate that the phosphatidylserine and phosphatidylethanolamine are localized almost exclusively in the inner monolayer of red cell membranes.

    Phospholipase degradation of phosphatidylcholine and sphingomyelin in intact cells indicates that they are present in the outer monolayer. This

---

*Problems with an asterisk (*) are answered in the Instructor's Manual.

conclusion depends on the red cell remaining intact during the treatment. In the case of sea snake venom, the absence of degradation of phosphatidylserine and phosphatidylethanolamine in intact cells provides an internal control. In the case of sphingomyelinase, there is no internal control, but the absence of lysis shows that the membrane is intact.

The results in Table 10–2 do not exclude the possibility that phosphatidylcholine and sphingomyelin are also located in the inner monolayer. However, the quantitation of sphingomyelin degradation by sphingomyelinase (provided in the body of the problem) indicates that sphingomyelin is localized almost entirely in the outer monolayer of the membrane. No such data are provided for phosphatidylcholine; thus, it is incorrect to conclude from the data given that phosphatidylcholine is located exclusively in the outer monolayer. However, other experiments not reported here do suggest that phosphatidylcholine is found almost entirely in the outer monolayer.

B. You chose red cells for these experiments because they contain no internal membranes. If the same experiments were performed on cells with internal membranes, it would have been impossible to measure directly the phospholipid composition of the inner monolayer, since phospholipids from the inner monolayer would have been hopelessly confused with those from internal membranes.

References: Bretscher, M. Asymmetrical lipid bilayer structure for biological membranes. *Nature New Biol.* 236:11–12, 1972.

Deenen, L.L.M.; DeGier, J. Lipids of the red cell membrane. In The Red Blood Cell (D. MacN. Surgenor, ed.), pp. 147–211. New York: Academic Press, 1974.

*10–5   **Reference:** Rousselet, A.; Guthmann, C.; Matricon, J.; Bienvenue, A.; Devaux, P.F. Study of the transverse diffusion of spin labeled phospholipids in biological membranes: 1. Human red blood cells. *Biochim. Biophys. Acta* 426:357–371, 1976.

10–6

A. One can estimate the half-time for flip-flop in these experiments by extending the curve in Figure 10–3 to the point at which 50% of the ESR signal is lost. For cells labeled in the inner monolayer, these data suggest a half-time for flip-flop of about 7 hours. For cells labeled in the outer monolayer, the half-time of flip-flop is much longer but cannot be estimated reliably. These data indicate that the rate of flip-flop of phospholipids between the two monolayers of the plasma membrane in red cells is extremely low. Similar experiments using synthetic bilayers have given even longer times; in fact, in the best experiments, when great care was taken not to allow oxidation or other damage to the lipids, the rate of flip-flop was immeasurably low (less than once per month).

B. Phospholipid 2 was used to label the inner monolayer, and phospholipid 1 was used to label the outer monolayer. As shown by the experiments in Figure 10–2B, phospholipid 2 in the inner monolayer is not reduced by the cytoplasm of red cells; when it is present in the outer monolayer, it can be reduced by ascorbate. Thus, phospholipid 2 is appropriate for measuring the rate of flip-flop from the inner to the outer monolayer. As shown by the experiments in Figure 10–2A, phospholipid 1 in the inner monolayer is reduced by red cell cytoplasm, but it is stable in the outer monolayer in the absence of ascorbate. Thus, phospholipid 1 is appropriate for measuring the rate of flip-flop from the outer to the inner monolayer.

C. One can make intact red cells with spin-labeled phospholipids exclusively in the inner monolayer by introducing phospholipid 2 into the membrane and then incubating the red cells for 1 hour in the presence of ascorbate. Ascorbate reduces the lipids in the outer monolayer, leaving red cells that are labeled only in the inner monolayer. Similarly, one can make intact red cells with spin-labeled phospholipids exclusively in the outer monolayer by introducing phospholipid 1 into the membrane and then incubating the red cells for 15 minutes in the absence of ascorbate. In this case the spin-labeled lipids in the inner monolayer are reduced by agents in the cytoplasm, leaving red cells that are labeled only in the outer monolayer.

Reference: Rousselet, A.; Guthmann, C.; Matricon, J.; Bienvenue, A.; Devaux, P.F. Study of the transverse diffusion of spin labeled phospholipids in biological membranes: 1. Human red blood cells. *Biochim. Biophys. Acta* 426:357–371, 1976.

Membrane Proteins

10–7

A. transmembrane
B. glycosylphosphatidylinositol (GPI) anchor
C. peripheral membrane, integral membrane
D. single-pass transmembrane, multipass transmembrane
E. porins
F. detergents
G. vectorial labeling
H. spectrin
I. band 3 protein
J. freeze-fracture
K. bacteriorhodopsin
L. rotational, lateral, flip-flop
M. heterocaryons
N. flourescence recovery after photobleaching (FRAP)
O. cell coat, glycocalyx
P. lectins
Q. selectins

10–8

A. True
B. False. This statement was once thought to be an accurate description of biological membranes, but it is now clear that many membrane proteins are inserted directly through the bilayer itself.
C. False. Such proteins are more likely to have their transmembrane segments arranged as α helices. An α-helical structure allows all of the hydrogen-bonding moieties along the peptide backbone to be satisfied, whereas a β-sheet structure can satisfy its hydrogen-bonding moieties only if it forms a closed barrel, which requires about eight strands.
D. True.
E. False. The reducing environment of the cytosol prevents formation of disulfide bonds on the cytosolic side of the membrane. Disulfide bonds form readily on the noncytosolic side.
F. True
G. False. Human red blood cells contain no internal membranes; at an early stage in their development they extrude their nuclei.
H. True
I. True
J. True
K. True
L. False. The apical and basolateral surfaces of epithelial cells, which are separated by intercellular tight junctions, have different lipid compositions.
M. True
N. False. The carbohydrate on internal membranes is directed away from the cytosol.

O. False. Although carbohydrate is attached mainly to integral membrane proteins, the glycocalyx can also contain adsorbed glycoproteins and proteoglycans.

10–9   Thus far, arrangements A, B, D, E, F, and I have been found in biological membranes. Arrangement C, which has carbohydrate on the cytoplasmic side of the membrane, does not seem to exist. Arrangements G and H, which show proteins completely buried or with only their tips embedded in the membrane, have not been found and are thought to be unlikely to occur on theoretical grounds.

*10–10

10–11

A. The elimination of sialic acid staining after sialidase treatment indicates that carbohydrate is exposed on the external surface. Because the carbohydrate is attached to glycophorin, it follows that glycophorin is also exposed on the external surface. This conclusion is supported by the results with pronase digestion, which eliminates sialic acid staining, presumably by clipping the peptide backbone. The results with pronase digestion indicate that band 3 is exposed to the external surface as well. In this case the appearance of the new protein band at about 70,000 daltons allows you to estimate that approximately 30,000 daltons of band 3 are exposed on the external surface. In neither digestion were the two spectrin bands affected, suggesting that spectrin is not exposed on the external surface.

B. One direct experimental approach for testing your colleague's objection is to break open the red cell ghosts before digesting them with pronase. If spectrin is resistant to pronase, its mobility on SDS polyacrylamide gels should be unaltered. However, if spectrin is located on the cytoplasmic surface, its mobility should be altered dramatically. These control experiments have been done; they show that spectrin is sensitive to pronase.

   Another approach is to make inside-out ghosts and see if it is possible to dissociate spectrin from the membrane by treatments that do not actually disrupt the membrane. This approach also has been successful, confirming that spectrin is on the cytoplasmic side and is not embedded in the membrane.

C. To determine which of the red cell proteins span the membrane using this basic experimental approach, it is necessary to prepare inside-out vesicles. Such vesicles can be prepared readily from red cell ghosts by disrupting them and allowing them to reseal under defined ionic conditions. When inside-out vesicles are treated with pronase, the mobilities and band 3 and glycophorin are altered. These results, along with the results above, indicate that band 3 and glycophorin are exposed on both surfaces of the red cell membrane and, therefore, must be transmembrane proteins.

References: Bennett, V.; Stenbuck, P.J. The membrane attachment protein for spectrin is associated with band 3 in human erythrocyte membranes. *Nature* 280:468–473, 1979.

Bennett, V.; Stenbuck, P.J. Association between ankyrin and the cytoplasmic domain of band 3 isolated from the human erythrocyte membrane. *J. Biol. Chem.* 255:6424–6432, 1980.

*10–12

10–13   The presence of both proteins in the pellet, as in mixture 3, 4, and 6 in Table 10–4, indicates an interaction between the proteins. The results with pairwise mixtures suggest that the four proteins are arranged as shown below:

band 3—ankyrin—spectrin—actin

An artist's conception of how these molecules are linked together to form the supporting meshwork on the cytoplasmic surface of red cells is shown in MBOC Figure 10–27. In reality, the interaction between actin and spectrin is too weak to be detected by this method, unless a third protein, band 4.1, is added.

*10–14

*10–15   **Reference:** Van Meer, G.; Simons, K. The function of tight junctions in maintaining differences in lipid composition between the apical and the basolateral cell surface domains of MDCK cells. *EMBO J.* 5:1455–1464, 1986.

Membrane Transport of Small Molecules and the Ionic Basis of Membrane Excitability

- **Principles of Membrane Transport**
- **Carrier Proteins and Active Membrane Transport**
- **Ion Channels and Electrical Properties of Membranes**

Principles of Membrane Transport

11–1

- A. membrane transport
- B. carrier, channel
- C. passive (facilitated diffusion), active
- D. electrochemical
- E. ionophores, mobile-ion, channel

11–2

- A. False. Lipid bilayers are impermeable to ions, but the plasma membrane contains specific ion channels and carriers that make it very permeable to particular ions under certain circumstances.
- B. True
- C. False. For carrier proteins to work like revolving doors would require that they flip-flop across the bilayer—which they do not.
- D. False. Such studies have revealed that membrane transport is mediated by a surprisingly *small* number of protein families, whose members have a common evolutionary origin. Each family, however, can contain a very *large* number of isoforms.
- E. True

11–3

- A. Cytochalasin B inhibits glucose transport competitively, suggesting that it binds at or near the site of D-glucose binding. If an excess of D-glucose is present, the binding site on the transporter will be occupied by D-glucose, preventing cytochalasin from binding and thereby interfering with cross-linking. On the other hand, L-glucose does not interfere with cross-linking, because it does not bind to the transporter and protect it from the binding of cytochalasin.
- B. Since treatment of the native glucose transporter with an enzyme that removes oligosaccharide side chains sharpens the electrophoretic band, the fuzziness must be due to heterogeneity of the carbohydrate moieties attached to the protein. Whether this heterogeneity represents variable occupancy of potential oligosaccharide addition sites on the protein or is due to actual variability in the length or sequence of the oligosaccharide side chains is not known. This degree of heterogeneity is unusual; most glycoproteins form much sharper bands on SDS polyacrylamide gels. Although there are about 350,000 molecules of this protein in each red cell (about the same as glycophorin and band 3), it went unnoticed for many years. In addition, there was considerable controversy about the molecular identity of the glucose transporter, with estimates of its molecular weight going as high as 200,000. All because it was a fuzzy band.

  Reference: Allard, W.J.; Lienhard, G.E. Monoclonal antibodies to the glucose transporter from human erythrocytes: identification of the transporter as a Mr = 55,000 protein. *J. Biol. Chem.* 260:8668–8675, 1985.

*11–4   **Reference:** Moriyoshi, K.; Masu, M.; Ishii, T.; Shigemoto, R.; Mizuno, N.; Nakanishi, S. Molecular cloning and characterization of the rat NMDA receptor. *Nature* 354:31–37, 1991.

Carrier Proteins and Active Membrane Transport

11–5

A. uniporters
B. symport
C. anion, antiport
D. Na⁺-K⁺ pump (Na⁺-K⁺ ATPase)
E. sarcoplasmic reticulum
F. secondary, primary
G. transcellular transport
H. microvilli
I. ABC transporter superfamily
J. multidrug resistance (MDR) protein

11–6

A. True
B. True
C. False. The Na⁺-K⁺ ATPase drives three Na⁺ ions out of the cell for every two K⁺ ions it pumps in. Therefore, it is electrogenic: it drives a net current across the membrane.
D. False. The flow of Ca²⁺ from the sarcoplasmic reticulum, where its concentration is high, to the cytosol, where its concentration is low, does not require an ATP-driven pump. The Ca²⁺ pump is responsible for the reverse process: pumping cytosolic Ca²⁺ back into the sarcoplasmic reticulum.
E. False. There is no direct connection between the light-activated proton pump (which indeed pumps protons out of the bacterium in sunlight) and the synthesis or hydrolysis of ATP. However, the bacteria do contain a separate membrane protein that can act as an ATP synthase when protons pass through it in an inward direction.
F. True
G. True

*11–7

11–8

A. If the entire free-energy change due to ATP hydrolysis ($\Delta G = -12$ kcal/mole) could be used to drive transport, then the maximum concentration gradient that could be achieved by ATP hydrolysis would have a free-energy change of +12 kcal/mole.

$$\Delta G_{in} = -2.3RT \log_{10} \frac{C_o}{C_i} + zFV$$

Rearranging the equation gives

$$\log_{10} \frac{C_o}{C_i} = \frac{-\Delta G_{in} + zFV}{2.3RT}$$

For an uncharged solute, the electrical term (zFV) drops to 0. Thus,

$$\log_{10} \frac{C_o}{C_i} = \frac{-\Delta G_{in}}{2.3RT}$$

Substituting for ΔG_{in}, R, and T gives

---

*Problems with an asterisk (*) are answered in the Instructor's Manual.

$$\log_{10}\frac{C_o}{C_i} = \frac{-12 \text{ kcal}/\text{mole}}{2.3 \times 1.98 \times 10^{-3} \text{ kcal}/{}^\circ\text{K mole} \times 310^\circ\text{K}}$$

$$\log_{10}\frac{C_o}{C_i} = -8.50$$

$$\log_{10}\frac{C_i}{C_o} = 8.50$$

$$\frac{C_i}{C_o} = 3.2 \times 10^8$$

Thus, for an uncharged solute a transport system that couples hydrolysis of 1 ATP to transport of 1 solute molecule could, in principle, drive a concentration difference across the membrane of more than eight orders of magnitude. Amazing!

B. If the entire free-energy change due to ATP hydrolysis ($\Delta G = -12$ kcal/mole) could be used to drive transport of Ca^{2+} out of the cell, then the maximum concentration gradient would yield a free-energy change of +12 kcal/mole.

$$\Delta G_{out} = 2.3RT \log_{10}\frac{C_o}{C_i} - zFV$$

Rearranging the equation gives

$$\log_{10}\frac{C_o}{C_i} = \frac{\Delta G_{out} + zFV}{2.3RT}$$

Since Ca^{2+} is charged, the electrical term must be included. Substituting for ΔG_{out}, R, T, z, F, and V, gives

$$\log_{10}\frac{C_o}{C_i} = \frac{12 \text{ kcal}/\text{mole} + (2 \times 23 \text{ kcal}/\text{V mole} \times -0.06 \text{ V})}{2.3 \times 1.98 \times 10^{-3} \text{ kcal}/{}^\circ\text{K mole} \times 310^\circ\text{K}}$$

$$\log_{10}\frac{C_o}{C_i} = 6.54$$

$$\frac{C_o}{C_i} = 3.5 \times 10^6$$

Thus a transport system that couples hydrolysis of 1 ATP to transport of 1 Ca^{2+} ion on the outside of the cell could, in principle, drive a concentration difference across the membrane of more than six orders of magnitude. Note, by comparison with uncharged solute, that pumping against the membrane potential reduces the theoretical limit by two orders of magnitude. The difference in Ca^{2+} concentration across a typical mammalian plasma membrane is more than four orders of magnitude, but well within the theoretical limit.

C. The free-energy change for transporting Na^+ out of the cell is

$$\Delta G_{out} = 2.3RT \log_{10}\frac{C_o}{C_i} - zFV$$

Substituting (with $2.3RT = 1.41$ kcal/mole),

$$\Delta G_{out} = \left(1.41\frac{\text{kcal}}{\text{mole}} \times \log_{10}\frac{145 \text{ mM}}{10 \text{ mM}}\right) - \left(1 \times \frac{23 \text{ kcal}}{\text{V mole}} \times -0.06 \text{ V}\right)$$

$$\Delta G_{out} = 3.0 \text{ kcal}/\text{mole Na}^+$$

$$\Delta G_{out} = 9.0 \text{ kcal}/3 \text{ mole Na}^+$$

The free-energy change for transporting K^+ into the cell is

$$\Delta G_{in} = -2.3RT \log_{10}\frac{C_o}{C_i} + zFV$$

Substituting (with $2.3 RT = 1.41$ kcal/mole)

$$\Delta G_{in} = -\left(1.41\frac{kcal}{mole} \times \log_{10}\frac{5\ mM}{140\ mM}\right) + \left(1 \times \frac{23\ kcal}{V\ mole} \times -0.06\ V\right)$$

$$\Delta G_{in} = 0.66\ kcal\ /\ mole\ K^+$$

$$\Delta G_{in} = 1.3\ kcal\ /\ 2\ mole\ K^+$$

The overall free-energy change for the Na$^+$-K$^+$ pump is

$$\Delta G = \Delta G_{out} + \Delta G_{in}$$

$$\Delta G = 9.0\ kcal\ /\ 3\ mole\ Na^+ + 1.3\ kcal\ /\ 2\ mole\ K^+$$

$$\Delta G = 10.3\ kcal\ /\ (3\ mole\ Na^+\ and\ 2\ mole\ K^+)$$

D. Since the hydrolysis at ATP provides 12 kcal/mole and the pump requires 10.3 kcal to transport 3 Na$^+$ out and 2 K$^+$ in, the efficiency of the Na$^+$-K$^+$ pump is

$$eff = \frac{10.3}{12.0}$$

$$= 86\%$$

Even with this remarkable efficiency the Na$^+$-K$^+$ pump typically accounts for a third of a mammalian cell's energy requirements and thus, presumably, a corresponding fraction of a mammal's total caloric intake.

*11–9

Ion Channels and Electrical Properties of Membranes

11–10

   A. channel, ion channels
   B. voltage changes, mechanical stimulation, ligand binding
   C. membrane potential, K$^+$ leak
   D. resting membrane potential, Nernst
   E. cell body, axon, dendrites
   F. action potential
   G. voltage-gated Na$^+$ channels, Na$^+$
   H. voltage-gated K$^+$ channels, K$^+$
   I. myelin sheath
   J. patch-clamp
   K. synapse, neurotransmitter, transmitter-gated
   L. excitatory neurotransmitters, inhibitory neurotransmitters
   M. neuromuscular junction
   N. spatial, temporal, grand
   O. axon hillock
   P. delayed K$^+$, early K$^+$
   Q. Ca^{2+}-activated K$^+$, adaptation
   R. long-term potentiation
   S. NMDA receptor

11–11

   A. True
   B. False. Carrier proteins *and* channel proteins saturate. It is thought that permeating ions have to shed most of their associated water molecules in order to pass, in single file, through the narrowest part of the channel; this limits their rate of passage. Thus, as ion concentrations are increased, the flux of ions through a channel increases proportionally but then levels off (saturates) at a maximum rate.

C. False. Although the Na$^+$-K$^+$ pump is electrogenic and makes a small (20%) direct contribution to the membrane potential because of the unequal stoichiometry of the exchange, the K$^+$ leak channel, which endows the membrane with a selective permeability to K$^+$, is responsible for the major portion of the membrane potential. (The Na$^+$-K$^+$ pump does make an important though indirect contribution to the membrane potential by maintaining the K$^+$ gradient across the membrane.)

D. True

E. True

F. True

G. False. Channels open in an all-or-nothing fashion. Thus the aggregate current does not indicate the degree to which individual channels are open but rather the total number of channels in the membrane that are open at any one time.

H. True

I. True

J. False. The concentration of Cl$^-$ is much higher outside the cell than inside. Thus, opening Cl$^-$ channels tends to increase the membrane potential, leading to hyperpolarization rather than depolarization.

K. True

L. True

M. False. Excitatory and inhibitory PSPs are summed spatially and temporally into a grand PSP. Only when the grand PSP is excitatory and of sufficient magnitude will the postsynaptic cell fire an action potential along its axon.

N. True

O. True

11–12

A. The expected membrane potential due to differences in K$^+$ concentration across the resting membrane is

$$V = 58 \text{ mV} \times \log_{10} \frac{C_o}{C_i}$$

$$V = 58 \text{ mV} \times \log_{10} \frac{9 \text{ mM}}{344 \text{ mM}}$$

$$V = -92 \text{ mV}$$

For Na$^+$, the equivalent calculation gives a value of +48 mV.

The assumption that the membrane potential is due solely to K$^+$ leads to a value near that of the resting potential. The assumption that the membrane potential is due solely to Na$^+$ leads to a value near that of the action potential.

These assumptions approximate the resting potential and action potential because K$^+$ *is* primarily responsible for the resting potential and Na$^+$ *is* responsible for the action potential. A resting membrane is 100-fold more permeable to K$^+$ than it is to Na$^+$ because of the presence of K$^+$ leak channels. The leak channel allows K$^+$ to leave the cell until the membrane potential rises sufficiently to oppose the K$^+$ concentration gradient. The theoretical maximum gradient (based on calculations like those above) is lowered somewhat by the entrance of Na$^+$, which carries positive charge into the cell (compensating for the positive charges on the exiting K$^+$). Were it not for the Na$^+$-K$^+$ pump, which continually removes Na$^+$, the resting membrane potential would be dissipated completely.

The action potential is due to a different channel, a voltage-gated Na$^+$ channel. These channels open when the membrane is stimulated, allowing Na$^+$ ions to enter the cell. The magnitude of the resulting membrane potential is limited by the difference in the Na$^+$ concentrations across the membrane.

The influx of Na^+ reverses the membrane potential locally, which opens adjacent Na^+ channels and ultimately causes an action potential to propagate away from the site of original stimulation.

B. The substitution of choline chloride for sodium chloride eliminates the action potential, as expected, since the action potential is due to specific Na^+ channels. As illustrated in the calculation above, the difference in concentrations of Na^+ across a membrane determines the magnitude of the action potential that results from Na^+ influx. Thus, if the Na^+ concentration outside the cell were reduced to half the normal value, the calculated membrane potential would be reduced to half the value calculated above. Measurements of the action potential for various mixtures of choline chloride and sodium chloride match these expectations.

Reference: Hille, B. Ionic Channels of Excitable Membranes, pp. 23–57. Sunderland, MA: Sinauer, 1984.

*11–13 Reference: Hille, B. Ionic Channels of Excitable Membranes, pp. 1–19. Sunderland, MA: Sinauer, 1984.

11–14

A. The resting potential in normal seawater is

$$V = 58\ \text{mV} \times \log_{10} \frac{9\ \text{mM}}{344\ \text{mM}} = -92\ \text{mV}$$

whereas the resting potential in seawater that is 60 mM KCl is

$$V = 58\ \text{mV} \times \log_{10} \frac{60\ \text{mM}}{344\ \text{mM}} = -44\ \text{mV}$$

Thus, the magnitude of the resting potential decreases by a factor of about 2.

B. The lack of activation in the absence of calcium suggests that activation requires an influx of Ca^{2+} into the cell from the extracellular fluid. The change in membrane potential in response to increased KCl opens a voltage-gated Ca^{2+} channel in the plasma membrane. The influx of calcium then initiates the other intracellular changes.

C. As you might expect from the answer to part B, addition of A23187 activates intracellular changes in the egg in regular seawater but not in calcium-free seawater.

   Not all eggs are like clam eggs, although many can be activated by some manipulation of their ionic environment. Thus, the sperm does more than just provide DNA. In case you were wondering, eggs activated by ionic changes rather than by sperm do not develop into embryos, although they often go through some abortive cell division cycles.

*11–15 Reference: Hille, B. Ionic Channels of Excitable Membranes, pp. 205–209. Sunderland, MA: Sinauer, 1984.

11–16

When acetylcholine is released from the synaptic vesicles of the neurons, some of the acetylcholine finds target receptors, some diffuses away, but most is rapidly hydrolyzed to acetate and choline, which are taken up by the nerve terminal. When the density of receptors is reduced, the probability diminishes that an acetylcholine molecule will find its receptor before it is hydrolyzed. The suboptimal transmission of the signal is responsible for the muscular weakness of myasthenic patients. One way to overcome their muscular weakness is to increase the concentration of acetylcholine in response to stimulation, in order to compensate for the reduced number of receptors. There are two general ways acetylcholine levels in the synaptic cleft might be increased: more acetylcholine could be released upon stimulation, or its hydrolysis could be inhibited. The drug neostigmine inhibits the enzyme acetylcholinesterase, which is responsible for hydrolysis of acetylcholine in the synaptic cleft, and thereby increases the efficiency of signal transmission across the synapse.

(You may have guessed that neostigmine binds to the acetylcholine receptor and stimulates muscle contraction directly. However, if it did, it would cause widespread contractions, not coordinated movements.)

Reference: Rowland, L.P. Diseases of chemical transmission at the nerve-muscular synapse: myasthenia gravis. In Principles of Neural Science, 2nd ed. (E.R. Kandel, J.H. Schwartz, eds.), pp. 147–185. New York: Elsevier, 1985.

*11–17 **Reference:** Sakmann, B. Elementary steps in synaptic transmission revealed by currents through single ion channels. *Science* 256:503–512, 1992.

11–18

A. The transmembrane segment M2 is responsible for the differences in conductance through the two types of acetylcholine-gated channels in young rat muscle. The results with chimeric cDNAs 3 and 4 indicate that the difference is in transmembrane segments M2, M3, and M4, but the result with chimeric cDNAs 1 and 2 rule out M4. Results with chimeric cDNAs 5 and 6 verify that the difference lies in segments M2 and M3. Finally, results with chimeric cDNAs 7 and 8 pinpoint the difference as transmembrane segment M2.

B. It is likely that the differences in channel conductance are due to the M2 transmembrane segment alone because that segment forms the actual lining of the pore through which the ions flow. The other transmembrane segments presumably serve to hold the M2 segment in the proper orientation. Since the 5 subunits of the acetylcholine-gated cation channel are homologous, it is thought that the pore is lined by the M2 transmembrane segments from each subunit.

C. As illustrated in Figure 11–8, glycine has a smaller side chain than threonine, while leucine has a larger side chain. If the size of the side chain determines the critical constriction in the pore, substitution of a glycine for threonine should increase the size of the pore and permit a freer flow of ions, hence increased current through the channel. By contrast, substitution of leucine for threonine should narrow the pore further and reduce the current.

The M2 transmembrane segments of the γ_1 and γ_2 subunits may differ in analogous ways. At some critical point in the M2 segment the γ_2 subunit may contain a bulkier amino acid side chain than the γ_1 subunit. Such a difference could account for the lower conductance of channels made with the γ_2 subunit relative to those made with the γ_1 subunit.

Reference: Sakmann, B. Elementary steps in synaptic transmission revealed by currents through single ion channels. *Science* 256:503–512, 1992.

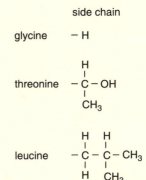

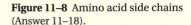

Figure 11–8 Amino acid side chains (Answer 11–18).

Intracellular Compartments and Protein Sorting

The Compartmentalization of Higher Cells

12–1

    A. organelles
    B. lumen
    C. sorting signals
    D. signal peptide, signal peptidase
    E. signal patches

12–2

    A. False. The membranes of intracellular compartments are selectively permeable (not impermeable). Selective permeability is conferred by transport proteins that help to establish the unique chemical identity of each compartment.

    B. True

    C. True

    D. False. The lumen of the ER is topologically equivalent to the outside of the cell, but the interior of the nucleus is topologically equivalent to the cytosol.

    E. True

    F. False. Signal peptides are used to direct proteins into the ER, mitochondria, chloroplasts, peroxisomes, and nucleus. Signal patches identify certain enzymes that are to be marked with specific sugar residues that then direct them from the Golgi apparatus into lysosomes; signal patches may also be used in other sorting steps that have been less well characterized.

    G. True

    H. False. The information required to construct a membrane-bounded organelle, such as the ER or Golgi, does not reside exclusively in the DNA that specifies the organelle proteins. Epigenetic information in the form of at least one distinct protein in the organelle membrane is also required. The epigenetic information is passed from parent cell to progeny cell in the form of the organelle itself.

12–3

    A. If the equivalent of one plasma membrane transits the ER every 24 hours and individual membrane proteins remain in the ER for 30 minutes (0.5 hr), then at any one time 0.021 (0.5 hr/24 hr) plasma membrane equivalents are present in the ER. Since the area of the ER membrane is 20 times greater than the area of the plasma membrane, the fraction of plasma membrane proteins in the ER is 0.021/20 = 0.001. Thus, the ratio of plasma membrane proteins to other membrane proteins in the ER is 1 to 1000. Out of every 1000 proteins in the ER membrane only 1 is in transit to the plasma membrane.

    B. In a cell that is dividing once per day the equivalent of one Golgi apparatus also must transit the ER every 24 hours. Thus, if the membrane of the Golgi apparatus is three times the area of the plasma membrane, three times as many Golgi apparatus membrane proteins will be present in the ER. Therefore, the ratio of Golgi apparatus membrane proteins to other membrane proteins in the ER is 3 to 1000.

C. If the areas of the membranes of all the rest of the compartments are equal to the area of the plasma membrane, then the ratio of membrane proteins bound for these compartments to the membrane proteins in the ER is 1 to 1000. Summing the contributions from all compartments, the ratio of membrane proteins, in transit, to proteins that are permanent residents of the ER membrane is 5 to 1000. Thus, 99.5% of the membrane proteins in the ER are permanent residents.

The Transport of Molecules into and out of the Nucleus

12–4

A. nuclear envelope
B. inner nuclear, nuclear lamina, outer nuclear
C. nuclear pores, nuclear pore complex
D. nuclear localization
E. nuclear lamins

12–5

A. True
B. False. The lipid bilayer of the inner and outer nuclear membranes are continuous with one another around the margin of each nuclear pore.
C. True
D. False. It seems likely that nuclear export, like nuclear import, also requires special recognition systems and a source of energy.
E. True
F. True

12–6

A. The portion of nucleoplasmin responsible for localization in the nucleus must reside in the tail. The nucleoplasmin head does not localize to the nucleus when injected into the cytoplasm, and it is the only injected fragment that is missing the tail.

B. These experiments suggest that the nucleoplasmin tail carries a nuclear localization signal and that accumulation in the nucleus is not the result of passive diffusion. The observations involving complete nucleoplasmin or fragments that retain the tail do not distinguish between the two models; they say only that the tail carries the important part of nucleoplasmin—be it a localization signal or a binding site. The key observations that argue against passive diffusion are the results with the nucleoplasmin head. Its lack of accumulation in the nucleus after injection into the cytoplasm could be rationalized (according to passive diffusion) on the basis of a missing binding site for some nuclear component. If the head is missing a binding site, however, it should not be retained after injection into the nucleus. Its retention in the nucleus suggests that the head is too large to pass through the nuclear pore. Since more massive forms of nucleoplasmin do pass through nuclear pores, passive diffusion of nucleoplasmin appears to be ruled out.
Reference: Dingwall, C.; Sharnick, S.V.; Laskey, R.A. A polypeptide domain that specifies migration of nucleoplasmin into the nucleus. *Cell* 30:449–458, 1982.

*12–7   **Reference:** Barnes, G.; Rine, J. Regulated expression of endonuclease EcoRI in *Saccharomyces cerevisiae*: nuclear entry and biological consequences. *Proc. Natl. Acad. Sci. USA* 82:1354–1358, 1985.

12–8

A. The critical flaw in your original protocol is that the killer protein, which was made at the high temperature but denied access to the nucleus in the

*Problems with an asterisk (*) are answered in the Instructor's Manual.

translocation mutants, will still be present when the cells are shifted to low temperature. When nuclear translocation resumes at the low temperature, the killer protein will be imported and the mutant cells will die. Your original protocol ensured that no new killer protein would be made (by shifting to low temperature in the presence of glucose) but did not take into account the killer protein that had been made in the presence of galactose before the temperature shift.

B. You need to modify your experimental protocol so that the previously made killer protein is rendered inactive before nuclear import is allowed to resume. There are many possible ways to accomplish this. For example, you might try leaving the cells at high temperature in the presence of glucose (so there is no new synthesis) for increasing periods of time to allow the previously made killer protein to be inactivated by normal degradation processes. You might also try to increase its rate of degradation by engineering its N-terminus so that it carries a destabilizing amino acid (MBOC, Chapter 5). Such a modification could turn the killer protein into a very short-acting molecule, which would disappear very rapidly in the absence of new synthesis. Alternatively, you might try to make a temperature-sensitive mutant of EcoRI that is active at the high temperature and inactive at the low temperature. Such a *cold-sensitive* protein would be active when nuclear transport was blocked in the import mutant and inactive when nuclear transport resumed.

C. There are several types of mutants that would be expected to survive the selection protocol without affecting the nuclear-transport machinery. For example, mutants that cannot transport galactose into the cell would be unable to turn on the hybrid gene and thus would survive the shift to high temperature in the presence of galactose. Other mutants in which EcoRI was not expressed from the hybrid gene would also survive the selection scheme. These mutants might include mutations in the promoter (which would prevent expression of the hybrid gene), mutations in the EcoRI portion of the hybrid gene (which would render EcoRI inactive), and mutations in the nuclear localization signal in the hybrid gene (which would prevent import of EcoRI into the nucleus).

Genetic selection schemes are rarely specific for the precise mutant you are seeking. In some cases, other classes can be predicted and screened out by an additional step. However, the real beauty of genetic selection schemes, the aspect that makes them like treasure hunts, is that they often give classes of mutants that you did not foresee. Sometimes these classes turn out to be trivial and uninteresting; however, they can also be extremely informative, leading you to insights that you did not anticipate.

12–9

A. Before phorbol ester treatment NF-κB is in the cytoplasm (Figure 12–3, lanes 5 and 6). After treatment NF-κB is found in the nucleus (lanes 3 and 4). Thus, NF-κB moves from the cytoplasm to the nucleus in response to phorbol ester treatment.

B. Even though treatment with phorbol ester activates protein kinase C, which can alter the activity of target proteins by phosphorylation, it is unlikely that NF-κB is directly activated by phosphorylation. The ability to activate NF-κB by treatment with mild denaturants suggests that inactive NF-κB does not differ from active NF-κB by covalent modification. It seems more likely that the activation of NF-κB is at least one step removed from the activity of protein kinase C.

C. One simple molecular mechanism to account for NF-κB activation by phorbol ester treatment of pre-B cells assumes that NF-κB is inactive because it is complexed with an inhibitor. Inactivation of NF-κB by a bound inhibitor is consistent with the activation of NF-κB by mild denaturants, which could exert their effect by causing the dissociation of the inhibitory subunit. The inhibitor could function by masking the enhancer-binding site of NF-κB and perhaps the nuclear localization signal as well. According to this model, treatment with phorbol ester would activate NF-κB by causing dissociation of the inhibitor. Since the usual effect of phorbol ester treatment is to activate protein kinase C, dissociation of the inhibitor might occur as a consequence of direct phosphorylation of the inhibitor or as a more indirect consequence of the phosphorylation of some other protein.

Reference: Baeuerle, P.A.; Baltimore, D. Activation of DNA-binding activity in an apparently cytoplasmic precursor of the NF-κB transcription factor. *Cell* 53:211–217, 1988.

The Transport of Proteins into Mitochondria and Chloroplasts

12–10

A. matrix space, inner membrane, intermembrane space, outer membrane
B. mitochondrial precursor
C. contact sites
D. thylakoid
E. stroma

12–11

A. False. Mitochondrially encoded proteins are located mostly in the inner mitochondrial membrane; however, the proteins encoded by chloroplasts are located mostly in the thylakoid membrane.
B. False. Mitochondrial signal peptides appear to be amphipathic helices with positively charged amino acids on one side and uncharged hydrophobic amino acids on the other side.
C. True
D. True
E. True
F. False. The ATP-driven cycle of binding and subsequent release of mitochondrial hsp70 is thought to pull the unfolded protein through a transmembrane channel into the matrix.
G. True
H. False. The pores in the outer mitochondrial membrane allow free passage of molecules less than 10,000 daltons. Since most proteins are larger than that, the outer membrane does form a permeability barrier to most proteins.
I. False. Mitochondria use the electrochemical proton gradient to help drive protein import. In chloroplasts, however, the electrochemical proton gradient is across the thylakoid membrane and, therefore, cannot aid import across the outer and inner membranes.

12–12

Normally, translation is much faster than mitochondrial import, so that proteins completely clear the ribosome before interacting with the mitochondrial membrane. By blocking protein synthesis with cycloheximide, you have made the rate of translation artificially slower than the rate of import. Since the signal peptide for protein import into mitochondria resides at the N-terminus, some of the partially synthesized mitochondrial proteins, which are still attached to ribosomes, will be able to interact with the mitochondrial membrane. The attempted import of even one such protein will attach the ribosome and the mRNA (and all other ribosomes translating the same mRNA molecule) to the mitochondrial membrane.

Reference: Kellems, R.E.; Allison, V.F.; Butow, R.A. Cytoplasmic type 80S ribosomes associated with yeast mitochondria. IV. Attachment of ribosomes to the outer membrane of isolated mitochondria. *J. Cell Biol.* 65:1–14, 1975.

A. Import of all four proteins into chloroplasts in your experiments is very efficient, as indicated by the small fraction of precursor (Figure 12–6B, upper band lane 2) that remains after incubation with the chloroplasts. This small fraction of precursor is still outside the chloroplast in all cases because it is digested by added protease (Figure 12–6B, lane 3).

B. Ferredoxin is localized to its normal compartment in your experiments, as indicated by its abundance in the stromal fraction (Figure 12–6B, FDFD lane 5) and its absence from the other fractions. Plastocyanin is also localized to its normal compartment—the thylakoid lumen, since the great majority is associated with thylakoids (Figure 12–6B, PCPC lane 6, lower band) in such a way that it cannot be digested with an added protease (Figure 12–6B, PCPC lane 7).

C. The hybrid proteins are imported into chloroplast compartments that are consistent with their N-terminal sequences. This result is most clear for FDPC, which is localized exclusively in the stromal fraction (Figure 12–6B, FDPC lane 5) just like ferredoxin. By contrast, only a very small fraction of PCFD makes it into the lumen of the thylakoid (Figure 12–6B, PCFD lane 6, lower band). Some seems to be bound to the thylakoid membrane, since it is digested when treated with a protease (Figure 12–6B, compare upper bands, lanes 6 and 7); however, the majority is found in the stromal fraction (Figure 12–6B, lane 5).

There are several reasons why import of PCFD to the thylakoid lumen may be so inefficient. (1) It may be difficult for the ferredoxin domain to cross the thylakoid membrane. However, it does manage to cross the outer and inner membranes and its overall charge is roughly the same as plastocyanin, which does cross the thylakoid membrane. (2) Ferredoxin normally picks up an iron-sulfur center at some point in its maturation. Thus, it is possible that it picks up the iron-sulfur center in the stroma and for that reason cannot cross the thylakoid membrane. (3) Ferredoxin normally functions in association with other proteins in the stroma. It may be that the ferredoxin domain in PCFD associates with these proteins and is, therefore, held in the stroma. These kinds of concerns often complicate otherwise straightforward mixing and matching experiments designed to elucidate protein import signals.

D. The multiple bands in your experiments are the key to understanding protein import into chloroplasts. The highest band for all proteins is the precursor—that is, the primary translation product, since it is present in the *in vitro* translation mixture in the absence of chloroplasts (Figure 12–6B, lane 1). The lower bands have had some portion of the protein removed. Although you anticipated that the protein would be removed from the N-terminus (which is why you set up the experiment as you did), your results confirm that the signal peptides are located at the N-terminus. The two-band pattern is associated with the N-terminal segment from ferredoxin (Figure 12–6B, FD and FDPC lane 2), whereas the three-band pattern is associated with the N-terminal segment from plastocyanin (Figure 12–6B, PC and PCFD, lane 2).

The two-band pattern associated with the ferredoxin signal peptide suggests that one cleavage event accompanies the import of ferredoxin to the stroma. The lower band corresponds to the functional form of ferredoxin in the stroma (Figure 12–6B, compare FD lanes 2 and 5). The three-band pattern associated with the plastocyanin signal peptide suggests that two cleavage events are involved in the import of plastocyanin to the thylakoid lumen. The lowest band corresponds to the functional form of plastocyanin in the thylakoid (Figure 12–6B, compare PC lanes 2 and 7). The intermediate band is probably an intermediate in import, suggesting that the signal peptide has two components that are removed sequentially. The N-terminal component of the signal peptide directs the protein into the stroma, and the C-terminal component directs the protein into the thylakoid lumen.

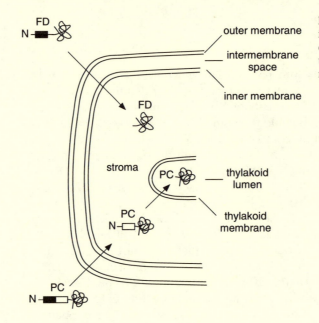

Figure 12–16 A schematic model illustrating the role of signal peptides in directing ferredoxin and plastocyanin to their appropriate chloroplast compartments (Answer 12–14).

E. A model that summarizes the role of the signal peptides in the import of ferredoxin and plastocyanin into their proper chloroplast compartments is illustrated in Figure 12–16. For each protein the signal peptide is shown as a box at the N-terminus of the protein. The shaded portion of the box corresponds to that portion of the signal peptide that is needed for import across the outer and inner chloroplast membranes. This signal peptide is removed as the proteins cross the membranes and enter the stroma. The unshaded segment of the box associated with plastocyanin is needed for import across the thylakoid membrane. It is removed during import into the thylakoid lumen.

Reference: Smeekens, S.; Bauerle, C.; Hageman, J.; Keegstra, K.; Weisbeek, P. The role of the transit peptide in the routing of precursors toward different chloroplast compartments. *Cell* 46:365–375, 1986.

Peroxisomes

12–15

A. peroxisomes
B. glyoxylate cycle, glyoxysomes

12–16

A. False. All eucaryotic cells contain peroxisomes.
B. True
C. False. New peroxisomes arise from preexisting ones by organelle growth and fission. All the membrane proteins and lipids of peroxisomes are imported from the cytosol.

*12–17 Reference: Osinga, K.A.; Swinkels, B.W.; Gibson, W.C.; Borst, P.; Veeneman, G.H.; Van Boom, J.H.; Michels, P.A.M.; Opperdoes, F.R. Topogenesis of microbody enzymes: sequence comparison of the genes for the glycosomal (microbody) and cytosolic phosphoglycerate kinases of *Trypanosoma brucei. EMBO J.* 4:3811–3817, 1985.

12–18

A. Since mRNA from normal cells and each of the mutant cell lines yield equal amounts of the 75-kd form but none of the 53-kd form, and the 53-kd form is present only in the normal cells, it is likely that the 53-kd form arises from the

75-kd form during the process of import into peroxisomes. This suggests that the 53-kd form is the active form of the enzyme. This conclusion is also supported by the observation that the mutant cells, which have no acyl CoA oxidase, have none of the 53-kd form.

B. The mutant cells have only the 75-kd form of the enzyme because their defective peroxisomes cannot import it and process it to the active 53-kd form. Because the 75-kd form disappears so quickly in the pulse-chase experiments in the mutant cells (without giving rise to the 53-kd form), the 75-kd form must be unstable in the cytosol and rapidly degraded. Catalase, by contrast, although prevented from entering the defective peroxisomes in the mutant cells, is stable in the cytosol. This accounts for the equal amounts of catalase activity in normal cells and the mutant cells.

C. The mutations in the mutant cell lines must affect different genes because when the cells are fused (Table 12–2, fusions 6 and 7) the heterocaryons have normal peroxisomes. This is a classic example of complementation. If the two mutant cell lines were defective in the same gene, the heterocaryon would be no better off than the original cells, and would still be peroxisome deficient.

D. The mutations in the two mutant cell lines must be recessive. When either of the mutant cell lines are fused to normal cells (Table 12–2, fusions 2 and 3), the heterocaryon has the phenotype of the normal cells—that is, they have functional peroxisomes. If either of the mutations was dominant, the resulting heterocaryon would still be peroxisome deficient.

References: Tsukamoto, T.; Yokota, S.; Fujiki, Y. Isolation of Chinese hamster ovary cell mutants defective in assembly of peroxisomes. *J. Cell. Biol.* 110:651–660, 1990.

Yajima, S.; Suzuki, Y.; Shimozawa, N.; Yamaguchi, S.; Orii, T.; Fujiki, Y.; Osumi, T.; Hashimoto, T.; Moser, H.W. Complementation study of peroxisome-deficient disorders by immunofluorescence staining and characterization of fused cells. *Human Genetics* 88:491–499, 1992.

The Endoplasmic Reticulum

12–19

A. endoplasmic reticulum (ER), ER lumen (ER cisternal space)
B. co-translationally, posttranslational
C. membrane-bound ribosomes, rough endoplasmic reticulum
D. ER signal peptide
E. polyribosome
F. microsomes
G. signal
H. signal recognition particle (SRP), SRP receptor
I. start-transfer signal
J. single-pass transmembrane, stop-transfer peptide
K. multipass transmembrane
L. binding protein (BiP)
M. glycoproteins
N. glycosylation, oligosaccharyl transferase
O. glycosylphosphatidylinositol (GPI) anchor
P. phospholipid translocators
Q. phospholipid exchange (phospholipid transfer)

12–20

A. True
B. True
C. False. The cytochrome P450 enzymes do not cleave drugs and metabolites into small pieces. Instead, they detoxify molecules by catalyzing the addition of hydroxyl groups. These groups serve as sites for addition of water-soluble

moieties (such as sulfate or glucuronic acid). Ultimately, the target molecule is rendered soluble enough so that it can be excreted in the urine.

D. False. Rough microsomes are more dense than smooth microsomes because of the attached ribosomes.

E. True

F. True

G. False. The binding of the signal recognition particle to the signal peptide causes a pause in protein synthesis. Synthesis resumes when the ribosomes carrying SRP bind to the SRP receptor, which is exposed on the cytosolic surface of the rough ER.

H. False. The import of proteins into the ER requires ATP hydrolysis but not ongoing protein synthesis.

I. True

J. True

K. True

L. True

M. False. The ER lumen does not contain reducing agents (they are in the cytosol) and, therefore, S—S bonds can form in the ER.

N. True

O. False. Proteins that are linked to glycosylphosphatidylinositol anchors are attached to the external surface of the plasma membrane. The attachment reaction occurs in the lumen of the ER, which is topologically equivalent to the outside of the cell.

P. True

Q. False. Transport vesicles carry new phospholipids to the plasma membrane, Golgi, and lysosomes. However, new phospholipids are transferred to mitochondria, plastids, and peroxisomes by phospholipid exchange proteins.

12–21

A. In the absence of microsomes a unique protein is synthesized (Figure 12–9, lane 1). This protein is accessible to protease digestion whether or not detergent is present (Figure 12–9, lanes 2 and 3), indicating that the protein is not protected by a membrane bilayer. Finally, treatment of the protein with endoglycosidase H does not alter its mobility (Figure 12–9, lane 4), indicating that the protein carries no *N*-linked sugars of the type added in the ER.

B. By the three criteria outlined in the problem (protease protection, *N*-linked sugars, and cleaved signal peptide), this protein is translocated across the microsomal membrane. (1) The protein is fully sensitive to protease in the presence of detergent (Figure 12–9, lane 7), but it is only partially sensitive to protease in the absence of detergent (Figure 12–9, lane 6). Thus, the protein is partially protected from protease in the presence of microsomes. (2) The rate of migration of the protein increases after treatment with endoglycosidase H (Figure 12–9, compare lanes 5 and 8), indicating that sugars are attached to the protein when it is translated in the presence of microsomes. (3) When the sugars are removed, the protein migrates faster than the precursor protein (Figure 12–9, compare lanes 1 and 8), indicating that a portion of the protein— presumably the signal peptide—is removed from the precursor when the protein is translated in the presence of microsomes.

C. The reduced size of the protein after protease treatment (Figure 12–9, lanes 5 and 6) indicates that a portion of the protein remains on the outside of the microsomes, accesible to protease. In combination with the results indicating cleavage of the signal peptide and addition of *N*-linked sugars, this result shows that the protein spans the membrane. Thus the protein is inserted only part way through the membrane and is presumably anchored in the membrane by a stop-transfer segment.

*12–22   **Reference:** Perara, E.; Rothman, R.E.; Lingappa, V.R. Uncoupling translocation from translation: implications for transport of proteins across membranes. *Science* 232:348–352, 1986.

12–23

The predicted arrangement of the proteins in the membrane of the ER and the identity of the membrane-spanning segments as start- or stop-transfer peptides are illustrated in Figure 12–17. The arrangement of the membrane-spanning segments for all these proteins is fixed by the orientation of the initial start-transfer peptide. For proteins A and D, which have cleavable signal peptides, the start-transfer peptide is oriented with its N-terminal end toward the cytosol. Cleavage of the signal peptide on the lumenal side of the ER exposes the new N-terminus of the protein in the lumen (Figure 12–17A and D). The internal start-transfer peptides in proteins B and C are oriented so that the positively charged end faces the cytosol. This orientation for protein C means that membrane-spanning segment 2 must be a start-transfer peptide.

*12–24   **Reference:** Manoil, C.; Beckwith, J. A genetic approach to analyzing membrane protein topology. *Science* 233:1403–1408, 1986.

12–25

A. Experiment 1 tests whether the acceptor membranes (red cell ghosts) are in excess. Since the PC exchange protein catalyzes an *exchange* reaction, there is a simple theoretical limit to how much transfer can occur at equilibrium. If the amount of donor and acceptor membranes were equal, for example, the limit of possible transfer would be 50%. Doubling the amount of acceptor membrane would raise the limit to 67% (a 2 to 1 ratio of donor to acceptor); tripling the acceptor membranes would raise the limit to 75% (a 3 to 1 ratio); and so on. Since adding more acceptor membranes made no difference, the red cell membranes must be in excess. Thus the 70% limit is not an equilibrium point for the exchange.

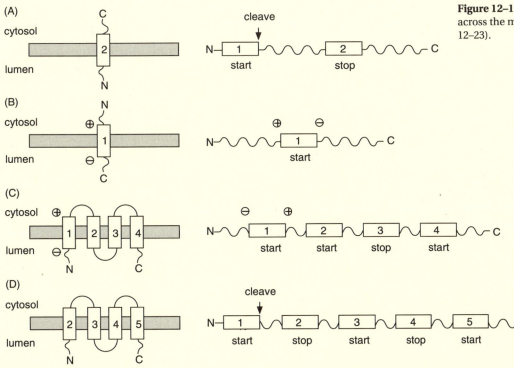

Figure 12–17 The arrangement of proteins across the membrane of the ER (Answer 12–23).

Experiment 2 rules out the possibility that the enzyme is inactivated during the reaction, since addition of fresh exchange protein causes no further exchange.

Experiment 3 eliminates the possibility that the starting labeled material was impure (that is, untransferable by the PC exchange protein, which is specific for PC) or was somehow altered during the course of the incubation.

B. The apparent explanation for the 70% limit is that the PC exchange protein transfers PC only from the outer monolayer of the vesicle bilayer. The area of the outside face of the donor vesicles is about 2.5 times the area of the inner face. The area of the surface of a sphere is $4/3\,\pi r^2$. Thus the ratio of the areas of the outer and inner faces of the donor vesicle is the ratio of squares of their radii, which is $10.5^2/(10.5-4.0)^2$ or 2.5. Since the outer surface is 2.5 times the inner surface, 71% (2.5/3.5) of the lipid is in the outer monolayer. Thus, 70% transfer is about the expected limit if the exchange protein can only exchange PC from the outer leaflet and PC does not flip-flop.

C. If the exchange protein exchanges PC only between outer leaflets, the label in the acceptor red cell membranes will all be in the outer leaflet and, therefore, all available for transfer. This result supports the idea that the PC exchange protein only transfers PC between outer monolayers.

Reference: Rothman, J.E.; Dawidowicz, E.A. Asymmetric exchange of vesicle phospholipids catalyzed by the phosphatidylcholine exchange protein. Measurement of inside-outside transitions. *Biochemistry* 14:2809–2816, 1975.

Vesicular Traffic in the Secretory and Endocytic Pathways

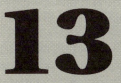

- **Transport from the ER Through the Golgi Apparatus**

- **Transport from the *Trans* Golgi Network to Lysosomes**

- **Transport from the Plasma Membrane via Endosomes: Endocytosis**

- **Transport from the *Trans* Golgi Network to the Cell Surface: Exocytosis**

- **The Molecular Mechanisms of Vesicular Transport and the Maintenance of Compartmental Diversity**

Transport from the ER Through the Golgi Apparatus

13–1

- A. Golgi apparatus
- B. Golgi stack, *cis* face, *trans* face
- C. *cis* Golgi, *trans* Golgi
- D. transitional elements
- E. brefeldin A
- F. *N*-linked
- G. high-mannose, complex
- H. *cis* compartment, *medial* compartment, *trans* compartment
- I. *O*-linked
- J. proteoglycans

13–2

- A. False. The number of Golgi stacks per cell varies greatly depending on the cell type: some animal cells contain one large stack, while certain plant cells contain hundreds of small ones.
- B. True
- C. False. It is thought that brefeldin A blocks forward transport from the ER through the Golgi without affecting the return transport from the Golgi to the ER. Blocking the forward pathway while leaving the return pathway intact causes the Golgi apparatus to empty into the ER.
- D. False. *N*-linked glycosylation begins in the ER with the addition of a single species of oligosaccharide to certain asparagine residues in target proteins. Some sugars are then removed from this oligosaccharide before it leaves the ER.
- E. True
- F. True
- G. True
- H. False. The function of *N*-linked oligosaccharides is unknown, but they evidently do not aid in transport.

13–3

- A. The radioactive label (GlcNAc) is added in the *medial* compartment, and the lectin precipitation depends on the presence of galactose, which is added in the *trans* compartment. Therefore, this experiment is following the movement of material between the *medial* and the *trans* compartments of the Golgi apparatus.
- B. If proteins moved through the Golgi apparatus by progression of the cisternae, then a protein that entered the Golgi in a mutant cell should remain with that stack and mature as the newly formed cisterna moves through the stack. Thus, the cisternal progression model predicts that none of the labeled G protein (which was labeled in the *medial* compartment of the Golgi apparatus in the mutant cell) should have galactose attached to it (which could only have been added in the Golgi apparatus from the wild-type cell). For this model the fusion of the infected mutant cells to uninfected wild-type cells (Table 13–1,

line 1) should be the same as the fusion of infected mutant cells to uninfected mutant cells (Table 13–1, line 2).

By contrast, if material moved through the Golgi apparatus by vesicle transport, there is the possibility that proteins might move between separated Golgi stacks inside transport vesicles. However, the frequency of movement of vesicles between different Golgi stacks is not addressed by the vesicle transport model. The vesicle transport model predicts only that some labeled G protein may acquire galactose in this way. For this model the fusion of infected mutant cells to uninfected wild-type cells (Table 13–1, line 1) should be more than for fusion of infected mutant cells to uninfected mutant cells (Table 13–1, line 2) but less than for fusion of infected wild-type cells to uninfected wild-type cells (Table 13–1, line 3).

C. The results in Table 13–1 support the vesicle transport model, since nearly half the labeled G protein acquired galactose. The extent of galactose addition is surprising because it suggests that once a vesicle leaves a cisterna, it has roughly an equal chance of fusing with a cisterna in the same or different Golgi stack. A number of other control experiments showed that the morphology of the Golgi stacks was unaltered by the fusion procedure, that the mutant and wild-type Golgi stacks remained distinct from one another, and that G protein did move into the wild-type Golgi stack.

Reference: Rothman, J.E.; Miller, R.L.; Urbani, L.J. Intercompartmental transport in the Golgi complex is a dissociative process: facile transfer of membrane protein between two Golgi populations. *J. Cell Biol.* 99:260–271, 1984.

*13–4

13–5

A. The altered G proteins with "membrane-spanning" segments that are 12, 8, or 0 amino acids long do not make it to the plasma membrane (Table 13–3). The presence of oligosaccharides indicates that each of these proteins is inserted into the ER membrane, which is expected since the signal peptide was not altered. The presence of the small C-terminal domain on the proteins with segments 12 and 8 amino acids long indicates that these proteins are anchored in the membrane much like the normal G protein. By contrast, the complete protease resistance of the G protein with a zero amino acid transmembrane segment indicates that it passed all the way through the ER membrane into the lumen. Thus, the G proteins with segments 12 and 8 amino acids long are in an internal membrane, but the G protein that is missing the membrane-spanning segment is in an internal lumen.

The partial endo-H resistance of the G protein with a membrane-spanning segment of 12 amino acids suggests that some fraction of this G protein makes it as far as the *medial* compartment of the Golgi, which is where the sugar modification occurs that renders the oligosaccharide endo-H resistant. The remainder of this protein is either in the membrane of the ER or the *cis* compartment of the Golgi. The endo-H sensitivity of the G proteins with 8 and 0 amino acid segments indicates that they do not make it past the *cis* compartment of the Golgi and may not make it out of the ER.

B. For the VSV G protein, the minimum length of the membrane-spanning segment appears to be 8 amino acid residues or less, since G proteins with modified membrane-spanning segments only 8 amino acids long are anchored in the membrane much like the normal G protein. This result is surprising since 8 amino acids arranged in an α helix are not thought to be long enough to span the membrane. There are several possibilities: the short membrane-spanning segments may be arranged as extended chains rather than as α helices; the membrane may be less than 3 nm thick at the point where these segments penetrate the membrane; and adjacent portions of the

---

*Problems with an asterisk (*) are answered in the Instructor's Manual.

G protein, including at least one basic amino acid (K or R) may be pulled into the membrane.

C. The minimum length of a membrane-spanning segment that is consistent with proper sorting of the G protein is 13 or 14 amino acids, since G proteins with segments 14 amino acids long are sorted to the plasma membrane like normal G proteins and G proteins with segments 12 amino acids long are not (Table 13–3). It is curious that shorter membrane-spanning segments anchor the protein in the membrane perfectly well but interfere with sorting. Two of several possibilities are (1) the folding of the cytoplasmic or the luminal domain is altered, thereby destroying the sorting signal and (2) the altered arrangement of amino acids in or near the membrane-spanning segment causes the protein to bind to a permanent resident of the ER or of the *cis* compartment of the Golgi.

Reference: Adams, G.A.; Rose, J.K. Structural requirements of a membrane-spanning domain for protein anchoring and cell surface transport. *Cell* 41:1007–1015, 1985.

Transport from the *Trans* Golgi Network to Lysosomes

13–6

A. lysosomes, acid hydrolases
B. vacuoles
C. autophagy
D. mannose 6-phosphate (M6P), M6P receptor
E. lysosomal storage

13–7

A. False. The proton pump in lysosomes pumps protons into the lysosome to maintain a low pH.
B. True
C. True
D. True
E. False. Proteins with KFERQ sequences on their surface are marked for uptake into lysosomes and degradation.
F. True
G. False. Addition of a weak base would cause M6P receptors to accumulate in late endosomes. M6P receptors, which bind lysosomal enzymes quite well at neutral pH, normally release bound enzymes at the lower pH of the late endosome and are then recycled to the Golgi. When the pH of the late endosome is raised, M6P receptors cannot release their bound enzymes, and because they cannot be recycled, they become trapped in the late endosome.
H. True
I. True

*13–8

13–9

A. In the hypothetical situation in which both the hydrolase and the M6P receptor are soluble, the rate of association will increase in direct proportion to the number of M6P groups. Each additional M6P group gives the hydrolase one additional way to bind to the receptor. Although the concentration of the hydrolase ([H]) remains the same, the concentration of M6P groups (which is what the receptor binds to) increases by a factor of four when the hydrolase has four M6P groups attached to it instead of one, thereby increasing the rate of association by a factor of four.

The rate of dissociation of a hydrolase from the receptor is the same whether the hydrolase has one M6P group or four. The rate of dissociation is related to the stability of the interaction between a single M6P group and the M6P receptor. That interaction is unaffected by other, unbound M6P groups on the hydrolase.

If the rate of association increases by a factor of four while the rate of dissociation remains unchanged, the affinity constant for the binding of a hydrolase with four M6P groups must be four times larger than the affinity constant for binding of a hydrolase with one M6P group.

B. If the first receptor is assumed to be locked in place and does not interfere with binding of a second receptor to other M6P groups on the hydrolase, the affinity constant for binding a second receptor will be three-quarters that of the affinity constant calculated in part A. The presence of one receptor on each hydrolase covers one M6P group, reducing by one-quarter the number available for subsequent binding to a second receptor. This would reduce the rate of association with the second receptor by one-quarter, giving an affinity constant that is also one-quarter less.

C. In the real situation with a soluble hydrolase and a membrane-bound receptor, binding to the first receptor would cause the hydrolase to become localized to a thin layer of the lumen adjacent to the membrane. This would have the effect of substantially increasing the concentration of the hydrolase in the immediate neighborhood of the M6P receptors. The increased local concentration would increase the rate of association with a second receptor correspondingly (but would not affect the rate of dissociation). As a result, the affinity constant for binding to a second receptor would increase substantially. The actual magnitude of the increase would depend on the volume of the lumen versus the volume of the thin layer adjacent to the membrane. (In Problem 15–24 we estimate the increase in concentration for a similar situation in which a protein is free in the cytosol versus bound to the plasma membrane.)

In framing this problem we have skirted several important issues that are essential to a detailed understanding of the true situation. For example, a hydrolase with four M6P groups can interact with as many as four M6P receptors (provided they do not interfere with one another), and at equilibrium (which the real system may never achieve) there would be hydrolases in the population with zero to four bound receptors, with the various forms interconvertible by appropriate rate constants. We have avoided this complexity to try to bring out two conceptual points. The presence of multiple M6P groups on lysosomal hydrolases increases their affinity for M6P receptors (and improves the efficiency of lysosomal targeting) in two distinct ways. First, multiple M6P groups increase the rate of association of the hydrolases with M6P receptors by providing more opportunities for binding. Second, multiple M6P groups increase the concentration of hydrolases near the membrane, giving rise to a much tighter overall binding than could be achieved if hydrolases had only a single M6P group. The situation is not unlike that of a climber on a sheer rock face: one toe- or finger-hold is good, but four are better.

13–10

A. The corrective factors are the lysosomal enzymes themselves. The enzyme missing in Hunter's syndrome is supplied by Hurler's cells and vice versa (that is, the enzyme missing in Hurler's syndrome is supplied by Hunter's cells). These enzymes are present in the medium because of a certain degree of inefficiency in sorting. Since they carry M6P, which normally should direct them to lysosomes, they presumably escaped capture by the lysosomal pathway and were secreted. They are taken into cells and delivered to lysosomes by a scavenger pathway, which operates due to a small number of M6P receptors on the cell surface. The degradative enzymes, bound to receptors, are taken up through coated pits into endosomes and are eventually delivered to lysosomes. Since lysosomes are the normal site of action for these degradative enzymes, the defect is corrected.

B. Protease treatment destroys the lysosomal enzymes themselves. Periodate treatment and alkaline phosphatase treatment both remove the M6P signal that is required for binding to the receptor, thus preventing the enzymes (which are still active) from entering the cell.

C. Such a scheme is unlikely to work for defects in cytosolic enzymes. External proteins normally do not cross membranes; thus, even when they are taken into cells, they remain in the lumen of a membrane-bounded compartment. In addition, foreign proteins are usually delivered to lysosomes and degraded.

   Reference: Kaplan, A.; Achord, D.T.; Sly, W.S. Phosphohexosyl components of a lysosomal enzyme are recognized by pinocytosis receptors on human fibroblasts. *Proc. Natl. Acad. Sci. USA* 74:2026–2030, 1977.

*13–11

13–12

Finding prelabeled lumenal material in the intracellular droplets indicates that the droplets are composed of lumenal material engulfed by the follicle cells. The progression of intracellular droplets from the periphery to the interior of the cell (presumably from early endosomes to late endosomes) is also more consistent with engulfment of lumenal material than with secretion. The presence of M6P on lumenal thyroglobulin suggests that the engulfed material is targeted to lysosomes where thyroglobulin is degraded to produce thyroxine. The mechanism for release of thyroxine into the bloodstream is unknown. The synthesis and release of thyroxine are summarized in Figure 13–12.

Note that the presence of M6P on thyroglobulin raises a very interesting cell biological problem. How does thyroglobulin manage to avoid entering the lysosome directly instead of being secreted? The answer is not known. The mannoses are phosphorylated in what appears to be the normal way. Perhaps other modifications, such as the sulfated tyrosines or high sialic acid content, mask the M6P signal, permitting the protein to be secreted. Other extracellular modifications, perhaps the iodination of tyrosines to form the thyroxine

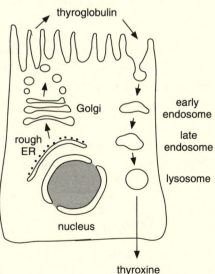

LUMEN OF FOLLICLE

Figure 13–12 Synthesis and release of thyroxine (Answer 13–12).

component of thyroglobulin, may activate the M6P signal for lysosomal targeting when thyroglobulin reenters the cell.

Reference: Herzog, V.; Neumuller, W.; Holzmann, B. Thyroglobulin, the major and obligatory exportable protein of thyroid follicle cells, carries the lysosomal recognition marker mannose 6-phosphate. *EMBO J.* 6:555–560, 1987.

Transport from the Plasma Membrane via Endosomes: Endocytosis

13–13

A. endocytosis
B. phagocytosis
C. macrophages, neutrophils
D. phagosomes
E. pinocytosis
F. endocytic-exocytic cycle
G. clathrin-coated pits, clathrin-coated vesicles
H. fluid-phase endocytosis
I. receptor-mediated endocytosis
J. low-density lipoproteins (LDL)
K. endosomal, early endosomes, late endosomes
L. transferrin
M. transcytosis

13–14

A. False. Not all particles that bind are ingested. Phagocytes have a variety of specialized surface receptors that are functionally linked to the phagocytic machinery of the cell. Only those particles that bind to these specialized receptors can be phagocytosed.
B. True
C. True
D. False. The LDL receptor and many other receptors enter coated pits irrespective of whether they have bound their specific ligands.
E. False. Many molecules that enter early endosomes are specifically diverted from the journey to late endosomes and lysosomes; they are recycled instead from early endosomes back to the plasma membrane via transport vesicles. Only those molecules that are not retrieved from endosomes are degraded.
F. True
G. True
H. False. During transcytosis, vesicles that form from either the apical or basolateral surface first fuse with early endosomes, where they are sorted into transport vesicles bound for the opposite surface.
I. True

13–15

A. HRP does not bind to a specific cellular receptor and is taken up only by fluid-phase endocytosis. Since endocytosis is a continuous (constitutive) process, HRP gets taken up steadily and its uptake does not saturate. By contrast, EGF binds to a specific EGF receptor and is internalized by receptor-mediated endocytosis. The limit to the amount of EGF that gets taken up is set by the number of EGF receptors on the cells; when the receptors are saturated, no further increase in uptake occurs (except at enormously high concentrations, where fluid-phase endocytosis becomes significant).
B. At saturation there are 4 pmol of EGF bound per 10^6 cells. This binding represents

$$\text{EGF bound} = \frac{4 \times 10^{-12} \text{ mole EGF}}{10^6 \text{ cells}} \times \frac{6 \times 10^{23} \text{ molecules}}{\text{mole}}$$
$$= 2.4 \times 10^6 \text{ EGF molecules/cell}$$

This number of receptors is about 10 times more than a typical cell would have. The cell line used for these studies was chosen for its high EGF receptor content, which may in part account for its cancerous nature.

C. An endocytic vesicle 20 nm (2×10^{-6} cm) in radius contains 3.3×10^{-17} ml of fluid.

$$\text{vesicle vol} = \tfrac{4}{3}\pi r^3$$
$$= \tfrac{4}{3} \times 3.14 \times \left(2 \times 10^{-6} \text{ cm}\right)^3$$
$$= 3.3 \times 10^{-17} \text{ cm}^3$$
$$= 3.3 \times 10^{-17} \text{ ml / vesicle}$$

The solution contains 1.5×10^{16} molecules/ml of HRP.

$$\text{HRP} = \frac{1 \text{ mg HRP}}{\text{ml}} \times \frac{\text{mmol}}{40{,}000 \text{ mg}} \times \frac{6.0 \times 10^{20} \text{ molecules}}{\text{mmol}}$$
$$= 1.5 \times 10^{16} \text{ molecules /ml}$$

Hence each vesicle contains 0.5 molecule of HRP.

$$\text{HRP} = \frac{1.5 \times 10^{16} \text{ molecules}}{\text{ml}} \times \frac{3.3 \times 10^{-17} \text{ ml}}{\text{vesicle}}$$
$$= 0.5 \text{ molecules /vesicle}$$

Since only half the vesicles contain HRP when the concentration is 1 mg/ml, it is not surprising that very few vesicles stained positively at a fiftyfold lower concentration of HRP.

D. These calculations, as alluded to by the authors, make the point that by having specific tight-binding receptors on the cell surface, cells can take up molecules from their surroundings at a much higher rate than they could simply by taking in fluid. Fishing provides an analogy. You could fish by taking random net-fulls from a stream, and occasionally you might catch a fish. But if you put bait where you cast your net, you increase your chances of success enormously. Each time a molecule of EGF hits a receptor, it sticks and subsequently makes its way to a coated pit to be internalized. If the EGF were simply trapped like HRP, its rate of uptake would be infinitesimal at the usual *in vivo* concentrations.

Reference: Haigler, H.T.; McKanna, J.A.; Cohen, S. Rapid stimulation of pinocytosis in human A-431 carcinoma cells by epidermal growth factor. *J. Cell Biol.* 83:82–90, 1979.

13–16

A. Binding of LDL by normal cells and JD's cells reaches a plateau because there are a limited number of LDL receptors per cell and they become saturated at high levels of LDL. The slope of the binding curve gives a measure of the binding affinity and the plateau gives a measure of the total number of binding sites (about 20,000 to 50,000, though you could not calculate this from the data shown here). JD has slightly fewer receptors on his cells, but they have an affinity similar to the normal cells.

Cells from patient FH bind essentially no LDL even at saturating external LDL levels. Either these cells completely lack the LDL receptor or the receptor is defective, so that its affinity for LDL is drastically reduced. It could also be that the cells do contain receptors, but for some reason they fail to appear on the surface of the cell.

B. Neither of the hypercholesterolemic patients' cells take up any LDL. Lack of entry is readily explained for patient FH because no LDL bound to the cells: no receptor, no uptake. This result indicates that the receptor is crucial for LDL-contained cholesterol to enter cells. Since LDL is not taken up by JD's cells, his LDL receptors must also be defective, in a different way than FH's LDL receptors. JD's cells bind LDL with the same affinity as normal and almost to the same level. Although his receptors are normal as far as LDL binding is concerned, the bound LDL does not get in. Thus, mere possession of a receptor on the cell surface is no guarantee of entry.

C. LDL must enter cells in order for the contained cholesterol esters to be released and hydrolyzed to cholesterol, which causes inhibition of cholesterol synthesis. In a normal person LDL enters the cells and inhibits cholesterol synthesis in the normal way. In the affected patients LDL does not enter the cells and, therefore, does not inhibit cholesterol synthesis.

D. If the defects in the hypercholesterolemic patients are due to defects in their LDL receptors, then free cholesterol should inhibit cholesterol synthesis in their cells as well as in normal cells. Free cholesterol does inhibit cholesterol synthesis in all these cells, strongly supporting the idea that the defects in the patients are due solely to problems with their LDL receptors.

Reference: Brown, M.S.; Goldstein, J.L. Receptor-mediated endocytosis: insights from the lipoprotein receptor system. *Proc. Natl. Acad. Sci. USA* 76:3330–3337, 1979.

*13–17 Reference: Brown, M.S.; Goldstein, J.L. Receptor-mediated endocytosis: insights from the lipoprotein receptor system. *Proc. Natl. Acad. Sci. USA* 76:3330–3337, 1979.

13–18

A. At 0°C endocytosis is blocked; therefore, the labeled transferrin receptors are trapped on the cell surface and are accessible to trypsin treatment. Most of the receptors in intact cells are not sensitive to trypsin because they are inside the cell (presumably in endosomes) and, therefore, are not accessible to the trypsin. When cells are incubated at 37°C, the labeled receptors are endocytosed and cycle through the endosomal compartment of the cell, thereby becoming inaccessible to trypsin.

B. Both trypsin treatment and antibody binding indicate that 30% of the total transferrin receptor is on the cell surface. When the transferrin receptors are allowed to recycle by incubation at 37°C, 30% is accessible to trypsin treatment of intact cells; therefore, 30% is on the surface. Similarly, antibody binds to 30% of the total receptor in the absence of detergent (0.54%/1.76% = 30%). Since recycling of transferrin receptors is very fast (see Problem 13–19), this distribution between the surface and internal compartments is the equilibrium distribution for transferrin receptors.

Reference: Bleil, J.D.; Bretscher, M.S. Transferrin receptor and its recycling in HeLa cells. *EMBO J.* 1:351–355, 1982.

*13–19 Reference: Bleil, J.D.; Bretscher, M.S. Transferrin receptor and its recycling in HeLa cells. *EMBO J.* 1:351–355, 1982.

Transport from the *Trans* Golgi Network to the Cell Surface: Exocytosis

13–20

A. exocytosis
B. constitutive, regulated
C. secretory vesicles
D. mast
E. synaptic vesicles

13–21

A. True
B. False. Secretory proteins, even those that are not normally expressed in a given secretory cell, are appropriately packaged into secretory vesicles. For this reason it is thought that the sorting signal, which is as yet undefined, is common to proteins in this class.
C. True
D. False. Once positioned beneath the plasma membrane, it waits until the cell receives an appropriate signal before fusing with the membrane and releasing its contents.
E. False. Localized stimulation of mast cells produces localized exocytosis. Thus, individual segments of the plasma membrane evidently can function independently.
F. True
G. True
H. True

13–22

A. Vesicles on the endocytic pathway will be labeled with colloidal gold; vesicles on the exocytic pathway will be labeled with ferritin.
B. Clathrin-coated vesicles are uncoated within a few seconds after they pinch off from the plasma membrane, so some will be caught with their coats off while others will still have their coats on.

13–23

The slow step in the constitutive secretion of transferrin relative to albumin occurs in the ER. The slow step in the constitutive secretion of albumin occurs in the Golgi. From Figure 13–8 it is clear that most of the transferrin in the cell is in the ER and most of the albumin is in the Golgi. The steady-state distribution of proteins along the constitutive pathway tells you where the proteins spend the majority of their time. As with any pathway, an accumulation occurs at the slow step. Therefore, the location of the majority of material corresponds to the slow step.

The constitutive secretion of transferrin is slow relative to albumin because it is delayed in the ER. This result appears to be general: if the constitutive secretion of a protein is slow, the protein is delayed for some reason in the ER.

References: Fries, E.; Gustafsson, L.; Peterson, P.A. Four secretory proteins synthesized by hepatocytes are transported from the endoplasmic reticulum to Golgi complex at different rates. *EMBO J.* 3:147–152, 1984.

Lodish, H.F.; Kong, N.; Snider, M.; Strous, G.J.A.M. Hepatoma secretory proteins migrate from rough endoplasmic reticulum to Golgi at characteristic rates. *Nature* 304:80–83, 1983.

*13–24 **References:** Gottlieb, T.A.; Beaudry, G.; Rizzolo, L.; Colman, A.; Rindler, M.J.; Adesnik, M.; Sabatini, D.D. Secretion of endogenous and exogenous proteins from polarized MDCK monolayers. *Proc. Natl. Acad. Sci. USA* 83:2100–2104, 1986.

Kondor-Koch, C.; Bravo, R.; Fuller, S.D.; Cutler, D.; Garoff, H. Protein secretion in the polarized epithelial cell line MDCK. *Cell* 43:297–306, 1985.

13–25

Antibodies specific for the cytoplasmic domain of synaptotagmin do not stain the nerve terminals because the cytoplasmic domain is never exposed on the outside of the cell. The luminal domain, however, is exposed to the outside of the cell when the synaptic vesicle fuses with the plasma membrane to release neurotransmitter molecules into the synaptic cleft. At that time the antibody can bind to the luminal domain of synaptotagmin. The membrane of the synaptic vesicle is quickly retrieved from the plasma membrane and reused to form new synaptic vesicles that contain bound antibodies within them. If the fusion of synaptic vesicles with the plasma membrane is stopped by lowering the temperature to 0°C, no labeling is observed.

Reference: Matteoli, M.; Takei, K.; Perin, M.S.; Südhof, T.C.; DeCamilli, P. Exo-endocytotic recycling of synaptic vesicles in developing processes of cultured hippocampal neurons. *J. Cell Biol.* 117:849–861, 1992.

The Molecular Mechanisms of Vesicular Transport and the Maintenance of Compartmental Diversity

13–26

A. coated vesicles
B. clathrin-coated, clathrin
C. adaptin
D. coatomer-coated
E. coatomer
F. ARF
G. SNAREs
H. Rab proteins

13–27

A. True
B. False. Coatomer-coated vesicles do mediate the outward vesicular transport from the ER and Golgi cisternae, but clathrin-coated vesicles mediate selective transport from the *trans* Golgi network to lysosomes and from the plasma membrane to endosomes. The counterflow of membrane from Golgi to the ER is mediated either by less well characterized vesicles or by means of elongated sacs or tubes of membrane that are dragged along microtubules.
C. True
D. False. Although the cage of clathrin itself seems to be the same in each clathrin-coated vesicle, the adaptins are different and mediate the capture of different types of cargo receptors.
E. True
F. True
G. True
H. True
I. False. Although Rab proteins were once thought to play this role because of their organelle-specific distribution, it is now thought that v-SNAREs and t-SNAREs account for the specific recognition of transport vesicles with their target membrane. The Rab proteins are now thought to control this recognition process by ensuring that the fit between a v-SNARE and a t-SNARE is correct before the vesicle is locked onto the target membrane.
J. True
K. True

13–28

Your experiments show that coatomer-coated vesicles transport G protein without concentrating their contents. The concentration of G protein in the cisternal space was actually slightly higher than in the vesicles and vesicle buds, as measured in two different ways. If the vesicles were transporting G protein in a selective way (like clathrin-coated vesicles), the concentration of G protein in the coatomer-coated vesicles should have been substantially higher than in the Golgi cisternae.

Reference: Orci, L.; Glick, B.S.; Rothman, J.E. A new type of coated vesicular carrier that appears not to contain clathrin: its possible role in protein transport within the Golgi stack. *Cell* 46:171–184, 1986.

*13–29   **References:** Donaldson, J.G.; Finazzi, D.; Klausner, R.D. Brefeldin A inhibits Golgi membrane-catalyzed exchange of guanine nucleotide into ARF protein. *Nature* 360:350–352, 1992.

Helms, J.B.; Rothman, J.E. Inhibition by brefeldin A of a Golgi membrane enzyme that catalyzes exchange of guanine nucleotide bound to ARF. *Nature* 360:352–354, 1992.

Orci, L.; Palmer, D.J.; Amherdt, M.; Rothman, J.E. Coated vesicle assembly in the Golgi requires only coatomer and ARF proteins from the cytosol. *Nature* 364:732–734, 1993.

13–30

A. Since vesicles form and accumulate when the function of Sec4 is impaired (as in *sec4^{ts}* and *sec4-Ile133*), the Sec4 protein cannot be involved in vesicle formation. Their accumulation in these mutants suggests that the vesicles can no longer deliver their cargo to the growing bud when the Sec4 protein is not working properly. Thus the Sec4 protein seems to be involved in vesicle targeting and fusion.

Functionally, the Sec4 protein resembles mammalian Rab proteins, which are also required for proper delivery of transport vesicles to their target membrane. Indeed, the Sec4 protein was the first identified member of the Rab family of proteins. The Sec4 protein is unlike mammalian ARF, which is required both for the formation of coatomer-coated vesicles and for their fusion to a target membrane.

B. From the description of the defects in the mutant Sec4 proteins and by analogy to Rab proteins, it is possible to outline the way the normal Sec4 protein functions in delivery of vesicles to the bud membrane (Figure 13–13).

The presence of some Sec4 protein (20% of total) in the cytosol of wild-type cells presumably represents Sec4 that is recycling after delivery of vesicles to the bud membrane. Removal of the C-terminal cysteines, by analogy to Ras proteins, presumably prevents attachment of a lipid that is essential for the binding of Sec4 to the forming vesicle. If Sec4 cannot bind to the vesicle, it cannot carry out its function.

In cells that express the Sec4-Ile133 protein, which is locked in its active conformation (even though no GTP is bound), essentially all of the mutant protein should be bound to vesicles. In its active form it cannot release itself

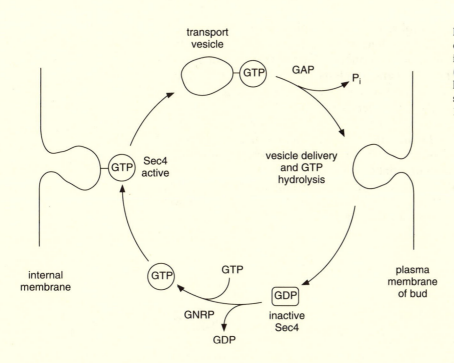

Figure 13–13 Outline of Sec4 function in delivery of transport vesicles from an internal membrane to the bud membrane (Answer 13–30). Additional proteins are likely to be involved in this process as shown for Rab proteins in MBOC Figure 13–57.

from vesicles and recycle. Thus there should be less in the cytosol than there is in wild-type cells.

C. The inhibitory properties of the "active" Sec4-Ile133 protein are very interesting and not altogether easy to interpret without further experimentation. Since Rab-like proteins are thought to function as monomers, it is unlikely that there is a direct effect of mutant Sec4-Ile133 protein on the normal Sec4 protein. More likely there is an indirect effect that prevents normal Sec4 from carrying out its function. There are several possibilities for such indirect effects. For example, if a vesicle component such as v-SNARE were present in limiting amounts, the accumulation of vesicles carrying abnormal Sec4 protein might deplete the supply and thereby interfere with the proper fusion of vesicles carrying the normal Sec4 protein. Alternatively, it may be that whenever an abnormal Sec4 protein interacts with the fusion machinery, it locks it into a nonfunctional form and reduces the availability of fusion components for interacting with vesicles carrying the normal Sec4 protein. Deciding which, if either, of these explanations is correct must await further experimentation.

Reference: Walworth, N.C.; Goud, B.; Kabcenell, A.K.; Novick, P.J. Mutational analysis of *SEC4* suggests a cyclical mechanism for the regulation of vesicular traffic. *EMBO J.* 8:1685–1693, 1989.

13–31

A. The use of the nonhydrolyzable analogue ATPγS allowed the complexes to form but prevented their dissociation. The ability to stabilize the complexes in this way allowed them to be extensively washed (also in the presence of ATPγS) to get rid of any proteins that were nonspecifically bound. The addition of ATP then allowed NSF to hydrolyze ATP and release the SNAPs and any associated SNAP receptors. This procedure offers two layers of biological specificity—assembly in the absence of ATP hydrolysis and disassembly with ATP hydrolysis—to ensure that the eluted proteins were biologically relevant.

B. The point of the experiment was to isolate SNAP receptors. If you had depended on the SNAPs in the membrane extract, they would have appeared in the eluted proteins and potentially confused the issue as to which proteins were SNAPs and which were SNAP receptors. The use of readily identifiable SNAPs removes this potential point of confusion. Another benefit in using known SNAPs is that it gives you an explicit experimental expectation, as discussed in part C.

C. If your ideas about how the complex assembles (which are illustrated schematically in Figure 13–10) are correct, then you would expect the total number of SNAP receptors to be about equal to about half the total number of SNAPs. This approximate ratio in the SNAPs and SNAP receptors among the proteins eluted from the column is what you expected and adds confidence that the proteins you isolated really are SNAP receptors.

D. Synaptic vesicles normally fuse with the presynaptic membrane. Thus syntaxins A and B, which are in the presynaptic target membrane, are potential t-SNAREs. Synaptobrevin-2, which is in synaptic vesicle membranes, and possibly SNAP-25 as well, are potential v-SNAREs.

Reference: Söllner, T.; Whiteheart, S.W.; Brunner, M.; Erdjument-Bromage, H.; Geromanos, S.; Tompst, P.; Rothman, J.E. SNAP receptors implicated in vesicle targeting and fusion. *Nature* 362:318–324, 1993.

Energy Conversion: Mitochondria and Chloroplasts

The Mitochondrion

14–1

    A. matrix space, intermembrane space
    B. outer
    C. respiratory chain (electron-transport chain), inner
    D. cristae
    E. acetyl coenzyme A (acetyl CoA)
    F. triacylglycerols (triglycerides)
    G. glycogen
    H. citric acid cycle (tricarboxylic acid cycle or Krebs cycle)
    I. oxidative phosphorylation
    J. electrochemical proton gradient
    K. proton-motive
    L. ATP synthase
    M. free-energy change (ΔG), standard free-energy change ($\Delta G°$)

14–2

    A. False. The intermembrane space is chemically equivalent to the cytosol with respect to small molecules because the outer mitochondrial membrane contains many copies of a transport protein that forms large aqueous channels. However, the composition of the matrix space is much more specialized because the inner mitochondrial membrane contains transport proteins that only allow passage of a restricted set of small molecules.
    B. True
    C. False. Animal cells store fuel in the form of fats (from fatty acids) and glycogen (from glucose).
    D. True
    E. False. During electron transport protons are pumped out of the mitochondrial matrix into the intermembrane space.
    F. True
    G. True
    H. False. ATP synthase is oriented in the inner membrane so that ATP is synthesized in the matrix space. It is transported out of the matrix by the ADP-ATP antiporter.
    I. False. Favored reactions have a negative ΔG; that is, they proceed with a *decrease* in free energy.
    J. True

*14–3

14–4

    A. The complete oxidation of citrate to CO_2 and H_2O occurs according to the balanced chemical reaction shown below.

$$C_6H_8O_7 + 4.5O_2 \rightarrow 6CO_2 + 4H_2O$$

Thus each molecule of citrate would require 4.5 molecules of oxygen for its complete oxidation.

*Problems with an asterisk (*) are answered in the Instructor's Manual.

The results in Table 14–1 were surprising to Krebs and others at the time because much more oxygen is consumed (40 mmol) than could be accounted for by oxidation of citrate itself. Only 13.5 mmol of oxygen would be required to oxidize 3 mmol of citrate completely (3×4.5). This calculation shows that citrate is acting catalytically in the oxidation of carbohydrates (which in these experiments were endogenous in the minced pigeon breasts). Although others were aware of the catalytic nature of other intermediates, Krebs was the first person to complete the circle of chemical reactions that constitute the citric acid cycle.

Krebs's experimental rationale is clearly laid out in the paper: "Since citric acid reacts catalytically in the tissue, it is probable that it is removed by a primary reaction but regenerated by a subsequent reaction. In the balance sheet no citrate disappears and no intermediate products accumulate. The first object of the study of intermediates is therefore to find conditions under which citrate disappears in the balance sheet."

B. The consumption of oxygen is low in the presence of the metabolic poisons because citrate is prevented from acting catalytically. The balanced equations for the conversion of citrate to α-ketoglutarate and succinate show that the amount of oxygen consumed is approximately what is expected.

For citrate conversion to α-ketoglutarate, half a molecule of oxygen is consumed.

$$C_6H_8O_7 + 0.5O_2 \rightarrow C_5H_6O_5 + CO_2 + H_2O$$

For citrate conversion to succinate, one molecule of oxygen is consumed.

$$C_6H_8O_7 + O_2 \rightarrow C_4H_6O_4 + 2CO_2 + H_2O$$

Thus the observed stoichiometry of oxygen consumption matches the expectations.

C. The absence of oxygen is crucial for demonstrating an accumulation of citrate from an intermediate in the cycle. In the presence of oxygen, citrate acts catalytically—is consumed and then regenerated—so that it does not accumulate no matter what intermediate is added. In the absence of oxygen, however, the conversion of citrate to α-ketoglutarate is blocked, since that conversion requires oxygen. Under these conditions citrate will accumulate if an appropriate intermediate is present. Of all the intermediates, only conversion of oxaloacetate to citrate does not require oxygen. The immediate precursor of oxaloacetate is malate. Since the conversion of malate to citrate requires oxygen, all other intermediates also must require oxygen to be converted to citrate. (The requirement for oxygen is indirect and is mediated through the cofactors NAD^+ and FAD; they accept electrons from the substrates and transfer them to the electron-transport chain and ultimately to oxygen.)

This reasoning might lead you to expect a quantitative conversion of oxaloacetate to citrate. However, in Krebs's experiments 300 μmol of oxaloacetate were added but only 13 μmol of citrate accumulated. What Krebs did not know was that citrate is generated by addition of acetyl CoA (undiscovered at the time) to oxaloacetate. The generation of acetyl CoA from its immediate precursor, pyruvate, is dependent on oxygen.

D. *E. coli* and yeast do indeed use the citric acid cycle. Krebs got this point wrong because he did not realize (nor did anyone for a long time) that citrate cannot get into these cells. Therefore, when he added citrate to intact *E. coli* and yeast, he found no stimulation of oxygen consumption. Passage of citrate across a membrane requires a membrane transporter, which is present in mitochondria but is absent in yeast and *E. coli*.

Reference: Krebs, H.A.; Johnson, W.A. The role of citric acid in intermediate metabolism in animal tissues. *Enzymologia* 4:148–156, 1937.

A. When the concentrations of the reactants and products are all 1 M, the reaction is at standard conditions and ΔG equals $\Delta G°$, which is –7.3 kcal/mole.

$$\Delta G = \Delta G° + 2.3\, RT \log_{10} \frac{[ADP][P_i]}{[ATP]}$$

$$= -7.3 \text{ kcal/mole} + 2.3\,(0.00198 \text{ kcal/°K mole})(310°\text{K}) \log_{10} \frac{1 \times 1}{1}$$

Since the $\log_{10}$ of 1 is 0,

$$\Delta G = -7.3 \text{ kcal/mole}$$

When the concentrations of the reactants and products are all 1 mM, ΔG equals –11.5 kcal/mole.

$$\Delta G = -7.3 \text{ kcal/mole} + (1.4 \text{ kcal/mole}) \log_{10} \frac{10^{-3} \times 10^{-3}}{10^{-3}}$$

$$= -7.3 \text{ kcal/mole} + (1.4 \text{ kcal/mole})\,(-3)$$

$$= -7.3 \text{ kcal/mole} - 4.2 \text{ kcal/mole}$$

$$\Delta G = -11.5 \text{ kcal/mole}$$

B. At the given concentrations of ATP, ADP, and P_i, the ΔG for ATP hydrolysis is –11.1 kcal/mole.

$$\Delta G = -7.3 \text{ kcal/mole} + (1.4 \text{ kcal/mole}) \log_{10} \frac{(0.001)(0.010)}{(0.005)}$$

$$= -7.3 \text{ kcal/mole} + (1.4 \text{ kcal/mole})\,(-2.7)$$

$$= -7.3 \text{ kcal/mole} - 3.8 \text{ kcal/mole}$$

$$\Delta G = -11.1 \text{ kcal/mole}$$

C. At equilibrium ΔG is 0. At equilibrium there is no longer any tendency for a reaction to proceed. If $[P_i]$ is 10 mM at equilibrium, then the ratio of [ATP] to [ADP] will be 6.1×10^{-8}.

$$0 = -7.3 \text{ kcal/mole} + (1.4 \text{ kcal/mole}) \log_{10} \frac{[ADP] \times (0.01)}{[ATP]}$$

$$7.3 \text{ kcal/mole} = (1.4 \text{ kcal/mole})\,(\log_{10} 0.01 + \log_{10}[ADP]/[ATP])$$

$$= (-2)\,(1.4 \text{ kcal/mole}) + (1.4 \text{ kcal/mole}) \log_{10}[ADP]/[ATP]$$

$$\log_{10}[ADP]/[ATP] = \frac{10.1 \text{ kcal/mole}}{1.4 \text{ kcal/mole}}$$

$$= 7.2$$

$$\log_{10}[ATP]/[ADP] = -7.2$$

$$[ATP]/[ADP] = 6.1 \times 10^{-8}$$

D. At a constant $[P_i]$, every tenfold change in the ratio of [ATP] to [ADP] will alter ΔG by 1.4 kcal/mole. As shown below, a tenfold increase in [ATP]/[ADP] will decrease ΔG by 1.4 kcal/mole.

$$\Delta G = \Delta G° + 1.4 \text{ kcal/mole} \log_{10} \frac{[ADP][P_i]}{[ATP]}$$

$$= \Delta G° + 1.4 \text{ kcal/mole} \log_{10}[P_i] + 1.4 \text{ kcal/mole} \log_{10} \frac{[ADP]}{[ATP]}$$

A tenfold increase in [ATP]/[ADP], which is equal to a tenfold decrease in [ADP]/[ATP], causes the $\log_{10}$ of the ratio in the expression above to decrease by −1. Thus each tenfold increase in the ratio causes 1.4 kcal/mole to be subtracted from the right-hand side, thereby decreasing ΔG by 1.4 kcal/mole. A 100-fold increase in the ratio of [ATP]/[ADP] decreases ΔG by 2.8 kcal/mole; a 1000-fold increase in the ratio decreases ΔG by 4.2 kcal/mole.

*14–6

14–7

A. The $\Delta G°$ for conversion of 3-phosphoglycerate (3PG) to pyruvate (PYR) and phosphate is the sum of $\Delta G°$ values for the individual steps in the reaction.

$$\Delta G°_{3PG\to PYR} = \Delta G°_{3PG\to PEP} + \Delta G°_{PEP\to PYR}$$
$$= 0.4 \text{ kcal/mole} + (-14.8 \text{ kcal/mole})$$
$$\Delta G°_{3PG\to PYR} = -14.4 \text{ kcal/mole}$$

B. The $\Delta G°$ for conversion of 3-phosphoglycerate to pyruvate and phosphate is independent of the pathway for the conversion. Thus, the $\Delta G°$ is −14.4 kcal/mole.

The $\Delta G°$ value for conversion of glycerate to pyruvate is obtained by subtracting $\Delta G°$ for 3-phosphoglycerate to glycerate (GLY) from the overall $\Delta G°$.

$$\Delta G°_{GLY\to PYR} = \Delta G°_{3PG\to PYR} - \Delta G°_{3PG\to GLY}$$
$$= -14.4 \text{ kcal/mole} - (-3.3 \text{ kcal/mole})$$
$$\Delta G°_{GLY\to PYR} = -11.1 \text{ kcal/mole}$$

C. The analysis above indicates that a very large standard free-energy change occurs between glycerate and pyruvate. Removal of water ($\Delta G° = -0.5$ kcal/mole) does not account for very much of this free-energy change. Thus it appears that the conversion of enolpyruvate to pyruvate is accompanied by a large standard free-energy change of around −10.6 kcal/mole. This reasoning suggests that the majority of the standard free-energy change associated with conversion of phosphoenolpyruvate to pyruvate (−10.6 kcal/mole out of −14.8 kcal/mole) comes from the conversion of enolpyruvate to pyruvate and not from the hydrolysis of the phosphate bond.

In fact, the standard free-energy change for phosphoenolpyruvate to pyruvate (−14.8 kcal/mole) is close to the sum of the enolpyruvate to pyruvate step (about −11 kcal/mole) and a normal standard free-energy change for hydrolysis of a simple phosphate ester bond (about −3.0 kcal/mole). Thus, the phosphate bond in phosphoenolpyruvate is a high-energy bond because its hydrolysis is linked to the very favorable conversion of enolpyruvate to pyruvate.

D. The $\Delta G°$ for the linked conversion of phosphoenolpyruvate to pyruvate and of ADP to ATP is −7.5 kcal/mole. The $\Delta G°$ for the linked reaction can be obtained by adding together the $\Delta G°$ values for the individual reactions. The individual reactions are

$$\begin{array}{ll} PEP \to PYR + P_i & \Delta G° = -14.8 \text{ kcal/mole} \\ ADP + P_i \to ATP & \Delta G° = 7.3 \text{ kcal/mole} \end{array}$$

NET: $PEP + ADP \to PYR + ATP$

$$\Delta G°_{PEP+ADP\to PYR+ATP} = \Delta G°_{PEP\to PYR} + \Delta G°_{ADP\to ATP}$$
$$= -14.8 \text{ kcal/mole} + 7.3 \text{ kcal/mole}$$
$$\Delta G°_{PEP+ADP\to PYR+ATP} = -7.5 \text{ kcal/mole}$$

Reference: Lipmann, F. Metabolic generation and utilization of phosphate bond energy. *Adv. Enzymol.* 1:99–162, 1941.

*14–8 **References:** Nicholls, D.G. Bioenergetics, pp. 159–164. London: Academic Press, 1982.

Tzagoloff, A. Mitochondria, pp. 212–213. New York: Plenum Press, 1982.

The Respiratory Chain and ATP Synthase

14–9

A. ATP synthase
B. cytochromes
C. iron-sulfur center
D. quinone (Q)
E. respiratory enzyme complexes
F. NADH dehydrogenase complex
G. cytochrome b-c_1 complex
H. cytochrome oxidase complex
I. conjugate redox pairs
J. redox potential (oxidation-reduction potential)
K. respiratory control

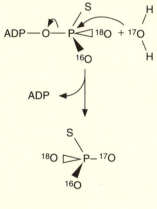

(A) ONE-STEP MECHANISM

inversion

14–10

A. True
B. True
C. False. ATP synthase is a reversible enzyme. Its direction of action depends on the balance between the steepness of the electrochemical proton gradient and the local D*G* for ATP hydrolysis.
D. True. (Because submitochondrial particles are inside-out, protons are pumped into the vesicle, causing the medium to become more basic.)
E. False. Although most proteins in the respiratory chain use iron atoms as electron carriers, one uses a flavin molecule and two use copper atoms as electron carriers.
F. True
G. False. The three respiratory enzyme complexes appear to exist as independent entities in the plane of the inner membrane and the ordered transfers of electrons is due entirely to the specificity of the functional interactions between the components of the chain.
H. True
I. True
J. False. Lipophilic weak acids act as uncoupling agents that dissipate the proton-motive force and stop ATP synthesis; however, they increase the flow of electrons through the respiratory chain by eliminating the respiratory control imposed by the electrochemical proton gradient.
K. True
L. True
M. True

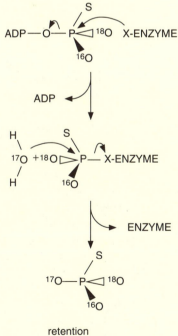

(B) TWO-STEP MECHANISM

retention

Figure 14–25 Stereochemical consequences of ATP hydrolysis by one-step (A) and two-step (B) mechanisms (Answer 14–11).

14–11

A. This experiment distinguishes very nicely between mechanisms involving a one-phosphate transfer and those involving two-phosphate transfers. Since each phosphate transfer results in inversion of the configuration around the phosphate atom, a one-transfer mechanism results in inversion and a two-transfer mechanism results in retention (inversion followed by inversion gives retention). As illustrated in Figure 14–25, direct attack of water on ATP to generate ADP and phosphate is a one-step mechanism and therefore produces inversion of configuration. Hydrolysis of ATP via an intermediate phosphorylated substance is a two-step mechanism, and therefore the configuration should be retained (Figure 14–25). (Note that the result does not, however, distinguish between a one-step mechanism and a mechanism involving three—or any odd number of—phosphate transfers.)

B. Inversion of configuration during the hydrolysis of ATP by ATP synthase indicates that the hydrolysis reaction does not occur through a two-transfer mechanism and therefore argues against the involvement of a single phosphorylated intermediate. Thus, hydrolysis of ATP by ATP synthase probably involves the direct attack of H_2O on ATP. If hydrolysis of ATP by ATP synthase is the reverse of the synthetic reaction (as it is thought to be), then synthesis of ATP from ADP and phosphate also occurs directly and not through a phosphorylated intermediate. (A mechanism involving three phosphate transfers is consistent with this analysis, but it is thought to be much less likely.)

Reference: Webb, M.R.; Grubmeyer, C.; Penefsky, H.S.; Trentham, D.R. The stereochemical course of phosphoric residue transfer catalyzed by beef heart mitochondrial ATPase. *J. Biol. Chem.* 255:11637–11639, 1980.

14–12

A. Oxygen accepts electrons from the electron-transport chain and is reduced to H_2O. Therefore, in the presence of oxygen the cytochromes would be drained of their electrons, that is, oxidized. Since the absorption bands do not show up in the presence of oxygen, the oxidized forms must not absorb light. The reduced forms of the cytochromes absorb light and are responsible for the characteristic absorption patterns. In the absence of oxygen the cytochromes pick up electrons from substrates (become reduced) but cannot get rid of them by transfer to oxygen. In the presence of oxygen, the electrons are transferred efficiently, leaving the cytochromes in their electron-deficient or oxidized state.

B. Keilin's observations indicate that the order of electron flow through the cytochromes is

$$\text{reduced substrates} \rightarrow \text{cytochrome b} \rightarrow \text{cytochrome c} \rightarrow \text{cytochrome a} \rightarrow O_2$$

This order can be deduced from Keilin's results. Since the bands become visible in the absence of oxygen, they represent the reduced (electron-rich) forms of the cytochromes. When oxygen is added, they are all converted to the oxidized (electron-poor) form. When cyanide is added, all the cytochromes are reduced, indicating that cyanide blocks the flow of electrons from the cytochromes to oxygen; that is, all the cytochromes are "upstream" of oxygen (in the sense of electron flow).

When urethane is added, cytochrome b remains reduced but cytochromes a and c become oxidized. Thus, urethane interrupts the flow of electrons from cytochrome b to cytochromes a and c, indicating that cytochrome b is "upstream" of cytochromes a and c.

These results indicate that either cytochrome a or c transfers electrons to oxygen. The inability of oxygen to oxidize a preparation of cytochrome c suggests, by elimination, that cytochrome a is responsible for transfer of electrons to oxygen. This ordering of cytochromes a and c is weak since it is based on a negative result (which could have other interpretations). Keilin himself confirmed this order by observing subtle spectral shifts in the cytochrome a band in the presence of cyanide under reducing conditions; he named the active component cytochrome a_3. We now know that cytochrome a is a large complex with several redox centers, one of which reacts with molecular oxygen.

C. The rapid oxidation of glucose to CO_2 prevents the disappearance of the absorption bands by providing a source of reduced substrates (ultimately NADH and $FADH_2$) that transfer electrons into the electron transport chain faster than oxygen can remove them. Under these conditions the cytochromes remain reduced (electron rich) and therefore continue to absorb light.

Reference: Keilin, D. The History of Cell Respiration and Cytochrome. Cambridge, UK: Cambridge University Press, 1966.

*14–13   References: Smith, H.T.; Ahmed, A.J.; Millet, F. Electrostatic interaction of cytochrome c with cytochrome c_1 and cytochrome oxidase. *J. Biol. Chem.* 256:4984–4990, 1981.

Capaldi, R.A.; Darley-Usmar, V.; Fuller, S.; Millet, F. Structural and functional features of the interaction of cytochrome c with complex III and cytochrome c oxidase. *FEBS Letters* 138:1–7, 1982.

14–14

A. The rate of oxygen consumption is determined by the rate of electron transport down the respiratory chain. Electron transport generates an electrochemical proton gradient, which opposes the flow of electrons. In the complete absence of a way to dissipate the gradient, the flow of electrons ultimately would stop when the electron pressure balances the opposing electrochemical proton gradient. In the experiment in Figure 14–9, the electrochemical proton gradient is dissipated at a slow background rate, which accounts for the slow background rate of oxygen consumption. Addition of ADP and its subsequent conversion to ATP allows protons to flow back into the mitochondria, dramatically reducing the electrochemical proton gradient and permitting the rapid transport of electrons to oxygen. The increased rate of electron transport produces an increased rate of oxygen consumption. When all the ADP is converted to ATP, proton flow again slows to the background rate, and the increased electrochemical proton gradient once again reduces the flow of electrons.

B. The slow background rate of oxygen consumption by mitochondria in the absence of added ADP indicates that electrons continue to flow down the electron-transport chain to oxygen in the absence of ATP synthesis. Such a flow can continue only if the electrochemical proton gradient is slowly being dissipated. If the mitochondrial inner membrane was completely impermeable to protons, the rate of oxygen consumption would drop to zero when proton pumping due to electron transport was balanced by the back pressure of the electrochemical proton gradient. Thus, the protons must be crossing the membrane in the absence of ATP synthesis.

   Several processes other than ATP synthesis from added ADP might account for the slow passage of protons across the membrane and the slow background rate of oxygen consumption. (1) The mitochondrial inner membrane is not completely impermeable to protons, which can slowly cross the membrane even in the absence of ATP synthesis. (2) The internal mitochondrial supply of ATP may be hydrolyzed to ADP and then reconverted to ATP using the proton-motive force. (3) If some mitochondria in the preparation are damaged so that their inner membranes are not intact, they will transport electrons to oxygen continuously because there will be no electrochemical proton gradient to oppose electron flow.

C. Since each pair of electrons that flows down the respiratory chain from NADH to oxygen reduces one oxygen atom, the $P/2e^-$ ratio is equivalent to the P/O ratio. The P/O ratio, as calculated below, is between 2.5 and 2.8 molecules of ATP per O atom. Uncertainty in the P/O ($P/2e^-$) ratio arises from the uncertainty in how much oxygen is consumed during conversion of 500 nmol ADP to ATP. If oxygen consumption is calculated as the difference between the horizontal gray lines in Figure 14–9, which is 100 nmol O_2, then the P/O ratio is 500 nmol ATP/200 nmol O, which is 2.5. On the other hand, if oxygen consumption is calculated as the difference between the slanted dotted lines in Figure 14–9, which is 90 nmol O_2, then the P/O ratio is 500 nmol ATP/180 nmol O, which is 2.8. The latter calculation makes the implicit assumption that the background rate of oxygen consumption continues during the conversion of ADP to ATP, which is a perfectly reasonable assumption. It turns out, however, that the natural slow flow of protons across intact inner membranes is quite sensitive to the size of the electrochemical proton gradient. The slight decline in the proton-motive force during ATP synthesis may reduce the leakage to nearly zero, in which case the larger value for oxygen consumption may be the more valid one (giving a P/O ratio of 2.5).

D. Several processes in these kinds of experiments, in addition to ATP synthesis, are driven by the electrochemical proton gradient. The uptake of substrate (D-hydroxybutyrate) into mitochondria may require symport with protons. The import of phosphate into mitochondria also requires symport with a proton. Finally, the exchange of internal ATP for external ADP is driven by the membrane potential, which is one component of the electrochemical proton gradient. Given that several processes are driven by the electrochemical proton gradient, it is not surprising that the P/O ratio is not an integer. Before the chemiosmotic theory, when chemical coupling hypotheses were fashionable, integral values were expected and values of 2.5 or 2.8 were assumed to "really" mean 3.

Reference: Nicholls, D.G. Bioenergetics. London: Academic Press, 1982.

*14–15   **Reference:** Nicholls, D.G. Bioenergetics, pp. 86, 110. London: Academic Press, 1982.

*14–16   **Reference:** Blaut, M.; Gottschalk, G. Evidence for a chemiosmotic mechanism of ATP synthesis in methanogenic bacteria. *Trends Biochem. Sci.* 10:486–489, 1985.

14–17

A. Each of the observations with ionophores is consistent with the idea that the movement of protons down the electrochemical proton gradient powers the flagella, as explained below for each observation.

1. During oxidation of glucose, bacteria pump protons out of the cell, establishing an electrochemical proton gradient, which is the sum of a proton gradient and a membrane potential. Addition of FCCP makes the membranes permeable to protons, thereby collapsing both the proton gradient and the membrane potential. In the absence of an electrochemical proton gradient to drive protons across the membrane, the flagellar motor cannot function.

2. In a medium contain K^+, valinomycin collapses the membrane potential specifically by allowing an influx of K^+ to balance the efflux of protons. Under these circumstances it is entirely the proton gradient (which is larger than usual because it is not opposed by the membrane potential) that drives the proton flux through the flagellar motor. The ability of bacteria to swim normally in the presence of the proton gradient alone is strong evidence that the flagellar motor is proton-powered.

3. In the absence of glucose (or any other substrate) for oxidation, there is no electrochemical proton gradient. In the presence of external K^+, valinomycin facilitates a flow of K^+ into the cell; this results in a membrane potential that is positive inside. Although there are protons available in the medium (from H_2O), this membrane potential is in the wrong orientation to drive the protons into the cell. As a result, the bacteria remain motionless.

4. In the absence of glucose, there is no electrochemical proton gradient. In the absence of external K^+ (that is, when Na^+ is in the medium), addition of valinomycin allows internal K^+ to move out of the cell (down its concentration gradient), creating a membrane potential that is positive outside. This membrane potential can drive protons into the cell for a while. Each proton that enters the cell lessens the membrane potential until the membrane potential is dissipated, at which point the cells stop swimming.

B. At first glance it seems peculiar that normal bacteria can swim in the absence of oxygen. In the absence of oxygen, there is no electron flow down the electron-transport chain and, therefore, no electron-transport-linked proton translocation across the membrane. What then is the source of protons to power the motor under anaerobic conditions? The mutant strain provides the essential clue. In the absence of the ATP synthase, bacteria cannot swim, which suggests that the ATP synthase in some way generates the proton gradient. In normal bacteria in the absence of oxygen, ATP that is generated anaerobically is used to drive the ATP synthase in *reverse*, causing protons to flow out of the cell. The resulting electrochemical proton gradient drives the protons back through the flagellar motor, allowing the bacteria to swim. The mutant bacteria cannot swim in the absence of glucose because they have no ATP synthase and, therefore, cannot create an electrochemical gradient in the absence of electron flow.

Reference: Manson, M.D.; Tedesco, P.; Berg, H.C.; Harold, F.M.; van der Drift, C. A protonmotive force drives bacteria flagella. *Proc. Natl. Acad. Sci. USA* 74:3060–3064, 1977.

Chloroplasts and Photosynthesis

14–18

A. plastid
B. stroma
C. thylakoids
D. photosynthetic electron-transfer (light), carbon-fixation (dark)
E. carbon-fixation (Calvin-Benson)
F. sucrose
G. starch
H. C_4, C_3
I. chlorophyll
J. antenna complex, photochemical reaction center
K. noncyclic photophosphorylation
L. Z scheme
M. cyclic photophosphorylation

14–19

A. True
B. False. The formation of O_2 requires light energy directly, whereas the fixation of CO_2 requires light energy only indirectly.
C. True
D. True
E. True
F. True
G. False. When an electron in a chlorophyll molecule is excited, it transfers its energy—not the electron—from one chlorophyll molecule to another by resonance energy transfer.
H. True
I. True
J. False. Each electron that is transferred from H_2O to $NADP^+$ requires two photons, one for each photosystem. Therefore, the reduction of $NADP^+$ to NADPH, which uses two electrons, requires four photons.
K. False. Cyclic photophosphorylation generates only ATP (not NADPH), and the balance between cyclic and noncyclic photophosphorylation is regulated by NADPH (not ATP).
L. True
M. True

The corn plant (C_4) eventually will kill the geranium (C_3). Because both plants fix CO_2, the concentration in the chamber falls. At low CO_2 concentration, the corn plant has a distinct advantage since the enzyme responsible for its initial carbon fixation has a high affinity for CO_2. By contrast, the geranium depends on ribulose bisphosphate carboxylase, which has a lower affinity for CO_2; furthermore, at low CO_2 concentrations O_2 competes with CO_2 for addition to ribulose 1,5-bisphosphate, ultimately liberating CO_2 in the process known as photorespiration. Not only does the geranium give up CO_2 in an abortive attempt at photosynthesis, it continues to respire (using its mitochondria), thereby providing even more CO_2 for the corn plant. The corn plant continues to fix CO_2 until the geranium wastes away and dies.

Reference: Becker, W.M. The World of the Cell, p. 282. Menlo Park, CA: Benjamin-Cummings, 1986.

*14–21

14–22

A. Starch formation requires light in the cactus and in C_4 plants (as well as C_3 plants). The synthesis of starch requires ATP and NADPH. These compounds are present in cells in only small amounts; they are not stored. During starch synthesis ATP and NADPH must be continuously regenerated in order for synthesis to continue. Regeneration of ATP and NADPH requires the photosynthetic electron-transfer reaction. Energy in sunlight energizes electrons in chlorophyll. Some electrons are passed to $NADP^+$ to generate NADPH; others are transferred along an electron-transport chain, generating an electrochemical proton gradient, which is coupled to ATP production.

B. CO_2 fixation in the cactus is outlined in Figure 14–26. Reactions shown with thick lines occur during the night; reactions shown with thin lines occur during the day. The carbon-fixation reactions in the cactus are essentially the same as those in C_4 plants. The key difference in the metabolic *pathways* of CO_2 fixation is that the cactus uses starch in the CO_2 pumping cycle. However, the common reactions of the pumping cycle are distributed differently in both space and time in C_4 plants and the cactus. In C_4 plants the reactions involve several cell types but occur all at the same time. By contrast, in the cactus they all occur in the same cell but at different times.

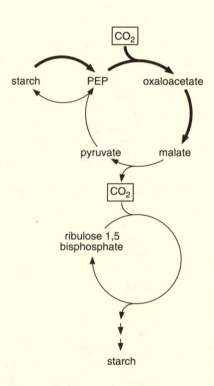

C. Starch is not required for CO_2 pumping in C_4 plants because the compounds that constitute the pump are used catalytically. In principle, a few molecules of phosphoenolpyruvate (PEP) could pump an unlimited amount of CO_2, because PEP is regenerated at the end of each pumping cycle. By contrast, the reactions in the cactus are stoichiometric: each CO_2 molecule that is stored as malate requires one molecule of PEP. Starch in the cactus is used as the source of PEP molecules (via glycolysis): the number of CO_2 molecules that can be fixed is limited by the amount of starch.

D. The principal advantage of this method of CO_2 fixation is that the cactus can seal itself off (close its stomata) during the heat of the day, thereby preventing water loss. Yet it can still provide a rich supply of CO_2 (from stored malate) for sugar synthesis during the day, when the production of ATP and NADPH are maximal due to photosynthetic electron-transfer reactions. At night, when there is less risk of water loss, it can open its stomata and fix CO_2.

Reference: Foyer, C.H. Photosynthesis, pp. 176–195. New York: Wiley, 1984.

*14–23 References: Curtis, H. Biology, 4th ed., p. 216. New York: Worth, 1983.

Figure 14–26 CO_2 Fixation in the cactus (Answer 14–22). Night reactions are shown as thick lines; day reactions are shown as thin lines. Only the carbon pathways are indicated: the cofactor requirements are not shown.

14–24

The burst of oxygen production when the illumination is switched to 650 nm suggests that this wavelength stimulates photosystem II, which accepts electrons directly from water and generates oxygen. Similarly, the dip in oxygen production when the illumination is switched to 700 nm suggests that this wavelength stimulates photosystem I, which accepts electrons from the electron-transport chain and, thus, is farther removed from the reactions that generate oxygen. This interpretation is supported by the more detailed analysis of the chromatic transients below.

The chromatic transients result because the two photosystems are out of balance with one another. Each separate wavelength preferentially (but not absolutely) stimulates one of the two photosystems. Thus, when photosystem I is stimulated (by 700-nm light), it pumps electrons out of the electron-transport chain that links the two photosystems, leaving them in a relatively oxidized state, primed to accept electrons from photosystem II. When the light is switched to 650 nm (which stimulates photosystem II), there is an initial rush of electrons (from H_2O) into the cytochrome chain that causes a burst of O_2 evolution. However, the flow of electrons through the cyctochromes is quickly limited by the electrons' ability to be transferred to photosystem I, which is suboptimally stimulated, and O_2 evolution slows.

When the light is switched back to 700 nm, the electron pressure from photosystem II (which is now suboptimally stimulated) is insufficient to push electrons into the relatively reduced (electron-rich) cytochromes. As a result, O_2 evolution is depressed transiently while electrons are bled off from the cytochromes. Once the cytochromes have been partially drained of their electrons, they can accept new electrons from photosystem II, thereby reestablishing the normal level of oxygen production.

References: Emerson, R. *Ann. Rev. Plant Physiol.* 9:1–24, 1958.

Lawlor, D.W. Photosynthesis: Metabolism, Control, and Physiology. New York: Wiley, 1987.

*14–25 **Reference:** Duysens, L.N.M.; Amesz, J.; Kamp, B.M. Two photochemical systems in photosynthesis. *Nature* 190:510–511, 1961.

14–26

A. These results support a gear-wheel connection between the abstraction of electrons from water and their activation in photosystem II reaction centers. The periodicity of O_2 evolution in response to flashes rules out the possibility that four photons must be delivered simultaneously to the reaction center. If four photons were needed simultaneously, then each flash should yield an equal burst of O_2.

The periodicity also argues against cooperation among four reaction centers to produce a molecule of O_2. At saturating light intensities, most of the reaction centers should be stimulated during each flash; if they could cooperate, they would generate O_2 at each flash. Furthermore, the results of the DCMU experiment definitely eliminate the possibility of cooperation. If four reaction centers were required to cooperate, one might expect a fourth-power dependence on the concentration of active centers. However, a thirtyfold reduction in active centers (DCMU inhibited 97% of the active centers) gave only a thirtyfold reduction in O_2 evolution (peaks of oxygen production were 3% of those in the absence of DCMU) instead of the enormous reduction (30^4) expected from a fourth-power dependence.

A periodicity in O_2 evolution is exactly what one would expect from a gear-wheel link between extraction of multiple electrons from water and photon excitation of single electrons in photosystem II reaction centers. Furthermore, each gear wheel must service a single reaction center. If one gear wheel could interact with four reaction centers, for example, then it could donate its four electrons from water during each flash, which would allow it to evolve O_2 during each flash, eliminating the periodicity.

B. The four-flash periodicity in the evolution of O_2 argues strongly that the gear wheel picks up four electrons at a time from two water molecules and passes them on to the photosystem II reaction center one at a time. The timing of the appearance of the first burst of O_2 says something about the dark-adapted state of the gear wheel, namely, that it holds three electrons. The first three flashes transfer those electrons. The gear wheel can then pick up four new electrons from water (in a reaction that depends on light), generating a molecule of oxygen in the process. (Actually, about a quarter of the gear wheels carry four electrons in the dark-adapted state, which is why there is significant oxygen evolution on the fourth flash.)

C. The periodicity is gradually damped out with increasing flash number because the multiple photosystems fall out of phase with one another. During a single flash most of the photosystem reaction centers capture one photon; however, some capture two photons, and some capture no photons. Those reaction centers that capture zero or two photons are out of step with the majority. After several flashes the number of out-of-step reaction centers increases sufficiently to obscure any periodicity. The period of dark adaptation at the beginning of the experiment is required to bring the majority of the reaction centers to the same state so that periodicity can be observed at all.

Reference: Forbush, B.; Kok, B.; McGloin, M. Cooperation of charges in photosynthetic oxygen evolution II. Damping of flash yield, oscillation and deactivation. *Photochem. Photobiol.* 14:307–321, 1971.

*14–27   Reference: Jagendorf, A.T.; Uribe, E. ATP formation caused by acid-base transition of spinach chloroplasts. *Proc. Natl. Acad. Sci. USA* 55:170–177, 1966.

14–28

A. The balanced equation for reduction of O_2 by Fe^{2+} is

$$4Fe^{2+} + O_2 + 4H^+ \rightarrow 4Fe^{3+} + 2H_2O$$

The two half-cell reactions are

$$4H^+ + O_2 + 4e^- \rightarrow 2H_2O \qquad E_o = 0.82 \text{ V}$$
$$Fe^{3+} + e^- \rightarrow Fe^{2+} \qquad E_o = 0.77 \text{ V}$$

In the balanced reaction Fe^{2+} is donating electrons, therefore

$$\Delta E_o = 0.82 \text{ V} - 0.77 \text{ V}$$
$$\Delta E_o = 0.05 \text{ V}$$

If the reaction occurs under standard conditions, $\Delta E = \Delta E_o$.
Using the relationship between ΔE and ΔG,

$$\Delta G = -nF \, \Delta E$$
$$= -4 \times 23 \text{ kcal/V mole} \times 0.05 \text{ V}$$
$$\Delta G = -4.6 \text{ kcal/mole}$$

or as it is sometimes stated,

$$\Delta G = -1.15 \text{ kcal/mole for each electron}$$

Thus the flow of electrons from Fe^{2+} to O_2 is thermodynamically favorable; the free-energy change for each electron, however, is fairly small. Fortunately, *T. ferrooxidans* does not depend on this redox reaction as a source of energy but rather as a way of detoxifying entering protons and as a source of electrons for reducing $NADP^+$.

B. The balanced reaction for reduction of $NADP^+ + H^+$ by Fe^{2+} is

$$NADP^+ + H^+ + 2Fe^{2+} \rightarrow NADPH + 2Fe^{3+}$$

The two half-cell reactions are

$$NADP^+ + H^+ + 2e^- \rightarrow NADPH \qquad E_o = -0.32\,V$$
$$Fe^{3+} + e^- \rightarrow Fe^{2+} \qquad E_o = 0.77\,V$$

In the balanced reaction Fe^{2+} is donating electrons, therefore

$$\Delta E_o = -0.32\,V - 0.77\,V$$
$$\Delta E_o = -1.09\,V$$

Under nonstandard conditions

$$\Delta E = \Delta E_o - \frac{2.3\,RT}{nF} \log_{10} \frac{[NADPH][Fe^{2+}]^2}{[NADP^+][Fe^{3+}]^2}$$

Since the concentrations of Fe^{2+} and Fe^{3+} are equal, they cancel out, and

$$\Delta E = -1.09\,V - \frac{2.3}{2} \times \frac{1.98 \times 10^{-3}\,kcal}{^\circ K\,mole} \times 310^\circ K \times \frac{V\,mole}{23\,kcal} \log_{10} \frac{10}{1}$$

$$= -1.09\,V - 0.03\,V$$

$$\Delta E = -1.12\,V$$

Under standard conditions $\Delta E = \Delta E_o$, and

$$\Delta G = -nF\,\Delta E$$

$$= -2 \times 23\,kcal/V\,mole \times (-1.09\,V)$$

$$\Delta G = 50.1\,kcal/mole\ (\text{or } 25\,kcal/mole\ \text{for each electron})$$

Under nonstandard conditions

$$\Delta G = -2 \times 23\,kcal/V\,mole \times (-1.12\,V)$$

$$\Delta G = 51.5\,kcal/mole\ (\text{or } 26\,kcal/mole\ \text{for each electron})$$

These calculations make it very clear that reduction of $NADP^+$ by Fe^{2+} is extremely unfavorable.

C. In the absence of a membrane potential, the free-energy change available from inward proton transport is

$$\Delta G = 2.3\,RT \log_{10} \frac{[H^+]_{in}}{[H^+]_{out}}$$

$$= 2.3 \times \frac{1.98 \times 10^{-3}\,kcal}{^\circ K\,mole} \times 310^\circ K \log_{10} \frac{10^{-6.5}}{10^{-2.0}}$$

$$= 2.3 \times \frac{1.98 \times 10^{-3}\,kcal}{^\circ K\,mole} \times 310^\circ K \log_{10} 10^{-4.5}$$

$$\Delta G = -6.4\,kcal/mole$$

If the ΔG for ATP synthesis is 11 kcal/mole, it would take at least two moles of protons (2×-6.4 kcal/mole $= -12.8$ kcal/mole), to drive ATP synthesis. Thus, each molecule of ATP would require that two protons be transported into the cell.

The energy of transport would have to be coupled to the synthesis of ATP, but thermodynamic calculations give no clue as to the actual mechanism of coupling. The protons could enter singly or together, depending on the mechanism. If they entered one at a time, the energy from the first proton would have to be stored in such a way that the energy of the second proton could be added to it.

D. If under standard conditions 50.1 kcal/mole are needed to reduce $NADP^+$, then a minimum of 8 moles of protons would have to be transported (8×-6.4 kcal/mole = -51.2 kcal/mole) to drive the electrons from Fe^{2+} to $NADP^+$.

Once again the thermodynamic calculations give no clue as to the actual mechanism by which electrons from Fe^{2+} are used to reduce $NADP^+$. The transport of protons could be coupled to the reverse flow of electrons in any number of ways. In principle, the transport of eight protons could be linked directly to the transfer of a pair of electrons from $2Fe^{2+}$ to $NADP^+$. Or, the electrons could be activated independently in a mechanism requiring the participation of four protons per electron. However, it seems more likely that the electrons are activated in increments, by being pushed from carrier to carrier up an electron-transport chain—much like the reverse of normal electron transport in mitochondria.

E. The fixation of each mole of CO_2 into glyceraldehyde 3-phosphate requires 3 moles of ATP and 2 moles of NADPH. The synthesis of these molecules requires transport of 22 moles of protons ($3 \times 2H^+$ for ATP + $2 \times 8H^+$ for NADPH). Therefore, 22 moles of Fe^{2+} must be oxidized to neutralize the transported protons. In addition, 2 moles of Fe^{2+} are required to provide electrons for each mole of NADPH. Thus, a total of 26 moles of Fe^{2+} must be oxidized for each mole of CO_2 fixed into glyceraldehyde 3-phosphate.

T. ferrooxidans, therefore, produces enormous quantities of Fe^{3+} during its normal growth. Were it not for the presence of other convenient reductants in the slag heaps, all the iron would be oxidized and the bacteria would stop growing.

Reference: Ingledew, J.W. *Thiobacillus ferrooxidans:* the bioenergetics of an acidophilic chemolithotroph. *Biochim. Biophys. Acta* 683:89–117, 1982.

The Evolution of Electron-Transport Chains

14–29

A. fermentation

14–30

A. True
B. True
C. True
D. True
E. True
F. True

The Genomes of Mitochondria and Chloroplasts

14–31

A. non-Mendelian inheritance (cytoplasmic inheritance)
B. mitotic segregation
C. maternal (uniparental)
D. cytoplasmic petite
E. urea cycle
F. endosymbiont hypothesis

14–32

A. False. Energy-converting organelles divide throughout interphase, out of phase with the division of the cell or with each other. Similarly, replication is not limited to S phase but occurs throughout the cell cycle. However, the process is regulated so that the total number of organelle DNA molecules doubles in every cell cycle.

B. True

C. False. Protein synthesis in both chloroplasts and mitochondria is much more like that in bacteria than that in the cytoplasm. It is true that the machinery in chloroplasts resembles bacterial machinery more closely than mitochondria, but both are clearly bacteriumlike.

D. True

E. False. The mitochondrial genetic code differs slightly from the nuclear code, but it also varies slightly from species to species.

F. True

G. True

H. False. The presence of introns in organelle genes is surprising precisely because corresponding introns have not been found in related bacterial genomes.

I. True

J. False. Variegated leaves are caused by the mitotic segregation of normal and defective chloroplasts.

K. True

L. False. Mitochondria from different tissues of the same organism often show characteristic tissue-specific differences in their content of nuclear proteins.

M. True

N. True

14–33

A. The results in Figure 14–20 are exactly what you would expect if mitochondrial DNA were replicated at random times throughout the cell cycle. Regardless of length of the chase, a constant fraction of the labeled DNA is triggered to replicate. Even the fraction of the DNA that is shifted (about 10%) is what you expect, since the labeling time with BrdU (2 hours) is about 10% of the cell cycle (20 hours).

If replication of mitochondrial DNA were confined to a specific part of the cell cycle, then the DNA that was labeled with ³H-thymidine in the first pulse would not be replicated a second time until the critical phase of the cell cycle came around again. As a result, very little of the labeled DNA should be shifted in density until that time. The critical time in the cell cycle would show up in this experiment as a high fraction of labeled DNA that was density shifted at a particular chase time.

B. It is true that the cell population is asynchronous, but asynchrony has no bearing on the interpretation of the experiment. If the mitochondrial DNA is replicating at random times, then the synchrony of the cell population is irrelevant. Your colleague's concern is directed at the possibility that an asynchronous cell population might obscure your ability to detect a timed replication of mitochondrial DNA. Your elegant experimental design, however, nicely gets around that potential objection. If mitochondria replicated at a specific time during the cell cycle, only those cells in that portion of the cycle would be labeled by the brief pulse of ³H-thymidine. Since in the remainder of the experiment you follow only the radioactive mitochondrial DNA molecules, the brief labeling period has, in essence, synchronized the cell population—you are blind to what happens in any cells that were outside the critical labeling period.

C. Your analysis of nuclear DNA should show a peak of density shifting between 15 and 20 hours, with very little shifting of radioactive DNA at shorter chase periods (Figure 14–27). Since nuclear DNA replicates in a specific phase of the cell cycle, only those cells in that portion of the cell cycle will be labeled with the pulse of ³H-thymidine. The nuclear DNA in the labeled cells will not replicate again until they pass through the entire cell cycle and arrive once

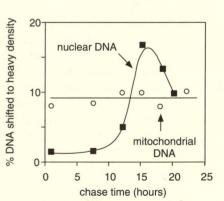

Figure 14–27 Peak of density-shifted nuclear DNA (Answer 14–33).

again in S phase. If the cells were radioactively labeled at the end of S phase initially, they will come into S phase after an additional 15 hours or so—at which point their density can be shifted by exposure to BrdU. If they were at the beginning of S phase when they were labeled, they will not enter S phase again for nearly 20 hours. A peak of density shifting of labeled nuclear DNA is indeed observed between 15 and 20 hours. The density-shifted DNA shows up as a peak and not a plateau because at times longer than 20 hours the labeled DNA passes out of S phase and once again becomes refractory to density shifting.

D. If mitochondrial DNA molecules were replicated at all times during the cell cycle, but individual molecules had to wait one cell cycle between replication events, there would be a peak of density shifting at 18 to 20 hours. Those molecules that were replicated during the pulse of ^{3}H-thymidine would be labeled. If these molecules had to wait one cell cycle to be replicated again, then they would not be subject to density shifting until one full cell cycle had passed. Thus, the mitochondrial DNA molecules labeled initially would not be density shifted until approximately 18 to 20 hours of chase. (The experimental results would resemble those for nuclear DNA replication, since nuclear DNA also has to wait one full cycle; however, the timing would be slightly different because mitochondrial DNA replicates in 2 hours whereas nuclear DNA takes 5 hours).

Reference: Bogenhagen, D.; Clayton, D.A. Mouse L cell mitochondrial DNA molecules are selected randomly for replication throughout the cell cycle. *Cell* 11:719–727, 1977.

*14–34 References: Montoya, J.; Ojala, D.; Attardi, G. Distinctive features of the 5′-terminal sequences of the human mitochondrial mRNAs. *Nature* 290:465–470, 1981.

Ojala, D.; Montoya, J.; Attardi, G. tRNA punctuation model of RNA processing in human mitochondria. *Nature* 290:470–474, 1981.

*14–35 Reference: Stern, D.B.; Palmer, J.D. Extensive and widespread homologies between mitochondrial DNA and chloroplast DNA in plants. *Proc. Natl. Acad. Sci. USA* 81:1946–1950, 1984.

14–36

The abnormal patterns of cytochrome absorption suggest that both *poky* and *puny* affect mitochondrial function. The genetic analysis is consistent with a cytoplasmic mode of inheritance for *poky*, but a nuclear mode of inheritance for *puny*.

Crosses 7, 8, and 9 in Table 14–4 are control crosses, which show that wild-type always yields fast-growing progeny and the mutants always yield slow-growing progeny.

Crosses 1 and 2 show the cytoplasmic mode of inheritance of *poky*. When *poky* was present in the protoperithecial parent (cytoplasmic donor), all the spores grew slowly (cross 1); when it was in the fertilizing parent, the spores grew rapidly (cross 2). This result is expected if the cytoplasmic donor determines the type of mitochondria present in the spores. In cross 1 *poky* was the cytoplasmic donor and the spores grew slowly. In cross 2 wild-type was the cytoplasmic donor and the spores grew rapidly.

Crosses 3 and 4 show the nuclear mode of inheritance of *puny*. In both crosses *puny* contributes a mutant gene to the fusion and wild-type contributes a normal gene to the fusion. These genes are divided up equally (in a Mendelian fashion) among the spores so that half the progeny grow rapidly and half the progeny grow slowly.

Crosses 5 and 6 are slightly more complicated because they involve the interplay of two mutations. In cross 5 *poky* is the cytoplasmic donor, and since all spores receive "*poky*" mitochondria, all spores grow slowly. Some spores (about half) will also carry the *puny* mutation in their nuclei (the other half will have wild-type nuclei—from *poky*), but it makes no difference whether the nuclei are normal or mutant because the mitochondria are already compromised by the *poky* mutation. In cross 6 *poky* is the nuclear donor (*puny* is the cytoplasmic donor); therefore, the *poky* mutation is present in *none* of the spores. Once again, half the spores will carry the *puny* mutation and half will be wild-type; however, in the absence of *"poky"* mitochondria, the nuclear phenotypes are expressed. Thus, half the spores will grow rapidly and half will grow slowly.

It was this sort of distortion from the expected Mendelian behavior of genes that led ultimately to the realization that mitochondria (and later chloroplasts) carried genetic material.

References: Mitchell, M.B.; Mitchell, H.K. A case of "maternal" inheritance in *Neurospora crassa*. *Proc. Natl. Acad. Sci. USA* 38:442–449, 1952.

Mitchell, M.B.; Mitchell, H.K.; Tissieres, A. Mendelian and non-Mendelian factors affecting the cytochrome system in *Neurospora crassa*. *Proc. Natl. Acad. Sci. USA* 39:605–613, 1953.

14–37

A. The deletions that suppress the *pet494* mutation do not affect the coding portion of the mRNA. Therefore, they do not alter the *coxIII* gene product and cannot affect its stability. By contrast, the replacement of the normal 5′ untranslated region with one from any of several other genes is perfectly consistent with an alteration in translation. These results suggest that the normal *PET494* gene product promotes translation of coxIII from the wild-type mitochondrial mRNA.

B. The rearrangements of the 5′ end of the *coxIII* gene result in deletion of essential mitochondrial genes, which normally would produce a cytoplasmic petite strain of yeast. Yet these strains have all the usual mRNAs and grow perfectly well. These observations suggest that the deleted DNA must be present somewhere else in the mitochondria. It turns out that the *pet494* suppressor strains contain both wild-type and deleted mitochondrial genomes. Although it is not uncommon for mitochondria to contain more than one DNA molecule, the mixture of DNAs is quite unusual. This so-called heteroplasmic state is normally quite unstable, and the individual mitochondrial genomes segregate rapidly. However, by demanding that the cells grow on glycerol, it is possible to maintain the heteroplasmic state indefinitely. In the presence of the nuclear *pet494* mutation, both mitochondrial genomes are required for growth on glycerol: the deleted genome provides translatable coxIII mRNA and the wild-type genome provides all other essential gene products.

Reference: Fox, T.D. Nuclear gene products required for translation of specific mitochondrially coded mRNAs in yeast. *Trends in Genetics* 2:97–100, 1986.

*14–38   **Reference:** Whatley, J.M.; John, P.; Whatley, F.R. From extracellular to intracellular: the establishment of mitochondria and chloroplasts. *Proc. R. Soc. Lond. B.* 204:165–187, 1979.

Cell Signaling

General Principles of Cell Signaling

15–1

- A. receptor
- B. local mediators, paracrine signaling
- C. neurotransmitter, synaptic signaling
- D. endocrine, hormones
- E. autocrine signaling
- F. eicosanoids
- G. gap junctions
- H. nitric oxide (NO)
- I. intracellular receptor superfamily
- J. steroid
- K. thyroid
- L. retinoids
- M. ion-channel-linked
- N. G-protein-linked
- O. enzyme-linked

15–2

- A. False. Signaling molecules that bind to intracellular receptors must be sufficiently small and hydrophobic to diffuse across the plasma membrane. There is no such restriction on the kinds of signaling molecules that bind to cell-surface receptors.
- B. False. The concentration of neurotransmitters in the synaptic cleft is much higher than, for example, the concentration of circulating hormones in the blood. Correspondingly, neurotransmitter receptors have a relatively low affinity for their ligand, which allows the neurotransmitter to dissociate rapidly from the receptor to terminate a response.
- C. True
- D. False. Gap junctions create narrow water-filled channels between cells that permit the exchange of small molecules but not proteins.
- E. True
- F. False. The same signaling molecule can bind to identical receptor proteins and yet produce very different responses in different types of target cells, reflecting differences in the internal machinery to which the receptors are coupled.
- G. True
- H. True
- I. True
- J. True
- K. False. Individual mammalian cells may contain more than 100 distinct kinds of protein kinases, most of which are serine/threonine kinases.

*15–3

---

*Problems with an asterisk (*) are answered in the Instructor's Manual.

*15–4   **Reference:** Yalow, R.S. Radioimmunassay: a probe for the fine structure of biologic systems. *Science* 200:1236–1245, 1978.

15–5

A. Since the half-life of radioactive iodine is 7 days, two atoms of radioactive iodine will give rise to one disintegration in a week. Thus, to determine the required number of picograms of labeled insulin, it is necessary to convert 1000 counts per minute to disintegrations per week (multiply by two to get the number of radioactive iodine atoms) and then to calculate the weight of the same number of insulin molecules.

1000 counts per minute is 2.0×10^7 disintegrations (disint) per week.

$$\frac{disint}{week} = \frac{1000 \ counts}{min} \times \frac{2 \ disint}{count} \times \frac{60 \ min}{hr} \times \frac{24 \ hr}{day} \times \frac{7 \ days}{week}$$
$$= 2.0 \times 10^7 \ disintegrations \ per \ week$$

Thus, 1000 cpm corresponds to 4.0×10^7 atoms or radioactive iodine. The weight of this number of insulin molecules is 0.76 pg (picograms).

$$insulin = 4.0 \times 10^7 \ molecules \times \frac{moles}{6 \times 10^{23} \ molecules} \times \frac{11,466 \ g}{mole} \times \frac{10^{12} \ pg}{g}$$

$$= 0.76 \ pg$$

B. For optimal sensitivity of the radioimmunoassay, the amounts of the tracer and unknown should be the same. Thus, given the starting assumptions, at optimum sensitivity you will be able to detect 0.76 pg of unlabeled insulin.

15–6

A. Both cell lines appear to contain a glucocorticoid receptor, based on the dexamethasone-induced increases in CAT activity. When transfected with the construct carrying the glucocorticoid-responsive enhancer on the viral DNA segment, cell line 1 showed a fivefold increase and cell line 2 showed an eightyfold increase in CAT activity in the presence of dexamethasone.

B. Cell line 1 differs significantly from cell line 2 in the level of CAT activity detected in the *absence* of dexamethasone after transfection of the construct containing the glucocorticoid-responsive enhancer. There are two reasonable explanations for this difference. (1) Cell line 1 contains a mutant glucocorticoid receptor that can partially activate the glucocorticoid enhancer in the viral segment in the absence of glucocorticoid. (2) Cell line 1 contains a tissue-specific protein that recognizes and stimulates a second (nonglucocorticoid-responsive) enhancer on the viral segment.

C. These potential explanations make different predictions for the outcome of experiments in which a variety of shorter viral segments are tested for CAT activity in the two cell lines. If the difference between the cell lines were due to different glucocorticoid receptors that recognize the same enhancer, then shorter pieces of the viral segments would all behave in one or the other of two ways: (1) they would not contain the glucocorticoid enhancer and would not be activated by dexamethasone in either cell line, or (2) they would contain the glucocorticoid enhancer and would show the same differential response to dexamethasone as the intact viral segment.

On the other hand, if the difference between the cell lines were due to different enhancer elements (glucocorticoid responsive and nonglucocorticoid responsive), then it should be possible to separate the different enhancers onto different DNA segments. Segments that contain only the glucocorticoid enhancer should give identical responses in the two cell lines; segments that contain only the nonglucocorticoid enhancer should give a stimulation in cell line 1 (that is independent of glucocorticoid) but no stimulation in cell line 2.

Actual experiments of this kind indicate that the viral segment contains a second enhancer that is specifically activated in cell line 1. Results of experiments such as illustrated in this problem suggest that genes may be controlled in a tissue-specific fashion by an interplay of multiple regulatory factors. Indeed, stimulation of gene expression by steroids normally requires that the responding cell have a steroid receptor and that the chromatin structure surrounding a potentially regulatable gene be open—an effect mediated by other transcription factors.

Reference: DeFranco, D.; Yamamoto, K.R. Two different factors act separately or together to specify functionally distinct activities at a single transcriptional enhancer. *Mol. Cell. Biol.* 6:993–1001, 1986.

*15–7   **References:** Payvar, F.; DeFranco, D.; Firestone, G.L.; Edgar, B.; Wrange, O.; Okret, S.; Gustafsson, J.-A.; Yamomoto, K.R. Sequence-specific binding of glucocorticoid receptor to MMTV DNA at sites within and upstream of the transcribed region. *Cell* 35:381–392, 1983.

Scheidereit, C.; Geisse, S.; Westphal, H.M.; Beato, M. The glucocorticoid receptor binds to defined nucleotide sequences near the promoter of mouse mammary tumor virus. *Nature* 304:749–752, 1983.

15–8

A. The cycloheximide-induced alteration of the puffing pattern is due to its effect on protein synthesis. The result indicates that a newly synthesized protein is required to turn off the early puffs and to turn on the late puffs. Presumably, the protein is synthesized from one of the early puffs.

The shut off of transcription from the intermolt puffs is insensitive to cycloheximide treatment. This observation suggests that the receptor-ecdysone complex turns off these puffs directly. (The shut off presumably is mediated by a protein because ecdysone is not a large enough molecule to have such specific interactions with naked DNA.)

B. The immediate regression of the early puffs upon ecdysone removal indicates that the ecdysone-receptor complex is required continuously to keep the genes turned on.

The premature activation of the late puffs under these conditions is unexpected. If activation of the late puffs depended only on a product of the early puffs, then they should be turned on at the same time (or even delayed due to a lower level of early product). The premature activation suggests that the receptor-ecdysone complex actually functions as an inhibitor, delaying activation until the concentration of the presumptive early-puff product reaches some critical level. Removal of ecdysone allows the puffs to be induced at a lower concentration of early product.

C. These experimental observations are summarized in the schematic diagram shown in Figure 15–20. The ecdysone-receptor complex binds to regulatory regions of intermolt, early, and late puffs. Binding at intermolt puffs turns them off, binding at early puffs turns them on, and binding at late puffs keeps them off. A product from one or more early puffs binds at the regulatory regions of early and late puffs, ultimately turning off the early puffs and turning on the late puffs.

Reference: Ashburner, M.; Chihara, C.; Meltzer, P.; Richards, G. Temporal control of puffing activity in polytene chromosomes. *Cold Spring Harbor Symp. Quant. Biol.* 38:655–662, 1973.

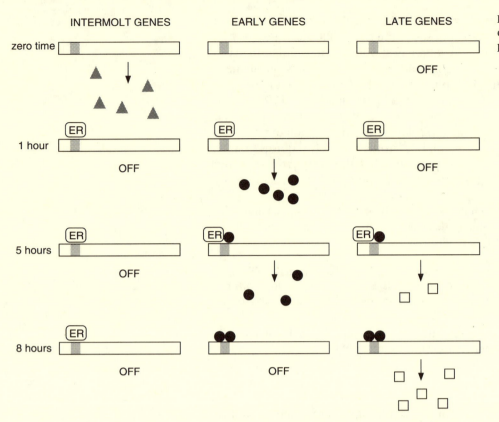

INTERMOLT GENES   EARLY GENES   LATE GENES

zero time

1 hour   OFF

5 hours   OFF

8 hours   OFF

OFF

OFF

OFF   OFF

OFF

Figure 15–20 Schematic diagram relating ecdysone-receptor (ER) binding to the pattern of gene activity (Answer 15–8).

Signaling via G-Protein-linked Cell-Surface Receptors

15–9

   A. trimeric GTP-binding proteins (G proteins)
   B. intracellular mediators (intracellular messengers, or second messengers)
   C. cyclic AMP, adenylyl cyclase, cyclic AMP phosphodiesterase
   D. stimulatory G protein (G_s)
   E. β-adrenergic receptors
   F. cholera toxin
   G. inhibitory G protein (G_i)
   H. pertussin toxin
   I. cyclic-AMP-dependent protein kinase (A-kinase)
   J. CRE-binding protein (CREB)
   K. serine/threonine phosphoprotein phosphatases
   L. phosphatidylinositol (PI)
   M. G_q, phospholipase C-β, inositol trisphosphate (IP_3), diacylglycerol
   N. Ca^{2+} oscillations
   O. protein kinase C (C-kinase, or PKC)
   P. calmodulin
   Q. Ca^{2+}/calmodulin-dependent protein kinases (CaM kinases)
   R. CaM-kinase II
   S. olfactory receptors
   T. cyclic GMP
   U. rod photoreceptors (rods), rhodopsin
   V. cyclic GMP phosphodiesterase

15–10

   A. True
   B. False. Extracellular signaling molecules control cyclic AMP levels by altering the activity of adenylyl cyclase.

C. True

D. True

E. False. It is the substrates for A-kinase, not A-kinase itself, that differ in different cell types.

F. True

G. True

H. False. In the nerve-specific pathway for Ca^{2+} signaling the cytosolic Ca^{2+} does come from outside the cell, but in the ubiquitous pathway it comes from the ER.

I. False. The phospholipid phosphatidylinositol is phosphorylated twice to make PIP_2. Cleavage of PIP_2 by phospholipase C-β then releases inositol trisphosphate.

J. True

K. True

L. False. The two branches of the inositol phospholipid signaling pathway often collaborate in producing a full cellular response. A number of cell types can be stimulated to proliferate in culture when treated with both a Ca^{2+} ionophore and a C-kinase activator but not when they are treated with either reagent alone.

M. False. Calmodulin is activated by binding Ca^{2+} but has no enzyme activity itself. Ca^{2+}/calmodulin acts by binding to other proteins to regulate their activity.

N. True

O. True

P. True

Q. False. In visual transduction, receptor activation leads to a fall in the level of cyclic GMP.

R. True

S. True

T. True

*15–11

15–12

A. The specific binding curve is obtained by subtracting the nonspecific curve from the total. As illustrated in Figure 15–21, the specific binding curve reaches a plateau above 4 nM alprenolol. Thus, the β-adrenergic receptors are saturated with alprenolol above this concentration.

B. There are 1500 β-adrenergic receptors per frog erythrocyte. Since one alprenolol binds per receptor, the number of bound alprenolol molecules is equal to the number of receptors. At saturation, 20,000 cpm of alprenolol binds per mg of erythrocyte membrane (Figure 15–21). Thus, the amount of bound alprenolol is

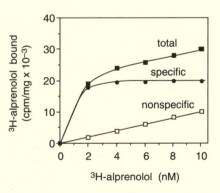

Figure 15–21 Specific binding of alprenolol to erythrocyte membranes. (Answer 15–12).

$$\text{bound alprenolol} = \frac{20 \times 10^3 \, \text{cpm}}{\text{mg}} \times \frac{\text{mmol}}{10^{13} \, \text{cpm}} \times \frac{6 \times 10^{20} \, \text{molecules}}{\text{mmol}}$$

$$\times \frac{\text{mg}}{8 \times 10^8 \, \text{erythrocyte}}$$

$$= 1500 \, \text{molecules per erythrocyte}$$

Since one molecule of alprenolol binds per β-adrenergic receptor, there are 1500 β-adrenergic receptors per erythrocyte.

Reference: Lefkowitz, R.J.; Limbird, L.E.; Mukherjee, C.; Caron, M.G. The β-adrenergic receptor and adenylate cyclase. *Biochim. Biophys. Acta* 457:1–39, 1976.

A. If A-kinase were essential in hamster cells, it would have been impossible to isolate mutants that lack the enzyme, since they could not survive. Thus A-kinase is not essential to these hamster cells (nor is it essential in several other cell lines in which such mutants have been isolated). One should not conclude, however, that A-kinase is not essential for the organism. There are many examples of enzyme defects that have minimal consequences for cells in culture but severely affect the intact organism.

B. Mutations that eliminate the catalytic subunit would be unresponsive to high levels of cyclic AMP and therefore resistant to cyclic AMP. These mutations would be recessive, for in the presence of the wild-type catalytic subunit, cyclic AMP responsiveness would be restored. (It might seem that mutations that eliminate the regulatory subunits would be the same. However, without the regulatory subunits A-kinase would be active at all times, which is a lethal condition—excess A-kinase activity is the reason that these cells are killed by high levels of cyclic AMP. Thus, cell lines without regulatory subunits would not have been isolated in the first place.)

Dominant mutations are somewhat more difficult to explain. In general, dominance indicates an altered activity rather than complete lack of activity. Dominant mutations have been found in both the regulatory and catalytic subunits. A possible dominant mutation in the regulatory subunit is one that increases its affinity for the catalytic subunit. Such a mutation might plausibly respond only to high levels of cyclic AMP (required to displace the tightly bound regulatory subunit), and it would be dominant because the mutant subunits bind the catalytic subunits at the low cyclic AMP concentrations that would displace the normal regulatory subunits.

Dominant mutations in the catalytic subunits are more difficult to explain. One possibility is that mutant catalytic subunits bind regulatory subunits more tightly, which would explain its altered responsiveness to cyclic AMP. Its dominance might be understood if the combination of a normal catalytic subunit rendered the heterodimer mutantlike in its binding to the regulatory subunits. If this were the case, an even mixture of mutant and normal catalytic subunits would be expected to have only one-quarter the normal A-kinase activity (only one-quarter of the catalytic dimers would have two wild-type subunits).

C. These experimental results generally support the contention that all cyclic AMP effects are mediated through A-kinase, but they fall short of proving the point. Similar studies in a variety of other cell lines have also identified mutants solely in the A-kinase pathway. Furthermore, cells that completely lack A-kinase show none of the effects normally associated with cyclic AMP. Consequently, there is no convincing evidence at present in eucaryotic cells that cyclic AMP has any target other than A-kinase.

Reference: Gottesman, M.M. Genetics of cyclic-AMP-dependent protein kinases. In Molecular Cell Genetics (M.M. Gottesman, ed.), pp. 711–743. New York: Wiley, 1985.

*15–14   References: Byers, D.; Davis, R.L.; Kiger, J.A. Defect in cyclic AMP phosphodiesterase due to the *dunce* mutation of learning in *Drosophila melanogaster*. *Nature* 289:79–81, 1981.

Chen, C.-N.; Denome, S.; Davis, R.L. Molecular analysis of cDNA clones and the corresponding genomic coding sequences of the *Drosophila* dunce[+] gene, the structural gene for cyclic AMP phosphodiesterase. *Proc. Natl. Acad. Sci. USA* 83:9313–9317, 1986.

15–15

A. Your experiments reproduce the paradoxical result you observed in brain slices. Cells transfected with the dopamine receptor, type II adenylyl cyclase, and α_s^* have high levels of cyclic AMP that are increased further by stimulating the dopamine receptor with quinpirole. Yet quinpirole has no effect on cyclic AMP levels in cells that are missing type II adenylyl cyclase or α_s^*.

B. Pertussis toxin eliminates the enhanced synthesis of cyclic AMP by quinpirole. Pertussis toxin is an enzyme that ADP ribosylates α_i so that G_i can no longer interact with its receptors. In the absence of this interaction the α_i subunit cannot exchange GTP for bound GDP and thus cannot dissociate from the $\beta\gamma$ subunit. Thus pertussis toxin blocks the enhancing signal from the quinpirole-activated dopamine receptor.

C. The simplest interpretation of your results is that type II adenylyl cyclase is maximally activated by a combination of α_s^* and $\beta\gamma$ subunits. The $\beta\gamma$ subunit is released from G_i upon activation of the dopamine receptor by quinpirole binding.

   The actual molecular explanation for the enhancing effects of $\beta\gamma$ subunits is unknown. Since free $\beta\gamma$ subunits have no effect on cyclic AMP levels in the absence of α_s^*, they do not stimulate type II adenylyl cyclase unless it has already bound an α_s subunit. One possible explanation is that the binding of an α_s to type II adenylyl cyclase causes a conformational change in the cyclase that exposes a binding site for the $\beta\gamma$ subunit. Binding of the $\beta\gamma$ subunit to that site could help lock the type II adenylyl cyclase into an active state, leading to enhanced or prolonged synthesis of cyclic AMP.

D. Expression of the cDNA for the α subunit of transducin eliminates the enhanced synthesis of cyclic AMP by quinpirole. Like pertussis toxin, the α subunit of transducin blocks the signal from the quinpirole-activated dopamine receptor; however, its effects are specific for the $\beta\gamma$ subunit. This was an important control experiment in the original studies because it pinpointed the $\beta\gamma$ subunit as the key component of G_i responsible for enhanced cyclic AMP levels in quinpirole-treated cells. With the results from pertussis toxin alone, it could be argued that the α_i subunit was having some anomalous enhancing effect in the presence of α_s^*.

   References: Federman, A.D.; Conklin, B.R.; Schrader, K.A.; Reed, R.R.; Bourne, H.R. Hormonal stimulation of adenylyl cyclase through G_i-protein $\beta\gamma$ subunits. *Nature* 356:159–161, 1992.

   Tang, W.-J.; Gilman, A.G. Type-specific regulation of adenylyl cyclase by G protein $\beta\gamma$ subunits. *Science* 254:1500–1503, 1991.

*15–16   Reference: Kurjan, I. Pheromone response in yeast. *Annu. Rev. Biochem.* 61:1097–1129, 1992.

*15–17   Reference: Burch, R.M.; Luini, A.; Axelrod, J. Phospholipase A_2 and phospholipase C are activated by distinct GTP-binding proteins in response to α_1-adrenergic stimulation in FRTL5 thyroid cells. *Proc. Natl. Acad. Sci. USA* 83:7201–7205, 1986.

15–18

A. The difference in the behavior of myosin light-chain kinase when purified in the presence and absence of protease inhibitors suggests that the regulatory domain of the kinase is sensitive to cleavage by proteases in the cell extract. Since the regulatory domain is expected to interact with calmodulin, it is not surprising that the enzyme lacking the regulatory domain is not retained on a calmodulin-affinity column. Activation of the kinase by proteolytic removal of the regulatory domain suggests that the domain in some way covers the active site. This regulatory "flap" moves out of the way upon Ca^{2+}/calmodulin binding. This motif of a regulatory flap covering an active site seems to be fairly common.

B. Upon entry into the cytoplasm, Ca^{2+} binds to calmodulin, causing a change in its conformation. The Ca^{2+}/calmodulin complex binds to the regulatory domain of myosin light-chain kinase, exposing the active site of the enzyme.

The activated kinase then phosphorylates myosin light chains. This alters their conformation so that they can interact with actin and thus initiate contraction. As you might predict, contraction is terminated through the action of phosphatases that remove the phosphate from the myosin light chains.

C. Calmodulin-affinity chromatography would probably not work very well as a first step in purification of the myosin kinase because cells contain an enormous amount of calmodulin (typically 1% of the total protein of a cell). In the presence of Ca^{2+} the myosin kinase would be bound to Ca^{2+}/calmodulin in the cell extract and thus would not bind efficiently to the calmodulin on the column. Therefore, the first two steps in your purification are necessary to remove the endogenous calmodulin so that the kinase will bind to the calmodulin-affinity column.

Reference: Adelstein, R.S.; Klee, C.B. Smooth muscle myosin light-chain kinase. In *Calcium and Cell Function* (W.Y. Cheung, ed.), Vol. 1, pp. 167–182. New York: Academic Press, 1980.

15–19

A. The activity of Ca^{2+} and that of diacylglycerol suggest that the normal sequence of events involves phospholipase C. Collagen fibers and thrombin stimulate a receptor on the surface of the platelet, which in turn activates phospholipase C, presumably through a G protein. Phospholipase C cleaves phosphatidylinositol bisphosphate to produce IP_3 and diacylglycerol. IP_3 mobilizes internal Ca^{2+} stores, which activates myosin light-chain kinase, resulting in the phosphorylation of the myosin light chain. This branch of the pathway can be stimulated by the calcium ionophore A23187. Diacylglycerol activates C-kinase, which phosphorylates the 40-kd protein. This branch of the pathway can be stimulated directly by diacylglycerol. These two individual pathways interact to stimulate serotonin release. The overall pathway for platelet activation is diagrammed in Figure 15–22.

B. Secretion of serotonin seems to require both calcium and diacylglycerol, since neither alone causes any secretion (Figure 15–10B). These experiments imply that the 40-kd protein is involved, although direct proof is lacking; its role is undefined. Calcium is thought to be more directly involved in secretion and in some way enables the fusion of membranes required for exocytosis.

Reference: Nishizuka, Y. Calcium, phospholipid turnover and transmembrane signaling. *Phil. Trans. R. Soc. Lond. (Biol.)* 302:101–112, 1983.

Figure 15–22 Overall pathway for platelet activation (Answer 15–19).

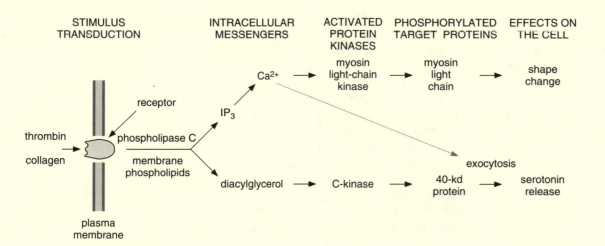

A. The opening of the K⁺ channel in the presence of GppNp and absence of acetylcholine may be somewhat surprising to you, since the release of GDP and the binding GTP by G proteins normally are stimulated by an activated receptor. Even in the absence of an activated receptor, however, G proteins exchange their bound nucleotides with nucleotides in the cytoplasm. Exchange is slow, and any bound GTP is quickly hydrolyzed in the absence of an activated receptor, thereby keeping the channel closed. The K⁺ channels open slowly when GppNp is present because, each time a GDP is released and a GppNp is bound, the G protein is locked into an active form. Over the course of a minute, enough G protein is activated in this way to open the K⁺ channels in the absence of acetylcholine.

B. The complete G protein does not activate the K⁺ channels in the absence of acetylcholine presumably because, like other trimeric G proteins, the active portion is inhibited by one of the subunits. The ability of the G_α subunit to open the K⁺ channel in the absence of acetylcholine and GTP suggests that it is the active portion of the G protein.

C. These experiments virtually rule out participation of an intracellular messenger in the activation of K⁺ channels. Since the buffer does not contain ATP or Ca^{2+}, neither of these potential messengers can be involved in opening the K⁺ channel. Nor is it likely that diglyceride or IP_3 participates, at least in their normal ways. Although either of these intermediates could be produced, neither could function as it usually does. IP_3 normally affects Ca^{2+} concentrations (which it cannot do here), and diglyceride normally activates protein kinase C (which requires ATP for its action). The simplest interpretation is that the G_α subunit activates the K⁺ channel directly, although an activation involving some other membrane component is not eliminated by these experiments.

D. A simple scheme for the G-protein-mediated activation of K⁺ channels by acetylcholine is shown in Figure 15–23.

Note: Although the original paper by Logothetis et al. concluded that it was $G_{\beta\gamma}$ that activated the K⁺ channel, subsequent studies have established that G_α is the active subunit. We have amended this problem to reflect the newer conclusion.

References: Birnbaumer, L.; Brown, A.M. G protein opening of K⁺ channels. *Nature* 327:21, 1987.

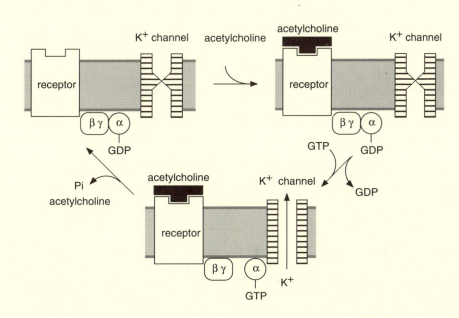

Figure 15–23 Diagram illustrating activation of K⁺ channels in heart by acetylcholine (Answer 15–20).

Logothetis, D.E.; Kurachi, Y.; Galper, J.; Neer, E.J.; Clapham, D.E. The βγ subunits of GTP-binding proteins activate the muscarinic K+ channel in heart. *Nature* 325:321–326, 1987.

Logothetis, D.E.; Kurachi, Y.; Galper, J.; Neer, E.J.; Clapham, D.E. G protein opening of K+ channels. *Nature* 327:22, 1987.

*15–21 **References:** Fung, B.K-K.; Stryer, L. Photolyzed rhodopsin catalyzes the exchange of GTP for bound GDP in retinal rod outer segments. *Proc. Natl. Acad. Sci. USA* 77:2500–2504, 1980.

Fung, B.K-K.; Hurley, J.B.; Stryer, L. Flow of information in the light-triggered cyclic nucleotide cascade of vision. *Proc. Natl. Acad. Sci. USA* 78:152–156, 1981.

Signaling via Enzyme-linked Cell-Surface Receptors

15–22

A. receptor guanylyl cyclases
B. receptor tyrosine kinases, protein kinases
C. SH2 domains
D. Ras superfamily, monomeric GTPases
E. Ras proteins
F. GTPase-activating proteins (GAPs), guanine nucleotide releasing proteins (GNRPs)
G. mitogen-activated protein (MAP) kinases
H. tyrosine-kinase-associated receptors
I. Src family
J. IL-2 receptor
K. protein tyrosine phosphatases
L. transforming growth factor-βs (TGF-βs)
M. receptor serine/threonine kinases
N. Notch

15–23

A. True
B. True
C. False. Ligand binding causes the receptor tyrosine kinase to assemble into dimers, which is the signal for activation of the kinase domain. In each case the receptors then phosphorylate themselves to initiate the intracellular signaling cascade.
D. True
E. True
F. True
G. False. Receptor tyrosine kinases and tyrosine-kinase-associated receptors activate some of the same signaling pathways, as might be expected since they both lead to phosphorylation of tyrosine residues.
H. False. When activated, receptor protein tyrosine phosphatases remove phosphate groups from tyrosine residues only on selected target proteins in the cell.
I. True
J. True
K. True

15–24

A. The concentrations of the two versions of RSV Src are inversely proportional to the volume in which they are distributed. Nonmyristoylated Src is distributed throughout the volume of the cell, which has a radius of 10 μm (10×10^{-6} m).

$$V_{cell} = \tfrac{4}{3}\pi r^3 = \tfrac{4}{3}\pi \left(10 \times 10^{-6}\, m\right)^3 = 4.1888 \times 10^{-15}\, m^3$$

Myristoylated Src is confined to a 4-nm layer beneath the plasma membrane. The volume of this layer is the volume of the cell minus the volume of a sphere with a radius 4 nm less than that of the cell.

$$\begin{aligned} V_{layer} &= V_{cell} - \tfrac{4}{3}\pi \left(r - 4\, nm\right)^3 \\ &= V_{cell} - \tfrac{4}{3}\pi \left(10 \times 10^{-6}\, m - 4 \times 10^{-9}\, m\right)^3 \\ &= V_{cell} - 4.1838 \times 10^{-15}\, m^3 \\ &= 0.0050 \times 10^{-15}\, m^3 \end{aligned}$$

Thus the volume of the cell is 834 times greater than the volume of a 4-nm layer beneath the membrane ($4.1888 \times 10^{-15}\, m^3 / 0.0050 \times 10^{-15}\, m^3$).

Even allowing for the interior regions of the cell from which the non-myristoylated Src is excluded (nucleus and lumena of organelles), it would still be more than two orders of magnitude less concentrated in the neighborhood of the membrane than myristoylated Src.

B. If a critical target for the Src protein tyrosine kinase is a membrane protein, the much lower concentration of nonmyristoylated Src in the neighborhood of the membrane could explain the lack of a transformed phenotype. The equilibrium constant (K) for association of Src with its membrane target (X) is

$$K = \frac{[\text{Src} \cdot \text{X}]}{[\text{Src}][\text{X}]}$$

Since the concentration of the target remains the same,

$$K[\text{X}] = \frac{[\text{Src} \cdot \text{X}]}{\text{Src}}$$

Thus, if the concentration of Src near the membrane falls by more than 100-fold, the concentration of the complex between Src and its membrane target must also decrease by more than 100-fold. Such a large decrease in the association of Src with its target could readily account for the lack of transformation by nonmyristoylated Src.

*15–25 **Reference:** Sadowski, H.B.; Shuai, K.; Darnell, J.E.; Gilman, M.Z. A common nuclear signal transduction pathway activated by growth factor and cytokine receptors. *Science* 261:1739–1744, 1993.

*15–26 **Reference:** Fantl, W.J.; Escobedo, J.A.; Martin, G.A.; Turck, C.W.; del Rosario, M.; McCormick, F.; Williams, L.T. Distinct phosphotyrosines on a growth factor receptor bind to specific molecules that mediate different signaling pathways. *Cell* 69:413–423, 1992.

15–27

A. The point of probing a duplicate filter with biotin-tagged GST is to enable you to eliminate false-positive clones that encode proteins that bind to the GST protein of the fusion protein or to the biotin tag. In the comparison of duplicate filters you look for colonies or phage plaques that bind to biotin-tagged GST-SH3, but do not bind to biotin-tagged GST.

B. This method could not be used as described for detecting proteins that bind to SH2 domains. The reason is that binding by SH2 domains requires that the target protein carries a phosphotyrosine residue. In this method the target protein is expressed in *E. coli*, which does not contain the tyrosine kinases required to make the required modification in the target protein. One way to make the screen work would be to incubate the filters first with a protein tyrosine kinase and ATP. Alternatively, a protein tyrosine might be engineered into *E. coli* so that it could be turned on at the same time the cDNA library was being expressed.

C. The main difference between protein-short-sequence interactions and the subunit-subunit interactions in multisubunit enzymes lies in their stability and reversibility, which in turn depends largely on the total number and aggregate strength of the weak bonds involved in their formations. Large contact surfaces such as those found among subunits in multisubunit enzymes make for very stable structures, whereas most of the examples of short patch recognition are more transient, and in some cases conditional, as in the interaction of SH2 domains with phosphotyrosine-containing proteins.

Reference: Ren, R.; Mayer, B.J.; Cicchetti, P.; Baltimore, D. Identification of a ten-amino-acid proline-rich SH3 binding site. *Science* 259:1157–1161, 1993.

*15–28 **References:** Fields, S.; Song, O. A novel genetic system to detect protein-protein interactions. *Nature* 340:245–246, 1989.

Vojtek, A.B.; Hollenberg, S.M.; Cooper, J.A. Mammalian Ras interacts directly with the serine/threonine kinase Raf. *Cell* 74:205–214, 1993.

15–29

A.
1. A VP16-cDNA might give rise to white colonies when grown in the presence of XGAL and absence of histidine if, for example, the cDNA encoded a protein that bound to a DNA sequence in front of the histidine gene. Such a DNA-binding protein with an attached VP16 activation domain could turn on transcription of the *HIS3* gene but not the *LacZ* gene.
2. A VP16-cDNA might give rise to blue colonies in the presence of XGAL and absence of histidine—even when no LexA-Ras plasmid is present in the cell—if, for example, the cDNA encoded a protein that bound a DNA sequence that appears in front of both the *HIS3* gene and the *LacZ* gene. Such a VP16-cDNA would turn on transcription of both genes.
3. A VP16-cDNA might give rise to blue colonies in the presence of XGAL and absence of histidine when expressed in a cell containing a LexA-lamin plasmid, if, for example, the cDNA encoded a protein that binds to the LexA domain of the LexA-Ras fusion protein.

B. As your friend points out, only one out of six cDNA inserts will be correctly fused to the *VP16* gene segment. In addition, many cDNAs that are correctly fused may not include enough of the coding sequence of the mRNA (many cDNAs are partial copies of the mRNA) to encode the portion of the protein that interacts with the LexA-Ras hybrid protein. But these are all concerns about efficiency and do not matter so long as a large enough population of cDNA is tested. In practice, it is best to prepare the cDNA library from a cell type or tissue in which the target protein (Ras in this case) is known to function. In that way it is likely that the mRNAs for any interacting proteins will be represented in the cDNA library.

C. Your advisor was wise to insist on confirmatory biochemical studies. The two-hybrid system, even with all the layers of selection and screening you built into it, can still give false positives. The 9 out of 19 clones with homology to Raf may seem very convincing given what you know about the biology of cell signaling. But it was still possible that the interaction detected by the two-hybrid system was an indirect interaction; for example, it could have been that there was a natural yeast protein that could bind to both Ras and Raf. Such a protein could bring the LexA-Ras and VP16-Raf hybrid proteins together to activate transcription in the absence of any direct interaction between Ras and Raf. The biochemical studies, which were carried out using proteins grown in *E. coli*, eliminated the possibility of this kind of indirect interaction.

References: Vojtek, A.B.; Hollenberg, S.M.; Cooper, J.A. Mammalian Ras interacts directly with the serine/threonine kinase Raf. *Cell* 74:205–214, 1993.

Zhang, X.-F.; Settleman, J.; Kyriakis, J.M.; Takenchi-Suzuki, E.; Elledge, S.J.; Marshall, M.S.; Bruder, J.T.; Rapp, U.R.; Avruch, J. Normal and oncogenic p21[ras] proteins bind to the amino-terminal domain of c-Raf-1. *Nature* 364:308–313, 1993.

Target-Cell Adaptation

15–30

A. adaptation or desensitization
B. receptor down-regulation
C. chemotaxis receptors

15–31

A. True
B. True
C. True
D. True
E. False. If the attractant level stays constant (regardlesss of the level), bacteria alternate tumbling and straight swimming at fairly rapid intervals.
F. False. The binding of a repellent activates the receptor and leads to tumbling; the binding of an attractant decreases the activity of the receptor and leads to smooth swimming. This makes sense from the bacterium's point of view: it wants to continue swimming toward an attractant, but it wants to tumble and change direction if it encounters a repellent.
G. True

15–32

A. The difference in binding of CGP-12177 and dihydroalprenolol to extracts of isoproterenol-treated cells suggests that the ligand-binding sites on some receptors are not directly exposed in the lysate but instead are enclosed by membrane. Dihydroalprenolol, being hydrophobic, can cross membranes and thus bind to all receptors. By contrast, CGP-12177, which is hydrophilic, cannot cross membranes and thus can only bind to exposed receptors. These results suggest that there are two populations of vesicles in the cell lysates: one with receptors facing outward and the other with receptors facing inward. The two populations are separated on sucrose-density gradients. The presence of 5′ nucleotidase in one population but not in the other indicates that the CGP-12177-binding population represents vesicles formed from fragments of the plasma membrane.

B. The parallel between CGP-12177 binding and hormone-dependent adenylyl cyclase activity suggests that the two are related; that is, the receptors that bind to CGP-12177 are the same ones that can activate adenylyl cyclase. This suggestion is supported by finding that CGP-12177 binding is specifically associated with vesicles formed from plasma membrane fragments. The other population of vesicles presumably represents internal vesicles formed by endocytosis, which would yield vesicles with the ligand-binding sites in the interior. These experiments suggest that isoproterenol-induced desensitization results from internalization of the receptors, making them unavailable for interaction with hormone and separating them from the proteins that couple them to adenylyl cyclase.

Reference: Hertel, C.; Muller, P.; Portenier, M.; Staehelin, M. Determination of the desensitization of β-adrenergic receptors by [³H]CGP-12177. *Biochem. J.* 216:669–674, 1983.

A. If phosphorylation of the two subunits occurs independently and at equal rates, four different types of receptor will exist: nonphosphorylated receptor, receptor phosphorylated on the γ subunit, receptor phosphorylated on the δ subunit, and receptor phosphorylated on both subunits. At 0.8 mole P/mole receptor each subunit would be 40% phosphorylated and 60% nonphosphorylated. Thus the ratio of the various receptor forms would be 36% with no phosphate (0.6 × 0.6), 24% with only the γ subunit phosphorylated (0.6 × 0.4), 24% with only the δ subunit phosphorlyated (0.6 × 0.4), and 16% with both subunits phosphorylated (0.4 × 0.4). At 1.2 mole P/mole receptor, the ratios would be: 16% with no phosphate, 24% with the γ subunit phosphorylated, 24% with the δ subunit phosphorylated, and 36% with both subunits phosphorylated.

B. These experiments suggest that desensitization requires only one phosphate per receptor and that phosphorylation of either the γ or the δ subunit is sufficient for desensitization. For both preparations, the fraction that behaves like the untreated receptor matches best the fraction calculated to carry no phosphate: 36% versus 36% at 0.8 mole P/mole receptor and 18% versus 16% at 1.2 mole P/mole receptor. This result suggests that phosphorylation of either subunit is sufficient to trigger desensitization. If a specific subunit were required to be phosphorylated, then the expected fractions behaving like the untreated receptor would have been 60% (24% + 36%) at 0.8 mole P/mole receptor and 40% (24% + 16%) at 1.2 mole P/mole receptor.

Reference: Huganir, R.L.; Delcour, A.H.; Greengard P.; Hess, G.P. Phosphorylation of the nicotine acetycholine receptor regulates its rate of desensitization. *Nature* 321:774–776, 1986.

*15–34 Reference: Manson, M.D.; Blank, V.; Brade, G.; Higgins, C.F. Peptide chemotaxis in *E. coli* involves the Tap signal transducer and the dipeptide permease. *Nature* 321:253–258, 1986.

15–35

A. The two cloned receptors, normal and truncated, both carry out signal transduction like the receptors in wild-type bacteria. Upon addition of aspartate, all three kinds of bacteria immediately suppress changes in direction of rotation. Thus the presence of the attractant (aspartate) in the medium is being communicated to the flagella in all three kinds of bacteria.

B. The adaptive properties of bacteria containing the cloned receptors are very different from wild-type bacteria. Wild-type bacteria return to their normal rate of tumbling (reversal of direction of rotation) within 3 minutes. Bacteria with the cloned normal receptor return to the normal rate of tumbling only after about 50 minutes, and bacteria with the truncated receptor do not begin to tumble even after more than 3 hours. Thus bacteria with the cloned normal receptor adapt more slowly than wild-type bacteria, whereas bacteria with the cloned truncated receptor evidently do not adapt.

C. The alterations in adaptation in bacteria with the cloned receptors suggest differences in the methylation rates or extents. The inability of the truncated receptor to be methylated provided a molecular basis for the inability of bacteria containing the receptor to adapt to a high-level aspartate. The molecular basis for the difference between wild-type bacteria and bacteria with the cloned normal receptor is more subtle. The difference in time of adaptation between the two kinds of bacteria is about fifteenfold (3 minutes versus 50 minutes), which is the same as the difference in numbers of receptors per cell. (Cloned genes expressed from plasmids are often overexpressed relative to their chromosomal counterparts.) Thus a reasonable explanation is that it takes the receptor methylase 15 times longer to methylate the more abundant normal receptor.

Reference: Russo, A.F.; Koshland, D.E. Separation of signal transduction and adaptation functions of the aspartate receptor in bacterial sensing. *Science* 220:1016–1020, 1983.

The Cytoskeleton

- The Nature of
  the Cytoskeleton
- Intermediate Filaments
- Microtubules
- Cilia and Centrioles
- Actin Filaments
- Actin-binding Proteins
- Muscle

The Nature of the Cytoskeleton

16–1

   A. cytoskeleton
   B. centrosome
   C. motor proteins

16–2

   A. True
   B. False. Although the minus end tends to lose subunits more readily than the plus end, it is normally stabilized by embedding it in the centrosome. The dynamic lengthening and shortening of microtubules occurs at the plus end. After growing for many minutes by adding subunits, its plus end may undergo a sudden transition that causes it to lose subunits, so that the microtubule shrinks rapidly and may disappear.
   C. True
   D. True
   E. True
   F. False. Only the centrosome in the cytotoxic T cells relocates to the zone of T-cell-target contact. The target cell remains unpolarized.
   G. True

*16–3

16–4

   A. A plausible model for control of pigment aggregation and dispersal is shown in Figure 16–26. The 57-kd protein changes its phosphorylation state according to the cyclic AMP level (which is regulated through hormone receptors in the cell membrane). In its phosphorylated form the 57-kd protein promotes dispersal, whereas in its nonphosphorylated form it promotes aggregation. This protein is phosphorylated by A-kinase, and it is dephosphorylated by a protein phosphatase.

     The regulatory scheme in Figure 16–26 accounts for all the observations except those with ATPγS. This analogue of ATP can be used as a substrate by

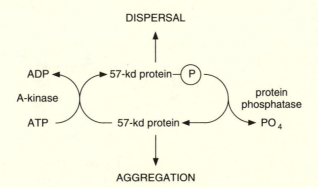

Figure 16–26 A plausible model for control of pigment aggregation and dispersal (Answer 16–4).

*Problems with an asterisk (*) are answered in the Instructor's Manual.

411

many protein kinases; presumably it serves as a substrate here as well. However, the slow rate of dispersal in the presence of ATPγS suggests either that the terminal thiolphosphate is transferred to the 57-kd protein more slowly or that the thiolphosphate group does not function as well on the 57-kd protein. The very slow reaggregation of pigment granules that are dispersed in the presence of ATPγS suggests that the thiolphosphate group is removed very slowly from the 57-kd protein by the phosphatase.

B. Although ATPγS is a reasonably good substrate for protein kinases, it is not a good substrate for enzymes, such as myosin light-chain kinase and ciliary dynein, that use its free energy of hydrolysis for movement in motile systems. Since dispersal of the pigment granules can occur in the presence of ATPγS, dispersal probably does not require ATP hydrolysis. Aggregation, by contrast, requires ATP hydrolysis and, therefore, is probably an active process.

Reference: Rozdzial, M.M.; Haimo, L.T. Bidirectional pigment granule movements of melanophores are regulated by protein phosphorylation and dephosphorylation. *Cell* 47:1061–1070, 1986.

*16–5   References: Chant, J.; Herskowitz, I. Genetic control of bud site selection in yeast by a set of gene products that constitute a morphogenetic pathway. *Cell* 65:1203–1212, 1991.

Chant, J.; Corrado, K.; Pringle, J.R.; Herskowitz, I. Yeast BUD5, encoding a putative GDP-GTP exchange factor, is necessary for bud site selection and interacts with bud formation gene *BEM1*. *Cell* 65:1213–1224, 1991.

Park, H.-O.; Chant, J.; Herskowitz, I. *BUD2* encodes a GTPase-activating protein for Bud1/Rsr1 necessary for proper bud-site selection in yeast. *Nature* 365:269–274, 1993.

Intermediate Filaments

16–6

A. keratins, keratin filaments
B. vimentin
C. desmin
D. glial fibrillary acidic protein
E. neurofilament proteins, neurofilaments
F. nuclear lamina, lamins

16–7

A. True
B. False. Intermediate filaments are very insoluble and make up the residue when cells are extracted with solutions containing high salt and nonionic detergents.
C. False. Intermediate filaments do not form polarized structures, as microtubules and actin filaments do. Intermediate filaments are constructed from symmetric tetramers whose ends are the same.
D. True
E. True
F. True
G. False. Some perfectly healthy living cells, such as the glial cells that make myelin in the central nervous system, completely lack cytoplasmic intermediate filaments. Even these cells, however, have nuclear lamins.
H. True

16–8

The coil 1A segment of nuclear lamin C matches the heptad repeat at 9 of 11 positions (Figure 16–27), which is very good. The match need not be perfect to

hydrophobic amino acids

```
 *    *    * * * * *    *    *    *  *  *        * *

DLQELNDRLAVYIDRVRSLETENAGLRLRITESEEVV

-A--D---A--D---A--D---A--D---A--D---A

 +    +    +   +    +    +    -    +    +    -    +
```

match with heptad repeat

Figure 16–27 Heptad repeat motif in the coil 1A region of lamin C (Answer 16–8). Hydrophobic amino acids are marked with an asterisk (*). When a hydrophobic amino acid occurs at an A or a D in the heptad repeat, it is assigned a +. The start of the heptad repeat was positioned to maximize matches.

allow formation of a coiled-coil. The matches to the heptad repeat in the other two marked segments (coil 1B and coil 2, Figure 16–4) are not as good, but they are still acceptable for formation of a coiled-coil.

Reference: McKeon, F.D.; Kirschner, M.W.; Caput, D. Homologies in both primary and secondary structure between nuclear envelope and intermediate filament proteins. *Nature* 319:463–468, 1986.

*16–9   **Reference:** Ottaviano, Y.; Gerace, L. Phosphorylation of the nuclear lamins during interphase and mitosis. *J. Biol. Chem.* 260:624–632, 1985.

Microtubules

16–10

A. plus end, minus end
B. aster
C. centrosome
D. pericentriolar material, centrosome matrix
E. dynamic instability
F. microtubule-associated proteins (MAPs)
G. cytoplasmic dyneins
H. kinesins

16–11

A. True
B. False. Colchicine binds to free tubulin subunits and prevents their assembly into microtubules. Because spindle microtubules are maintained through a chemical equilibrium with free tubulin subunits, the removal of usable free subunits by binding to colchicine leads to disassembly of the microtubules.
C. True
D. True
E. False. Although centrosomes, the major microtubule organizing centers in almost all animal cells, do contain centrioles, a number of microtubule organizing centers in plants, animals, and fungi do not. The common feature of all microtubule organizing centers is an electron-dense matrix that usually contains γ-tubulin and other centrosome-specific proteins.
F. True
G. True
H. True
I. False. Acetylation and detyrosination occur most completely on stable microtubules and are thought to further stabilize mature microtubules, as well as provide binding sites for microtubule-associated proteins.
J. True
K. False. The microtubules in axons are all oriented with their plus ends pointing away from the cell body, but the microtubules in dendrites have mixed polarity: some have their plus ends pointing away from the cell body, while others have their plus ends pointing toward the cell body.
L. True
M. True

16–12

A. Centrosomes lower the critical concentration by providing nucleation sites for microtubule growth. Nucleation sites make it easier to start new microtubules; moreover, they protect the bound end from disassembly. Thus, once started, a microtubule is more likely to persist. In the absence of such a nucleation site, it is much more difficult to start a microtubule and both ends serve as sites for disassembly.

B. The shapes of the curves in the presence and absence of centrosomes differ because of the nature of the assays used to detect polymerization. In the absence of centrosomes (Figure 16–6A), the assay was for total polymer formed, which depends only on the concentration of added tubulin. Thus it increases indefinitely in a linear fashion as long as the concentration of tubulin is increased. In the presence of centrosomes (Figure 16–6B) the assay was for the number of microtubules per centrosome. Since each centrosome has only a limited number of nucleation sites (about 60 for the centrosomes used in this experiment), the measurement must reach a plateau at high tubulin concentrations.

C. A concentration of tubulin dimers of 1 mg/ml corresponds to 9.1 μM.

$$[\text{tubulin}] = \frac{1 \text{ mg tubulin}}{\text{ml}} \times \frac{\text{mmol tubulin}}{1.1 \times 10^5 \text{ mg tubulin}} \times \frac{1000 \text{ ml}}{\text{L}}$$
$$= 9.1 \times 10^{-3} \text{ mmol / L}$$
$$= 9.1 \times 10^{-3} \text{ mM}$$
$$[\text{tubulin}] = 9.1 \text{ μM}$$

This value is below the critical concentration for microtubule assembly in the absence of centrosomes. Thus, without a nucleation site for growth, which is provided by the centrosome, a cell would have no microtubules. This simple consideration probably explains why the majority of microtubules originate from centrosomes in animal cells.

Reference: Mitchison, T.; Kirschner, M. Microtubule assembly nucleated by isolated centrosomes. *Nature* 312:232–237, 1984.

*16–13   Reference: Mitchison, T.; Kirschner, M.W. Properties of the kinetochore *in vitro*. I. Microtubule nucleation and tubulin binding. *J. Cell Biol.* 101:755–765, 1985.

16–14

A. The two ends of an individual microtubule appear to behave independently of one another. One end can grow while the other shrinks, and both ends can grow or shrink at the same time. Furthermore, the transitions between growth states at the two ends do not correlate with one another in any obvious way.

B. The simple GTP-cap hypothesis predicts that the faster-growing end, which has the longer GTP cap, should be more stable than the slower-growing end, which has a shorter GTP cap. Thus, according to the simple GTP-cap hypothesis, a fast-growing end should persist in a growth state longer than a slow-growing end; that is, a fast-growing end should switch from a growth state to a shrinking state less frequently than a slow-growing end. (The hypothesis says nothing about how frequently a shrinking end, which does not have a cap, will be converted into a growing end.)

The experimental results appear, if anything, to run counter to the predictions of the simple GTP-cap hypothesis. The growth periods at the "active" ends do not seem to be significantly longer (they actually appear somewhat shorter) than the growth periods at the "inactive" ends. Thus, these results do not support the simple GTP-cap hypothesis. In cells proteins other than tubulin may bind to GTP caps and help to stabilize fast-growing ends.

C. Since centrosomes nucleate growth of microtubules by binding to the minus end, all the free ends are plus ends. As a consequence, only one type of behavior of ends ("active" or "inactive") should be observed. The "active" end corresponds to the plus end.

Since MAPs tend to stabilize microtubules against disassembly, they would be expected to reduce the frequency of switching between the two growth states and extend the length of time they remain in the growing state. This result is observed. In fact, the switches in the growth state are abolished and growth is smooth and continuous until the steady-state length is reached, after which the length remains constant.

Reference: Horio, T.; Hotami, H. Visualization of the dynamic instability of individual microtubules by dark-field microscopy. *Nature* 321:605–607, 1986.

*16–15 **Reference:** Svoboda, K.; Schmidt, C.F.; Schnapp, B.J.; Block, S.M. Direct observation of kinesin stepping by optical trapping interferometry. *Nature* 365:721–727, 1993.

Cilia and Centrioles

16–16

A. cilia
B. flagella
C. axoneme
D. ciliary dynein

16–17

A. True
B. True
C. False. Ciliary bending occurs due to sliding between adjacent outer doublet microtubules.
D. True
E. False. When centrioles duplicate, the two members of a pair first separate and then a daughter centriole is formed perpendicular to each original centriole. One daughter centriole and one original centriole form the new pair of centrioles.

*16–18

16–19

Adjacent cilia in the electron micrograph in Figure 16–11 are oriented identically with reference to the central pair of microtubules. The identical orientation suggests that some structural feature of the axoneme constrains the direction of bending, that is, that cilia are designed to bend in a particular way. The orientation of the central pair of microtubules correlates with the direction of bending, which is always in a plane drawn between the two central microtubules. To make this relationship clearer, imagine that the axoneme shown in cross-section in Figure 16–10 extends straight up out of the page; this axoneme could bend toward the top or bottom of the page, but not to the left or right. (In addition, in many axonemes one adjacent set of outer doublet microtubules—specifically, the pair at the bottom of Figure 16–10—are cross-linked so that they cannot slide relative to one another. This modification further confines the plane of bending to the one observed.)

If the microtubules of the central pair are aligned parallel to one another throughout the length of the cilium, then all bending in the power stroke and in the return stroke will be in one plane. However, it is not uncommon for the central pair of microtubules to be twisted around one another, in which case the direction of bending rotates around the axis of the cilium as the bend propagates up the cilium. The consequence of this arrangement is a unidirectional power stroke (which depends only on the orientation of the central pair at the base of the cilium, where bending is initiated) and a helical return stroke as the bend moves to the tip of the cilium.

*16–20   **References:** Brokaw, C.J.; Luck, D.J.L.; Huang, B. Analysis of the movement of *Chlamydomonas* flagella: the function of the radial-spoke system is revealed by comparison of wild-type and mutant flagella. *J. Cell Biol.* 92:722–732, 1982.

Gibbons, I.R. Cilia and flagella of eukaryotes. *J. Cell Biol.* 91:107s–124s, 1981.

16–21

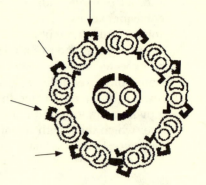

(A) UPWARD BEND

One pattern of dynein activity that could account for the planar bending of an axoneme is depicted in Figure 16–28. If the dynein arms on the left half of the axoneme are active (arrows in Figure 16–28A) and the ones on the right half are passive, the cilium will bend upward. This is difficult to imagine in three dimensions, but take it slowly. First, the orientation of the axonemes shown in Figure 16–28 is the standard one, that is, with the tip of the axoneme *below* the plane of the page. Second, the dynein arms push their neighbor doublets toward the *tip* of the axoneme, so the doubles are being pushed *below* the plane of the page. Third, the doublets at the *top* of the diagram in Figure 16–28A will be pushed the farthest below the page because its total displacement is the sum of incremental displacements produced by all four active dynein arms. Fourth, the doublet that moves the farthest defines the "inside" of the bend (see Figure 16–13). Therefore, since the top doublet moves the farthest, the axoneme will bend upward (toward the top of the page) when the dynein arms on the left half of the axoneme are active.

The same reasoning argues that the axoneme will bend downward (toward the bottom of the page) if the dynein arms on the right half of the axoneme are active and the ones on the left half are passive (Figure 16–28B).

The actual pattern of dynein activity that gives rise to planar bending is not yet understood. The two central singlet microtubules are natural candidates for regulatory elements: they are surrounded by nonidentical proteins; they contact different subsets of outer doublets; and they are linked (indirectly) to the two sets of dynein arms used in the model proposed above.

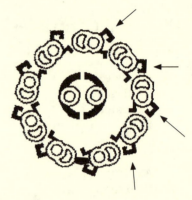

(B) DOWNWARD BEND

Figure 16–28 One possible pattern of dynein activity that could produce planar bending of an axoneme (Answer 16–21). (A) Upward bend. (B) Downward bend. Arrows indicate active dynein arms.

16–22

A. If the affected gene controlled the synthesis of the missing proteins in the mutant flagella, then none of those proteins would be present in the gametes from the mutant. After fusion the mutant flagella would be repaired by addition of unlabeled components from the nonradioactive wild-type gametes, since all protein synthesis was inhibited. As a result, the *autoradiograph* of the electrophoretic pattern would look exactly like the mutant alone.

By contrast, if the affected gene encoded a protein whose assembly into the axoneme must precede the addition of other proteins, all the proteins except the defective one would be present in functional form in gametes from the mutant. Since the mutant was grown in radioactive medium, these proteins would be labeled. Thus, after fusion the mutant flagella would be repaired by addition of a mixture of labeled components from the mutant gametes and unlabeled components from the wild-type gametes. The only exception would be the protein encoded by the defective gene: it would come entirely from the wild-type gametes and, therefore, would be unlabeled. Under these circumstances the autoradiograph of the electrophoretic pattern would look like that from the wild-type gametes with a single missing spot. The missing spot would correspond to the product of the defective gene.

B. Analysis of second-site, intragenic revertants also can distinguish between the possibilities. If the affected gene encoded a product that is part of the assembled axoneme, then some revertants (those with an altered number of charged amino acids) should produce patterns with one spot at a new location (usually slightly left or right of the normal position, due to an effect on the isoelectric point of the protein). If the affected gene controlled the synthesis of the 17 missing proteins, then none of the revertants would have an altered pattern (since none of the component proteins are directly affected by mutation). This approach is less satisfactory than the first, since an unaltered pattern does not allow one to conclude that the defective gene controls synthesis—it may be that not enough revertants were examined.

In conjunction, these two methods for analyzing flagellar mutants have proven enormously powerful. Most mutants, *pf*14 included, affect assembly directly as judged by both assays; that is, they encode a protein that forms part of the axoneme structure. Moreover, the two methods agree on which protein is encoded by the defective gene: the unlabeled spot in dikaryon analysis corresponds to the shifted spot in revertent analysis.

Reference: Luck, D.J.L. Genetic and biochemical dissection of the eucaryotic flagellum. *J. Cell Biol.* 98:789–794, 1984.

Actin Filaments

16–23

A. actin filaments, actin
B. thymosin
C. profilin
D. lamellipodium
E. Rho, Rac

16–24

A. True
B. False. The ATP bound to the actin monomer is hydrolyzed after polymerization, but the ADP is trapped in the polymer. The hydrolysis of ATP during actin polymerization is analogous to the GTP hydrolysis that accompanies microtubule assembly.
C. True
D. True
E. False. The even distribution of actin between filaments and monomers is surprising given that the monomer concentration (typically 50–200 μM) is much higher than the critical concentration for pure actin (less than 1 μM).
F. True
G. False. Actin is continuously polymerizing near the tip of the leading edge and continuously depolymerizing at more internal sites, giving the impression that the leading edge is propelling itself forward by pushing actin filaments backward.
H. True
I. True
J. True
K. True

16–25

A. Although the end points for polymerization and ATP hydrolysis were the same, the initial rate of ATP hydrolysis was less than the initial rate of polymerization. (Compare the slopes of the two curves in Figure 16–15 at short times.) At the time when all the actin was polymerized (about 30 seconds), less than half the ATP was hydrolyzed. It is the difference in initial rates that your advisor noticed, and, as he said, it proves that actin polymerization can occur in the absence of ATP hydrolysis.

B. Since the rate of polymerization is faster than the rate of ATP hydrolysis, newly added actin subunits must still retain bound ATP. Since the bound ATP is not hydrolyzed until sometime after assembly, growing actin filaments have ATP "caps." Once an ATP-actin monomer has bound to a filament, the ATP can be hydrolyzed.

    Reference: Carlier, M.-F.; Pantaloni, D.; Korn, E.D. Evidence for an ATP cap at the ends of actin filaments and its regulation of the F-actin steady state. *J. Biol. Chem.* 259:9983–9986, 1984.

*16–26   Reference: Tilney, L.G.; Inoue, S. Acrosomal reaction of *Thyone* sperm. II. The kinetics and possible mechanism of acrosomal process elongation. *J. Cell Biol.* 93:820–827, 1982.

16–27

A. The electron microscopic examination reveals that actin monomers normally add four times faster to the plus ends of the decorated actin filaments than to the minus ends. The different rates of assembly at the ends can be estimated by measuring the lengths (counting the subunits in Figure 16–18) of newly assembled actin at the plus and minus ends.

B. As is evident from the micrographs, cytochalasin B interferes with filament assembly by stopping actin polymerization at the plus end, which is normally the preferred end for addition of monomers. One plausible mechanism to explain this inhibition is that cytochalasin B binds to the plus end of the actin filament and physically blocks addition of new actin monomers.

    This mechanism can also account for the viscosity measurements. Since growth at the minus end is unaffected, the filaments continue to grow, but much more slowly. The slower growth rate explains the slower increase in viscosity in the presence of cytochalasin B. The lower viscosity at the plateau indicates that the actin filaments are shorter in the presence of cytochalasin B. Why is it that the filaments do not ultimately grow to the same length even though their growth rate is slower? The filaments are shorter when they are growing only from the minus ends because the critical concentration for assembly at the minus end is higher than the critical concentration for assembly at the plus end. This is another way of saying that the equilibrium for assembly at the minus end is shifted more toward the free subunits than the equilibrium for assembly at the plus end. The difference in equilibrium constants means that filaments that are growing at the plus end (or both ends) will be longer than filaments growing only at the minus end.

C. An actin filament normally grows at different rates at the plus and minus ends. This observation indicates that the monomer probably undergoes a conformational change upon addition to an actin filament. If all subunits, assembled and free, were identical in conformation, the rates of growth at the two ends should be the same. Since the subunits are joined to the polymer by identical sets of interactions at the two ends, their on and off rates should be the same. Hence, the rates of growth should be the same. (See Panel 16–1 in MBOC, pp. 824–825.)

    The apparent binding of cytochalasin B to the plus end of a filament, but not to an actin monomer, is also consistent with a conformational change upon actin addition. If an actin monomer exists in two conformations, it might reasonably bind cytochalasin B in one conformation but not in the other. This argument is weaker than the one above because cytochalasin B could have the same effect by binding to a site on the polymer that spans adjacent subunits (a site present only on the polymer). These two models make different predictions about how many cytochalasin B molecules should be bound per filament. Although this number is difficult to measure accurately, the estimates favor a single cytochalasin B molecule per filament.

    Reference: MacLean-Fletcher, S.; Pollard, T.D. Mechanism of action of cytochalasin B on actin. *Cell* 20:329–341, 1980.

Actin-binding Proteins

16–28

    A. actin-binding proteins
    B. spectrin, ankyrin
    C. fimbrin
    D. α-actinin
    E. filamin
    F. gelsolin
    G. myosins
    H. myosin-I, myosin-II
    I. contractile ring
    J. stress fibers
    K. focal contacts
    L. microvilli
    M. villin
    N. tropomyosins

16–29

    A. True
    B. False. The parallel bundles found in microspikes and filopodia have all their actin filaments oriented with the same polarity. Bundles in stress fibers and in the contractile ring have filaments arranged with opposite polarity. The mixture of polarities allows myosin motors to cause such bundles to contract.
    C. True
    D. True
    E. False. The statement is true for myosin-II molecules but not for myosin-I molecules. The tails of the various myosin-I molecules differ from those of myosin-II molecules and from each other. Myosin-I molecules do not polymerize into bipolar fragments. The short tails of myosin-I molecules contain sites that bind to other actin filaments or to membranes.
    F. True
    G. True
    H. True
    I. True
    J. True
    K. False. Remarkably, *Dictyostelium* cells without myosin-II can still move over the substratum and respond chemotactically to a source of cyclic AMP. Thus myosin-II is not absolutely essential for cell locomotion.

16–30

The observation that defects in either α-actinin or gelation factor cause no problems in motility or development, whereas defects in both cause profound problems, suggests that each protein alone can substitute for the functions normally provided by the other. Although α-actinin and gelation factor presumably serve somewhat different roles in *Dictyostelium*, they also provide a mutual backup system. This appears to be common in higher organisms. As reverse genetics is used to knock out more and more genes, it is becoming clear that many are redundant, that is, alternative backup functions are provided by other genes.

    One peculiar observation in these experiments is that the amoebae of the doubly mutant strains move perfectly well, even though the two major actin-binding proteins are defective. This may mean that there are still other actin-binding proteins in *Dictyostelium* that function sufficiently well to allow amoebae to move but not well enough to permit the cell movements associated with development.

Reference: Witke, W.; Schleicher, M.; Noegel, A.A. Redundancy in the microfilament system: abnormal development of *Dictyostelium* cells lacking two F-actin cross-linking proteins. *Cell* 68:53–62, 1992.

*16–31   **Reference:** Broschat, K.O.; Stidwell, P.R.; Burgess, D.R. Phosphorylation controls brush border motility by regulating myosin structure and association with the cytoskeleton. *Cell* 35:561–571, 1983.

Muscle

16–32

A. myofibrils
B. sarcomeres
C. sarcoplasmic reticulum
D. heart muscle
E. smooth muscle

16–33

A. False. Z discs mark the ends of each sarcomere and link sarcomeres into a long chain, which is the myofibril.
B. True
C. True
D. False. The action potential triggered by a nerve impulse affects a voltage-sensitive protein in the plasma membrane that is linked to Ca^{2+} release channels in the sarcoplasmic reticulum. When activated by the incoming action potential, the voltage-sensitive proteins trigger the Ca^{2+} release channels to open, probably by direct mechanical coupling. Once open, the Ca^{2+} release channels allow Ca^{2+} ions to flow from the lumen of the sarcoplasmic reticulum into the cytosol.
E. True
F. False. Although nebulin does act as a molecular ruler to specify the length of actin filaments, titin is a springlike molecule that keeps myosin thick filaments centered in the sarcomere.
G. True
H. False. In smooth muscle cells the phosphorylation of myosin-II light chains is catalyzed by myosin light-chain kinase, whose action requires the binding of a Ca^{2+}/calmodulin complex. As a result, contraction is controlled by the level of cytosolic Ca^{2+}, as in cardiac and skeletal muscle.

16–34

A. The locations of the striated muscle components in the electron micrograph are illustrated schematically in Figure 16–29A. α-actinin is a component of the Z disc; titin links the myosin-II filaments to the Z disc; and nebulin binds along the length of each actin filament.
B. The micrograph in Figure 16–20B shows a hypercontracted muscle. Thus the light band has entirely disappeared, and a new band, caused by the overlap of actin filaments, has appeared in the middle of the sarcomere. The schematic relationship of the two electron micrographs in Figure 16–20 is shown in Figure 16–29B.

16–35

A. Two helices of myosin-II molecules would dovetail perfectly (Figure 16–30). Using the numbering scheme in Figure 16–21, the individual myosin molecules at the ends of each helix would abut end to end as follows: 1 with 6′, 2 with 5′, 3 with 4′, 4 with 3′, 5 with 2′, and 6 with 1′ (where the numbers with primes indicate a second copy of the helix).

(A) RELAXED

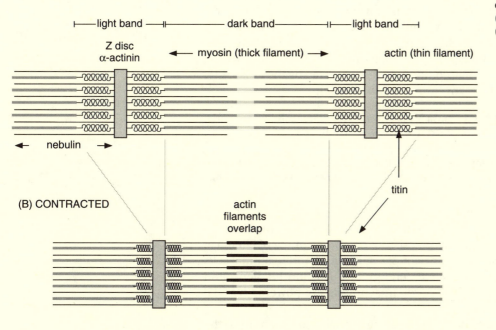

|— light band —|—————— dark band ——————|— light band —|

Z disc
α-actinin   ←—— myosin (thick filament) ——→   actin (thin filament)

←— nebulin —→

titin

(B) CONTRACTED

actin
filaments
overlap

Figure 16–29 Schematic diagrams of electron micrographs in Figure 16–20 (Answer 16–34). (A) Relaxed muscle. (B) Contracted muscle.

If you tried to match the two ends in your imagination, you undoubtedly discovered what a difficult mental feat this is. However, it can be readily demonstrated with a concrete object. For example, arrange six pencils around a central one and secure the bundle with a rubber band. Recess successive pencils equally around the cylinder to produce helically staggered ends. Two identical bundles will fit together perfectly. To become familiar with the geometric principles of protein assembly, one must play with real objects; the ability to visualize will improve accordingly.

B. The bare zone is determined by the distance between the first myosin heads on the oppositely oriented helices (Figure 16–30). The closest heads (using the numbering in Figure 16–21) would be the ones on myosins 1 and 1'. Their separation would be equal to the length of the myosin rod (150 nm) plus one-sixth that amount (25 nm) due to the helical stagger (myosin 1' abuts myosin 6). This theoretical estimate of the length of the bare zone (175 nm) compares reasonably well with the measured length of the bare zone, which is 160 nm.

C. Tapering of the myosin-II filament at its ends is expected if the filament is a helix. Successive cross-sections through the end would show six myosin molecules, then five, then four, and so on to one—a natural thinning of material at the end (Figure 16–30). In electron micrographs the tapering is apparent over the last 100 to 150 nm of the filament, as expected according to this explanation.

D. There is nothing inherent in the structure of a helix composed of subunits such as myosin-II that defines (or dictates) its length. The average length of a helix is determined by the overall concentration of subunits and by the relative rates of assembly and disassembly at its ends; but even when conditions are constant, considerable variation in length will occur. The remarkable uniformity of length of myosin-II thick filaments from striated muscle demands something in addition to the myosin-II molecules themselves. Thus, there may be a length-determining molecule that is associated with myosin thick filaments. In fact, some other biological helices that have defined lengths are known to be associated with length-determining molecules. The lengths of the tobacco mosaic virus capsid and of the bacteriophage lambda tail, which are both helices, are set by length-determining molecules that stretch from one end of the helix to the other.

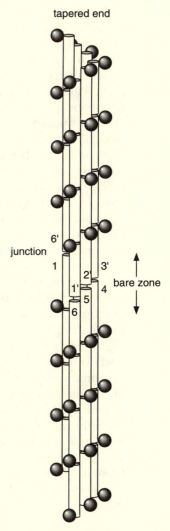

tapered end

junction

6'
1
2' 3'
1'
5   4   bare zone
6

tapered end

Figure 16–30 A myosin-II thick filament (Answer 16–35).

As we have tried to illustrate in this problem, an understanding of the principles of protein assembly generates important insights into biological structure. Such an understanding can tell you when a structural feature is a reasonable consequence of protein assembly (as in the case of the bare zone and the tapering at the tip). Equally important, it can tell you when a structural feature demands some additional component (as suggested by the uniformity of length of myosin thick filament from striated muscle).

*16–36   Reference: Goldman, Y.E.; Hibberd, M.G.; McCray, J.A.; Trentham, D.R. Relaxation of muscle fibers by photolysis of caged ATP. *Nature* 300:701–705, 1980.

*16–37   Reference: Gordon, A.M.; Huxley, A.F.; Julian, F.J. The variation in isometric tension with sarcomere length in vertebrate muscle fibres. *J. Physiol.* 184:170–192, 1966.

16–38

A. In each cycle the chemical free energy that drives the cycle is provided by hydrolysis of ATP. Although ATP hydrolysis is a common source of chemical free energy, it is not the only one. For example, sugar transport in animal cells is usually powered directly by the free energy in a Na^+ ion gradient, and movements on the ribosome during protein synthesis are powered by GTP hydrolysis.

The mechanical work accomplished during muscle contraction is the motion of actin thin filaments relative to myosin thick filaments. The mechanical work done during active transport of Ca^{2+} is the pumping of ions against their concentration gradient from the inside of the cell to the outside.

B. Actin is alternately bound tightly and then released in each cross-bridge cycle during muscle contraction; Ca^{2+} is alternately bound tightly and then released during its active transport.

In the diagram in Figure 16–25A, actin is tightly bound to myosin at each point where the two are in contact. The binding of ATP to the myosin head converts it to a weak binding form, allowing it to detach from actin. (Although each of these steps is shown separately in the diagram, the binding of ATP is thought to initiate a conformational change, which in turn reduces the affinity of myosin for actin, thereby promoting detachment of actin and the completion of the conformational change.)

In the diagram in Figure 16–25B, Ca^{2+} is tightly bound to the transport protein when it is on the inside of the cell (upper drawing) but only weakly bound when it faces the outside of the cell (lower drawing). Although the tightness of binding is not immediately apparent in the diagrammatic representation, if follows from the concept of active transport. Since the pump is transporting Ca^{2+} against its concentration gradient, the pump must have a high affinity for Ca^{2+} on the inside of the cell (so that Ca^{2+} can be bound effectively at the low intracellular concentration of Ca^{2+}) and a low affinity for Ca^{2+} on the outside of the cell (so that Ca^{2+} can be released effectively at the high external concentration of Ca^{2+}).

C. In both cycles the "power stroke" is the conformational change indicated on the right side of the cycles as drawn in Figure 16–25. The "return stroke" in each case is the conformational change indicated on the left side of the drawings in Figure 16–25.

Reference: Eisenberg, E.; Hill, T.L. Muscle contraction and free energy transduction in biological systems. *Science* 227:999–1006, 1985.

The Cell-Division Cycle

The General Strategy of the Cell Cycle

17–1

A. cell-division cycle
B. cell-cycle control
C. mitosis
D. cytokinesis, M phase
E. S phase
F. G_1 phase, G_2 phase
G. cyclin-dependent protein kinase (Cdk)
H. mitotic cyclins, G_1 cyclins

17–2

A. True
B. False. For an asynchronously growing population of cells, the fraction of cells in S phase is equal to the fraction of the cell cycle that is devoted to S phase. To know the actual length of S phase requires knowledge of the overall length of the cell cycle.
C. True
D. False. Although it was once thought that the cell cycle might operate in this way, it is now clear that the cell cycle control system is regulated by brakes that can operate at specific checkpoints in response to feedback signals that reflect the progress of the downstream processes.
E. False. In unfavorable circumstances higher eucaryotic cells usually arrest the cell cycle at the G_1 checkpoint, before the cell commits its resources to chromosome duplication.
F. False. Cyclins bind to Cdk to stimulate their protein kinase activity and control phosphorylation of target proteins. Cyclins are rapidly degraded only after the active Cdk-cyclin complex has triggered the transition through a cell-cycle checkpoint.

*17–3

17–4

A. The overall length of the cell cycle is equivalent to the time it takes for the entire population of cells to double in number. To find the length of the cell cycle, select any two points on the graph in Figure 17–1 between which the number of cells has doubled. The time separating those two points is the length of the cell cycle. For example, the first two data points in Figure 17–1 are at 3×10^5 cells (10 hours) and 6×10^5 cells (30 hours). Since the population of mouse L cells doubled in 20 hours (30 hours – 10 hours), the length of the cell cycle is 20 hours.
B. In outline, the length of G_2 can be derived from the data in Figure 17–2A, the length of S from the data in Figure 17–2B, and the length of G_1 from the overall length of the cell cycle minus (M + S + G_2).

   ^{3}H-thymidine is incorporated only during S phase. Thus no label will be present in cells undergoing mitosis until the cells that were at the very tail end

---

*Problems with an asterisk (*) are answered in the Instructor's Manual.

423

of S phase when the label was added have traversed the G_2 portion of the cell cycle. The appearance of the first labeled mitotic cells at 3 hours after ^{3}H-thymidine addition (Figure 17–2A) suggests that G_2 is 3 hours long. The majority of mitotic cells, however, did not become labeled until 4 hours after addition of label (and a few did not become labeled until 5 hours after addition of label). This variation suggests that there is some variability in the length of G_2: if the length of G_2 were precisely defined, 100% of mitoses would be labeled when labeling was first observed. Thus, the length of G_2 in mouse L cells is 3 to 4 hours.

The length of S phase can be deduced from the number of silver grains over labeled mitotic cells (Figure 17–2B). Cells that were at the very end of S phase when label was added will have incorporated very little ^{3}H-thymidine and thus will have very few silver grains, whereas cells that were at the beginning of S phase will have incorporated much more label and thus will have many more silver grains. The important realization is that cells at the beginning of S phase and cells that were in G_1 will have the same number of silver grains, since they incorporated label for the same length of time (that is, throughout S phase). Thus the beginning of S phase is the point in Figure 17–2B at which the number of silver grains per labeled cell reaches a plateau, which is about 10 hours before mitosis. Since G_2 is about 3 hours long, S phase must be about 7 hours long.

Given that M phase is 1 hour, S phase is 7 hours, G_2 is 3 to 4 hours, and the overall length of the cell cycle is 20 hours, G_1 must be 8 to 9 hours long.

Reference: Stanners, C.P.; Till, J.E. DNA synthesis in individual L-strain mouse cells. *Biochim. Biophys. Acta* 37:406–419, 1960.

*17–5   **References:** Bootsma, D.; Budke, L.; Vos, O. Studies on synchronous division of tissue culture cells initiated by excess thymidine. *Exp. Cell Res.* 33:301–309, 1964.

Bostock, C.J.; Prescott, D.M.; Kirkpatrick, J.B. An evaluation of the double thymidine block for synchronizing mammalian cells at the G_1-S border. *Exp. Cell Res.* 68:163–168, 1971.

Rao, P.N.; Johnson, R.T. Mammalian cell fusion: I. Studies on the regulation of DNA synthesis and mitosis. *Nature* 225:159–164, 1970.

Xeros, N. Deoxyriboside control and synchronization of mitosis. *Nature* 194:682–683, 1962.

17–6

The constancy of the length of S phase in haploid versus diploid and diploid versus tetraploid organisms may not be so surprising. If one assumes that particular chromosomes and regions within chromosomes have a defined order of replication, then halving or doubling the number of chromosomes would not be expected to alter the schedule. Moreover, the ratio of genes encoding the replication machinery (DNA polymerases, helicases, initiation factors, etc.) to the amount of DNA would not change. By contrast, the DNA of different organisms might well be expected to have different ratios of critical genes to DNA content, which could account for the correlation seen in Table 17–1.

Reference: Prescott, D.M. Reproduction of Eucaryotic Cells, pp. 85–86. New York: Academic Press, 1975.

The Early Embryonic Cell Cycle and the Role of MPF

17–7

A. early embryonic cell cycles
B. maturation-promoting factor (MPF)

C. cyclin

D. re-replication block

17–8

A. True

B. True

C. True

D. False. Although synthesis of new cyclin protein is required for each division cycle, synthesis of cyclin mRNA is not required: cyclin synthesis occurs even in the absence of a nucleus.

E. True

F. False. Although accumulation of cyclin is required for MPF activity, the sudden, all-or-none activation of MPF is a consequence of an autocatalytic process in which active MPF generates additional active MPF.

G. True

H. False. Unlike the standard cell cycle, no feedback system operates to detect incompletely replicated chromosomes in the cleaving embryo. The time required for replication is invariant due to the stockpiles of nutrients in the egg and the protected environment of the embryo.

I. False. The G_2 nucleus does not resume DNA synthesis because in some way or other its passage through S creates a block to re-replication.

J. True

17–9

A. Since each transfer accomplishes a twentyfold dilution (50 nl/1000 nl), 10 transfers yield a dilution factor of 20^{10}, which is equal to 10^{13}. It is unreasonable for a molecule to have an undiminished biological effect over this range of dilution.

B. The appearance of MPF activity in the absence of protein synthesis suggests that an inactive precursor of MPF is being activated. In principle, activation could involve one of several kinds of posttranslational modifications, such as protease cleavage or phosphorylation. MPF is activated by phosphorylation.

C. In order for MPF to propagate its activated state through serial transfers, it must be able to activate itself. If it were a protease, for example, active MPF might activate its inactive precursor by cleavage, such as trypsin-mediated cleavage of trypsinogen to produce more trypsin. In the case of MPF, however, active MPF functions as a protein kinase that activates its inactive precursor by phosphorylation. Note that it is not necessary that MPF activates itself; for example, it could activate another protein kinase, which in turn activates the precursor to MPF. Nevertheless, the principle is the same.

D. Since there is no detectable MPF activity in an immature oocyte, MPF cannot be the source of the original activation event. Presumably, a protein synthesized in response to progesterone stimulation (therefore cycloheximide sensitive) is responsible, directly or indirectly, for the initial activation of MPF.

Reference: Wasserman, W.J.; Masui, Y. Effects of cycloheximide on a cytoplasmic factor initiating meiotic maturation in *Xenopus* oocytes. *Exp. Cell Res.* 91:381–388, 1975.

17–10

A. Colchicine has no apparent effect on the synthesis of cyclin, since the amount of cyclin continues to increase in colchicine-treated cells. Colchicine does, however, have a dramatic effect on cyclin destruction. Colchicine strongly delays the destruction of cyclin but does not completely block it. Two observations in these experiments point to the continued destruction of cyclin, albeit with significantly altered kinetics. As shown in Figure 17–4, the amount of cyclin levels off in the presence of colchicine, suggesting that the rate of synthesis is balanced by the rate of destruction. As shown in Figure

17–5, when cyclin synthesis is blocked by emetine, cyclin is eventually destroyed in the presence of colchicine.

    The stabilization of cyclin certainly could account for the metaphase arrest produced by colchicine. As long as cyclin levels stay high, MPF will also stay high and keep the cell in M phase.

B. Inhibition of protein synthesis blocks the synthesis of new cyclin. Because colchicine delays cyclin destruction but does not prevent it, cyclin is eventually destroyed. When cyclin levels fall, MPF activity declines as well. In the absence of MPF activity, cells exit mitosis. In the presence of colchicine the exit from mitosis is rather peculiar: the absence of microtubules prevents normal chromosome segregation and cell division.

    Reference: Hunt, T.; Luca, F.C.; Ruderman, J.V. The requirement for protein synthesis and degradation, and the control of destruction of cyclins A and B in the meiotic and mitotic cell cycles of the clam embryo. *J. Cell Biol.* 116:707–724, 1992.

*17–11   **Reference:** Holloway, S.L.; Glotzer, M.; King, R.W.; Murray, A.W. Anaphase is initiated by proteolysis rather than by the inactivation of maturation-promoting factor. *Cell* 73:1393–1402, 1993.

Yeasts and the Molecular Genetics of Cell-Cycle Control

17–12

A. budding yeast
B. fission yeast
C. *cdc* (cell-division cycle)
D. *wee*
E. *cdc2*
F. Start
G. Start kinase
H. p53

17–13

A. True
B. False. In budding yeast and mammalian cells the G_1 checkpoint is more prominent. The G_2 checkpoint is most important in fission yeast.
C. False. Yeasts, like all fungi, keep their nuclear envelope intact throughout mitosis: a mitotic spindle forms inside the nucleus, and after chromosome segregation has been completed, the nucleus pinches in half.
D. True
E. True
F. False. The complex of cyclin and Cdc2 is inactive until it is phosphorylated by an activating kinase, and the inhibitory phosphate added by Wee1 kinase is removed by Cdc25 phosphatase.
G. True
H. False. The protein kinase component, Cdc2, in Start kinase and MPF is identical. The differing specificities are due presumably to their different cyclin components.
I. True
J. True
K. False. Inhibitors of DNA replication delay entry into mitosis in mammalian cells but not in early frog embryos.
L. True
M. False. Feedback controls seem to be based on negative signals that arrest the control system. Negative signals that arrest the cycle are inherently more sensitive ways to detect completion of incremental processes such as DNA replication and chromosome attachment to spindles.

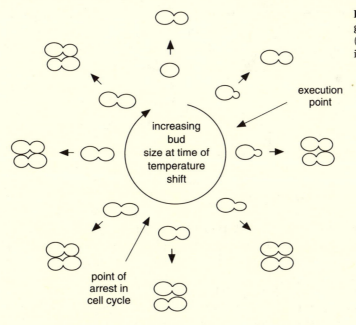

Figure 17–16 The execution point for the gene product affected by your mutant (Answer 17–15). The execution point is indicated by the arrow.

increasing bud size at time of temperature shift

execution point

point of arrest in cell cycle

*17–14 **Reference:** Hartwell, L.H. Cell division from a genetic perspective. *J. Cell Biol.* 77:627–637, 1978.

17–15

A. The execution point for your temperature-sensitive mutant is marked on Figure 17–16. Cells that were at a point in the cell cycle before the execution point when the temperature was raised grow to the characteristic landmark morphology but do not divide. Cells that were beyond the execution point at the time when the temperature was raised divide and then stop at the landmark morphology during the next cell-division cycle.

B. The characteristic landmark morphology defines the time at which the cell stops its progress through the cell cycle, as indicated for your mutant in Figure 17–16. The landmark morphology is clearly different from the morphology at the execution point. Therefore, the execution point and the point at which growth is arrested do not correspond in your mutant.

At first glance it may seem odd that the execution point and the point of growth arrest do not coincide. An analogy may make the situation clearer. The addition of engine mounts to the chassis is an early step in the assembly of an automobile. Without engine mounts the engine cannot be added and a complete car cannot be built. However, in the absence of engine mounts assembly of other parts of the automobile can continue until a point is reached at which all further assembly depends on the engine mounts. In this case the normal execution point for engine-mount addition is early, but the arrest point for assembly is relatively late, with a characteristic landmark morphology that resembles a complete automobile (until one looks under the hood).

Reference: Hartwell, L.H. Cell division from a genetic perspective. *J. Cell Biol.* 77:627–637, 1978.

17–16

A. The state of phosphorylation of Wee1 and Cdc25 is the result of the balance between the protein kinase and protein phosphatase activities that regulate them. By inhibiting the protein phosphatases, okadaic acid causes Wee1 and Cdc25 to accumulate in their phosphorylated forms as shown in Figure 17–17. Since this change activates MPF, Wee1 and Cdc25 must have originally been present in the extract in their nonphosphorylated forms. Thus active Wee1

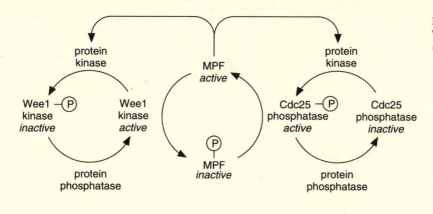

kinase is nonphosphorylated as is inactive Cdc25 phosphatase (Figure 17–17). Knowing which forms are phosphorylated allows you to label the arrows that correspond to the kinases and phosphatases that control Wee1 and Cdc25 phosphorylation (Figure 17–17).

B. The protein kinases and phosphatases that control phosphorylation of Wee1 and Cdc25 must be specific for serine/threonine residues because they are affected by okadaic acid, which is specific for serine/threonine phosphatases.

C. Okadaic acid has no direct effect on Cdc2 phosphorylation because it is phosphorylated on a tyrosine residue. Tyrosine phosphatases are unaffected by okadaic acid. The decrease in Cdc2 phosphorylation is a consequence of the change in activation of Wee1 kinase and Cdc25 phosphatase.

D. As soon as some active MPF appears it would begin to phosphorylate Wee1 and Cdc25, inactivating the kinase and activating the phosphatase. The resultant decrease in Wee1 kinase activity and increase in Cdc25 phosphatase activity would lead to dephosphorylation (and activation) of more MPF. This in turn would further decrease the activity of Wee1 kinase and further increase the activity of Cdc25 phosphatase, leading to still more MPF activity. Thus the initial appearance of a little MPF activity would rapidly lead to its complete activation.

This sort of activation is referred to as a positive feedback loop. It is a common means of regulation when it is advantageous for a system to flip rapidly from one state to another without lingering in the intermediate states.
Reference: Kumagai, A.; Dunphy, W.G. Regulation of the Cdc25 protein during the cell cycle in *Xenopus* extracts. *Cell* 70:139–151, 1992.

17–17

A. Careful examination of the time-lapse pictures in Figure 17–10 shows that all the cells without buds arrested at the dumbbell stage, whereas all of the cells with buds formed viable colonies. The appearance of a bud corresponds with the beginning of S phase. Haploid cells that have partially or fully replicated their genomes are more resistant to x-ray induced breaks because a break in one chromosome can be repaired by recombination with the intact homologue. Haploid cells in G_1 are especially sensitive to breaks because they contain no second intact copy of the chromosome with which to recombine. After replication such a cell will contain two copies of the chromosome, but both will be broken at the same position. Thus even in G_2, a haploid cell that suffers a break in G_1 will not have an intact chromosome with which to repair itself by homologous recombination.

B. The observation that dumbbell-stage cells have a single nucleus and no spindle indicates that the cells are arrested in G_2 prior to mitosis. The presence of a bud indicates that the cells have already passed the G_1 checkpoint and such cells will complete S phase.

C. Half the wild-type cells temporarily arrest in the dumbbell stage while they wait for damage to be repaired. The mitotic entry checkpoint senses damaged

DNA and halts the cell cycle until the damage is repaired. Once the damage is repaired the cells enter mitosis and divide to produce viable microcolonies.

D. *rad*52 cells remain arrested at the dumbbell stage because they are incapable of repairing their damaged chromosomes. The continued signal from the damaged DNA prevents the cells from passing the mitotic entry checkpoint.

E. Very few *rad*9 mutant cells arrest at the dumbbell stage because they are defective in their ability to sense DNA damage. In these cells the mitotic entry checkpoint does not function. Division in the absence of repair leads to haploid cells that have broken chromosomes and cells that are missing pieces of chromosomes. Both situations lead to nonviable cells. Only a small fraction of cells (30%) manage to repair their chromosomes in the absence of a checkpoint delay.

F. If *rad*9 cells are artificially held in G_2, the number of viable cells increases. The artificial delay allows the cells time to repair their damaged chromosomes so that they then can enter mitosis with an intact genome. The important point is that *rad*9 cells contain all the necessary enzymes required for DNA repair; they are defective only in sensing DNA damage.

References: Hartwell, L.H.; Weinert, T.A. Checkpoints: controls that ensure the order of cell cycle events. *Science* 246:629–634, 1989.

Weinert, T.A.; Hartwell, L.H. The *rad*9 gene controls the cell cycle response to DNA damage in *Saccharomyces cerevisiae*. *Science* 241:317–322, 1988.

*17–18   References: Murray, A.; Hunt, T. The Cell Cycle: An Introduction, pp. 143–144. New York: W.H. Freeman, 1993.

Cell-Division Controls in Multicellular Animals

17–19

A. growth factors
B. PDGF (platelet-derived growth factor)
C. oncogene, proto-oncogene
D. tumor-suppressor genes
E. early-response, delayed-response
F. retinoblastoma (Rb)
G. cell senescence

17–20

A. True
B. False. Both yeasts and higher animals have multiple cyclins, but unlike yeasts higher animals also have multiple cyclin-dependent kinases. It seems, therefore, that separate mammalian Cdk proteins perform the various functions that in yeast can be carried out by a single one.
C. True
D. False. Proliferation of most cell types in an intact animal depends on a specific combination of growth factors rather than a single growth factor.
E. True
F. False. There is no rigid rule in mammalian cells. Cell growth and cell division can be independently stimulated by growth factors to give differently specialized cells that vary enormously in their ratio of cytoplasm to DNA.
G. True
H. False. Serum-deprived cells continue through the current cell cycle until they reach the G_1 checkpoint where they enter G_0.
I. True
J. False. The cell population density at which cell proliferation ceases in the confluent monolayer increases with increasing concentration of growth factors in the medium.

K. True
L. True
M. True
N. True
O. True
P. True

*17–21 **Reference:** O'Keefe, E.J.; Pledger, W.J. A model of cell-cycle control: sequential events regulated by growth factors. *Mol. Cell. Endocrinol.* 31:167–186, 1983.

17–22

A. Experiment 1 shows that the 170-kd protein has a high-affinity binding site for EGF and therefore is most likely the EGF receptor. The control experiment of incubating the membrane preparation in the presence of excess unlabeled EGF demonstrates that EGF binding to the 170-kd protein is specific, not random.

B. Experiments 2, 3, and 4 show that the 170-kd protein (the EGF receptor) becomes phosphorylated with radioactive phosphate in the presence of γ-^{32}P-ATP. Transfer of the phosphate from the γ position in ATP demonstrates that the EGF receptor is a substrate for a protein kinase. Since the labeling intensity of the 170-kd protein is increased in the presence of EGF, the activity of the protein kinase is stimulated by EGF.

C. Experiments 3 and 4 show that the EGF receptor can transfer phosphate from ATP to protein; thus, it is a protein kinase. Experiment 3 is less convincing than experiment 4. In experiment 3, although the antibody is specific for the EGF receptor, it is not unreasonable to question whether other proteins, perhaps including a protein kinase, might have been trapped within the antibody precipitate. In experiment 4 the EGF receptor is first separated by molecular weight from other proteins that might contaminate the antibody precipitate. Only in the unlikely event that the contaminating protein kinase was also a 170-kd protein would experiment 4 lead you astray.

D. With the caveat mentioned in part C, experiment 4 indicates convincingly that the EGF receptor is a substrate for its own protein kinase activity.

Reference: Cohen, S.; Ushiro, H.; Stocheck, C.; Chinkers, M. A native 170,000 epidermal growth factor receptor-kinase complex from shed plasma membrane vesicles. *J. Biol. Chem.* 257:1523–1531, 1982.

17–23

These results suggest that density-dependent inhibition of cell growth can be completely ascribed to changes in cell shape. This correlation becomes apparent when the ^{3}H-thymidine incorporation is expressed in a way that takes into account the difference in the number of cells per plate; for example, as cpm/1000 cells (Table 17–6). Note, however, that these data do not distinguish whether cell shape "controls" cell proliferation or whether both parameters are dependent on a third factor related in some way to the substrate. The disappearance of bundles of microfilaments, the decreased attachment points to the substratum, or the reduction in the rate of nutrient uptake are all possibilities.

Reference: Folkman, J.; Moscona, A. Role of cell shape in growth control. *Nature* 273:345–349, 1978.

*17–24 **References:** Fung, Y-K.T.; Murphree, A.L.; T'Ang, A.; Qian, J.; Hinrichs, S.H.; Benedict, W.F. Structural evidence for the authenticity of the human retinoblastoma gene. *Science* 236:1657–1661, 1987.

Knudson, A.G. Mutation and cancer: statistical study of retinoblastoma. *Proc. Natl. Acad. Sci. USA* 68:820–823, 1971.

Table 17–6 Incorporation of ^{3}H-Thymidine by Cells Grown on Normal Dishes and on Poly(HEMA)-treated Dishes (Answer 17–23)

Type of Dish	Cell Density (cells/dish)	Confluency	Cell Height (µm)	^{3}H Incorporation (cpm/1000 cells)
Normal	60,000	subconfluent	6	253
Normal	200,000	confluent	15	55
Normal	500,000	confluent	22	7
Poly(HEMA)	30,000	sparse	6	250
Poly(HEMA)	30,000	sparse	15	50
Poly(HEMA)	30,000	sparse	22	7

17–25

A. As discussed in Problem 17–24, the absence of a shoulder on any of the three curves suggests that in all cases only a single event is needed to trigger tumor production in mice that are already expressing one or both oncogenes.

B. Although the rate of tumor production is much higher in mice with both oncogenes, activation of the cellular *ras* proto-oncogene cannot be a required event in the production of tumors in mice that are already expressing the MMTV-regulated *myc* oncogene. Nor can activation of the cellular *myc* proto-oncogene be a required event in triggering tumor formation in mice that are already expressing the MMTV-regulated *ras* oncogene. As indicated in part A, even when mice contain both *myc* and *ras*, some additional event is required to produce a tumor. If *myc* plus *ras* were sufficient for tumor formation, then all mice would develop tumors as soon as they passed through puberty.

C. The rate of tumor production in mice with both oncogenes is much higher than expected if the effects of the individual oncogenes were additive. Thus the two oncogenes together have a synergistic effect on the rate of tumor production. However, as argued in part B, activation of both oncogenes is not sufficient to generate a tumor. Thus, the two oncogenes acting together must open up a pathway to tumor production that can be triggered by any one of several low-frequency events or that can be triggered by one very common event. The nature of the activating events is unclear for any of these transgenic mice.

Reference: Sinn, E.; Muller, W.; Pattengale, P.; Tepler, I.; Wallace, R.; Leder, P. Coexpression of MMTV/v-Ha-*ras* and MMTV/c-*myc* genes in transgenic mice: synergistic action of oncogenes *in vivo*. *Cell* 49:465–475, 1987.

The Mechanics
of Cell Division

- An Overview of M Phase
- Mitosis
- Cytokinesis

An Overview of M Phase

18–1

A. chromosome condensation
B. mitosis, cytokinesis
C. mitotic spindle
D. contractile ring
E. centrosome, centrioles

18–2

A. True
B. False. Bacteria, which generally have only one chromosome, segregate the replicated copies to daughter cells without special condensation by a mechanism that involves chromosome attachment to the bacterial plasma membrane.
C. True
D. True
E. False. The five stages of mitosis occur in strict sequential order, but cytokinesis begins during anaphase and continues through the end of M phase.
F. True
G. True

*18–3

18–4

The patterns of centriole duplication and splitting that account for the mercaptoethanol-induced abnormalities in cell division are diagrammed in Figure 18–14. In essence, the treatment with mercaptoethanol allows the centrosome to split a second time without an intervening duplication of the centrioles (Figure 18–14A). (Centrosomes split the first time as the egg entered mitosis.) As a consequence, each of the spindle poles of the tetrapolar spindle has only one centriole rather than the normal pair of centrioles. Evidently, a centrosome with a single centriole is a perfectly adequate spindle pole.

The daughter cells from the four-way division of the egg receive a centrosome with a single centriole. During the next cell cycle the centriole is duplicated to form a normal centrosome with a centriole pair (Figure 18–14A). However, the usual form of the centrosome upon entry into mitosis has two centriole pairs instead of one. Thus the centrosome in the daughter cells looks like a single spindle pole and indeed forms a monopolar spindle (Figure 18–14B). Most commonly, cell division is aborted, and the cell traverses the cell cycle again, allowing the centrosome to be duplicated a second time so that it possesses two centriole pairs (Figure 18–14B, top). This second duplication puts the centrosome cycle back in step with the cell cycle and further cell divisions occur normally (Figure 18–14B, top). In the rarer cases in which the daughter cells form a bipolar spindle, the centrosomes split to form two centrosomes each with a single centriole (Figure 18–14B, bottom). Although the splitting allows cell division to occur, it presents the daughter cells with the

*Problems with an asterisk (*) are answered in the Instructor's Manual.

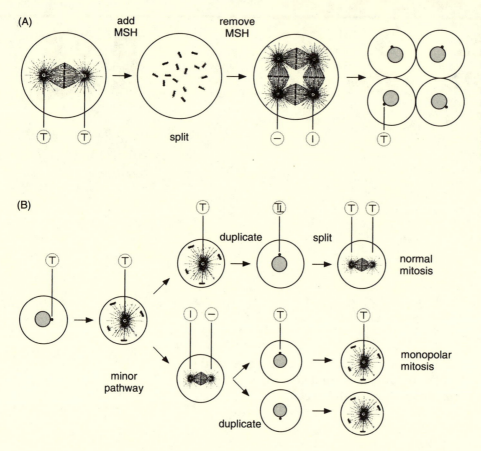

(A) add MSH → remove MSH → split

Figure 18–14 Duplication and splitting of centrosomes in mercaptoethanol-treated sea urchin eggs (Answer 18–4). The mercaptoethanol-induced splitting of centrosomes to form spindle poles with single centrioles is shown in (A). The aborted cell division allowing two rounds of centriole duplication to generate a normal centrosome is shown in the upper pathway in (B). The splitting of a centrosome with a single pair of centrioles is shown in the lower pathway in (B).

same problem as the parent: a centrosome with a single centriole (Figure 18–14B, bottom). In this case the centrosome cycle remains out of step with the cell-division cycle.

Note that although the cell divisions can be restored to normal, eggs that undergo a four-way division develop abnormally because none of the cells receives a full complement of chromosomes.

References: Maizia, D.; Harris, P.J.; Bibring, T. The multiplicity of mitotic centers and the time-course of their duplication and separation. *J. Biophys. Biochem. Cytol.* 7:1–20, 1960.

Sluder, G.; Rieder, C.L. Centriole number and the reproductive capacity of spindle poles. *J. Cell Biol.* 100:887–896, 1985.

18–5

A. The formation of the three types of patches observed in speckled kernels is shown in Figure 18–15. The formation of a colorless, nonwaxy (*c-Wx*) patch results from a breakage that eliminates the dominant color (*C*) allele (Figure 18–15A). In the absence of the dominant allele the color of the patch is determined by the recessive colorless (*c*) allele on the normal chromosome (which is not shown in the figure).

The formation of colorless, waxy (*c-wx*) spots in a colorless, nonwaxy (*c-Wx*) patch is due to a second breakage event that eliminates the dominant nonwaxy (*Wx*) allele (Figure 18–15B). In the absence of the dominant allele the spot is waxy (*wx*) due to the recessive allele on the normal chromosome (not shown).

The formation of an intensely colored patch is due to a breakage event that leads to a dicentric chromosome with multiple copies of the dominant color allele (Figure 18–15C). Thus the genetic constitution of the intensely colored patch is *C-C-Wx*.

434   Chapter 18 : The Mechanics of Cell Division

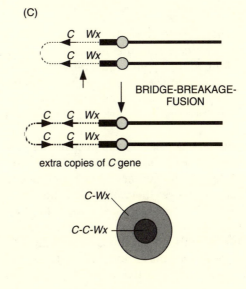

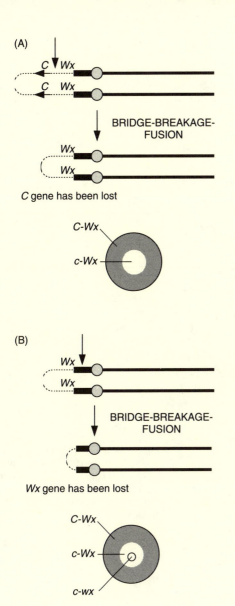

Figure 18–15 Formation of three different types of patches observed in speckled kernels (Answer 18–5). Vertical arrows pointing to the dicentric chromosomes show the position of the breaks that lead to formation of the patches. In each case the upper half of the starting dicentric chromosome is arbitrarily shown to give rise to the product dicentric chromosome. The formation of a colorless, waxy (*c-wx*) spot in a colorless, nonwaxy (*c-Wx*) patch requires the bridge-breakage-fusion shown in (A) followed by the bridge-breakage-fusion shown in (B).

B. You would never expect to see a colored spot within a colorless patch because, once eliminated, the dominant color (*C*) allele cannot be regained by further bridge-breakage-fusion cycles.

C. You would expect to see colorless spots within an intensely colored patch because the dominant color (*C*) allele could be lost by subsequent bridge-breakage-fusion cycles.

The demonstration by McClintock of bridge-breakage-fusion cycles in plants was one of the earliest indications that the broken ends of chromosomes are in some way "sticky"—entirely different from natural chromosome ends. It is clear now that most cells have an active repair pathway for joining broken DNA ends together as a defense against potentially lethal double-strand breaks. So long as breaks are rare, the correct ends are joined. But when multiple breaks are present, the wrong partners can be joined, leading to translocations or other genetic rearrangements. In humans such rearrangements are often associated with cancers.

Reference: McClintock, B. The behavior of successive nuclear divisions of a chromosome broken at meiosis. *Proc. Natl. Acad. Sci. USA* 25:405–416, 1939.

A. The simplest way to make this calculation is to draw all the possible arrangements of three chromosomes on the three mitotic spindles of tripolar eggs and on the four mitotic spindles of tetrapolar eggs, as shown schematically in Figure 18–16. Upon separation of the sister chromatids of each chromosome followed by cell division, some of the arrangements yield cells that do not contain at least one chromosome, as indicated by shaded regions in Figure 18–16.

For tripolar eggs there are 10 possible arrangements, 7 of which give rise to three cells that each has at least one chromosome. Thus 70% of tripolar eggs will have three cells containing at least one chromosome.

For tetrapolar eggs there are twenty possible arrangements, eight of which give rise to four cells that each have at least one chromosome. Thus 40% of tetrapolar eggs will have four cells containing at least one chromosome.

B. If the total number of chromosomes is the critical factor, more than 70% of tripolar eggs and more than 40% of tetrapolar eggs would be expected to develop into normal plutei. If the distribution of chromosomes is the critical factor, then $(0.7)^9$, or 4%, of tripolar eggs and $(0.4)^9$, or 0.03%, of tetrapolar eggs would be expected to develop into normal plutei.

Boveri found that 58 out of 695, or 8%, of tripolar eggs developed into normal plutei and that zero out of 1170, or less than 0.09%, of tetrapolar eggs developed into normal plutei. These results agree remarkably well with

(A) TRIPOLAR EGGS

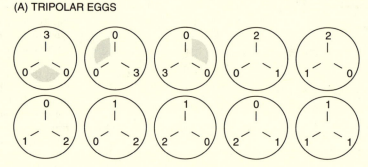

(B) TETRAPOLAR EGGS

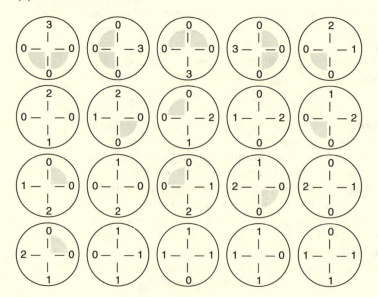

Figure 18–16 All possible arrangements of three chromosomes on the mitotic spindles of tripolar eggs (A) and tetrapolar eggs (B) (Answer 18–6). Shaded portions indicate cells that will not have at least one chromosome.

expectations based on the idea that individual chromosomes carry only a portion of the total genetic information.

References: Baltzer, F. Theodor Boveri: Life and Work of a Great Biologist. Berkeley: University of California Press, 1967.

Boveri, T. Über mehrpolige mitosen als mittel zur analyse des zellkerns. *Verh. d. phys.-med. Ges. Würzburg, N.F.* 35:67–90, 1902. (Available in English translation, Foundations of Experimental Embryology, ed. by B. Willier and J. Oppenheimer. Englewood Cliffs, NJ: Prentice-Hall, 1964.)

Boveri, T. Zellenstudien VI: Die Entwicklung dispermer Seeigeleier. Ein Beitrag zur Befruchtungslehre und zur Theorie des Kernes. *Jena. Zeitschr. Naturw.* 43:1–292, 1907.

Mitosis

18–7

A. polar, kinetochore, astral
B. centromere
C. kinetochores
D. prometaphase
E. metaphase
F. anaphase
G. anaphase A, anaphase B
H. telophase

18–8

A. False. Most of the microtubules that assemble on centrosomes are less stable than interphase microtubules.
B. True
C. False. Even before anaphase, while the sister chromatids are still linked, there is already a tension generated in the kinetochore microtubules, tugging outward on the chromosome and inward on the spindle poles.
D. False. The DNA sequences at the centromere are binding sites for proteins in both yeasts and higher eucaryotes. In neither case do they encode mRNA or proteins.
E. True
F. True
G. True
H. False. Equal and opposite pulls would tend to position chromosomes at any position between the poles. Some imbalance in the pulls or an additional force must act to position the chromosomes on the metaphase plate.
I. True
J. False. It is thought that sister chromatids are held together along their length by chromosomal proteins that must be altered in some way at anaphase to permit the chromatids to be pulled apart.
K. True
L. True
M. False. Sixty to 80% occurs at the kinetochore.
N. True
O. False. No special DNA sequences are required for formation of a nucleus.

18–9

A. Dicentric plasmids are stable in bacteria because bacteria use a completely different mechanism to segregate their chromosomes. Bacterial chromosomes are attached directly to specialized regions of the cell membrane that are gradually separated by the growth of membrane between them. Fission occurs between the two attachment sites, so that each daughter receives one chromosome. Thus bacteria are indifferent to the presence of centromeric sequences on the plasmid DNA—which is fortunate for scientists who want to clone centromeric DNA.

B. Dicentric plasmids are unstable in yeasts for the same reason that dicentric chromosomes are unstable in higher eucaryotes. If the two centromeric sequences attach to opposite poles, the spindle apparatus can exert enough pull on the DNA molecule to break its phosphodiester backbone. Roughly half the time a plasmid would be expected to orient itself on the spindle so that its two centromeres are attached to opposite poles. Thus there is a very high probability that a plasmid will be broken at each cell division, hence the instability.

C. Since monocentric plasmids are very stable, it seems most likely that the mechanism for deletion of centromeric sequences from dicentric plasmids relates to the breakage they suffer during mitosis. As illustrated in Figure 18–17, a circular plasmid must suffer two breaks to permit the centromeres to separate during mitosis. This breakage naturally separates the centromeres from one another onto linear fragments of the original plasmid. If the ends of a fragment join to make a circle, the resulting plasmid will contain a single centromeric sequence (Figure 18–17). However, only those fragments that contain the yeast origin of replication (*ARS1*) and the selected marker (*TRP1*) can continue to grow in future generations.

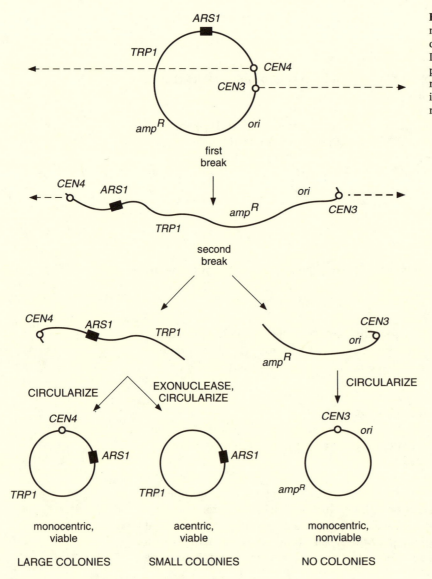

Figure 18–17 A mechanism for generating monocentric and acentric plasmids from a dicentric plasmid in yeast (Answer 18–9). Dashed arrows indicate the direction of pull toward the spindle poles. Viability refers to the ability of the plasmid to grow in yeast under selective conditions (which require *ARS1* and *TRP1*).

This mechanism does not readily account for the loss of both centromeric sequences; rather, it predicts that one centromeric sequence will be retained. Once the dicentric plasmid is reduced to a monocentric plasmid, it should be stable. The loss of both centromeric sequences probably involves a process other than simple breakage. One likely possibility (especially given a background knowledge of yeast that you were not provided with in the problem) is that the broken ends are digested by exonucleases, which occasionally remove the remaining centromeric sequence before the fragment circularizes (Figure 18–17).

Reference: Mann, C.; Davis, R.W. Instability of dicentric plasmids in yeast. *Proc. Natl. Acad. Sci. USA* 80:228–232, 1983.

*18–10   Reference: Murray, A.W.; Szostoak, J.W. Pedigree analysis of plasmid segregation in yeast. *Cell* 34:961–970, 1982.

*18–11   Reference: Mitchison, T.J.; Kirschner, M.W. Properties of the kinetochore *in vitro*. II. Microtubule capture and ATP-dependent translocation. *J. Cell. Biol.* 101:766–777, 1985.

18–12

A. If the second centromere were active, it would indeed destabilize the plasmid as described in Problem 18–9. In your experiments, however, growth on galactose keeps the introduced *CEN6* inactive by promoting transcription toward it.

B. Transcription requires that RNA polymerase be able to unwind DNA and separate the two strands over a short distance as it moves along the DNA. The stable assembly of a kinetochore evidently is not compatible with these activities of RNA polymerase. This has ample precedent in the activity of some gene regulatory proteins, which bind tightly to DNA sequences in the path of RNA polymerase and thereby block its progression along the DNA.

It is not so clear how transcription inactivates a kinetochore. Transcription apparently does not disturb the special arrangement of nucleosomes around centromeric DNA, and the majority of transcripts terminate at the border of the *CEN* sequence. It may be that the approach of the transcriptional apparatus close to the edge of the centromere destabilizes microtubule attachment just enough to kill the kinetochore effectively.

C. Low fidelity of chromosome transmission can arise in several ways other than by disruption of kinetochore function. Mutations in tubulin genes and in microtubule-based motor proteins, which can interfere with attachment to kinetochores or with spindle function, also show elevated rates of chromosome loss. Mutations that affect DNA metabolism can also lead to chromosome loss by interfering with proper chromosome replication.

Reference: Doheny, K.F.; Sorger, P.K.; Hyman, A.A.; Tugendreich, S.; Spencer, F.; Hieter, P. Identification of essential components of the *S. cerevisiae* kinetochore. *Cell* 73:761–774, 1993.

18–13

A. For the sectoring assay to be meaningful, it is essential that all cells start with the same genetic endowment. The use of a conditional centromere on minichromosome A makes this possible. Growing the cells on galactose keeps the second centromere inactive so that minichromosome A is faithfully transmitted at each cell division. When the cells are spread on agar plates containing glucose, however, the second centromere is activated because transcription from the *GAL10* promoter stops.

B. Sectoring results when the minichromosome is broken. A dicentric minichromosome is more apt to break when it is pulled toward opposite spindle poles by attachment to two strong centromeres (minichromosome A) than when pulled by attachments to one strong and one weak centromere (minichromosome B). The weak centromere, which is known to reduce centromere function by an order of magnitude, either lets go in the tug-of-war or perhaps does not attach in the first place. In either case the result is less sectoring than observed for minichromosome A, which has two strong centromeres.

C. The decrease in sectoring observed when minichromosome A is introduced into *ctf13* yeast as compared with wild-type yeast is fully consistent with the idea that the *ctf13* gene encodes a kinetochore protein. Presumably the defective kinetochore protein encoded by *ctf13* weakens the kinetochore sufficiently so that when the dicentric minichromosome is stretched between the spindle poles, one of the protein-DNA connections will break before the DNA breaks. In the wild-type strain, where the protein-DNA connection is stronger, the DNA breaks more often, resulting in a higher frequency of sectoring.

D. The *ctf13* mutation enhances the loss of normal chromosomes and stabilizes dicentric minichromosomes for the same reason: it makes a slightly defective kinetochore. The defect in the kinetochore allows normal chromosomes to detach from the spindle microtubules occasionally and be lost. The defect also allows one centromere to detach when a dicentric minichromosome is pulled toward both spindle poles, thereby preventing breakage of the minichromosome and allowing it to be transmitted faithfully at a higher frequency than in a wild-type strain.

Reference: Doheny, K.F.; Sorger, P.K.; Hyman, A.A.; Tugendreich, S.; Spencer, F.; Hieter, P. Identification of essential components of the *S. cerevisiae* kinetochore. *Cell* 73:761–774, 1993.

*18–14   **Reference:** Bloom, K. The centromere frontier: kinetochore components, microtubule-based motility, and the CEN-value paradox. *Cell* 73:621–624, 1993.

Cytokinesis

18–15

A. cytokinesis
B. furrowing
C. contractile ring
D. midbody
E. cell plate
F. phragmoplast
G. preprophase band

18–16

A. True
B. False. The position of the mitotic spindle anticipates the position of the cleavage furrow: it is positioned centrally for symmetric cleavage and asymmetrically for asymmetric cleavage.
C. True
D. True
E. True
F. True

18–17

A. The chromosome-signaling hypothesis predicts that furrows will form only where chromosomes have been aligned. This prediction matches the result of the first division but not the result of the second division, where three furrows are formed instead of the expected two furrows.

B. The polar-relaxation hypothesis predicts that there should be two furrows at the first division instead of one as is observed. In the toroidal egg the spindles should relax a ring of the cell surface on either side of the equatorial plane. As a consequence, there should be two regions where the cell surface furrows: at the equatorial plane and opposite it on the other side of the torus. Note that the events of the second division match the expectations of the relaxation hypothesis.

C. The aster-stimulation hypothesis predicts that there should be a single furrow at the first cell division and three furrows at the second cell division, which matches the experimental observations. A single furrow is expected at the first division because the spindle fibers from the two poles interact only at the equatorial plane. At the second division, however, the spindle fibers interact not only at the two equatorial positions, but also at the position of the third furrow.

Reference: Rappaport, R. Establishment of the mechanism of cytokinesis in animal cells. *Int. Rev. Cytol.* 105:245–281, 1986.

Cell Junctions, Cell Adhesion, and the Extracellular Matrix

Cell Junctions

19–1

A. connective tissues
B. epithelium
C. occluding, anchoring, communicating
D. tight
E. apical, basolateral
F. cell-cell adherens
G. adhesion belt (zonula adherens)
H. cell-matrix adherens
I. focal contacts (adhesion plaques)
J. septate
K. desmosomes
L. hemidesmosomes
M. gap junction
N. plasmodesmata

19–2

A. False. Tight junctions provide molecule-tight seals between cells.
B. True
C. False. The intestinal epithelium is 10,000 times more leaky to ions than is the bladder epithelium.
D. False. Gap junctions are communicating junctions.
E. True. Desmosomes and adhesion belts link cells to other cells, while hemidesmosomes and focal contacts link cells to the extracellular matrix.
F. False. Despite the differences between various connexin proteins, their basic structure and function are so highly conserved that functional gap junctions can often form between cells expressing different connexins, even if the two cells are from different vertebrates.
G. False. The permeability of gap junctions is regulated by intracellular calcium and pH. The extracellular calcium concentration and pH are closely regulated by physiological mechanisms and do not normally change very much.
H. True

19–3

As shown in Figure 19–19, the three protein monomers have distinctly different assembly properties because of the placement of complementary binding domains on their surfaces. Monomer B can assemble into a long chain, as would be expected for one sealing strand of a tight junction. Monomer A could assemble into a large two-dimensional aggregate, as is characteristic of desmosomes. Monomer C could assemble only into a tetramer and thus would be unable (in the absence of additional binding domains) to assemble into a chain or a large aggregate.

*19–4   **Reference:** Van Meer, G.; Simons, K. The function of tight junctions in maintaining differences in lipid composition between the apical and the basolateral cell surface domains of MDCK cells. *EMBO J.* 5:1455–1464, 1986.

*Problems with an asterisk (*) are answered in the Instructor's Manual.

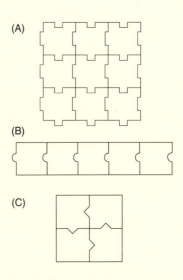

Figure 19–19 Assembly of protein monomers (Answer 19–3).

Your measurements support the two-state model of resistance of tight junctions because the resistance of an epithelium is logarithmically related to the number of sealing strands in the junction, as shown by the straight line in Figure 19–20. The line your data points define can be described by

$$R = R_{min} P^{-n}$$

where

R = specific resistance of the juntion
R_{min} = minimum resistance of the juntion (where there are no sealing strands in the tight junction)
P = probability that a given strand is open
n = number of strands in the tight junction

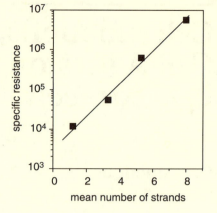

Figure 19–20 A plot of specific resistance versus the number of strands in the tight junction (Answer 19–5).

The value of P from your data is about 0.4. Thus, an individual sealing strand has a relatively high probability of being in the open state. However, if there are enough strands, the probability of their all being open at once is very low, so the overall junction is very tight.

The morphological basis for the closed and open states is unknown. There are, however, three types of protein-protein contacts in a tight junction: (1) contacts between the individual subunits to form a sealing strand in one cell, (2) contacts between individual sealing strands on the same cell to form the characteristic anastomosing network of strands, and (3) contacts between sealing strands on different cells to seal the space between the cells. If any contacts were broken, a pathway would open for the passage of small molecules across an individual sealing strand.

Reference: Claude, P. Morphological factors influencing transepithelial permeability: a model for the resistance of the zonula occludens. *J. Memb. Biol.* 39:219–232, 1978.

*19–6   **Reference:** Lawrence, T.S.; Beers, W.H.; Gilula, N.B. Transmission of hormonal stimulation by cell-to-cell communication. *Nature* 272:501–506, 1978.

19–7

A. HRP and fluorescein enter both cells at the early two-cell stage (but not at the late two-cell stage) because the cells are still connected by cytoplasmic bridges, which allow passage of large molecules.

B. Fluorescein enters all cells in the compacted eight-cell embryo because the cells are connected by gap junctions. Since gap junctions permit passage only of molecules less than 1000 daltons or so, HRP is confined to the cell into which it was initially injected. The different result before and after compaction indicates that formation of gap junctions is associated with compaction.

C. If you inject current from the HRP injection electrode, you would detect it in the fluorescein electrode only in the early two-cell embryo and the compacted eight-cell embryo. Only at these two stages are the adjacent cells electrically coupled. At the two-cell stage, the coupling is mediated by the cytoplasmic bridge remaining from cell division; at the eight-cell stage, the coupling is mediated by gap junctions.

Reference: Lo, C.W.; Gilula, N.B. Gap junctional communication in the preimplantation mouse embryo. *Cell* 18:399–409, 1979.

19–8

In invertebrates sheets of cells are often connected by structures known as septate junctions as well as by tight junctions. Micrograph (A) shows the characteristic appearance of septate junctions in section, and micrograph (B), the regular lines of particles that are typical of these junctions when seen in freeze fracture. Septate junctions serve primarily as adhering junctions.

A. septate junction (transmission electron micrograph)
B. septate junction from an insect gut (freeze fracture)
C. gap junction (*en face* view of a heavy metal-infiltrated section)
D. gap junction (freeze fracture)

Cell-Cell Adhesion

19–9

A. cell-cell adhesion molecules (CAMs)
B. cadherins
C. E-cadherin
D. homophilic, heterophilic
E. selectins
F. neural cell adhesion molecule (N-CAM)
G. fasciclin II

19–10

A. False. Although the quote is probably true in spirit, it is incorrect in detail. Warren Lewis was a pioneer in cell biology who was trying to draw attention to the importance of the adhesive properties of cells in tissues at a time when the problem had been largely ignored by the biologists of the day. Much of our bodies, however, is connective tissue whose integrity depends on the quality of the matrix rather than of the cells that inhabit it. It is not at all easy to dissociate cells from tissues, as anyone who has eaten a tough piece of steak can testify.

B. True

C. False. The development of tissues does not normally depend on the sorting of randomly mixed cell types.

D. False. Antibodies directed against most cell-surface components do not block cell adhesion. Thus far, antibodies against cell-surface glycoproteins involved in the adhesion process are the only ones that actually inhibit adhesion.

E. False. Identical cadherins on different cells bind to one another directly, thus the mechanism of binding is homophilic.

F. True. (It is somewhat counterintuitive that cells, which stick together very tightly, are bound by weaker interactions than those between a hormone and its receptor, which form a relatively transient complex. The strength of the cell-cell binding derives from multiple weak interactions, which when summed, are very strong.)

G. True

H. True

*19–11 **Reference:** Bonner, J.T.; Savage, L.J. Evidence for the formation of cell aggregates by chemotaxis in the development of the slime mold *Dictyostelium discoideum. J. Exp. Zool.* 106:1–26, 1947.

19–12

A. The fraction of the phage population that will be attached by at least one tail fiber at any one instant is equal to one minus the fraction not attached by any tail fibers, which is $(0.5)^{12} = 0.00024$ for wild-type bacteria and $(0.5)^6 = 0.016$ for *ompC⁻* bacteria. Thus, at any instant 99.98% of the phage population will be attached to wild-type bacteria and 98.4% will be attached to *ompC⁻* bacteria.

B. The very small difference in the fraction of the phage population attached to wild-type and *ompC⁻* bacteria at first seems too little to account for the 1000-fold difference in infectivity. However, since T4 must wander around the surface of a bacterium to find an appropriate place to attach its baseplate, the instantaneous calculation is misleading. If, for example, T4 must stay bound to the bacterial surface for 500 "instants" during its wandering, then $(0.9998)^{500}$ = 90% will remain attached to wild-type bacteria, but only $(0.984)^{500}$ = 0.03% will remain attached to *ompC⁻* bacteria. This difference would be more than enough to account for the 1000-fold difference in infectivity.

By associating with the bacterial surface through multiple weak interactions, bacteriophage T4 can wander around the surface without falling off. This allows a search for relatively rare injection sites, which are at points of connection between the inner and outer membranes. Multiple weak adhesive interactions between cells presumably are important for similar reasons: they allow a cell to search its neighbors without getting stuck prematurely.

Reference: Goldberg, E. Recognition, attachment, injection. In Bacteriophage T4 (C.K. Mathews, E.M. Kutter, G. Mosig, P.B. Berget, eds.), pp. 32–39. Washington, DC: American Society for Microbiology, 1983.

*19–13   Reference: Bischoff, R. Rapid adhesion of nerve cells to muscle fibers from adult rats is mediated by a sialic-acid-binding receptor. *J. Cell Biol.* 102:2273–2280, 1986.

19–14

The different sensitivities of type 5 and type 21 cells to trypsin and mercaptoethanol argue that two different cell-adhesion molecules are interacting. The dissociation of the factor from type 5 cells into two components upon treatment with mercaptoethanol suggests that one of these components is a linker molecule that is attached to type 5 cells by a disulfide bond. Alternatively, it could mean that the factor from type 5 cells is a two-component cell-adhesion molecule.

Reference: Crandall, M.A.; Brock, T.D. Molecular basis of mating in the yeast *Hansenula wingei. Bact. Rev.* 32:139–163, 1968.

The Extracellular Matrix of Animals

19–15

A. extracellular matrix
B. connective tissue
C. fibroblasts
D. glycosaminoglycans (GAGs)
E. hyaluronan
F. proteoglycans
G. collagens
H. collagen fibrils
I. fibril-associated collagens
J. elastin
K. fibronectin
L. RGD sequence, type III fibronectin
M. type IV
N. basal lamina
O. laminin

19–16

A. False. The extracellular matrix plays an active role influencing the development, migration, proliferation, shape, and metabolism of cells that contact it.
B. True
C. False. The core proteins are also highly diverse.
D. True
E. True
F. True

G. False. The elasticity of elastin fibers derives from their lack of secondary structure; elastin forms random coils. The hydrogen bonds that stabilize the α helix are too strong to be disrupted by the kinds of forces that deform elastin.

H. True

I. True

J. False. The basal lamina survives and provides a scaffolding along which regenerating cells can migrate.

K. True

19–17

A. Cycloheximide is an inhibitor of protein synthesis. It causes a reduction in $^{35}SO_4$ incorporation in the absence of ONP-β-D-xyloside because it blocks the synthesis of the core protein to which the polysaccharide chains—the substrate for $^{35}SO_4$ incorporation—are added. Since $^{35}SO_4$ incorporation is not sensitive to cycloheximide in the presence of ONP-β-D-xyloside, that incorporation must not depend on protein synthesis. The similarity of ONP-β-D-xyloside to the serine-xylose linkage to the protein suggests (correctly) that ONP-β-D-xylose serves as a primer for formation of polysaccharide chains. (Remember that in all cases the $^{35}SO_4$ counts were shown to be in glycosaminoglycans.)

B. In the presence of ONP-β-D-xyloside large amounts of $^{35}SO_4$-labeled material are released into the medium because they cannot be incorporated into the extracellular matrix of the tissue. An individual polysaccharide chain, even if full size, which these are not, is quite small relative to the proteoglycan to which it is normally attached.

C. ONP-β-D-xyloside serves as a primer for sugar addition, whereas ONP-α-D-xyloside does not, as indicated by its inability to block incorporation of counts into tissue or to stimulate counts in the medium (Table 19–5). Presumably, the sugar transferase responsible for adding the next sugar in the chain is specific for the β linkage between xylose and its attached group.

D. ONP-β-D-xyloside might block development of salivary glands in tissue culture in two ways. First, by competing for sugar addition to the core protein (note decrease in tissue counts in Table 19–5), it may alter the proteoglycan content of the extracellular matrix in such a way that the salivary gland cells cannot develop. Alternatively, the increase in the concentration of free polysaccharide chains could complete for proteoglycan binding sites on the surfaces of the salivary gland cells.

   Reference: Thompson, H.A.; Spooner, B.S. Inhibition of branching morphogenesis and alteration of glycosaminoglycan biosynthesis in salivary glands treated with β-D-xyloside. *Dev. Biol.* 89:417–424, 1982.

*19–18   Reference: Sykes, B. The molecular genetics of collagen. *Bioessays* 3:112–117, 1985.

19–19

A. The threefold increase in molecular weight when the disulfide bonds were left intact indicates that the resistant peptides in each partially reassembled collagen molecule are held together by disulfide bonds.

B. Since type III pN-collagen and type III collagen produced identical patterns after trypsin treatment, their resistant peptides almost certainly are identical. Thus, if all the resistant peptides are linked by disulfide bonds, the relevant bonds must be those at the C-terminus, that is, the ones that are present in type III collagen.

C. If reassembly began at random sites, there would be a more complicated fragment pattern. More importantly, however, some of the resistant peptides would not contain the site for the disulfide bond in the C-terminal region. As a consequence, the molecular weight of those peptides would not increase because they would not contain a disulfide bond. Since all the resistant peptides do contain the disulfide bond, reassembly is not initiated at random sites. Thus reassembly seems to initiate at a specific site that must be near the position of the disulfide bonds at the C-terminus.

D. The orderly progression from small to large fragments with increasing time indicates a zipperlike mechanism beginning, as argued above, from the C-terminal region (Figure 19–21). An all-or-none mechanism would not be expected to show any intermediate stages in the reassembly process.

The presence of discrete bands, which differ in size by several hundred amino acids, requires some additional explanation. If the zippering process was absolutely smooth and all trypsin-cleavage sites were equally sensitive, then a whole series of bands would be expected, each differing from the last by an increment reflecting the spacing of the trypsin-sensitive bonds (lysines and arginines). The absence of this finely spaced ladder of bands suggests either (1) that the cleavage sites are not equally sensitive (the trypsin digestions were very brief in these experiments) or (2) that zippering is not smooth but, rather, pauses at characteristic points, representing more difficult regions to assemble.

It is important to realize that the reassembly results tell you nothing directly about the natural assembly of type III collagen. It could be argued reasonably that the disulfide bonds at the C-terminus provide an artificial nucleation site for reassembly that is not present naturally. Nevertheless, these reassembly results are consistent with other *in vivo* observations. For example, point mutants of collagen typically are better assembled on the C-terminal side of the mutation than on the N-terminal side. This observation suggests that assembly is initiated toward the C-terminus and progresses zipperlike toward the N-terminus. A zipperlike assembly from one end also makes good theoretical sense: in such a highly repetitious molecule as collagen, random internal initiation would rarely lead to correctly aligned polypeptides.

All these considerations, taken together, from the basis for the current belief that fibrillar collagens assemble zipperlike from the C-terminus and that a primary function of the C-terminal propeptide is to register the polypeptides for proper assembly.

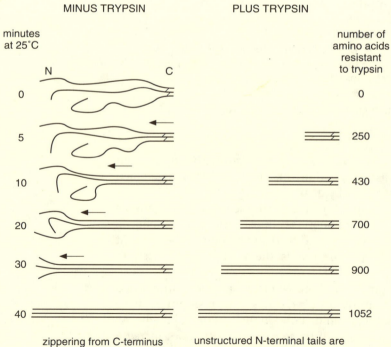

MINUS TRYPSIN   PLUS TRYPSIN

minutes
at 25°C

N   C

number of
amino acids
resistant
to trypsin

minutes at 25°C	number of amino acids resistant to trypsin
0	0
5	250
10	430
20	700
30	900
40	1052

zippering from C-terminus   unstructured N-terminal tails are
progresses with time of incubation   removed by trypsin treatment

Figure 19–21 Zipperlike assembly of type III collagen molecules, initiating from the C-terminal end (Answer 19–19). Results of trypsin digestion of partially assembled molecules is indicated on the right. The trypsin-resistant peptides are the ones that are visualized after electrophoresis.

Reference: Bachinger, H.P.; Bruckner, P.; Timpl, R.; Prockop, D.J.; Engel, J. Folding mechanism of the triple helix in type III collagen and type III pN-collagen. *Eur. J. Biochem.* 106:619–632, 1980.

*19–20 References: Pierschbacher, M.D.; Ruoslahti, E. Cell attachment activity of fibronectin can be duplicated by small synthetic fragments of the molecule. *Nature* 309:30–33, 1984.

Pytela, R.; Pierschbacher, M.D.; Ruoslahti, E. Identification and isolation of a 140 kd cell surface glycoprotein with properties expected of a fibronectin receptor. *Cell* 40:191–198, 1985.

Ruoslahti, E.; Pierschbacher, M.D. Arg-Gly-Asp: a versatile cell recognition signal. *Cell* 44:517–518, 1986.

19–21

A. Since the α1(I) collagen probe and your probe give identical patterns of mRNA, your gene is very likely to be the α1(I) collagen gene. You, of course, would want to confirm this by showing that your probe hydridized directly to the α1(I) probe.

   The identity of the affected gene is somewhat surprising because the α1(I) collagen gene encodes a component of type I collagen, which is found primarily in skin, tendons, bones, ligaments, and internal organs. The principal collagen in blood vessels is type III collagen. (See MBOC Table 19–4, p. 980). It seems that type I collagen is important for maintaining blood vessel integrity, at least early in development.

B. Matings between heterozygotes will follow simple Mendelian patterns. In the embryos before the lethality is expressed, one-fourth will be homozygous for the unaffected gene (pattern 1), one-fourth will be homozygous for the affected gene (pattern 3), and one-half will be heterozygous (pattern 2). Since pattern 3 does not appear in live births, one-third of newborns will be homozygous for the unaffected gene (pattern 1) and two-thirds will be heterozygous (pattern 2).

C. Cloning the unaffected gene that corresponds to the gene carrying the retroviral insertion is a straightforward and very powerful approach. The general procedure is outlined below.

   1. Prepare a genomic DNA library from a mouse that carries the retrovirus insert. Use a restriction enzyme that cuts outside the retrovirus sequences.
   2. Using retrovirus DNA as a probe, screen the library to find clones that carry retrovirus sequences.
   3. Subclone a segment of the cell DNA flanking the retrovirus sequences. (It is essential that this DNA not contain repetitive DNA sequences.) The DNA flanking the retrovirus contains the gene of interest.
   4. Use the subcloned cell DNA to screen a genomic DNA library (and/or cDNA library) prepared from normal mice. These clones contain unaltered versions of the gene of interest.

   References: Jaenisch, R.; Harbers, K.; Schnieke, A.; Lohler, J.; Chumakov, I.; Jahner, D.; Grotkopp, D.; Hoffmann, E. Germline integration of Moloney murine leukemia virus at the Mor13 locus leads to recessive lethal mutation and early embryonic death. *Cell* 32:209–216, 1983.

   Schnieke, A.; Harbers, K.; Jaenisch, R. Embryonic lethal mutation in mice induced by retrovirus insertion into the α1(I) collagen gene. *Nature* 304:315–320, 1983.

Extracellular Matrix Receptors on Animal Cells: The Integrins

19–22

    A. integrins

    B. stress fibers

19–23

    A. True

    B. False. The same integrin molecule can have different ligand-binding activities in different cells, suggesting that additional cell-type-specific factors can interact with integrins to modulate their binding activity.

    C. True

    D. False. Extracellular fibronectin filaments assemble at the surface of cultured fibroblasts in alignment with intracellular stress fibers, which are bundles of actin filaments.

    E. True

    F. True

19–24

    A. Several of the experiments in Table 19–8 indicate that the stimulated binding due to the anti-β_1 antibody occurs by the interaction of fibronectin with $\alpha_5\beta_1$ integrin. The antibody against the α_5 chain blocks fibronectin binding in the presence or absence of the anti-β_1 antibody. Anti-fibronectin antibody interferes with anti-β_1 stimulated binding. And most conclusively, a peptide containing the RGD motif (through which integrins bind fibronectin) blocks the stimulated binding, whereas a similar peptide without the complete RGD motif has no effect.

    B. Synthesis of new integrin molecules or transfer from an intracellular compartment would require ATP. Experiments that show that metabolic poisons have no effect on anti-β_1 stimulated binding eliminate these possibilities. Thus, the increased affinity of $\alpha_5\beta_1$ integrin for fibronectin seems to be due to some effect of the anti-β_1 antibody on the existing population of integrins.

    C. The experiments in Figure 19–17 combined with the results with metabolic poisons indicate that $\alpha_5\beta_1$ integrin can exist in a low-affinity and a high-affinity state. Apparently, the anti-β_1 antibody by binding to a specific site on the β_1 chain induces a conformational change in the $\alpha_5\beta_1$ integrin that flips it into its high-affinity state. Only binding at select sites would be expected to cause such conformational changes, which correlates with the rarity of monoclonal antibodies that have the properties of this particular anti-β_1 antibody. As more and more integrins are studied, it appears that low- and high-affinity states are common. In many cases the affinity can be altered by events occurring within the cell.

        Reference: Faull, R.J.; Kovach, N.L.; Harlan, J.M.; Ginsberg, M.H. Affinity modulation of integrin $\alpha_5\beta_1$: regulation of the functional response by soluble fibronectin. *J. Cell Biol.* 121:155–162, 1993.

*19–25   **Reference:** O'Toole, T.E.; Mandelman, D.; Forsyth, J.; Shattil, S.J.; Plow, E.F.; Ginsberg, M.H. Modulation of the affinity of integrin $\alpha_{IIb}\beta_3$ (GPIIb-IIIa) by the cytoplasmic domain of α_{IIb}. *Science* 254:845–847, 1991.

The Plant Cell Wall

19–26

    A. primary cells walls, secondary cell wall
    B. turgor pressure
    C. cellulose microfibrils
    D. hemicelluloses

19–27

    A. False. A mature cell may simply retain the primary cell wall or, far more commonly, produce a rigid secondary cell wall either by thickening the primary wall or by depositing new layers with a different composition underneath the old ones.
    B. True
    C. False. Although the major components of plant cell walls are polysaccharides, there is a variable contribution from structural proteins, which are thought to strengthen the wall.
    D. False. In the absence of cortical microtubules, new cellulose microfibrils are deposited in the same orientation as the preexisting microfibrils. Intact microtubules seem to be required for any change in the orientation of cellulose microfibrils.
    E. True

*19–28

19–29

    A. The accepted explanation for the ability of an antisense RNA to block expression of the normal gene is that the two RNAs—the antisense RNA and the normal RNA—will hybridize to make a double-stranded RNA that cannot be translated. This would effectively block synthesis of the ACC synthase enzyme and prevent formation of ethylene. But this may not be the true mechanism. In some plants a phenomenon called RIPing pairs duplicated sequences in meiosis and introduces mutations into both. Thus, it may be that the normal ACC synthase gene is inactivated in your transgenic tomatoes.
    B. In all likelihood!
    Reference: Oeller, P.W.; Min-Wong, L.; Taylor, L.P.; Pike, D.A.; Theologis, A. Reversible inhibition of tomato fruit senescence by antisense RNA. *Science* 254:437–439, 1991.

References

Adams, G.A.; Rose, J.K. Structural requirements of a membrane-spanning domain for protein anchoring and cell surface transport. *Cell* 41:1007–1015, 1985. (Problem 13–5)

Adelstein, R.S.; Klee, C.B. Smooth muscle myosin light-chain kinase. In Calcium and Cell Function (W.Y. Cheung, ed.), Vol. 1, pp. 167–182. New York: Academic Press, 1980. (Problem 15–18)

Alberts, B.M.; Frey, L. T4 bacteriophage gene 32: a structural protein in the replication and recombination of DNA. *Nature* 227:1313–1318, 1970. (Problem 6–33)

Allard, W.J.; Lienhard, G.E. Monoclonal antibodies to the glucose transporter from human erythrocytes: identification of the transporter as a Mr = 55,000 protein. *J. Biol. Chem.* 260:8668–8675, 1985. (Problem 11–3)

Anderson, J.E. Restriction endonucleases and modification methylases. *Curr. Opin. Struct. Biol.* 3:24–30, 1993. (Problem 7–6)

Andreasen, P.H.; Dreisig, H.; Kristiansen, K. Unusual ciliate-specific codons in *Tetrahymena* mRNAs are translated correctly in a rabbit reticulocyte lysate supplemented with a subcellular fraction from *Tetrahymena*. *Biochem. J.* 244:331–335, 1987. (Problem 6–5)

Ashburner, M.; Chihara, C.; Meltzer, P.; Richards, G. Temporal control of puffing activity in polytene chromosomes. *Cold Spring Harbor Symp. Quant. Biol.* 38:655–662, 1973. (Problem 15–8)

Bachinger, H.P.; Bruckner, P.; Timpl, R.; Prockop, D.J.; Engel, J. Folding mechanism of the triple helix in type III collagen and type III pN-collagen. *Eur. J. Biochem.* 106:619–632, 1980. (Problem 19–19)

Baeuerle, P.A.; Baltimore, D. Activation of DNA-binding activity in an apparently cytoplasmic precursor of the NF-κB transcription factor. *Cell* 53:211–217, 1988. (Problem 12–9)

Baltzer, F. Theodor Boveri: Life and Work of a Great Biologist. Berkeley: University of California Press, 1967. (Problem 18–6)

Barnes, G.; Rine, J. Regulated expression of endonuclease EcoRI in *Saccharomyces cerevisiae*: nuclear entry and biological consequences. *Proc. Natl. Acad. Sci. USA* 821:1354–1358, 1985. (Problem 12–7)

Becker, W.M. The World of the Cell, p. 282. Menlo Park, CA: Benjamin-Cummings, 1986. (Problem 14–20)

Bender, J.; Kleckner, N. Genetic evidence that Tn10 transposes by a nonreplicative mechanism. *Cell* 45:801–815, 1986. (Problem 6–43)

Bennett, V.; Stenbuck, P.J. Association between ankyrin and the cytoplasmic domain of band 3 isolated from the human erythrocyte membrane. *J. Biol. Chem.* 255:6424–6432, 1980. (Problem 10–11)

Bennett, V.; Stenbuck, P.J. The membrane attachment protein for spectrin is associated with band 3 in human erythrocyte membranes. *Nature* 280:468–473, 1979. (Problem 10–11)

Berget, S.M.; Berk, A.J.; Harrison, T.; Sharp, P.A. Spliced segments at the 5′ termini of adenovirus-2 late mRNA: a role for heterogeneous nuclear RNA in mammalian cells. *Cold Spring Harbor Symp. Quant. Biol.* 42:523–529, 1977. (Problem 8–24)

Birnbaumer, L.; Brown, A.M. G protein opening of K⁺ channels. *Nature* 327:21, 1987. (Problem 15–20)

Bischoff, R. Raid adhesion of nerve cells to muscle fibers from adult rats is mediated by a sialic-acid-binding receptor. *J. Cell Biol.* 102:2273–2280, 1986. (Problem 19–13)

Blackburn, E.H.; Gall, J.G. A tandemly repeated sequence at the termini of the extrachromosomal ribosomal RNA genes in *Tetrahymena*. *J. Mol. Biol.* 120:33–53, 1978. (Problem 8–4)

Blaut, M.; Gottschalk, G. Evidence for a chemiosmotic mechanism of ATP synthesis in methanogenic bacteria. *Trends Biochem. Sci.* 10:486–489, 1985. (Problem 14–16)

Bleil, J.D.; Bretscher, M.S. Transferrin receptor and its recycling in HeLa cells. *EMBO J.* 1:351–355, 1982. (Problems 13–18, 13–19)

Bloom, K. The centromere frontier: kinetochore components, microtubule-based motility, and the CEN-value paradox. *Cell* 73:621–624, 1993. (Problem 18–14)

Bloom, K.S.; Carbon, J. Yeast centromere DNA is in a unique and highly ordered structure in chromosomes and small circular minichromosomes. *Cell* 29:305–317, 1982. (Problem 8–8)

Boeke, J.D.; Garfinkel, D.J.; Styles, C.A.; Fink, G.R. Ty elements transpose through an RNA intermediate. *Cell* 40:491–500, 1985. (Problem 8–35)

Bogenhagen, D.; Clayton, D.A. Mouse L cell mitochondrial DNA molecules are selected randomly for replication throughout the cell cycle. *Cell* 11:719–727, 1977. (Problem 14–33)

Bonner, J.T.; Savage, L.J. Evidence for the formation of cell aggregates by chemotaxis in the development of the slime mold *Dictyostelium discoideum*. *J. Exp. Zool.* 106:1–26, 1947. (Problem 19–11)

Bootsma, D.; Budke, L.; Vos, O. Studies on synchronous division of tissue culture cells initiated by excess thymidine. *Exp. Cell Res.* 33:301–309, 1964. (Problem 17–5)

Bostock, C.J.; Prescott, D.M.; Kirkpatrick, J.B. An evaluation of the double thymidine block for synchronizing mammalian cells at the G_1-S border. *Exp. Cell Res.* 68:163–168, 1971. (Problem 17–5)

Boveri, T. Über mehrpolige Mitosen als Mittel zur Analyse de Zellkerns. *Verh. d. phys.-med. Ges. Würzburg, N.F.* 35:67–90, 1902. (Available in English translation, Foundations of Experimental Embryology, ed. by B. Willier and J. Oppenheimer. Englewood Cliffs, NJ: Prentice-Hall, 1964.) (Problem 18–6)

Boveri, T. Zellenstudien VI: Die Entwicklung dispermer Seeigeleier. Ein Beitrag zur Befruchtungslehre und zur Theorie des Kernes. *Jena. Zeitschr. Naturw.* 43:1–292, 1907. (Problem 18–6)

Bretscher, M. Asymmetrical lipid bilayer structure for biological membranes. *Nature New Biol.* 236:11–12, 1972. (Problem 10–4)

Brewer, B.J.; Fangman, W.L. The localization of replication origins on ARS plasmids in *S. cerevisiae*. *Cell* 51:463–471, 1987. (Problem 8–16)

Brokaw, C.J.; Luck, D.J.L.; Huang, B. Analysis of the movement of *Chlamydomonas* flagella: the function of the radial-spoke system is revealed by comparison of wild-type and mutant flagella. *J. Cell Biol.* 92:722–732, 1982. (Problem 16–20)

Broschat, K.O.; Stidwell, P.R.; Burgess, D.R. Phosphorylation controls brush border motility by regulating myosin structure and association with the cytoskeleton. *Cell* 35:561–571, 1983. (Problem 16–31)

Brown, C.J.; Ballabio, A.; Rupert, J.L.; Lafreniere, R.G.; Grompe, M.; Tonlorenzi, R.; Willard, H.F. A gene from the region of the human X inactivation centre is expressed exclusively from the inactive X chromosome. *Nature* 349:38–44, 1993. (Problem 7–14)

Brown, M.S.; Goldstein, J.L. Receptor-mediated endocytosis: insights from the lipoprotein receptor system. *Proc. Natl. Acad. Sci. USA* 76:3330–3337, 1979. (Problems 13–16, 13–17)

Brutlag, D.; Kornberg, A. Enzymatic synthesis of deoxyribonucleic acid. 36. A proofreading function for the 3′ to 5′ exonuclease activity in deoxyribonucleic acid polymerases. *J. Biol. Chem.* 247:241–248, 1972. (Problem 6–24)

Burch, R.M.; Luini, A.; Axelrod, J. Phospholipase A_2 and phospholipase C are activated by distinct GTP-binding proteins in response to α_1-adrenergic stimulation in FRTL5 thyroid cells. *Proc. Natl. Acad. Sci. USA* 83:7201–7205, 1986. (Problem 15–17)

Byers, D.; Davis, R.L.; Kiger, J.A. Defect in cyclic AMP phosphodiesterase due to the *dunce* mutation of learning in *Drosophila melanogaster. Nature* 289:79–81, 1981. (Problem 15–14)

Callan, H.G. The nature of lampbrush chromosomes. *Int. Rev. Cytol.* 15:1–34, 1963. (Problem 8–11)

Capaldi, R.A.; Darley-Usmar, V.; Fuller, S.; Millet, F. Structural and functional features of the interaction of cytochrome c with complex III and cytochrome c oxidase. *FEBS Letters* 138:1–7, 1982. (Problem 14–13)

Carlier, M.-F.; Pantaloni, D.; Korn, E.D. Evidence for an ATP cap at the ends of actin filaments and its regulation of the F-actin steady state. *J. Biol. Chem.* 259:9983–9986, 1984. (Problem 16–25)

Cha, T.A.; Alberts, B.M. Studies of the DNA helicase-RNA primase unit from bacteriophage T4. A trinucleotide sequence on the DNA template starts RNA primer synthesis. *J. Biol. Chem.* 261:7001–7010, 1986. (Problem 6–25)

Chamberlain, J.S.; Gibbs, R.A.; Ranier, J.E.; Caskey, C.T. Multiplex PCR for the diagnosis of Duchenne muscular dystrophy. In PCR Protocols (M.A. Innis, D.H. Gelfand, J.J. Sninsky, T.J. White, eds.), pp. 272–281. San Diego, CA: Academic Press, 1990. (Problem 7–26)

Chant, J.; Corrado, K.; Pringle, J.R.; Herskowitz, I. Yeast BUD5, encoding a putative GDP-GTP exchange factor, is necessary for bud site selection and interacts with bud formation gene *BEM1. Cell* 65:1213–1224, 1991. (Problem 16–5)

Chant, J.; Herskowitz, I. Genetic control of bud site selection in yeast by a set of gene products that constitute a morphogenetic pathway. *Cell* 65:1203–1212, 1991. (Problem 16–5)

Chen, C.-N.; Denome, S.; Davis, R.L. Molecular analysis of cDNA clones and the corresponding genomic coding sequences of the *Drosophila* dunce+ gene, the structural gene for cyclic AMP phosphodiesterase. *Proc. Natl. Acad. Sci. USA* 83:9313–9317, 1986. (Problem 15–14)

Claude, P. Morphological factors influencing transepithelial permeability: a model for the resistance of the zonula occlude *J. Memb. Biol.* 39:219–232, 1978. (Problem 19–5)

Cohen, S.; Ushiro, H.; Stocheck, C.; Chinkers, M. A native 170,000 epidermal growth factor receptor-kinase complex from shed plasma membrane vesicles. *J. Biol. Chem.* 257:1523–1531, 1982. (Problem 17–22)

Conway, L.; Wickens, M. Analysis of mRNA 3′-end formation by modification interference: the only modifications which prevent processing lie in AAUAAA and the poly (A) site. *EMBO J.* 6:4177–4184, 1987. (Problem 8–25)

Cox, M.; Lehman, I.R. The polarity of the recA protein-mediated branch migration. *Proc. Natl. Acad. Sci. USA* 78:6023–6027, 1981. (Problem 6–34)

Craigen, W.J.; Cook, R.G.; Tate, W.P.; Caskey, C.T. Bacterial peptide chain release factors: conserved primary structure and possible frameshift regulation of release factor 2. *Proc. Natl. Acad. Sci. USA* 82:3616–3620, 1985. (Problem 6–8)

Crandall, M.A.; Brock, T.D. Molecular basis of mating in the yeast *Hansenula wingei. Bact. Rev.* 32:139–163, 1968. (Problem 19–14)

Crick, F.H.; Brenner, S. The absolute sign of certain phase-shift mutants in bacteriophage T4. *J. Mol. Biol.* 26:361–363, 1967. (Problem 6–41)

Curtis, H. Biology, 4th ed., p. 216. New York: Worth, 1983. (Problem 14–23)

Davis, R.L.; Weintraub, H.; Lassar, A.B. Expression of a single transfected cDNA converts fibroblasts to myoblasts. *Cell* 51:987–1000, 1987. (Problem 9–29)

Deenen, L.L.M.; DeGier, J. Lipids of the red cell membrane. In The Red Blood Cell (D. MacN. Surgenor, ed.), pp. 147–211. New York: Academic Press, 1974. (Problem 10–4)

DeFranco, D.; Yamamoto, K.R. Two different factors act separately or together to specify functionally distinct activities at a single transcriptional enhancer. *Mol. Cell. Biol.* 6:993–1001, 1986. (Problem 15–6)

Dingwall, C.; Sharnick, S.V.; Laskey, R.A. A polypeptide domain that specifies migration of nucleoplasmin into the nucleus. *Cell* 30:449–458, 1982. (Problem 12–6)

Dodson, M.; Dean, F.B.; Bullock, P.; Echols, H.; Hurwitz, J. Unwinding of duplex DNA from the SV40 origin of replication by T-antigen. *Science* 238:964–967, 1987. (Problem 8–17)

Doheny, K.F.; Sorger, P.K.; Hyman, A.A.; Tugendreich, S.; Spencer, F.; Hieter, P. Identification of essential components of the *S. cerevisiae* kinetochore. *Cell* 73:761–774, 1993. (Problems 18–12, 18–13)

Donaldson, J.G.; Finazzi, D.; Klausner, R.D. Brefeldin A inhibits Golgi membrane-catalyzed exchange of guanine nucleotide into ARF protein. *Nature* 360:350–352, 1992. (Problem 13–29)

Dunn, T.M.; Hahn, S.; Ogden, S.; Schleif, R.F. An operator at −280 base pairs that is required for repression of *araBAD* operon promoter: addition of DNA helical turns between the operator and promoter cyclically hinders repression. *Proc. Natl. Acad. Sci. USA* 81:5017–5020, 1984. (Problem 9–12)

Duysens, L.N.M.; Amesz, J.; Kamp, B.M. Two photochemical systems in photosynthesis. *Nature* 190:510–511, 1961. (Problem 14–25)

Edelman, P.; Gallant, J. Mistranslation in *E. coli. Cell* 10:131–137, 1977. (Problem 6–9)

Eisenberg, E.; Hill, T.L. Muscle contraction and free energy transduction in biological systems. *Science* 227:999–1006, 1985. (Problem 16–38)

Elledge, S.J.; Mulligan, J.T.; Ramer, S.W.; Spottswood, M.; Davis, R.W. λYES: a multifunctional cDNA expression vector for the isolation of genes by complementation of yeast and *Escherichia coli* mutations. *Proc. Natl. Acad. Sci. USA* 88:1731–1735, 1991. (Problem 7–23)

Emerson, R. *Ann. Rev. Plant Physiol.* 9:1–24, 1958. (Problem 14–24)

Fantl, W.J.; Escobedo, J.A.; Martin, G.A.; Turck, C.W.; del Rosario, M.; McCormick, F.; Williams, L.T. Distinct phosphotyrosines on a growth factor receptor bind to specific molecules that mediate different signaling pathways. *Cell* 69:413–423, 1992. (Problem 15–26)

Farrell, P.J.; Balkow, K.; Hunt, T.; Jackson, R.J.; Trachsel, H. Phosphorylation of initiation factor eIF-2 and the control of reticulocyte protein synthesis. *Cell* 11:187–200, 1977. (Problem 9–34)

Faull, R.J.; Kovach, N.L.; Harlan, J.M.; Ginsberg, M.H. Affinity modulation of integrin $\alpha_5\beta_1$: regulation of the functional response by soluble fibronectin. *J. Cell Biol.* 121:155–162, 1993. (Problem 19–24)

Federman, A.D.; Conklin, B.R.; Schrader, K.A.; Reed, R.R.; Bourne, H.R. Hormonal stimulation of adenylyl cyclase through G$_i$-protein $\beta\gamma$ subunits. *Nature* 356:159–161, 1992. (Problem 15–15)

Fields, S.; Song, O. A novel genetic system to detect protein-protein interactions. *Nature* 340:245–246, 1989. (Problem 15–28)

Folkman, J.; Moscona, A. Role of cell shape in growth control. *Nature* 273:345–349, 1978. (Problem 17–23)

Forbush, B.; Kok, B.; McGloin, M. Cooperation of charges in photosynthetic oxygen evolution II. Damping of flash yield, oscillation and deactivation. *Photochem. Photobiol.* 14:307–321, 1971. (Problem 14–26)

Fox, T.D. Nuclear gene products required for translation of specific mitochondrially coded mRNAs in yeast. *Trends in Genetics* 2:97–100, 1986. (Problem 14–37)

Foyer, C.H. Photosynthesis, pp. 176–195. New York: Wiley, 1984. (Problem 14–22)

Friedberg, E.C. DNA repair. New York: W.H. Freeman, 1984. (Problem 6–17)

Fries, E.; Gustafsson, L.; Peterson, P.A. Four secretory proteins synthesized by hepatocytes are transported from the endoplasmic reticulum to Golgi complex at different rates. *EMBO J.* 3:147–152, 1984. (Problem 13–23)

Fung, B.K-K.; Hurley, J.B.; Stryer, L. Flow of information in the light-triggered cyclic nucleotide cascade of vision. *Proc. Natl. Acad. Sci. USA* 78:152–156, 1981. (Problem 15–21)

Fung, B.K-K.; Stryer, L. Photolyzed rhodopsin catalyzes the exchange of GTP for bound GDP in retinal rod outer segments. *Proc. Natl. Acad. Sci. USA* 77:2500–2504, 1980. (Problem 15–21)

Fung, Y-K.T.; Murphree, A.L.; T'Ang, A.; Qian, J.; Hinrichs, S.H.; Benedict, W.F. Structural evidence for the authenticity of the human retinoblastoma gene. *Science* 236:1657–1661, 1987. (Problem 17–24)

Gall, J.G. Personal communication. (Problem 8–11)

Gartenberg, M.R.; Crothers, D.M. DNA sequence determinants of CAP-induced bending and protein binding affinity. *Nature* 333:824–829, 1988. (Problem 9–5)

Gay, D.A.; Yen, T.J.; Lau, J.T.Y.; Cleveland, D.W. Sequences that confer β-tubulin autoregulation through modulated mRNA stability reside within exon 1 of a β-tubulin mRNA. *Cell* 50:671–679, 1987. (Problem 9–35)

Gelehrter, T.D.; Collins, F.S. Principles of Medical Genetics, p. 80. Baltimore: Williams & Wilkins, 1990. (Problem 7–17)

Gibbons, I.R. Cilia and flagella of eukaryotes. *J. Cell Biol.* 91:107s–124s, 1981. (Problem 16–20)

Godowski, P.J.; Rusconi, S.; Miesfeld, R.; Yamamoto, K.R. Glucocorticoid receptor mutants that are constitutive activators of transcriptional enhancement. *Nature* 325:365–385, 1987. (Problem 9–17)

Goldberg, E. Recognition, attachment, injection. In Bacteriophage T4 (C.K. Mathews, E.M. Kutter, G. Mosig, P.B. Berget, eds.), pp. 32–39. Washington, DC: American Society for Microbiology, 1983. (Problem 19–12)

Goldman, Y.E.; Hibberd, M.G.; McCray, J.A.; Trentham, D.R. Relaxation of muscle fibers by photolysis of caged ATP. *Nature* 300:701–705, 1980. (Problem 16–36)

Gordon, A.M.; Huxley, A.F.; Julian, F.J. The variation in isometric tension with sarcomere length in vertebrate muscle fibres. *J. Physiol.* 184:170–192, 1966. (Problem 16–37)

Gottesman, M.M. Genetics of cyclic-AMP-dependent protein kinases. In Molecular Cell Genetics (M.M. Gottesman, ed.), pp. 711–743. New York: Wiley, 1985. (Problem 15–13)

Gottlieb, T.A.; Beaudry, G.; Rizzolo, L.; Colman, A.; Rindler, M.J.; Adesnik, M.; Sabatini, D.D. Secretion of endogenous and exogenous proteins from polarized MDCK monolayers. *Proc. Natl. Acad. Sci. USA* 83:2100–2104, 1986. (Problem 13–24)

Goutte, C.; Johnson, A.D. a1 protein alters the DNA-binding specificity of α2 repressor. *Cell* 52:875–882, 1988. (Problem 9–25)

Haigler, H.T.; McKanna, J.A.; Cohen, S. Rapid stimulation of pinocytosis in human A-431 carcinoma cells by epidermal growth factor. *J. Cell Biol.* 83:82–90, 1979. (Problem 13–15)

Harland, R.M.; Laskey, R.A. Regulated replication of DNA microinjected into eggs of *Xenopus laevis*. *Cell* 21:761–771, 1980. (Problem 8–20)

Hartwell, L.H. Cell division from a genetic perspective. *J. Cell Biol.* 77:627–637, 1978. (Problems 17–14, 17–15)

Hartwell, L.H.; Weinert, T.A. Checkpoints: controls that ensure the order of cell cycle events. *Science* 246:629–634, 1989. (Problem 17–17)

Helms, J.B.; Rothman, J.E. Inhibition by brefeldin A of a Golgi membrane enzyme that catalyzes exchange of guanine nucleotide bound to ARF. *Nature* 360:352–354, 1992. (Problem 13–29)

Hertel, C.; Muller, P.; Portenier, M.; Staehelin, M. Determination of the desensitization of β-adrenergic receptors by [^{3}H]CGP-12177. *Biochem. J.* 216:669–674, 1983. (Problem 15–32)

Herzog, V.; Neumuller, W.; Holzmann, B. Thyroglobulin, the major and obligatory exportable protein of thyroid follicle cells, carries the lysosomal recognition marker mannose 6-phosphate. *EMBO J.* 6:555–560, 1987. (Problem 13–12)

Higuchi, R. Recombinant PCR. In PCR Protocols (M.A. Innis, D.H. Gelfand, J.J. Sninsky, T.J. White, eds.), pp. 177–183. San Diego, CA: Academic Press, 1990. (Problem 7–35)

Hille, B. Ionic Channels of Excitable Membranes. Sunderland, MA: Sinauer, 1984. (Problems 11–12, 11–13, 11–15)

Holloway, S.L.; Glotzer, M.; King, R.W.; Murray, A.W. Anaphase is initiated by proteolysis rather than by the inactivation of maturation-promoting factor. Cell 73:1393–1402, 1993. (Problem 17–11)

Horio, T.; Hotami, H. Visualization of the dynamic instability of individual microtubules by dark-field microscopy. *Nature* 321:605–607, 1986. (Problem 16–14)

Horowitz, S.; Gorovsky, M.A. An unusual genetic code in nuclear genes of *Tetrahymena*. *Proc. Natl. Acad. Sci. USA* 82:2452–2455, 1985. (Problem 6–5)

Huberman, J.A.; Riggs, A.D. On the mechanism of DNA replication in mammalian chromosomes. *J. Mol. Biol.* 32:327–341, 1968. (Problem 8–18)

Huganir, R.L.; Delcour, A.H.; Greengard, P.; Hess, G.P. Phosphorylation of the nicotine acetycholine receptor regulates its rate of desensitization. *Nature* 321:774–776, 1986. (Problem 15–33)

Hunt, T.; Luca, F.C.; Ruderman, J.V. The requirement for protein synthesis and degradation, and the control of destruction of cyclins A and B in the meiotic and mitotic cell cycles of the clam embryo. *J. Cell Biol.* 116:707–724, 1992. (Problem 17–10)

Ingledew, J.W. *Thiobacillus ferrooxidans:* the bioenergetics of an acidophilic chemolithotroph. *Biochim. Biophys. Acta* 683:89–117, 1982. (Problem 14–28)

Inman, R.B.; Schnos, M. Structure of branch points in replicating DNA: presence of single-stranded connections in lambda DNA branch points. *J. Mol. Biol.* 56:319–325, 1971. (Problem 6–23)

Jacks, T.; Varmus, H. Expression of the Rous sarcoma virus *pol* gene by ribosomal frameshifting. *Science* 230:1237–1242, 1985. (Problem 6–8)

Jaenisch, R.; Harbers, K.; Schnieke, A.; Lohler, J.; Chumakov, I.; Jahner, D.; Grotkopp, D.; Hoffmann, E. Germline integration of Moloney murine leukemia virus at the Mor13 locus leads to recessive lethal mutation and early embryonic death. *Cell* 32:209–216, 1983. (Problem 19–21)

Jagendorf, A.T.; Uribe, E. ATP formation caused by acid-base transition of spinach chloroplasts. *Proc. Natl. Acad. Sci. USA* 55:170–177, 1966. (Problem 14–27)

Johnson, P.J.; Kooter, J.M.; Borst, P. Inactivation of transcription by UV irradiation of *T. brucei* provides evidence for a multicistronic transcription unit including a VSG gene. *Cell* 51:273–281, 1987. (Problem 8–23)

Kaplan, A.; Achord, D.T.; Sly, W.S. Phosphohexosyl components of a lysosomal enzyme are recognized by pinocytosis receptors on human fibroblasts. *Proc. Natl. Acad. Sci. USA* 74:2026–2030, 1977. (Problem 13–10)

Karrer, K.M.; Gall, J.G. The macronuclear ribosomal DNA of *Tetrahymena pyriformis* is a palindrome. *J. Mol. Biol.* 104:421–453, 1976. (Problem 7–9)

Keilin, D. The History of Cell Respiration and Cytochrome. Cambridge, UK: Cambridge University Press, 1966. (Problem 14–12)

Kellems, R.E.; Allison, V.F.; Butow, R.A. Cytoplasmic type 80S ribosomes associated with yeast mitochondria. IV. Attachment of ribosomes to the outer membrane of isolated mitochondria. *J. Cell Biol.* 65:1–14, 1975. (Problem 12–12)

Knudson, A.G. Mutation and cancer: statistical study of retinoblastoma. *Proc. Natl. Acad. Sci. USA* 68:820–823, 1971. (Problem 17–24)

Koleske, A.J.; Buratowski, S.; Nonet, M.; Young, R.A. A novel transcription factor reveals a functional link between the RNA polymerase II CTD and TFIID. *Cell* 69:883–894, 1992. (Problem 9–16)

Kondor-Koch, C.; Bravo, R.; Fuller, S.D.; Cutler, D.; Garoff, H. Protein secretion in the polarized epithelial cell line MDCK. *Cell* 43:297–306, 1985. (Problem 13–24)

Konkel, D.A.; Maizel, J.V.; Leder, P. The evolution and sequence comparison of two recently diverged mouse chromosomal β-globin genes. *Cell* 18:865–873, 1979. (Problem 8–5)

Krause, M.; Hirsh, D. A trans-spliced leader sequence on actin mRNA in *C. elegans*. *Cell* 49:753–761, 1987. (Problem 8–28)

Krebs, H.A.; Johnson, W.A. The role of citric acid in intermediate metabolism in animal tissues. *Enzymologia* 4:148–156, 1937. (Problem 14–4)

Kuhne, T.; Wieringa, B.; Reiser, J.; Weissmann, C. Evidence against a scanning model for RNA splicing. *EMBO J.* 2:727–733, 1983. (Problem 8–29)

Kumagai, A.; Dunphy, W.G. Regulation of the Cdc25 protein during the cell cycle in *Xenopus* extracts. *Cell* 70:139–151, 1992. (Problem 17–16)

Kurjan, I. Pheromone response in yeast. *Annu. Rev. Biochem.* 61:1097–1129, 1992. (Problem 15–16)

Landegren, U.; Kaiser, R.; Sanders, J.; Hood, L. A ligase-mediated gene detection technique. *Science* 241:1077–1080, 1988. (Problem 7–16)

Lawlor, D.W. Photosynthesis: Metabolism, Control, and Physiology. New York: Wiley, 1987. (Problem 14–24)

Lawrence, T.S.; Beers, W.H.; Gilula, N.B. Transmission of hormonal stimulation by cell-to-cell communication. *Nature* 272:501–506, 1978. (Problem 19–6)

LeBowitz, J.H.; McMacken, R. The *Escherichia coli* DnaB replication protein is a DNA helicase. *J. Biol. Chem.* 261:4738–4748, 1986. (Problem 6–26)

Leff, S.E.; Evans, R.M.; Rosenfeld, M.G. Splice commitment dictates neuron-specific alternative RNA processing in calcitonin/CGRP gene expression. *Cell* 48:517–524, 1987. (Problem 9–32)

Lefkowitz, R.J.; Limbird, L.E.; Mukherjee, C.; Caron, M.G. The β-adrenergic receptor and adenylate cyclase. *Biochim. Biophys. Acta* 457:1–39, 1976. (Problem 15–12)

Liebold, E.A.; Munro, H.N. Cytoplasmic protein binds *in vitro* to a highly conserved sequence in the 5′ untranslated region of ferritin heavy- and light-subunit mRNAs. *Proc. Natl. Acad. Sci. USA* 85:2171–2175, 1988. (Problem 9–33)

Lindahl, T.; Demple, B.; Robins, P. Suicide inactivation of the *E. coli.* O^6-methylguanine-DNA methyltransferase. *EMBO J.* 1:1359–1363, 1982. (Problem 6–19)

Lipmann, F. Metabolic generation and utilization of phosphate bond energy. *Adv. Enzymol.* 1:99–162, 1941. (Problem 14–7)

Lo, C.W.; Gilula, N.B. Gap junctional communication in the preimplantation mouse embryo. *Cell* 18:399–409, 1979. (Problem 19–7)

Lodish, H.F.; Kong, N.; Snider, M.; Strous, G.J.A.M. Hepatoma secretory proteins migrate from rough endoplasmic reticulum to Golgi at characteristic rates. *Nature* 304:80–83, 1983. (Problem 13–23)

Logothetis, D.E.; Kurachi, Y.; Galper, J.; Neer, E.J.; Clapham, D.E. The βγ subunits of GTP-binding proteins activate the muscarinic K⁺ channel in heart. *Nature* 325:321–326, 1987. (Problem 15–20)

Logothetis, D.E.; Kurachi, Y.; Galper, J.; Neer, E.J.; Clapham, D.E. G protein opening of K⁺ channels. *Nature* 327:22, 1987. (Problem 15–20)

Luck, D.J.L. Genetic and biochemical dissection of the eucaryotic flagellum. J. Cell Biol. 98:789–794, 1984. (Problem 16–22)

MacLean-Fletcher, S.; Pollard, T.D. Mechanism of action of cytochalasin B on actin. *Cell* 20:329–341, 1980. (Problem 16–27)

MacNicol, A.; Muslin, A.J.; Williams, L.T. Raf-1 kinase is essential for early *Xenopus* development and mediates the induction of mesoderm by FGF. *Cell* 73:571–583, 1993. (Problem 7–36)

Maizia, D.; Harris, P.J.; Bibring, T. The multiplicity of mitotic centers and the time-course of their duplication and separation. *J. Biophys. Biochem. Cytol.* 7:1–20, 1960. (Problem 18–4)

Mann, C.; Davis, R.W. Instability of dicentric plasmids in yeast. *Proc. Natl. Acad. Sci. USA* 80:228–232, 1983. (Problem 18–9)

Manoil, C.; Beckwith, J. A genetic approach to analyzing membrane protein topology. *Science* 233:1403–1408, 1986. (Problem 12–24)

Manson, M.D.; Blank, V.; Brade, G.; Higgins, C.F. Peptide chemotaxis in *E. coli* involves the Tap signal transducer and the dipeptide permease. *Nature* 321:253–258, 1986. (Problem 15–34)

Manson, M.D.; Tedesco, P.; Berg, H.C.; Harold, F.M.; van der Drift, C. A protonmotive force drives bacteria flagella. *Proc. Natl. Acad. Sci. USA* 74:3060–3064, 1977. (Problem 14–17)

Matteoli, M.; Takei, K.; Perin, M.S.; Südhof, T.C.; DeCamilli, P. Exoendocytotic recycling of synaptic vesicles in developing processing of cultured hippocampal neurons. *J. Cell Biol.* 117:849–861, 1992. (Problem 13–25)

McClintock, B. The behavior of successive nuclear divisions of a chromosome broken at meiosis. *Proc. Natl. Acad. Sci. USA* 25:405–416, 1939. (Problem 18–5)

McKeon, F.D.; Kirschner, M.W.; Caput, D. Homologies in both primary and secondary structure between nuclear envelope and intermediate filament proteins. *Nature* 319:463–468, 1986. (Problem 16–8)

Miller, J.H. Mutagenic specificity of ultraviolet light. *J. Mol. Biol.* 182:45–65, 1985. (Problem 6–16)

Mitchell, M.B.; Mitchell, H.K. A case of "maternal" inheritance in *Neurospora crassa. Proc. Natl. Acad. Sci. USA* 38:442–449, 1952. (Problem 14–36)

Mitchell, M.B.; Mitchell, H.K.; Tissieres, A. Mendelian and non-Mendelian factors affecting the cytochrome system in *Neurospora crassa. Proc. Natl. Acad. Sci. USA* 39:605–613, 1953. (Problem 14–36)

Mitchison, T.; Kirschner, M. Microtubule assembly nucleated by isolated centrosomes. *Nature* 312:232–237, 1984. (Problem 16–12)

Mitchison, T.; Kirschner, M.W. Properties of the kinetochore *in vitro*. I. Microtubule nucleation and tubulin binding. *J. Cell Biol.* 101:755–765, 1985. (Problem 16–13)

Mitchison, T.J.; Kirschner, M.W. Properties of the kinetochore *in vitro*. II. Microtubule capture and ATP-dependent transloca-tion. *J. Cell Biol.* 101:766–777, 1985. (Problem 18–11)

Monod, J. The phenomenon of enzymatic adaptation. Growth Symposium XI:223–289, 1947. [Reprinted in Selected Papers in Molecular Biology by Jacques Monod (A. Lwoff, A. Ullmann, eds.), pp. 68–134, New York: Academic Press, 1947]. (Problem 9–10)

Montoya, J.; Ojala, D.; Attardi, G. Distinctive features of the 5'-terminal sequences of the human mitochondrial mRNAs. *Nature* 290:465–470, 1981. (Problem 14–34)

Moriyoshi, K.; Masu, M.; Ishii, T.; Shigemoto, R.; Mizuno, N.; Nakanishi, S. Molecular cloning and characterization of the rat NMDA receptor. *Nature* 354:31–37, 1991. (Problem 11–4)

Mowry, K.L.; Steitz, J.A. Identification of human U7 snRNP as one of several factors involved in the 3'-end maturation of histone premessenger RNAs. *Science* 238:1682–1687, 1987. (Problem 8–26)

Murray, A.; Hunt, T. The Cell Cycle: An Introduction, pp. 143–144. New York: W.H. Freeman, 1993. (Problem 17–18)

Murray, A.W.; Szostak, J.W. Pedigree analysis of plasmid segrega-tion in yeast. *Cell* 34:961–970, 1982. (Problem 18–10)

Murray, E.J.; Grosveld, F. Site-specific demethylation in the promoter of human γ-globin does not alleviate methylation mediated suppression. *EMBO J.* 6:2329–2335, 1987. (Problem 9–28)

Nathans, J.; Piantanida, T.P.; Eddy, R.L.; Shows, T.B.; Hogness, D.S. Molecular genetics of inherited variation in human color vision. *Science* 232:203–210, 1986. (Problems 8–33, 8–34)

Nathans, J.; Thomas, D.; Hogness, D.S. Molecular genetics of human color vision: the genes encoding blue, green, and red pigments. *Science* 232:193–202, 1986. (Problem 8–33)

Nicholls, D.G. Bioenergetics. London: Academic Press, 1982. (Problems 14–8, 14–14, 14–15)

Ninfa, A.J.; Reitzer, L.J.; Magasanik, B. Initiation of transcription at the bacterial glnAp2 promoter by purified E. coli components is facilitated by enhancers. Cell 50:1039–1046, 1987. (Problem 9–11)

Nishizuka, Y. Calcium, phospholipid turnover and transmembrane signaling. *Phil. Trans. R. Soc. Lond. (Biol.)* 302:101–112, 1983. (Problem 15–19)

Norman, C.; Runswick, M.; Pollock, R.; Treisman, R. Isolation and properties of cDNA clones encoding SRF, a transcription factor that binds to the *c-fos* serum response element. *Cell* 55:989–1003, 1988. (Problem 9–6)

Ochman, H.; Medhora, M.M.; Garza, D.; Hartl, D.L. Amplification of flanking sequences by inverse PCR. In PCR Protocols (M.A. Innis, D.H. Gelfand, J.J. Sninsky, T.J. White, eds.), pp. 219–227. San Diego, CA: Academic Press, 1990. (Problem 7–28)

Oeller, P.W.; Min-Wong, L.; Taylor, L.P.; Pike, D.A.; Theologis, A. Reversible inhibition of tomato fruit senescence by antisense RNA. *Science* 254:437–439, 1991. (Problem 19–29)

Ojala, D.; Montoya, J.; Attardi, G. tRNA punctuation model of RNA processing in human mitochondria. *Nature* 290:470–474, 1981. (Problem 14–34)

O'Keefe, E.J.; Pledger, W.J. A model of cell-cycle control: sequential events regulated by growth factors. *Mol. Cell. Endocrinol.* 31:167–186, 1983. (Problem 17–21)

Orci, L.; Glick, B.S.; Rothman, J.E. A new type of coated vesicular carrier that appears not to contain clathrin: its possible role in protein transport within the Golgi stack. *Cell* 46:171–184, 1986. (Problem 13–28)

Orci, L.; Palmer, D.J.; Amherdt, M.; Rothman, J.E. Coated vesicle assembly in the Golgi requires only coatomer and ARF proteins from the cytosol. *Nature* 364:732–734, 1993. (Problem 13–29)

Orr-Weaver, T.L.; Spradling, A.D. *Drosophila* chorion gene amplification requires an upstream region regulating s18 transcription. *Mol. Cell. Biol.* 6:4624–4633, 1986. (Problem 8–19)

Osinga, K.A.; Swinkels, B.W.; Gibson, W.C.; Borst, P.; Veeneman, G.H.; Van Boom, J.H.; Michels, P.A.M.; Opperdoes, F.R. Topogenesis of microbody enzymes: sequence comparison of the genes for the glycosomal (microbody) and cytosolic phosphoglycerate kinases of *Trypanosoma brucei. EMBO J.* 4:3811–3817, 1985. (Problem 12–17)

O'Toole, T.E.; Mandelman, D.; Forsyth, J.; Shattil, S.J.; Plow, E.F.; Ginsberg, M.H. Modulation of the affinity of integrin $\alpha_{IIb}\beta_3$ (GPIIb-IIIa) by the cytoplasmic domain of α_{IIb}. *Science* 254:845–847, 1991. (Problem 19–25)

Ottaviano, Y.; Gerace, L. Phosphorylation of the nuclear lamins during interphase and mitosis. *J. Biol. Chem.* 260:624–632, 1985. (Problem 16–9)

Park, H.-O.; Chant, J.; Herskowitz, I. *BUD2* encodes a GTPase-activating protein for Bud1/Rsr1 necessary for proper bud-site selection in yeast. *Nature* 365:269–274, 1993. (Problem 16–5)

Payvar, F.; DeFranco, D.; Firestone, G.L.; Edgar, B.; Wrange, O.; Okret, S.; Gustafsson, J.-A.; Yamomoto, K.R. Sequence-specific binding of glucocorticoid receptor to MMTV DNA at sites within and upstream of the transcribed region. *Cell* 35:381–392, 1983. (Problem 15–7)

Perara, E.; Rothman, R.E.; Lingappa, V.R. Uncoupling translocation from translation: implications for transport of proteins across membranes. *Science* 232:348–352, 1986. (Problem 12–22)

Pierschbacher, M.D.; Ruoslahti, E. Cell attachment activity of fibronectin can be duplicated by small synthetic fragments of the molecule. *Nature* 309:30–33, 1984. (Problem 19–20)

Pollock, R.; Treisman, R. A sensitive method for the determination of protein-DNA binding specificities. *Nucleic Acids Res.* 18:6197–6204, 1990. (Problem 7–27)

Ponticelli, A.S.; Schultz, D.W.; Taylor, A.F.; Smith, G.R. Chi-dependent DNA strand cleavage by recBCD enzyme. *Cell* 41:145–151, 1985. (Problem 6–32)

Potter, H.; Dressler, D. On the mechanism of genetic recombinaton: electron microscopic observation of recombination intermediates. *Proc. Natl. Acad. Sci. USA* 73:3000–3004, 1976. (Problem 6–36)

Powell, L.M., et al. A novel form of tissue-specific RNA processing produces apolipoprotein-B48 in intestine. *Cell* 50:831–840, 1987. (Problem 9–36)

Prescott, D.M. Reproduction of Eucaryotic Cells, pp. 85–86. New York: Academic Press, 1975. (Problem 17–6)

Prunell, A.; Kornberg, R.D.; Lutter, L.; Klug, A.; Levitt, M.; Crick, F.H. Periodicity of deoxyribonuclease I digestion of chromatin. *Science* 204:855–858, 1979. (Problem 8–6)

Pytela, R.; Pierschbacher, M.D.; Ruoslahti, E. Identification and isolation of a 140 kd cell surface glycoprotein with properties expected of a fibronectin receptor. *Cell* 40:191–198, 1985. (Problem 19–20)

Rao, P.N.; Johnson, R.T. Mammalian cell fusion: I. Studies on the regulation of DNA synthesis and mitosis. *Nature* 225:159–164, 1970. (Problem 17–5)

Rappaport, R. Establishment of the mechanism of cytokinesis in animal cells. *Int. Rev. Cytol.* 105:245–281, 1986. (Problem 18–17)

Ren, R.; Mayer, B.J.; Cicchetti, P.; Baltimore, D. Identification of a ten-amino-acid proline-rich SH3 binding site. *Science* 259:1157–1161, 1993. (Problem 15–27)

Rothman, J.E.; Dawidowicz, E.A. Asymmetric exchange of vesicle phospholipids catalyzed by the phosphatidylcholine exchange protein. Measurement of inside-outside transitions. *Biochemistry* 14:2809–2816, 1975. (Problem 12–25)

Rothman, J.E.; Miller, R.L.; Urbani, L.J. Intercompartmental transport in the Golgi complex is a dissociative process: facile transfer of membrane protein between two Golgi populations. *J. Cell Biol.* 99:260–271, 1984. (Problem 13–3)

Rousselet, A.; Guthmann, C.; Matricon, J.; Bienvenue, A.; Devaux, P.F. Study of the transverse diffusion of spin labeled phospholipids in biological membranes: 1. Human red blood cells. B*iochim. Biophys. Acta* 426:357–371, 1976. (Problems 10–5, 10–6)

Rowland, L.P. Diseases of chemical transmission at the nerve-muscular synapse: myasthenia gravis. In Principles of Neural Science, 2nd ed. (E.R. Kandel, J.H. Schwartz, eds.), pp. 147–185. New York: Elsevier, 1985. (Problem 11–16)

Rozdzial, M.M.; Haimo, L.T. Bidirectional pigment granule movements of melanophores are regulated by protein phosphorylation and dephosphorylation. *Cell* 47:1061–1070, 1986. (Problem 16–4)

Ruffner, D.E.; Sprung, C.N.; Minghetti, P.P.; Gibbs, P.E.M.; Dugaiczyk, A. Invasion of the human albumin-α-fetoprotein gene family by *Alu, Kpn,* and two novel repetitive DNA elements. *Mol. Biol. Evol.* 4:1–9, 1987. (Problem 8–36)

Ruoslahti, E.; Pierschbacher, M.D. Arg-Gly-Asp: a versatile cell recognition signal. *Cell* 44:517–518, 1986. (Problem 19–20)

Russo, A.F.; Koshland, D.E. Separation of signal transduction and adaptation functions of the aspartate receptor in bacterial sensing. *Science* 220:1016–1020, 1983. (Problem 15–35)

Sadowski, H.B.; Shuai, K.; Darnell, J.E.; Gilman, M.Z. A common nuclear signal transduction pathway activated by growth factor and cytokine receptors. *Science* 261:1739–1744, 1993. (Problem 15–25)

Safer, B.; Kemper, W.; Jagus, R. Identification of a 48S preinitiation complex in reticulocyte lysate. *J. Biol. Chem.* 253:3384–3386, 1978. (Problem 6–10)

Sakmann, B. Elementary steps in synaptic transmission revealed by currents through single ion channels. *Science* 256:503–512, 1992. (Problems 11–17, 11–18)

Sanger, F., et al. The nucleotide sequence of bacteriophage φX174. *J. Mol. Biol.* 125:225–246, 1978. (Problem 6–44)

Sawadogo, M.; Roeder, R.G. Factors involved in specific transcription by human RNA polymerase II: analysis by a rapid and quantitative *in vitro* assay. *Proc. Natl. Acad. Sci. USA* 82:4394–4398, 1985. (Problems 9–13, 9–14)

Sawadogo, M.; Roeder, R.G. Interaction of a gene-specific transcription factor with the adenovirus major late promoter upstream of the TATA box region. *Cell* 43:165–175, 1985. (Problem 9–15)

Scheidereit, C.; Geisse, S.; Westphal, H.M.; Beato, M. The glucocorticoid receptor binds to defined nucleotide sequences near the promoter of mouse mammary tumor virus. *Nature* 304:749–752, 1983. (Problem 15–7)

Schleif, R.F. Genetics and Molecular Biology, Chap. 13. Reading, MA: Addison Wesley, 1986. (Problem 9–9)

Schnieke, A.; Harbers, K.; Jaenisch, R. Embryonic lethal mutation in mice induced by retrovirus into the α1(I) collagen gene. *Nature* 304:315–320, 1983. (Problem 19–21)

Sinn, E.; Muller, W.; Pattengale, P.; Tepler, I.; Wallace, R.; Leder, P. Coexpression of MMTV/v-Ha-*ras* and MMTV/c-*myc* genes in transgenic mice: synergistic action of oncogenes *in vivo. Cell* 49:465–475, 1987. (Problem 17–25)

Sluder, G.; Rieder, C.L. Centriole number and the reproductive capacity of spindle poles. *J. Cell Biol.* 100:887–896, 1985. (Problem 18–4)

Smeekens, S.; Bauerle, C.; Hageman, J.; Keegstra, K.; Weisbeek, P. The role of the transit peptide in the routing of precursors toward different chloroplast compartments. *Cell* 46:365–375, 1986. (Problem 12–14)

Smith, H.T.; Ahmed, A.J.; Millet, F. Electrostatic interaction of cytochrome c with cytochrome c₁ and cytochrome oxidase. *J. Biol. Chem.* 256:4984–4990, 1981. (Problem 14–13)

Söllner, T.; Whiteheart, S.W.; Brunner, M.; Erdjument-Bromage, H.; Geromanos, S.; Tompst, P.; Rothman, J.E. SNAP receptors implicated in vesicle targeting and fusion. *Nature* 362:312–324, 1993. (Problem 13–31)

Spierer, A.; Spierer, P. Similar levels of polyteny in bands and interbands of *Drosophila* giant chromosomes. *Nature* 307:176–178, 1984. (Problem 8–12)

Staden, R. An interactive graphics program for comparing and aligning nucleic acid and amino acid sequences. *Nucleic Acids Res.* 10:2951–2961, 1982. (Problem 8–5)

Stanners, C.P.; Till, J.E. DNA synthesis in individual L-strain mouse cells. *Biochim. Biophys. Acta* 37:406–419, 1960. (Problem 17–4)

Stanojevic, D.; Small, S.; Levine, M. Regulation of a segmentation stripe by overlapping activators and repressors in the *Drosophila* embryo. *Science* 254:1385–1387, 1991. (Problem 9–18)

Stern, D.B.; Palmer, J.D. Extensive and widespread homologies between mitochondrial DNA and chloroplast DNA in plants. *Proc. Natl. Acad. Sci. USA* 81:1946–1950, 1984. (Problem 14–35)

Svoboda, K.; Schmidt, C.F.; Schnapp, B.J.; Block, S.M. Direct observation of kinesin stepping by optical trapping interferometry. *Nature* 365:721–727, 1993. (Problem 16–15)

Sykes, B. The molecular genetics of collagen. *Bioessays* 3:112–117, 1985. (Problem 19–18)

Szostak, J.W.; Blackburn, E.H. Cloning yeast telomeres on linear plasmid vectors. *Cell* 19:245–255, 1982. (Problem 8–3)

Tang, W.-J.; Gilman, A.G. Type-specific regulation of adenylyl cyclase by G protein βγ subunits. *Science* 254:1500–1503, 1991. (Problem 15–15)

Teo, I.; Sedgwick, B.; Kilpatrick, M.W.; McCarthy, T.V.; Lindahl, T. The intracellular signal for induction of resistance to alkylating agents in E. *coli. Cell* 45:315–324, 1986. (Problem 6–18)

Thompson, C.M.; Koleske, A.J.; Chao, D.M.; Young, R.A. A multisubunit complex associated with the RNA polymerase II CTD and TATA-binding protein in yeast. *Cell* 73:1361–1375, 1993. (Problem 9–16)

Thompson, H.A.; Spooner, B.S. Inhibition of branching morphogenesis and alteration of glycosaminoglycan biosynthesis in salivary glands treated with β-D-xyloside. *Dev. Biol.* 89:417–424, 1982. (Problem 19–17)

Tilney, L.G.; Inoue, S. Acrosomal reaction of *Thyone* sperm. II. The kinetics and possible mechanism of acrosomal process elongation. *J. Cell Biol.* 93:820–827, 1982. (Problem 16–26)

Treisman, R. Identification and purification of a polypeptide that binds to the c-fos serum response element. *EMBO J.* 6:2711–2717, 1987. (Problem 9–6)

Treisman, R. Transient accumulation of *c-fos* RNA following serum stimulation requires a conserved 5′ element and *c-fos* 3′ sequences. *Cell* 42:889–902, 1985. (Problem 9–37)

Tsukamoto, T.; Yokota, S.; Fujiki, Y. Isolation of Chinese hamster ovary cell mutants defective in assembly of peroxisomes. *J. Cell Biol.* 110:651–660, 1990. (Problem 12–18)

Tzagoloff, A. Mitochondria, pp. 212–213. New York: Plenum Press, 1982. (Problem 14–8)

van Arsdell, S.W.; Weiner, A.M. Human genes for U2 small nuclear RNA are tandemly repeated. *Mol. Cell. Biol.* 4:492–499, 1984. (Problem 8–32)

Van Meer, G.; Simons, K. The function of tight junctions in maintaining differences in lipid composition between the apical and the basolateral cell surface domains of MDCK cells. *EMBO J.* 5:1455–1464, 1986. (Problems 10–15, 19–4)

Vojtek, A.B.; Hollenberg, S.M.; Cooper, J.A. Mammalian Ras interacts directly with the serine/threonine kinase Raf. *Cell* 74:205–214, 1993. (Problems 15–28, 15–29)

Vollrath, D.; Nathans, J.; Davis, R.W. Tandem array of human visual pigment genes at Xq28. Science 240:1669–1671, 1988. (Problem 8–33)

Walworth, N.C.; Goud, B.; Kabcenell, A.K.; Novick, P.J. Mutational analysis of *SEC4* suggests a cyclical mechanism for the regulation of vesicular traffic. *EMBO J.* 8:1685–1693, 1989. (Problem 13–30)

Wasserman, W.J.; Masui, Y. Effects of cycloheximide on a cytoplasmic factor initiating meiotic maturation in *Xenopus* oocytes. *Exp. Cell Res.* 91:381–388, 1975. (Problem 17–9)

Webb, M.R.; Grubmeyer, C.; Penefsky, H.S.; Trentham, D.R. The stereochemical course of phosphoric residue transfer catalyzed by beef heart mitochondrial ATPase. *J. Biol. Chem.* 255:11637–11639, 1980. (Problem 14–11)

Weinert, T.A.; Hartwell, L.H. The *rad*9 gene controls the cell cycle response to DNA damage in *Saccharomyces cerevisiae. Science* 241:317–322, 1988. (Problem 17–17)

Weintraub, H.; Groudine, M. Chromosomal subunits in active genes have an altered conformation. *Science* 193:848–856, 1976. (Problem 8–13)

Whatley, J.M.; John, P.; Whatley, F.R. From extracellular to intracellular: the establishment of mitochondria and chloroplasts. *Proc. R. Soc. Lond. B.* 204:165–187, 1979. (Problem 14–38)

Wilson, T.; Treisman, R. Removal of poly(A) and consequent degradation of *c-fos* mRNA facilitated by 3′ AU-rich sequences. *Nature* 336:396–399, 1988. (Problem 9–37)

Witke, W.; Scheicher, M.; Noegel, A.A. Redundancy in the microfilament system: abnormal development of *Dictyostelium* cells lacing two F-actin cross-linking proteins. *Cell* 68:53–62, 1992. (Problem 16–30)

Wolffe, A.P.; Brown D.D. DNA replication *in vitro* erases a *Xenopus* 5S RNA gene transcription complex. *Cell* 47:217–227, 1986. (Problem 9–27)

Workman, J.L.; Roeder, R.G. Binding of transcription factor TFIID to the major late promoter during *in vitro* nucleosome assembly potentiates subsequent initiation by RNA polymerase II. *Cell* 51:613–622, 1987. (Problem 9–21)

Xeros, N. Deoxyriboside control and synchronization of mitosis. *Nature* 194:682–683, 1962. (Problem 17–5)

Yajima, S.; Suzuki, Y.; Shimozawa, N.; Yamaguchi, S.; Orii, T.; Fujiki, Y.; Osumi, T.; Hashimoto, T.; Moser, H.W. Complementation study of peroxisome-deficient disorders by immunofluorescence staining and characterization of fused cells. *Human Genetics* 88:491–499, 1992. (Problem 12–18)

Yalow, R.S. Radioimmunassay: a probe for the fine structure of biologic systems. *Science* 200:1236–1245, 1978. (Problem 15–4)

Yen, T.J.; Machlin, P.S.; Cleveland, D.W. Autoregulated instability of β-tubulin mRNAs by recognition of the nascent amino terminus of β-tubulin. *Nature* 334:580–585, 1988. (Problem 9–35)

Zahringer, J.; Baliga, B.S.; Munro, H.N. Novel mechanism for translational control in regulation of ferritin synthesis by iron. *Proc. Natl. Acad Sci. USA* 73:857–861, 1976. (Problem 9–33)

Zhang, X.-F.; Settleman, J.; Kyriakis, J.M.; Takenchi-Suzuki, E.; Elledge, S.J.; Marshall, M.S.; Bruder, J.T.; Rapp, U.R.; Avruch, J. Normal and oncogenic p21^ras proteins bind to the amino-terminal domain of c-Raf-1. *Nature* 364:308–313, 1993. (Problem 15–29)

Zieg, J.; Silverman, M.; Hilmen, M.; Simon, M. Recombinational switch for gene expression. *Science* 196:170–172, 1977. (Problem 9–24)

Zinkel, S.S.; Crothers, D.M. DNA bend direction by phase sensitive detection. *Nature* 328:178–181, 1987. (Problem 9–5)

Cited Researchers

Gall, J.G. (Problems 7–9, 8–4, 8–11)
Gallant, J. (Problems 6–9, 15–20)
Galper, J. (Problem 15–20)
Garfinkel, D.J. (Problem 8–35)
Garoff, H. (Problem 13–24)
Gartenberg, M.R. (Problem 9–5)
Garza, D. (Problem 7–28)
Gay, D.A. (Problem 9–35)
Geisse, S. (Problem 15–7)
Gelehrter, T.D. (Problem 7–17)
Gerace, L. (Problem 16–9)
Geromanos, S. (Problem 13–31)
Gibbons, I.R. (Problem 16–20)
Gibbs, P.E.M. (Problem 8–36)
Gibbs, R.A. (Problem 7–26)
Gibson, W.C. (Problem 12–17)
Gilman, A.G. (Problem 15–15)
Gilman, M.Z. (Problem 15–25)
Gilula, N.B. (Problem 19–6, 19–7)
Ginsberg, M.H. (Problems 19–24, 19–25)
Glick, B.S. (Problem 13–28)
Glotzer, M. (Problem 17–11)
Godowski, P.J. (Problem 9–17)
Goldberg, E. (Problem 19–12)
Goldman, Y.E. (Problem 16–36)
Goldstein, J.L. (Problems 13–16, 13–17)
Gordon, A.M. (Problem 16–37)
Gorovsky, M.A. (Problem 6–5)
Gottesman, M.M. (Problem 15–13)
Gottlieb, T.A. (Problem 13–24)
Gottschalk, G. (Problem 14–16)
Goud, B. (Problem 13–30)
Goutte, C. (Problem 9–25)
Greengard, P. (Problem 15–33)
Grompe, M. (Problem 7–14)
Grosveld, F. (Problem 9–28)
Grotkopp, D. (Problem 19–21)
Groudine, M. (Problem 8–13)
Grubmeyer, C. (Problem 14–11)
Gustafsson, J.-A. (Problem 15–7)
Gustafsson, L. (Problem 13–23)
Guthmann, C. (Problems 10–5, 10–6)

Hageman, J. (Problem 12–14)
Hahn, S. (Problem 9–12)
Haigler, H.T. (Problem 13–15)
Haimo, L.T. (Problem 16–4)
Harbers, K. (Problem 19–21)
Harlan, J.M. (Problem 19–24)
Harland, R.M. (Problem 8–20)
Harold, F.M. (Problem 14–17)
Harris, P.J. (Problem 18–4)
Harrison, T. (Problem 8–24)
Hartl, D.L. (Problem 7–28)
Hartwell, L.H. (Problems 17–14, 17–15, 17–17)
Hashimoto, T. (Problem 12–18)
Helms, J.B. (Problem 13–29)
Herskowitz, I. (Problem 16–5)
Hertel, C. (Problem 15–32)
Herzog, V. (Problem 13–12)
Hess, G.P. (Problem 15–33)
Hibberd, M.G. (Problem 16–36)
Hieter, P. (Problems 18–12, 18–13)
Higgins, C.F. (Problem 15–34)
Higuchi, R. (Problem 7–35)

Hill, T.L. (Problem 16–38)
Hille, B. (Problems 11–12, 11–13, 11–15)
Hilmen, M. (Problem 9–24)
Hirsh, D. (Problem 8–28)
Hoffmann, E. (Problem 19–21)
Hogness, D.S. (Problems 8–33, 8–34)
Hollenberg, S.M. (Problems 15–28, 15–29)
Holloway, S.L. (Problem 17–11)
Holzmann, B. (Problem 13–12)
Hood, L. (Problem 7–16)
Horio, T. (Problem 16–14)
Horowitz, S. (Problem 6–5)
Hotami, H. (Problem 16–14)
Huang, B. (Problem 16–20)
Huberman, J.A. (Problem 8–18)
Huganir, R.L. (Problem 15–33)
Hunt, T. (Problems 9–34, 17–10, 17–18)
Hurley, J.B. (Problem 15–21)
Hurwitz, J. (Problem 8–17)
Huxley, A.F. (Problem 16–37)
Hyman, A.A. (Problems 18–12, 18–13)

Ingledew, J.W. (Problem 14–28)
Inman, R.B. (Problem 6–23)
Inoue, S. (Problem 16–26)
Ishii, T. (Problem 11–4)

Jacks, T. (Problem 6–8)
Jackson, R.J. (Problem 9–34)
Jaenisch, R. (Problem 19–21)
Jagendorf, A.T. (Problem 14–27)
Jagus, R. (Problem 6–10)
Jahner, D. (Problem 19–21)
John, P. (Problem 14–38)
Johnson, A.D. (Problem 9–25)
Johnson, P.J. (Problem 8–23)
Johnson, R.T. (Problem 17–5)
Johnson, W.A. (Problem 14–4)
Julian, F.J. (Problem 16–37)

Kabcenell, A.K. (Problem 13–30)
Kaiser, R. (Problem 7–16)
Kamp, B.M. (Problem 14–25)
Kaplan, A. (Problem 13–10)
Karrer, K.M. (Problem 7–9)
Keegstra, K. (Problem 12–14)
Keilin, D. (Problem 14–12)
Kellems, R.E. (Problem 12–12)
Kemper, W. (Problem 6–10)
Kiger, J.A. (Problem 15–14)
Kilpatrick, M.W. (Problem 6–18)
King, R.W. (Problem 17–11)
Kirkpatrick, J.B. (Problem 17–5)
Kirschner, M.W. (Problems 16–8, 16–12, 16–13, 18–11)
Klausner, R.D. (Problem 13–29)
Kleckner, N. (Problem 6–43)
Klee, C.B. (Problem 15–18)
Klug, A. (Problem 8–6)
Knudson, A.G. (Problem 17–24)
Kok, B. (Problem 14–26)
Koleske, A.J. (Problem 9–16)
Kondor-Koch, C. (Problem 13–24)
Kong, N. (Problem 13–23)

Konkel, D.A. (Problem 8–5)
Kooter, J.M. (Problem 8–23)
Korn, E.D. (Problem 16–25)
Kornberg, A. (Problem 6–24)
Kornberg, R.D. (Problem 8–6)
Koshland, D.E. (Problem 15–35)
Kovach, N.L. (Problem 19–24)
Krause, M. (Problem 8–28)
Krebs, H.A. (Problem 14–4)
Kristiansen, K. (Problem 6–5)
Kuhne, T. (Problem 8–29)
Kumagai, A. (Problem 17–16)
Kurachi, Y. (Problem 15–20)
Kurjan, I. (Problem 15–16)
Kyriakis, J.M. (Problem 15–29)

Lafreniere, R.G. (Problem 7–14)
Landegren, U. (Problem 7–16)
Laskey, R.A. (Problems 8–20, 12–6)
Lassar, A.B. (Problem 9–29)
Lau, J.T.Y. (Problem 9–35)
Lawlor, D.W. (Problem 14–24)
Lawrence, T.S. (Problem 19–6)
LeBowitz, J.H. (Problem 6–26)
Leder, P. (Problems 8–5, 17–25)
Leff, S.E. (Problem 9–32)
Lefkowitz, R.J. (Problem 15–12)
Lehman, I.R. (Problem 6–34)
Levine, M. (Problem 9–18)
Levitt, M. (Problem 8–6)
Liebold, E.A. (Problem 9–33)
Lienhard, G.E. (Problem 11–3)
Limbird, L.E. (Problem 15–12)
Lindahl, T. (Problems 6–18, 6–19)
Lingappa, V.R. (Problem 12–22)
Lipmann, F. (Problem 14–7)
Lo, C.W. (Problem 19–7)
Lodish, H.F. (Problem 13–23)
Logothetis, D.E. (Problem 15–20)
Lohler, J. (Problem 19–21)
Luca, F.C. (Problem 17–10)
Luck, D.J.L. (Problems 16–20, 16–22)
Luini, A. (Problem 15–17)
Lutter, L. (Problem 8–6)

Machlin, P.S. (Problem 9–35)
MacLean-Fletcher, S. (Problem 16–27)
MacNicol, A. (Problem 7–36)
Magasanik, B. (Problem 9–11)
Maizel, J.V. (Problem 8–5)
Maizia, D. (Problem 18–4)
Mandelman, D. (Problem 19–25)
Mann, C. (Problem 18–9)
Manoil, C. (Problem 12–24)
Manson, M.D. (Problems 14–17, 15–34)
Marshall, M.S. (Problem 15–29)
Martin, G.A. (Problem 15–26)
Masu, M. (Problem 11–4)
Masui, Y. (Problem 17–9)
Matricon, J. (Problems 10–5, 10–6)
Matteoli, M. (Problem 13–25)
Mayer, B.J. (Problem 15–27)
McCarthy, T.V. (Problem 6–18)
McClintock, B. (Problem 18–5)
McCormick, F. (Problem 15–26)

McCray, J.A. (Problem 16–36)
McGloin, M. (Problem 14–26)
McKanna, J.A. (Problem 13–15)
McKeon, F.D. (Problem 16–8)
McMacken, R. (Problem 6–26)
Medhora, M.M. (Problem 7–28)
Meltzer, P. (Problem 15–8)
Michels, P.A.M. (Problem 12–16)
Miesfeld, R. (Problem 9–17)
Miller, J.H. (Problem 6–16)
Miller, R.L. (Problem 13–3)
Millet, F. (Problem 14–13)
Minghetti, P.P. (Problem 8–36)
Min-Wong, L. (Problem 19–29)
Mitchell, H.K. (Problem 14–36)
Mitchell, M.B. (Problem 14–36)
Mitchison, T. (Problems 16–12, 16–13, 18–11)
Mizuno, N. (Problem 11–4)
Monod, J. (Problem 9–10)
Montoya, J. (Problem 14–34)
Moriyoshi, K. (Problem 11–4)
Moscona, A. (Problem 17–23)
Moser, H.W. (Problem 12–18)
Mowry, K.L. (Problem 8–26)
Mukherjee, C. (Problem 15–12)
Muller, P. (Problem 15–32)
Muller, W. (Problem 17–25)
Mulligan, J.T. (Problem 7–23)
Munro, H.N. (Problem 9–33)
Murphree, A.L. (Problem 17–24)
Murray, A.W. (Problems 17–11, 17–18, 18–10)
Murray, E.J. (Problem 9–28)
Muslin, A.J. (Problem 7–36)

Nakanishi, S. (Problem 11–4)
Nathans, J. (Problems 8–33, 8–34)
Neer, E.J. (Problem 15–20)
Neumuller, W. (Problem 13–12)
Nicholls, D.G. (Problems 14–8, 14–14, 14–15)
Ninfa, A.J. (Problem 9–11)
Nishizuka, Y. (Problem 15–19)
Noegel, A.A. (Problem 16–30)
Nonet, M. (Problem 9–16)
Norman, C. (Problem 9–6)
Novick, P.J. (Problem 13–30)

Ochman, H. (Problem 7–28)
Oeller, P.W. (Problem 19–29)
Ogden, S. (Problem 9–12)
Ojala, D. (Problem 14–34)
O'Keefe, E.J. (Problem 17–21)
Okret, S. (Problem 15–7)
Opperdoes, F.R. (Problem 12–16)
Orci, L. (Problems 13–28, 13–29)
Orii, T. (Problem 12–18)
Orr-Weaver, T.L. (Problem 8–19)
Osinga, K.A. (Problem 12–17)
Osumi, T. (Problem 12–18)
O'Toole, T.E. (Problem 19–25)
Ottaviano, Y. (Problem 16–9)

Palmer, D.J. (Problem 13–29)
Palmer, J.D. (Problem 14–35)
Pantaloni, D. (Problem 16–25)
Park, H.-O. (Problem 16–5)
Pattengale, P. (Problem 17–25)
Payvar, F. (Problem 15–7)
Penefsky, H.S. (Problem 14–11)
Perara, E. (Problem 12–22)
Perin, M.S. (Problem 13–25)
Peterson, P.A. (Problem 13–23)
Piantanida, T.P. (Problems 8–33, 8–34)
Pierschbacher, M.D. (Problem 19–20)
Pike, D.A. (Problem 19–29)
Pledger, W.J. (Problem 17–21)
Plow, E.F. (Problem 19–25)
Pollard, T.D. (Problem 16–27)
Pollock, R. (Problems 7–27, 9–6)
Ponticelli, A.S. (Problem 6–32)
Portenier, M. (Problem 15–32)
Potter, H. (Problem 6–36)
Powell, L.M., et al. (Problem 9–36)
Prescott, D.M. (Problems 17–5, 17–6)
Pringle, J.R. (Problem 16–5)
Prockop, D.J. (Problem 19–19)
Prunell, A. (Problem 8–6)
Pytela, R. (Problem 19–20)

Ramer, S.W. (Problem 7–23)
Ranier, J.E. (Problem 7–26)
Rao, P.N. (Problem 17–5)
Rapp, U.R. (Problem 15–29)
Rappaport, R. (Problem 18–17)
Reed, R.R. (Problem 15–15)
Reiser, J. (Problem 8–29)
Reitzer, L.J. (Problem 9–11)
Ren, R. (Problem 15–27)
Richards, G. (Problem 15–8)
Rieder, C.L. (Problem 18–4)
Riggs, A.D. (Problem 8–18)
Rindler, M.J. (Problem 13–24)
Rine, J. (Problem 12–7)
Rizzolo, L. (Problem 13–24)
Robins, P. (Problem 6–19)
Roeder, R.G. (Problems 9–13, 9–14, 9–15, 9–21)
Rose, J.K. (Problem 13–5)
Rosenfeld, M.G. (Problem 9–32)
Rothman, J.E. (Problems 12–25, 13–3, 13–28, 13–29, 13–31)
Rothman, R.E. (Problem 12–22)
Rousselet, A. (Problems 10–5, 10–6)
Rowland, L.P. (Problem 11–16)
Rozdzial, M.M. (Problem 16–4)
Ruderman, J.V. (Problem 17–10)
Ruffner, D.E. (Problem 8–36)
Runswick, M. (Problem 9–6)
Ruoslahti, E. (Problem 19–20)
Rupert, J.L. (Problem 7–14)
Rusconi, S. (Problem 9–17)
Russo, A.F. (Problem 15–35)

Sabatini, D.D. (Problem 13–24)
Sadowski, H.B. (Problem 15–25)
Safer, B. (Problem 6–10)
Sakmann, B. (Problems 11–17, 11–18)

Sanders, J. (Problem 7–16)
Sanger, F. (Problem 6–44)
Savage, L.J. (Problem 19–11)
Sawadogo, M. (Problems 9–13, 9–14, 9–15)
Scheicher, M. (Problem 16–30)
Scheidereit, C. (Problem 15–7)
Schleif, R.F. (Problems 9–9, 9–12)
Schmidt, C.F. (Problem 16–15)
Schnapp, B.J. (Problem 16–15)
Schnieke, A. (Problem 19–21)
Schnos, M. (Problem 6–23)
Schrader, K.A. (Problem 15–15)
Schultz, D.W. (Problem 6–32)
Sedgwick, B. (Problem 6–18)
Settleman, J. (Problem 15–29)
Sharnick, S.V. (Problem 12–6)
Sharp, P.A. (Problem 8–24)
Shattil, S.J. (Problem 19–25)
Shigemoto, R. (Problem 11–4)
Shimozawa, N. (Problem 12–18)
Shows, T.B. (Problems 8–33, 8–34)
Shuai, K. (Problem 15–25)
Silverman, M. (Problem 9–24)
Simon, M. (Problem 9–24)
Simons, K. (Problems 10–15, 19–4)
Sinn, E. (Problem 17–25)
Sluder, G. (Problem 18–4)
Sly, W.S. (Problem 13–10)
Small, S. (Problem 9–18)
Smeekens, S. (Problem 12–14)
Smith, G.R. (Problem 6–32)
Smith, H.T. (Problem 14–13)
Snider, M. (Problem 13–23)
Söllner, T. (Problem 13–31)
Song, O. (Problem 15–28)
Sorger, P.K. (Problems 18–12, 18–13)
Spencer, F. (Problems 18–12, 18–13)
Spierer, A. (Problem 8–12)
Spierer, P. (Problem 8–12)
Spooner, B.S. (Problem 19–17)
Spottswood, M. (Problem 7–23)
Spradling, A.D. (Problem 8–19)
Sprung, C.N. (Problem 8–36)
Staden, R. (Problem 8–5)
Staehelin, M. (Problem 15–32)
Stanners, C.P. (Problem 17–4)
Stanojevic, D. (Problem 9–18)
Steitz, J.A. (Problem 8–26)
Stenbuck, P.J. (Problem 10–11)
Stern, D.B. (Problem 14–35)
Stidwell, P.R. (Problem 16–31)
Stocheck, C. (Problem 17–22)
Strous, G.J.A.M. (Problem 13–23)
Stryer, L. (Problem 15–21)
Styles, C.A. (Problem 8–35)
Südhof, T.C. (Problem 13–25)
Suzuki, Y. (Problem 12–18)
Svoboda, K. (Problem 16–15)
Swinkels, B.W. (Problem 12–17)
Sykes, B. (Problem 19–18)
Szostak, J.W. (Problems 8–3, 18–10)

Takei, K. (Problem 13–25)
Takenchi-Suzuki, E. (Problem 15–29)
T'Ang, A. (Problem 17–24)
Tang, W.-J. (Problem 15–15)

Index

Prefixes

Symbol	Name	Value
E-	exa-	10^{18}
P-	peta-	10^{15}
T-	tera-	10^{12}
G-	giga-	10^{9}
M-	mega-	10^{6}
k-	kilo-	10^{3}
h-	hecto-	10^{2}
da-	deca-	10^{1}
d-	deci-	10^{-1}
c-	centi-	10^{-2}
m-	milli-	10^{-3}
μ-	micro-	10^{-6}
n-	nano-	10^{-9}
p-	pico-	10^{-12}
f-	femto-	10^{-15}
a-	atto-	10^{-18}

Radioactive Isotopes

Isotope	Emission	Half-Life	Counting Efficiency[a]	Maximum Specific Activity[b]
^{14}C	beta	5730 years	96%	62 mCi/mmol
^{3}H	beta	12.3 years	65%	29 Ci/mmol
^{35}S	beta	87.4 days	97%	1490 Ci/mmol
^{125}I	gamma, auger, and conversion electrons	60.3 days	78%	2400 Ci/mmol
^{32}p	beta	14.3 days	100%	9120 Ci/mmol
^{131}I	beta and gamma	8.04 days	100%	16,100 Ci/mmol

[a] Maximum efficiency for an unquenched sample in a liquid scintillation counter. Most real samples are quenched to some extent.

[b] This value assumes one atom of radioisotope per molecule. If there are two radioactive atoms per molecule, the specific activity will be twice as great, and so on.